Lecture Notes in Computer Science 11355

Commenced Publication in 1973
Founding and Former Series Editors:
Gerhard Goos, Juris Hartmanis, and Jan van Leeuwen

More information about this series at http://www.springer.com/series/7407

Gautam K. Das · Partha S. Mandal
Krishnendu Mukhopadhyaya · Shin-ichi Nakano (Eds.)

WALCOM: Algorithms and Computation

13th International Conference, WALCOM 2019
Guwahati, India, February 27 – March 2, 2019
Proceedings

 Springer

Editors
Gautam K. Das (iD)
Indian Institute of Technology Guwahati
Guwahati, India

Krishnendu Mukhopadhyaya (iD)
Indian Statistical Institute
Kolkata, India

Partha S. Mandal (iD)
Indian Institute of Technology Guwahati
Guwahati, India

Shin-ichi Nakano (iD)
Gunma University
Kiryu, Japan

ISSN 0302-9743   ISSN 1611-3349   (electronic)
Lecture Notes in Computer Science
ISBN 978-3-030-10563-1   ISBN 978-3-030-10564-8   (eBook)
https://doi.org/10.1007/978-3-030-10564-8

Library of Congress Control Number: 2018965171

LNCS Sublibrary: SL1 – Theoretical Computer Science and General Issues

This Springer imprint is published by the registered company Springer Nature Switzerland AG
The registered company address is: Gewerbestrasse 11, 6330 Cham, Switzerland

Preface

WALCOM, the Workshop on Algorithms and Computations, made a humble beginning in Dhaka in 2007. The idea was to create an international platform of high quality that would be easy to attend for researchers from South Asian countries. It has come a long way since then. This proceedings volume contains papers presented at WALCOM 2019, the 13th International Conference and Workshop on Algorithms and Computations, held from February 27 to March 2, 2019, at the Indian Institute of Technology, Guwahati (IIT Guwahati), India. The topics covered included diverse areas such as computational geometry, combinatorial algorithms, approximation algorithms, graph algorithms, graph drawing, parallel and distributed algorithms. The conference was organized by the Department of Mathematics, IIT Guwahati, as a part of the Silver Jubilee celebration of the Institute.

Initially, WALCOM was held in Bangladesh and India on alternate years. In 2016 it made the first step outside and was hosted in Nepal. The 2017 edition in Taiwan was a grand success. After another two rounds in Dhaka, Bangladesh, in 2018 and Guwahati, India, this year, it is again set to go out to Singapore in 2020. Currently WALCOM enjoys a high degree of respect from the research community. A large portion of the credit goes to the eminent scientists in the Steering Committee from Bangladesh, Germany, India, Japan, Korea, and the UK. The Program Committee for 2019 had 30 distinguished members from 17 different countries such as Australia, Austria, Bangladesh, Brazil, Canada, France, Germany, Greece, India, Israel, Italy, Japan, The Netherlands, Oman, South Korea, Taiwan, and the USA.

The growing popularity of WALCOM is reflected in the fact that we had 100 submissions by 240 authors from 21 different countries. Each paper was reviewed by at least three experts. Finally, 30 papers were selected for presentation at the conference. Continuing with the tradition of WALCOM, The Best Paper and The Best Student Paper were awarded. The decision was announced during the conference. We acknowledge the continued support from Springer in publishing the proceedings in the prestigious LNCS series. We are happy to announce that like previous years, this year too, two special issues — one of *Journal of Graph Algorithms and Applications* and one of *Theoretical Computer Science* — are planned for extended and upgraded versions of selected papers from WALCOM 2019.

The rich tradition of WALCOM was continued by three invited talks by three very distinguished scientists, namely, Professor Mark de Berg, Department of Computer Science, TU Eindhoven, The Netherlands, Professor David Peleg, Department of Computer Science and Applied Mathematics, Weizmann Institute of Science, Israel, and Professor Saket Saurabh, Theoretical Computer Science Group, Institute of Mathematical Sciences, India. We are grateful to the speakers for taking time from their busy schedules and delivering excellent and illuminating lectures. We express our gratitude to all the members of the Program Committee and the external reviewers for their in depth reviews. We thank all the authors who submitted their valuable work to the conference. We are happy to note that we were able to meet all the deadlines without extensions. Our sincere appreciation to the members of the Steering Committee for their guidance and advice. Special thanks are due the director and the administration of IIT Guwahati for hosting the event and all other help. The EasyChair platform made our life so much simpler. We greatly acknowledge the assistance received from the Science and Engineering Research Board, Government of India and Capillary Technologies. It is difficult to get financial support for a conference in theoretical computer science. Finally, hearty congratulations to the Organizing Committee of WALCOM 2019 for successfully organizing the event.

February 2019

Gautam K. Das
Partha S. Mandal
Krishnendu Mukhopadhyaya
Shin-ichi Nakano

Organization

Steering Committee

Kyung-Yong Chwa	KAIST, South Korea
Costas S. Iliopoulos	KCL, UK
M. Kaykobad (Chair)	BUET, Bangladesh
Petra Mutzel	TU Dortmund, Germany
Shin-ichi Nakano	Gunma University, Japan
Subhas Chandra Nandy	Indian Statistical Institute, Kolkata, India
Takao Nishizeki	Tohoku University, Japan
C. Pandu Rangan	IIT Madras, India
Md. Saidur Rahman	BUET, Bangladesh

Program Committee

Sang Won Bae	Kyonggi University, South Korea
Arnab Bhattacharyya	Indian Institute of Science, India
Hans Bodlaender	Utrecht University, The Netherlands
Prosenjit Bose	Carleton University, Canada
Paz Carmi	Ben-Gurion University, Israel
Marek Chrobak	University of California Riverside, USA
Anirban Dasgupta	IIT Gandhinagar, India
Stephane Durocher	University of Manitoba, Canada
Rudolf Fleischer	GUtech, Oman
Guilherme Dias da Fonseca	Université Clermont Auvergne, France
Arijit Ghosh	Indian Statistical Institute, India
Seok-Hee Hong	University of Sydney, Australia
Giuseppe Liotta	University of Perugia, Italy
Matúš Mihalák	Maastricht University, The Netherlands
Debajyoti Mondal	University of Saskatchewan, Canada
Krishnendu Mukhopadhyaya (Co-chair)	Indian Statistical Institute, India
Petra Mutzel	TU Dortmund, Germany
Shin-ichi Nakano (Co-chair)	Gunma University, Japan
Martin Nöllenburg	TU Wien, Austria
Leonidas Palios	University of Ioannina, Greece
Vangelis Paschos	University of Paris-Dauphine, France
Rossella Petreschi	University of Rome la Sapienza, Italy
Md. Saidur Rahman	BUET, Bangladesh
M. Sohel Rahman	BUET, Bangladesh
Kunihiko Sadakane	The University of Tokyo, Japan
Etsuji Tomita	The University of Electro-Communications, Japan

Csaba D. Toth   California State University Northridge, USA
Jan Vahrenhold   University of Münster, Germany
Sue Whitesides   University of Victoria, Canada
Hsu-Chun Yen   National Taiwan University, Taiwan

Organizing Committee

Swaroop Nandan Bora   IIT Guwahati, India
Durga Charan Dalal   IIT Guwahati, India
Gautam Kumar Das (Co-chair)   IIT Guwahati, India
Sangram Kishor Jena   IIT Guwahati, India
Kalpesh Kapoor   IIT Guwahati, India
K. V. Krishna   IIT Guwahati, India
Partha Sarathi Mandal   IIT Guwahati, India
  (Co-chair)
Debasish Pattanayak   IIT Guwahati, India
Himadri Sekhar Paul   TCS Innovation Labs, India
S. V. Rao   IIT Guwahati, India
Arnab Sarkar   IIT Guwahati, India
N. Selvaraju (HOD)   IIT Guwahati, India
Vinay Wagh   IIT Guwahati, India

Additional Reviewers

Abu-Affash, Karim   Furini, Fabio
Ahmed, Abu Reyan   Galesi, Nicola
Akitaya, Hugo   Ganian, Robert
Aono, Yoshinori   Gerard, Yan
Bahoo, Yeganeh   Ghosh, Sasthi Charan
Belmonte, Rémy   Giannakos, Aristotelis
Bhagat, Subhash   Grilli, Luca
Calamoneri, Tiziana   Hasunuma, Toru
Chanda, Bhabatosh   Kamali, Shahin
Chang, Yi-Jun   Kather, Philipp
Conte, Alessio   Katsikarelis, Ioannis
Crombez, Loic   Katz, Matya
Darryl, Hill   Kijima, Shuji
De Carufel, Jean-Lou   Kim, Sang-Sub
Di Giacomo, Emilio   Klawitter, Jonathan
Di Pierro, Alessandra   Klute, Fabian
Dublois, Louis   Lampis, Michael
Dütsch, Fabian   Maheshwari, Anil
Favreau, Jean-Marie   Massini, Annalisa
Ferdous, S. M.   Mehrabi, Saeed
Fujiwara, Hiroshi   Migler-Vondollen, Theresa

Montecchiani, Fabrizio
Morin, Pat
Mukhopadhyaya, Srabani
Navarra, Alfredo
Nederlof, Jesper
Pereira de Sá, Vinícius G.
Pritam, Siddharth
Sahu, Aryabartta
Schultz Xavier Da Silveira,
  Luís Fernando
Sen, Sagnik

Sharma, Gokarna
Sinaimeri, Blerina
Smid, Michiel
Stamoulis, Georgios
Tappini, Alessandra
Upadrasta, Ramakrishna
van der Zanden, Tom
Wintraecken, Mathijs
Yamauchi, Yukiko
Yanhaona, Muhammad Nur
Zohora, Fatema Tuz

ETH-Tight Exact Algorithms for Hard Geometric Problems Using Geometric Separators (Abstract of Invited Talk)

Mark de Berg

Department of Computer Sciece, TU Eindhoven
M.T.d.Berg@tue.nl

Many well-known optimization problems on graphs, including INDEPENDENT SET, HAMILTONIAN CYCLE, and the TRAVELING SALESMAN PROBLEM (TSP) are NP-hard. Hence, we do not expect to have polynomial-time algorithms for solving these problems exactly. However, we may still be able to solve them in so-called *sub-exponential time*, that is, with an algorithm whose running time is of the form $2^{o(n)}$, where n is the input size. This turns out to be the case for many problems—including the ones mentioned above—when the input graph is planar. In particular, INDEPENDENT SET and HAMILTO-NIAN CYCLE can be solved in $2^{O(\sqrt{n})}$ time on n-vertex planar graphs. The fact that many problems on planar graphs admit algorithms with $2^{O(\sqrt{n})}$ running time has been dubbed the *square-root phenomenon*. A main tool behind this phenomenon is the famous *Planar Separator Theorem*, which states that for any planar graph $\mathcal{G} = (V, E)$ there is a subset $S \subset V$ of $O(\sqrt{n})$ vertices whose removal splits $\mathcal{G}$ into connected components of size at most $2n/3$.

In the first part of my talk I will discuss some recent work [1] that extends these results to certain classes of *(geometric) intersection graphs*. The intersection graph induced by a set V of objects in the plane (or in some higher-dimensional space) is the graph $\mathcal{G} = (V, E)$ whose vertices correspond to the objects in V and where $E = \{(o, o') \in V \times V : o \cap o' \neq \emptyset\}$. In other words, there is an edge between two objects if and only if they intersect each other. Intersection graphs are a generalization of planar graphs, because any planar graph can be realized as the intersection graph of a set of disks—actually, even as the intersecting graph of a set of disks with disjoint interiors. Intersection graphs can have arbitrarily large cliques and so they do not have small separators. Still, as I will explain in the talk, there is a "clique-based" separator for intersection graphs of disks (and, more generally, of so-called *fat objects*) that makes it possible to solve many problems on such graphs in sub-exponential time. When the disks (or: fat objects) are similar in size, then this approach even leads to algorithms with running time $2^{O(\sqrt{n})}$, which is *ETH-tight*: unless the Exponential-Time Hypothesis fails, there can be no algorithm that solves these problems on unit-disk graphs in $2^{o(\sqrt{n})}$ time.

This work was supported by the NETWORKS project, funded by the Netherlands Organization for Scientific Research NWO under project no. 024.002.003.

In the second part of the talk I will focus on EUCLIDEAN TSP, where we want to find a shortest tour visiting a given set P of n points in the plane (or in some higher-dimensional space). The celebrated Help-Karp dynamic-programming algorithm solves TSP on general weighted graphs in $O(n^2 2^n)$ time, but no sub-exponential algorithms are known for this case. For the EUCLIDEAN TSP, however, there are algorithms with $n^{O(\sqrt{n})} = 2^{O(\sqrt{n} \log n)}$ running time. I will explain a recent result [2] which improves this to $2^{O(\sqrt{n})}$, which is ETH-tight. The algorithm is based on a new "distance-based" separator theorem for point sets.

References

1. de Berg, M., Bodlaender, H.L., Kisfaludi-Bak, S., Marx, D., van der Zanden, T.: A framework for ETH-tight algorithms and lower bounds in geometric intersection graphs. In: 50th ACM Symposium on Theory of Computing (STOC 2018), pp. 574–586 (2018)
2. de Berg, M., Bodlaender, H.L., Kisfaludi-Bak, S., Kolay, S.: An ETH-tight exact algorithm for Euclidean TSP. In: 59th Annual IEEE Symposium on Foundations of Computer Science, pp. 450–461 (2018)

Contents

Miscellaneous

Data Structures

Parallel and Distributed Algorithms

Packing and Covering

Invited Talks

Graph Profile Realizations
and Applications to Social Networks

Amotz Bar-Noy[1], Keerti Choudhary[2], David Peleg[2(✉)], and Dror Rawitz[3]

[1] City University of New York (CUNY), New York City, USA
amotz@sci.brooklyn.cuny.edu
[2] Weizmann Institute of Science, Rehovot, Israel
{keerti.choudhary,david.peleg}@weizmann.ac.il
[3] Bar Ilan University, Ramat-Gan, Israel
dror.rawitz@biu.ac.il

Abstract. The social standing of individuals in a social network is typically determined locally according to the individual's neighborhood or by a comparison between the individual and its neighbors. In this paper, we consider various criteria that measure social status and the extent in which individuals are satisfied with their social status. We study these criteria from the point of view of network realization: given a satisfaction specification, decide whether there exists a network realizing this specification.

1 Introduction

Consider a society consisting of a population V of n individuals with id's $V = \{1, \ldots, n\}$, connected by a social network. The *status* of individuals in such a society reflects their wealth, power or social influence. This paper focuses on questions related to possible *local status relations* of individuals in the network.

We consider two main settings. In the first, the individuals are fully ranked by their status. For simplicity, we assume that the status of an individual is represented by its id, namely, $i + 1$ has a higher status than i. Note that this ranking is independent of the structure of the social network. In the second setting, the status of an individual reflects the number of social connections it has in the network, namely, its degree. Naturally, other settings may be considered, e.g., one where individuals are grouped into (ranked) social classes, or one where individuals are grouped into unordered classes, say, ones reflecting opinions or political affinity.

We are interested in the way individuals view their social status, the criteria they apply to evaluate their social status, and the extent to which they are satisfied (or unsatisfied) with their relative social status. Status criteria may be *absolute* or *relative*, namely, dependent on comparisons with the status of other individuals in the nearby vicinity. Various notions by which an individual may compare itself with its neighbors were considered in the literature. We may say that a vertex i is *satisfied* if its rank or degree satisfies a certain (absolute or relative) condition.

G. K. Das et al. (Eds.): WALCOM 2019, LNCS 11355, pp. 3–14, 2019.
https://doi.org/10.1007/978-3-030-10564-8_1

In this context, one may study the relation between the *structure* of a given social network and the *satisfaction* of its members. The straightforward direction of this study involves looking at some given social networks and analyzing their *satisfaction profile*. Our focus here, however, is on the interesting *dual* problem where, rather than being given the *social network*, we are given a *satisfaction specification*, namely, a description of the *desired* pattern of satisfaction, and are asked whether there is a network *realizing* this specification, namely, a graph whose satisfaction profile conforms to the given specification.

In this paper, we introduce this line of study via a number of simple examples, illustrating different aspects of the problem.

Related Work. Dual problems of the type studied in this paper, sometimes referred to as *graph realization* or *graph construction* problems, were considered in a variety of contexts. The most well-studied problem of this type is that of *degree sequences*, a.k.a. *graphic sequences*. An n-element integer sequence σ is graphic if there exists an n-vertex graph whose degree sequence equals σ. Conditions for a sequence to be graphic and algorithms for deciding if a sequence is graphic and constructing a realizing graph were studied in [5,8,10–12,15–19], and some of them turned out to be relevant in the context of social networks, cf. [3,6,13]. A recent generalization of the problem, named the *graphic deviation* problem, concerns finding, for a given sequence σ that is not graphic, the graph whose degree sequence is closest to that of the given sequence [4]. Sampling questions on regular graphs were studied, e.g., in [20]. Other types of graph realizations are, e.g., the neighborhood list problem [1], the related shotgun assembly problem [14], and a number of other problem discussed in [2].

2   Rank-Based Notions of satisfaction

Let us start with a simple *relative* rank-based notion of satisfaction.

Example 1: $\mathcal{H}^{R>}$. Under this satisfaction measure, the profile is an n-entry vector $\langle s_1, \ldots, s_n \rangle$, where $0 \leq s_i \leq n-1$ for every i, and the requirement is that each vertex i has exactly s_i neighbors whose id is smaller than i.[1]

Observation 1. *The specification $\mathcal{H}^{R>}$ is realizable if and only if $s_i \leq i-1$ for every $1 \leq i \leq n$.*

Proof: Let us first suppose that the required condition holds. Then we construct G by adding, for every vertex $2 \leq i \leq n$, exactly s_i edges, connecting it to the vertices $\{i-1, \ldots, i-s_i\}$. (All the ranks $i-x$, for $1 \leq x \leq s_i$, do exist, since $i - x \geq i - s_i \geq i - i + 1 = 1$ by the required condition.)

Conversely, suppose $s_i \geq i$ for some $1 \leq i \leq n$. Then it is impossible to connect vertex i to s_i neighbors of rank lower than its own, since there are only $i - 1 < s_i$ such vertices altogether.   ∎

---

[1] A similar example can be studied where each vertex i has *at least* s_i such neighbors.

Next we turn to *local-patriotic* notions of satisfaction. The previous example compared the status of an individual to others in its neighborhood. One may consider also satisfaction notions based on the quality of one's neighborhood. This may be relevant in societies where being connected to the high echelons of society is of value even when one's own rank is not particularly high.

Example 2: $\mathcal{H}^{hs}(k, \epsilon)$. Here i is satisfied if at least an ϵ-fraction of its neighbors belong to "high society"[2], namely, have rank higher than k.

In this context, it makes little sense to consider isolated vertices. In fact, if isolated vertices are allowed, then the empty graph is a valid realization. We therefore examine what happens when every vertex must have at least one neighbor.

Let $A = \{i \mid i > k\}$ be the set of "highly ranked" vertices (of rank higher than k), and let $B = V \setminus A$ be the remaining vertices. By definition, $|A| = n - k$ and $|B| = k$.

We show that if the class of "highly ranked" vertices is not too small, i.e., not smaller than $\Theta(\sqrt{n})$, then there exists a realization. The next observation gives the exact bound. Let $\delta = \epsilon/(1 - \epsilon)$.

Lemma 1. *The specification $\mathcal{H}^{hs}(k, \epsilon)$ is realizable if and only if*

$$n - k - 1 \geq \delta \cdot \left\lceil \frac{k}{n - k} \right\rceil. \tag{1}$$

Proof: We first show that Condition (1) is sufficient. Given n and k, construct a split graph $G_{n,k}$ as follows. Connect the vertices of the set A by a clique and connect each vertex of B as a leaf to one vertex of A, in an evenly distributed way. We now show that if Condition (1) holds, then $G_{n,k}$ is a realizing graph for the specification $\mathcal{H}^{hs}(k, \epsilon)$.

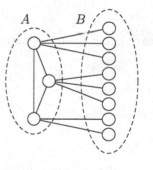

Let d_i^A and d_i^B denote the number of neighbors of i in G that belong to the sets A and B, respectively. It follows that $d_i^A = 1$ and $d_i^B = 0$ for $i \in B$, and that $d_i^A = |A| - 1 = n - k - 1$ and $d_i^B \in \{\lfloor k/(n - k) \rfloor, \lceil k/(n - k) \rceil\}$ for $i \in A$. Clearly, every vertex of B is satisfied. A vertex $i \in A$ is satisfied if $d_i^A \geq \epsilon \cdot \frac{d_i^A}{d_i^A + d_i^B}$, or, if $d_i^A \geq \delta \cdot d_i^B$. Condition (1) implies that $d_i^A = n - k - 1 \geq \delta \cdot \lceil k/(n - k) \rceil \geq \delta \cdot d_i^B$, hence the specification $\mathcal{H}^{hs}(k, \epsilon)$ is realized by $G_{n,k}$.

---

[2] One may consider also a non-uniform version of this example, where different vertices desire different percentiles or different thresholds.

Next we prove that Condition (1) is necessary, by showing that if a specification is realizable, then the construction $G_{n,k}$ realizes it as well. Consider a realizable specification $\mathcal{H}^{hs}(k,\epsilon)$, and let G be the realizing graph. One may modify G in three stages as follows. First, add all missing edges between vertices in A, and remove any edge between two vertices in B, producing a split graph. Notice that the modification may only increase d_i^A for $i \in A$ and decrease d_j^B for $j \in B$. Hence, the resulting graph G_1 still realizes the specification. Next, for every $j \in B$, remove all edges from j to A but one. As $d_j^B = 0$ for every $j \in B$, j remains satisfied. Also, for $i \in A$, d_i^B may only decrease and thus i remains satisfied. It follows that the resulting G_2 realizes the specification. Finally, balance the degrees of vertices in A, by evenly distributing the leaves of B among them. Notice that the resulting graph is $G_{n,k}$. Since $\max_{i \in A} d_i^B$ may only decrease during this stage, this graph still realizes the specification $\mathcal{H}^{hs}(k,\epsilon)$. Note that in $G_{n,k}$, there must exist at least one vertex $i \in A$ such that $d_i^B = \lceil k/(n-k) \rceil$. Also $d_i^A = n - k - 1$. Since $G_{n,k}$ realizes the specification $\mathcal{H}^{hs}(k,\epsilon)$, it follows that i is satisfied, so $n - k - 1 = d_i^A \geq \delta \cdot d_i^B = \delta \cdot \lceil k/(n-k) \rceil$, implying Condition (1). Hence the condition is necessary.   ∎

Corollary 1. *(1) The condition $n-k \geq \sqrt{\delta}\sqrt{n}+1$ is sufficient for the realizability of specification $\mathcal{H}^{hs}(k,\epsilon)$; (2) The condition $n - k \geq \sqrt{\delta}\sqrt{n} - \delta$ is necessary for the realizability of specification $\mathcal{H}^{hs}(k,\epsilon)$.*

Proof: To prove Part (1), suppose $n - k \geq \sqrt{\delta}\sqrt{n} + 1$ holds. Rearranging and squaring, we get $(n - k - 1)^2 \geq \delta n$, hence also

$$(n - k - 1)(n - k) \geq \delta n.$$

Rearranging again, we have

$$n - k - 1 \geq \delta \cdot \frac{n}{n-k} = \delta \cdot \left(\frac{k}{n-k} + 1\right),$$

which implies that

$$n - k - 1 \geq \delta \cdot \left\lceil \frac{k}{n-k} \right\rceil.$$

By Lemma 1, the specification is realizable.

To prove Part (2), suppose the specification is realizable. Therefore, Condition (1) of Lemma 1 holds, implying that $n - k - 1 \geq \delta \cdot k/(n-k)$. Rewriting and squaring, we get that

$$(n - k)^2 \geq (\delta k/(n-k) + 1)(n-k) = \delta k + n - k = \delta n + (1 - \delta)(n - k),$$

or equivalently,

$$(n - k)^2 + (\delta - 1)(n - k) \geq \delta n.$$

Therefore also

$$(n - k + \delta)^2 = (n - k)^2 + 2\delta(n - k) + \delta^2 \geq \delta n.$$

Taking the square root and rearranging again, we get that $n - k \geq \sqrt{\delta}\sqrt{n} - \delta$, establishing the claim. ∎

Corollary 2. *For constant ϵ, the specification is realizable if and only if the number of highly ranked vertices in A satisfies $n - k = \Omega(\sqrt{n})$.*

In particular, note that $\delta = 1$ for $\epsilon = 1/2$. Corollary 1 then implies that in order to ensure that at least half the neighbors of every vertex i are of rank k or higher, the number of vertices of rank k or higher must be at least $\sqrt{n} - 1$, and $\sqrt{n} + 1$ such vertices suffice.

3 Degree-Based Notions of satisfaction

The satisfaction notions discussed next refer, e.g., to connections and influence, and hence rely on the vertex degrees, rather than on their id's. Our first example considers an *absolute* degree-based notion of satisfaction.

Example 3: $\mathcal{H}^D(k)$. Here, a vertex i is satisfied if its degree is at least k. As the ranks do not play a role in the definition of satisfaction, the profile can be condensed into a pair $\langle n, \ell \rangle$, for $n \geq 2$ and $0 \leq \ell \leq n$, specifying that exactly ℓ of the n vertices in the graph are satisfied.

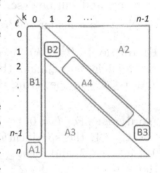

Analysis: For an n-vertex specification, the range $0 \leq k \leq n - 1$ and $0 \leq \ell \leq n$ can be classified into two categories, the realizable and non-realizable specifications. Each category can be partitioned further into a number of sub-ranges as follows (see figure for a schematic illustration of the partition).

Category A: Realizable specifications.

Case A1: $k = 0$ and $\ell = n$. This case is realizable by the empty graph.
Case A2: $k \geq 1$ and $\ell \leq k - 1$.

A possible realization is a split graph consisting of a clique K_ℓ of ℓ vertices and an independent set I of $n - \ell$ vertices, where all edges in $K_\ell \times I$ are contained in the edge set. In this case, the degree of each of the ℓ vertices in K_ℓ is $n - 1 \geq k$ and the degree of each vertex in I is $\ell < k$. See the figure, in which the dotted lines represent edges.

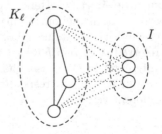

Case A3: $k \geq 1$ and $\ell \geq k+1$. A possible realization is a split graph consisting of a clique K_ℓ of ℓ vertices and an independent set I of $n - \ell$ vertices, where none of the edges in $K_\ell \times I$ are contained in the edge set. In this case, the degree of each of the ℓ vertices in K_ℓ is $\ell - 1 \geq k$ and the degree of each vertex in I is $0 < k$. See the previous figure, omitting the dotted lines.

Case A4: $2 \leq k = \ell \leq n - 2$.

A possible realization is a split graph consisting of a clique K_ℓ of ℓ vertices, an independent set I of $n - \ell - 2$ vertices, where none of the edges in $K_\ell \times I$ exist, and two vertices u and v such that every vertex of K_ℓ is connected to exactly one of u or v, while the degree of u and v is at least 1. In this case, the degree of each of the ℓ vertices in K_ℓ is $\ell = k$, the degree of each vertex in I is 0, and the degree of both u and v is less than $\ell = k$. See figure (where $n = 6$ and $\ell = 3$).

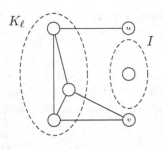

Category B: Non-realizable specifications. The three easily verified cases are **(B1)** $k = 0$ and $\ell < n$, **(B2)** $k = \ell = 1$, and **(B3)** $k = \ell = n - 1$.

We remark that for the measure $\mathcal{H}^D(k)$ we have also a full characterization for the special cases where the realizing graph must be connected and when it is required to be a forest or a tree. These characterizations are omitted for lack of space, and will be described in the full paper.

Next we consider *relative* degree-based notions of satisfaction.

Example 4: $\mathcal{H}^{DN}(k)$. A vertex i is satisfied if its degree is greater than at least k of its neighbors. Again, the profile can be represented by a pair $\langle n, \ell \rangle$, for $n \geq 2$ and $0 \leq \ell \leq n$, specifying that exactly ℓ of the n vertices are satisfied.

Analysis: For $k = 0$, the only realizable specification is clearly $\langle n, n \rangle$. For $k = 1$, the specification is realizable for $n \in \{2, 3, 4, 5, 6\}$ if and only if $0 \leq \ell \leq n - 2$, and for $n \geq 7$ if and only if $0 \leq \ell \leq n - 1$ (see [2] for details). Hereafter, we consider $2 \leq k \leq n - 1$.

It is easy to verify that for $\ell \geq n - k + 1$, the specification $\langle n, \ell \rangle$ is non-realizable, since the k lowest degree vertices in the graph must be unsatisfied.

For $\ell \leq n - k$, the specification is realizable by a split graph construction similar to that of case A2 in the previous Example 3.

Example 5: $\mathcal{H}^{Dav}$. Here i is satisfied if its degree is greater than the average degree of its neighbors. Conversely, i is unsatisfied if its degree is less than the average degree of its neighbors. A vertex that is neither satisfied nor unsatisfied is referred to as *indifferent*. (Note that a vertex without neighbors is considered indifferent.) In this case, the profile is represented by a triple $\langle n, h, s \rangle$, where h, s and $n - h - s$, respectively, are the number of satisfied, unsatisfied, and indifferent vertices.

Analysis: The specifications can be classified into two categories, the realizable specifications and the non-realizable ones. Each category can be partitioned further into a number of sub-ranges as follows (see figure for a schematic illustration of the partition.)

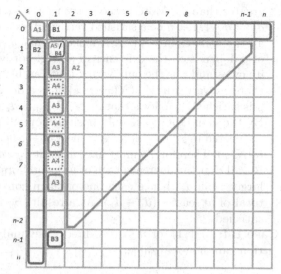

Category A: Realizable specifications.

Case A1: $\langle n, 0, 0 \rangle$: This specification is realizable by a complete graph K_n.

Case A2: $\langle n, h, s \rangle$, for $h \geq 1$ and $s \geq 2$: This case is realizable using a clique K_{n-h-s} on $n - h - s$ vertices and a split graph where one side is a clique K_h and the other is an independent set I_s. The edge set of the split graph contains all possible edges between the two sides.

Case A3: $\langle n, h, 1 \rangle$, where $2 \leq h \leq n-3$ is even. This case is realizable as follows. First construct a clique K_{h+2} over the vertices $\{1, \ldots, h+2\}$. Remove from it the matching $\{(1, 2), (3, 4), \ldots, (h-1, h)\}$ (of $h/2$ edges), and connect the vertices $\{1, 2, \ldots, h\}$ to an additional new vertex $h+3$. Finally, add a separate clique K_{n-3-h} on the vertices $\{h+4, \ldots, n\}$.

Case A4: $\langle n, h, 1 \rangle$, where $3 \leq h \leq n-5$ is odd. This case is realizable as follows. First construct a clique K_{h+4} over the vertices $\{1, \ldots, h+4\}$. Remove from it a cycle on the h vertices $\{1, 2, \ldots, h\}$ and the two edges $\{(h+1, h+2), (h+3, h+4)\}$ on the remaining four vertices. Connect the cycle vertices to the vertex $h+5$. Finally, add a separate clique K_{n-h-5} on the vertices $\{h+6, \ldots, n\}$.

Case A5: $\langle n, 1, 1 \rangle$, where $n \geq 6$. This case is realizable using a construction similar to that of the figure in case A6 of Example 6.

Category B: Non-realizable specifications.

Case B1: $\langle n, 0, s \rangle$, for $s > 0$: If there were a realization, then $\sum_i \sum_{j \in N(i)} d_i > \sum_i d_i^2$, yielding a contradiction (as these sums are always equal in graphs).

Case B2: $\langle n, h, 0 \rangle$, for $h > 0$: If there were a realization, then $\sum_i \sum_{j \in N(i)} d_j < \sum_i d_i^2$, again yielding a contradiction.

Case B3: $\langle n, n-1, 1 \rangle$: We show that there is no realization in this case. Assume, toward contradiction, that G is a realization with $n-1$ satisfied vertices and one unsatisfied vertex. Let x be the only unsatisfied vertex and denote its degree by d_x. Observe that all the other degrees in the graph are strictly greater than d_x. Moreover, observe that $V \setminus (N(x) \cup \{x\}) \neq \emptyset$, since otherwise $d_x = n - 1$. Let y be a satisfied vertex whose degree d_y is the minimum among all the vertices in $V \setminus \{x\}$. It must be that $y \in N(x)$, because the satisfaction of y depends on the existence of a neighbor whose degree is strictly less than its own (and x is the only vertex that can fit this requirement). It also follows that any vertex in $V \setminus (N(x) \cup \{x\})$ has a degree strictly larger than d_y. Since y is satisfied, it follows that $d_y^2 > \sum_{z \in N(y)} d_z$, or $S_y = \sum_{z \in N(y)} (d_y - d_z) > 0$. By the choice of y, the only positive contribution to S_y is $d_y - d_x$. Now y is connected to x and can have at most $d_x - 1$ neighbors in $N(x)$, since it belongs to $N(x)$. Therefore, y must have at least $d_y - d_x$ neighbors in $V \setminus (N(x) \cup \{x\})$. Recall that the degree of each of y's neighbors from $V \setminus (N(x) \cup \{x\})$ is strictly larger than d_y, hence each one of them contributes at least -1 to S_y, for a total of at least $-(d_y - d_x)$. As a result, S_y is non-positive and y cannot be satisfied.

Case B4: $\langle n, 1, 1 \rangle$ for $n \leq 5$: A direct case analysis reveals that these specifications are non-realizable.

We suspect that the remaining unclassified cases, namely, $\langle n, n - 4, 1 \rangle$ for odd n, $\langle n, n - 3, 1 \rangle$ for even n, and $\langle n, n - 2, 1 \rangle$ for any n, are also non-realizable.

The satisfaction profile based on the measure $\mathcal{H}^{Dav}$ is related to the well-known *friendship paradox*, by which in most social networks, most vertices are unsatisfied, in the sense that their neighbors have more neighbors (on average) than themselves [7,9].

Note that there are no realizations with $n - 1$ satisfied vertices, but one can have $n - 1$ unsatisfied vertices. This asymmetry may possibly be viewed as another by-product of the Friendship Paradox, in which high degree vertices influence the situation more than low degree vertices.

Finally, we consider a *homophilic* degree-based notion of satisfaction. *Homophily* refers to the tendency for people to have (non-negative) ties with people who are similar to themselves in socially significant ways. In social networks (graphs), homphily is usually measured by the similarities among the degrees. We say that a vertex is close to regularity if its degree is "similar" to the degrees of its neighbors. In its extreme manifestation, hereafter referred to as *regularity*, the degree of a regular vertex must be equal to the degrees of all of its neighbors[3]. By definition, if all the vertices are regular then the graph is a regular graph. Moreover, under most definitions for regularity, all the vertices of a regular graph are regular.

In a social network, an interesting realization problem may be to explore constructions that guarantee homophily for a specified number of vertices. This motivates the following example.

Example 6: $\mathcal{H}^{DH}$. Vertex i is satisfied if it is regular, namely, its degree is the same as the degrees of all of its neighbors.

---

[3] A more relaxed version requires the vertex degree to be roughly equal to the average degree of its neighbors. See the previous section.

Again, the specification $\langle n, \ell \rangle$ requires that exactly ℓ of the n vertices in the graph be regular.

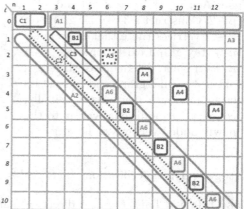

Analysis: We classify the specifications $\langle n, \ell \rangle$ of $\mathcal{H}^{DH}$ into three categories: the specifications realizable by a connected graph, those that are realizable by a disconnected graph (where an isolated vertex is considered to be regular) but not by a connected graph, and the non-realizable specifications.

These categories can be partitioned further into a number of sub-ranges as follows (see above figure for a schematic illustration of the partition.)

Category A: Specifications $\langle n, \ell \rangle$ realizable by a connected graph.

Case A1: $n \geq 3$ and $\ell = 0$. The specifications are realizable by a star graph with $n - 1$ leaves.

Case A2: $n \geq 1$ and $\ell = n$. These specifications are realizable by a clique K_n.

Case A3: $n \geq 5$ and $1 \leq \ell \leq n-3$, where $\ell \neq n/2 - 1$.

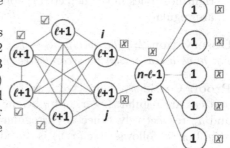

These specifications are realizable as follows. Start with a clique of size $\ell + 2$ and a star with root s and $n - \ell - 3$ leaves. Erase an arbitrary edge (i, j) from the clique and connect s to i and j, as in the figure (where the number in each vertex is its degree and there is an x next to the irregular vertices).

Case A4: $n \geq 8$, n even, and $\ell = n/2 - 1$. The construction of **case A3** fails when $\ell + 1 = n - \ell - 1$, since in this case the degree of the star root s equals those of the clique vertices, so the above graph satisfies the specification $(n, \ell+3)$. However, noting that when $n \geq 8$ the original star has $n - \ell - 3 \geq 2$ leaves, the above construction can be modified by disconnecting one of the leaves of the star from the root s and connecting it to one of the other leaves of the star.

Case A5: $n = 6$ and $\ell = 2$. Here, the 6-vertex path is a realization.

Case A6: $n \geq 6$, n even, and $\ell = n-2$. Construct a 3-regular graph on $n-2$ vertices (note that a 3-regular n-vertex graph exists for every even $n \geq 4$.) Next, delete one of the edges of the 3-regular graph and connect its endpoints to a new vertex. Finally, connect one additional vertex to the newly added vertex (see figure).

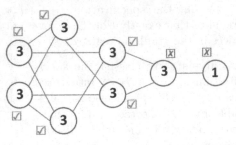

Category B: Specifications $\langle n, \ell \rangle$ realizable only by a disconnected graph.

Case B1: $n = 4$ and $\ell = 1$. A direct case analysis reveals that this specification is unrealizable by a connected graph. A disconnected realization consists of a 3-vertex path and an isolated vertex.

Case B2: $n \geq 7$, n odd, and $\ell = n - 2$. This case can be realized by an n-vertex disconnected graph composed of the 6-vertex graph depicted in the adjoining figure and a clique on $n - 6$ vertices. All vertices but the two rightmost ones in the depicted graph are regular.

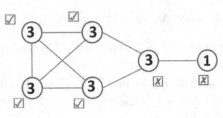

Observation 2: *The specification $\mathcal{H}^{DH}$ for $n \geq 6$, n odd, and $\ell = n - 2$, does not have a realization by a connected graph.*

Proof: Consider a connected n-vertex graph G, $n \geq 6$, in which exactly $n - 2$ vertices are regular. Let x and y be the irregular vertices, whose degrees are d_x and d_y respectively. Since any irregular vertex must have at least one irregular neighbor, it follows that (x, y) is an edge in G. Moreover, the edge (x, y) is a bridge in G, because all the rest of the vertices in the graphs have degree d_x or d_y and there is no edge between a vertex of degree d_x and a vertex of degree d_y. It follows that $d_x \neq d_y$, since otherwise the entire graph G is regular, in contradiction to the assumption that G has two irregular vertices.

Omit the edge (x, y) from G to get two disjoint graphs, an n_x-vertex graph G_x and an n_y vertex graph G_y. The degree of x in G_x is $d_x - 1$, therefore the sum of the degrees in G_x is $n_x \cdot d_x - 1$. This sum must be even, hence n_x is odd. Similarly n_y is odd. As a result, $n = n_x + n_y$ is an even number. ∎

Category C: Non-realizable specifications $\langle n, \ell \rangle$.

Case C1: $n \in \{1, 2\}$ and $\ell = 0$. The specification cannot be realized for $n \in \{1, 2\}$ with $\ell = 0$ since any graph of one or two vertices is regular.

Case C2: $n \geq 2$ and $\ell = n - 1$. The specification cannot be realized for $n \geq 3$ and $\ell = n - 1$. To see this, suppose that there exists a realizing graph G, and w.l.o.g. let n be the irregular vertex. If its degree is $d_n = 0$, then n is regular too, a contradiction. Therefore, $d_n = k > 0$ and let $i_1, \ldots, i_k$ be its neighbors. Since each such i_j is regular, it must satisfy $d_{i_j} = d_n = k$. Hence, the degree of vertex n is equal to the degrees of all of its neighbors, which makes n regular, a contradiction.

Case C3: $n \in \{3, 4, 5\}$ and $\ell = n - 2$. In each of these cases, it can be verified by direct case analysis that the specification cannot be realized.

Finally, let us remark that for the measure $\mathcal{H}^{DH}$ as well we have also a full characterization for the special case where the realizing graph is required to be a forest or a tree. This characterization is again deferred to the full paper for lack of space.

References

1. Aigner, M., Triesch, E.: Realizability and uniqueness in graphs. Discret. Math. **136**, 3–20 (1994)
2. Bar-Noy, A., Choudhary, K., Peleg, D., Rawitz, D.: Structural information and communication complexity. In: Lotker, Z., Patt-Shamir, B. (eds.) SIROCCO 2018. LNCS, vol. 11085, pp. 3–13. Springer, Heidelberg (2018). https://doi.org/10.1007/978-3-030-01325-7_1
3. Blitzstein, J.K., Diaconis, P.: A sequential importance sampling algorithm for generating random graphs with prescribed degrees. Internet Math. **6**(4), 489–522 (2011)
4. Broom, M., Cannings, C.: Graphic deviation. Discret. Math. **338**, 701–711 (2015)
5. Choudum, S.A.: A simple proof of the Erdös-Gallai theorem on graph sequences. Bull. Aust. Math. Soc. **33**(1), 67–70 (1991)
6. Cloteaux, B.: Fast sequential creation of random realizations of degree sequences. Internet Math. **12**(3), 205–219 (2016)
7. Eom, Y.-H., Jo, H.-H.: Generalized friendship paradox in complex networks: the case of scientific collaboration. Sci. Rep. **4**(4603), 1–6 (2014)
8. Erdös, P., Gallai, T.: Graphs with prescribed degrees of vertices [Hungarian]. Matematikai Lapok **11**, 264–274 (1960)
9. Feld, S.L.: Why your friends have more friends than you do. Am. J. Sociol. **96**, 1464–1477 (1991)
10. Hakimi, S.L.: On realizability of a set of integers as degrees of the vertices of a linear graph - I. SIAM J. Appl. Math. **10**(3), 496–506 (1962)
11. Havel, V.: A remark on the existence of finite graphs [in Czech]. Casopis Pest. Mat. **80**, 477–480 (1955)
12. Kelly, P.J.: A congruence theorem for trees. Pacific J. Math. **7**, 961–968 (1957)
13. Mihail, M., Vishnoi, N.: On generating graphs with prescribed degree sequences for complex network modeling applications. In: 3rd Workshop on Approximation and Randomization Algorithms in Communication Networks (2002)
14. Mossel, E., Ross, N.: Shotgun assembly of labeled graphs. CoRR, abs/1504.07682 (2015)

15. O'Neil, P.V.: Ulam's conjecture and graph reconstructions. Am. Math. Mon. **77**, 35–43 (1970)
16. Sierksma, G., Hoogeveen, H.: Seven criteria for integer sequences being graphic. J. Graph Theory **15**(2), 223–231 (1991)
17. Tripathi, A., Tyagi, H.: A simple criterion on degree sequences of graphs. Discret. Appl. Math. **156**(18), 3513–3517 (2008)
18. Ulam, S.M.: A Collection of Mathematical Problems. Wiley, Hoboken (1960)
19. Wang, D.L., Kleitman, D.J.: On the existence of n-connected graphs with prescribed degrees ($n > 2$). Networks **3**, 225–239 (1973)
20. Wormald, N.C.: Models of random regular graphs. Surv. Comb. **267**, 239–298 (1999)

Parameterized Computational Geometry via Decomposition Theorems

Fahad Panolan[1], Saket Saurabh[2(✉)], and Meirav Zehavi[3]

[1] University of Bergen, Bergen, Norway
fahad.panolan@uib.no
[2] The Institute of Mathematical Sciences, HBNI, Chennai, India
saket@imsc.res.in
[3] Ben-Gurion University, Beersheba, Israel
meiravze@bgu.ac.il

Abstract. Parameterized complexity is one of the most established algorithmic paradigms to deal with computationally hard problems. In the first two decades, the field largely focused on problems arising from studies of graphs and networks. However, lately the focus has changed substantially and it has started to permeate into other fields such as computational geometry, and computational social choice theory. In this article, we will survey some exciting developments in the emerging field of parameterized computational geometry through our contributions. We will focus on designing efficient parameterized algorithms on unit-disk graphs via new graph decomposition theorems.

1 Introduction

While many interesting graph problems remain NP-complete even when restricted to geometric graphs such as planar graphs and unit-disk graphs, the restriction of a problem to geometric graphs is usually considerably more tractable algorithmically than the problem on general graphs. Over the last four decades, it has been proved that many graph problems on planar graphs admit subexponential time algorithms [16,22,29], subexponential time parameterized algorithms [1,28,30], linear kernels [2,4,6] and (Efficient) Polynomial Time Approximation Schemes ((E)PTAS) [3,10,17,18,23,24,27].

It is very natural to repeat the algorithmic successes on planar graphs and more generally on graphs excluding a fixed graph H as a minor on other geometric graphs. In last few years, there has been successful attempt in extending the methods and techniques used for designing efficient algorithms (approximation, exact or parameterized) on planar graphs to geometric graphs such as unit-disk graphs and map graphs. In this article we will focus only on unit-disk graphs and describe various new tools, primarily graph decomposition theorems for unit-disk graphs and show its applicability in designing efficient parameterized algorithms.

We start by giving two well known methods for designing efficient parameterized algorithms, namely, the theory of bidimensionality [11–13] and Baker style

G. K. Das et al. (Eds.): WALCOM 2019, LNCS 11355, pp. 15–27, 2019.
https://doi.org/10.1007/978-3-030-10564-8_2

vertex/edge decomposition theorems [3] (Sect. 3). In Sect. 4 we extend the ideas of bidimensionality to unit-disk graphs [21]. However, this method only works when the problem in question is easily solvable on cliques. So to extend the applicability of these ideas, we define a notion of clique-grid graphs [20] and show its utility in Sect. 5. In particular we do as follows. If an input n-vertex unit-disk graph G contains a clique of size $\mathsf{poly}(k)$ (such a clique can be found in polynomial time), then we have a trivial YES-instance or NO-instance, depending on the problem. Otherwise, we show that the unit disk graph G in a YES-instance of the problem admits, sometimes after a polynomial time preprocessing, a specific type of (ω, Δ, τ)-decomposition, where the meaning of ω, Δ and τ is as follows. The vertex set of G is partitioned into cliques $C_1, \ldots, C_d$, each of size at most $\omega = k^{\mathcal{O}(1)}$. We also require that after contracting each of the cliques C_i to a single vertex, the maximum vertex degree Δ of the obtained graph $\tilde{G}$ is $\mathcal{O}(1)$, while the treewidth τ of $\tilde{G}$ is $\mathcal{O}(\sqrt{k})$. Moreover, the corresponding tree decomposition of $\tilde{G}$ can be constructed efficiently. We use the tree decomposition of $\tilde{G}$ to construct a tree decomposition of G by "uncontracting" each of the contracted cliques C_i. While the width of the obtained tree decomposition of G can be of order $\omega \cdot \tau = k^{\mathcal{O}(1)}$, we show that each of our parameterized problems can be solved in time $f(\Delta) \cdot \omega^{f(\Delta) \cdot \tau}$. Here we use dynamic programming over the constructed tree decomposition of G, however there is a twist from the usual way of designing such algorithms. This part of the algorithm is problem-specific—in order to obtain the claimed running time, we have to establish a very specific property for each of the problems. Finally, in Sect. 6, we give a Baker style contraction decomposition on unit-disk graphs and show its utility in designing faster parameterized algorithm for MINIMUM BISECTION. The final results are based on [19].

2   Preliminaries

In this section we give some important notations. All graphs considered in this paper are finite, undirected and simple. For the terms which are not explicitly defined here, we use standard notations from [15]. For a graph G, its vertex set is denoted by $V(G)$ and its edge set is denoted by $E(G)$. For a vertex $v \in V(G)$, its (open) *neighbourhood* $N_G(v)$ is the set of all vertices adjacent to it and its *closed neighborhood* is the set $N_G(v) \cup \{v\}$. Given an edge $e = xy$ of a graph G, the graph G/e is obtained from G by contracting e. That means that the endpoints x and y are replaced by a new vertex $v_{x,y}$ which is adjacent to the old neighbors of x and y (except for x and y). A graph H obtained by a sequence of edge-contractions is said to be a *contraction* of G. A graph H is a *minor* of a graph G if H is the contraction of some subgraph of G. Let G, H be two graphs. A subgraph G' of G is said to be a *minor-model* of H in G if G' contains H as a minor. We say that a graph G is H-*minor-free* when it does not contain H as a minor. We also say that a graph class $\mathcal{G}$ is H-*minor-free*. A graph class $\mathcal{G}$ is said to be *minor-closed/contraction-closed* if every minor/contraction of a graph in $\mathcal{G}$ also belongs to $\mathcal{G}$.

Given a positive integer t, we denote by $\boxplus_t$ the $t \times t$ grid. Formally, for a positive integer t, a $t \times t$ *grid* $\boxplus_t$ is a graph with vertex set $\{(x,y) \; : \; x,y \in \{1,\ldots,t\}\}$. Thus $\boxplus_t$ has exactly t^2 vertices. Two different vertices (x,y) and (x',y') are adjacent if and only if $|x - x'| + |y - y'| = 1$.

Planar Graphs. A *drawing* of a graph G on the plane is the mapping of each vertex in $V(G)$ to a point in the plane, and of each edge $\{u,v\} \in E(G)$ to a plane curve whose extreme points are the points mapped to u and v. An *embedding* of a graph G on the plane is a drawing of G on the plane such that any two distinct plane curves of the drawing can intersect only at their endpoints. A *planar graph* is a graph that can be embedded on the plane, and a *plane graph* is a planar graph with a fixed embedding.

Unit Disk Graphs. Given a set of geometric objects O, the *intersection graph* of O is the graph G with the vertex set $V(G) = O$ and the edge set $E(G) = \{\{u,v\} \; : \; u,v \in O, \; u \cap v \neq \emptyset\}$. That is, every geometric object is represented by a vertex, and two vertices are adjacent if and only if the objects that they represent intersect.

Let $P = \{p_1 = (x_1,y_1), p_2 = (x_2,y_2), \ldots, p_n = (x_n,y_n)\}$ be a set of points in the (Euclidean) plane. In the *unit disk graph (UDG) model*, for every $i \in [n]$, we let d_i denote the disk of radius 1 whose centre is p_i. Accordingly, we denote $D = \{d_1, d_2, \ldots, d_n\}$. Then, the *UDG of D* is the intersection graph of D.

Next we define the notion of a *tree-decomposition* of a graph G.

Definition 1. *A* tree decomposition *of a graph G is a pair $\mathcal{T} = (T,\beta)$, where T is a tree and β is a function from $V(T)$ to $2^{V(G)}$, that satisfies the following conditions.*

- $\bigcup_{x \in V(T)} \beta(x) = V(G)$.
- *For every edge $\{u,v\} \in E(G)$ there exists $x \in V(T)$ such that $\{u,v\} \subseteq \beta(x)$.*
- *For every vertex $v \in V(G)$, if $v \in \beta(x) \cap \beta(y)$ for some $x,y \in V(T)$, then $v \in \beta(z)$ for all z on the unique path between x and y in T.*

The width *of $\mathcal{T}$ is $\max_{x \in V(T)} |\beta(x)| - 1$. Each $\beta(x)$ is called a* bag. *The* treewidth *of G, denoted by $\mathbf{tw}(G)$, is the minimum width over all tree decompositions of G.*

3   Two Tools on Planar Graphs

In this section we give two well-known methods in designing efficient parameterized algorithms on planar graphs. The first one generally yields a parameterized subexponential algorithms.

3.1   Grid Theorems and Bidimensionality

Most of the modern results on planar graph use the notion of treewidth in crucial ways. Let $\mathcal{G}_t$ be a graph class comprising graphs of treewidth at most t.

Treewidth is a minor-closed parameter and hence the class $\mathcal{G}_t$ comprising graphs of treewidth at most t is closed under taking minors. From the Graph Minors theorem it follows that the property of having treewidth at most t can be characterized by a finite set of forbidden minors $\mathrm{Forb}(\mathcal{G}_t)$. That is, a graph G has treewidth at most t if and only if it does not contain any graph $H \in \mathrm{Forb}(\mathcal{G}_t)$ as a minor. These forbidden graphs give us algorithmic handle on problems we want to solve on planar graphs. But what do graphs of $\mathrm{Forb}(\mathcal{G}_t)$ look like? We apparently do not know the answer to this question. However, we would like to get some "approximate characterization" for graphs having treewidth at most t that is more tractable. Grids form the desired obstacle for planar graphs. The following theorem is due to Robertson, Seymour and Thomas; we present here a version with refined constants due to Gu and Tamaki.

Theorem 1 (Planar excluded grid theorem, [25,32]). *Let t be a nonnegative integer. Then every planar graph G of treewidth at least $9t/2$ contains $\boxplus_t$ as a minor. Furthermore, for every $\epsilon > 0$ there exists an $\mathcal{O}(n^2)$ algorithm that, for a given n-vertex planar graph G and integer t, either outputs a tree decomposition of G of width at most $(9/2 + \epsilon)t$, or constructs a minor model of $\boxplus_t$ in G.*

The planar excluded grid theorem (Theorem 1) provides a powerful tool for designing parametrized subexponential time algorithms on planar graphs. In all these algorithms we use the win/win approach. We first approximate the treewidth of a given planar graph. If the treewidth turns out to be small, we use standard dynamic programing to find a solution. Otherwise, we know that our graph contains a large grid as a minor, and using this we should be able to conclude the right answer to the instance.

Let us see an example of this strategy. Let us first look at PLANAR VERTEX COVER, i.e., for a given planar graph G and parameter k, we need to determine whether there exists a vertex cover of G of size at most k. We need to answer the following three simple questions.

(i) How small can be a vertex cover of $\boxplus_t$? It is easy to check that $\boxplus_t$ contains a matching of size $\lfloor t^2/2 \rfloor$, and hence every vertex cover of $\boxplus_t$ is of cardinality at least $\lfloor t^2/2 \rfloor$.

(ii) Given a tree decomposition of width t of G, how fast can we solve VERTEX COVER? This can be done in time $2^t \cdot t^{\mathcal{O}(1)} \cdot n$ [8].

(iii) Is VERTEX COVER minor-closed? In other words, is it true that for every minor H of graph G, the vertex cover of H does not exceed the vertex cover of G?

The class of graphs with vertex cover at most k is minor-closed, i.e., a graph G has a vertex cover of size at most k, then the same holds for every minor of G. Thus, if G contains $\boxplus_t$ as a minor for some $t \geq \sqrt{2k+2}$, then by (i) G has no vertex cover of size k. By the planar excluded grid theorem, this means that the treewidth of a planar graph admitting a vertex cover of size k is smaller than $\frac{9}{2}\sqrt{2k+2}$.

We summarize the above discussion with the following algorithm. For $t = \lceil \sqrt{2k+2} \rceil$ and some $\epsilon > 0$, by making use of the constructive part of Theorem 1 we either compute in time $\mathcal{O}(n^2)$ a tree decomposition of G of width at most $(\frac{9}{2}+\epsilon)t$, or we conclude that G has $\boxplus_t$ as a minor. In the second case, we infer that G has no vertex cover of size at most k. However, if a tree decomposition has been constructed, then by (ii) we can solve the problem in time $2^{(\frac{9}{2}+\epsilon)\lceil\sqrt{2k+2}\rceil} \cdot k^{\mathcal{O}(1)} \cdot n = 2^{\mathcal{O}(\sqrt{k})} \cdot n$. The total running time of the algorithm is hence $2^{\mathcal{O}(\sqrt{k})} \cdot n + \mathcal{O}(n^2)$.

Let us extract the properties of PLANAR VERTEX COVER which were essential for obtaining a subexponential parameterized algorithm.

(P1) The size of any solution in $\boxplus_t$ is of order $\Omega(t^2)$.

(P2) Given a tree decomposition of width t, the problem can be solved in time $2^{\mathcal{O}(t)} \cdot n^{\mathcal{O}(1)}$.

(P3) The problem is minor-monotone, i.e., if G has a solution of size at most k, then every minor of G also has a solution of size at most k.

The above method of designing subexponential parameterized algorithms is known as bidimensionality and we refer to [11–13] and book chapter in [8] for further information.

3.2   Shifting Technique and Decomposition Theorems

We now present another technique for obtaining fixed-parameter tractable algorithms for problems on planar graphs using treewidth. The main idea of the approach originates in the work on approximation schemes on planar graphs, pioneered in the 1980s by Baker [3]. The methodology is widely used in modern approximation algorithms, and is called the *shifting technique*, or simply *Baker's technique*. In this section we present a parameterized counterpart of this framework.

We will exhibit the method with the MINIMUM BISECTION problem. For a given n-vertex graph G and integer k, the task is to decide whether there exists a partition of $V(G)$ into sets A and B, such that $\lfloor n/2 \rfloor \leq |A|, |B| \leq \lceil n/2 \rceil$ and the number of edges with one endpoint in A and the second in B is at most k. In other words, we are looking for a balanced partition (A, B) with an (A, B)-cut of size at most k. Such a partition (A, B) will be called a *k-bisection* of G. For this problem we will use the following decomposition theorem.

Lemma 1 ([8]). *Let G be a planar graph and k be a nonnegative integer. Then the edge set of G can be partitioned into $k + 1$ sets such that after contracting edges of any of these sets, the resulting graph admits a tree decomposition of width at most ck, for some constant $c > 0$. Moreover, such a partition, together with tree decompositions of width at most ck of respective graphs, can be found in polynomial time.*

We remark here that on planar graph there are forms of Lemma 1, where on the place of edge contractions, we do edge deletions. Also, there is vertex version, where we partition the vertex set and the operation is vertex deletion. See [8, Chap. 7] for further information.

It will be convenient to work with a slightly more general variant of the MINIMUM BISECTION problem. In the following, we will assume that G can be a multigraph, i.e., it can have multiple edges between the same pair of vertices. Moreover, the graph comes together with a weight function $\mathbf{w} : V(G) \to \mathbb{Z}_{\geq 0}$ on vertices, and from a k-bisection (A, B) we will require that $\lfloor \mathbf{w}(V(G))/2 \rfloor \leq \mathbf{w}(A), \mathbf{w}(B) \leq \lceil \mathbf{w}(V(G))/2 \rceil$. Of course, by putting unit weights we arrive at the original problem.

Theorem 2 ([8]). MINIMUM BISECTION *on planar graphs can be solved in time* $2^{\mathcal{O}(k)} \cdot W \cdot n^{\mathcal{O}(1)}$, *where W is the maximum weight of a vertex.*

Proof. We use Lemma 1 to partition the set of edges of the input planar graph G into sets $S_0 \cup \cdots \cup S_k$. Note here that Lemma 1 is formally stated only for simple graphs, but we may extend it to multigraphs by putting copies of the same edge always inside the same set S_j.

Suppose that there exists a k-bisection (A, B) of G, and let F be the set of edges between A and B. Then at least one of the sets S_j is disjoint from F. Let us contract all the edges of S_j, keeping multiple edges but removing created loops. Moreover, whenever we contract some edge uv, we define the weight of the resulting vertex as $\mathbf{w}(u) + \mathbf{w}(v)$. Let G_j be the obtained multigraph. Since F is disjoint from S_j, during this contraction we could have just contracted some parts of $G[A]$ and some parts of $G[B]$. Therefore, the new multigraph G_j also admits a k-bisection (A', B'), where A', B' comprise vertices originating in subsets of A and B, respectively.

On the other hand, if for any G_j we find some k-bisection (A', B'), then uncontracting the edges of S_j yields a k-bisection (A, B) of G. These two observations show that G admits a k-bisection if and only if at least one of the multigraphs G_j does. However, Lemma 1 provided us a tree decomposition of each G_j of width $\mathcal{O}(k)$. Hence, we can apply a standard dynamic programming algorithm to each G_j, and thus solve the input instance (G, k) in time $2^{\mathcal{O}(k)} \cdot W \cdot n^{\mathcal{O}(1)}$. Note here that the maximum weight of a vertex in each G_j is at most nW. □

4 Bidimensionality on Unit-Disk Graphs

In this section we extend the bidimensionality framework to classes of geometric graphs, in particular on unit-disk graphs. The key ingredient of the framework on planar graphs was Planar excluded grid theorem (Theorem 1). We need a analogue of this result for unit-disk graphs. However, a family of unit-disk graphs contains arbitrarily large cliques and hence such linear excluded grid theorem is not possible. We overcome this obstacle by showing that cliques are the only pathological case. In particular, we show that the treewidth of every unit-disk graph excluding a clique of constant size as a subgraph and excluding a $k \times k$ grid as a minor, is $\mathcal{O}(k)$. Let $\mathcal{G}_U^t$ be the class of unit-disk graphs, not containing clique K_t on t vertices as a subgraph. We refer to such graphs as K_t-free graphs. In such graphs we have the following result.

Lemma 2 ([21]). *Any unit disk graph G with maximum vertex degree Δ contains a $\lfloor \frac{\mathbf{tw}(G)}{144\Delta^3} \rfloor \times \lfloor \frac{\mathbf{tw}(G)}{144\Delta^3} \rfloor$ grid as a minor.*

Lemma 2 is extremely useful for problems where cliques are easy to handle. We show the utility of Lemma 2 by designing a parameterized subexponential time algorithms for VERTEX COVER.

Theorem 3 ([21]). VERTEX COVER *admits a parameterized subexponential time $2^{\mathcal{O}(k^{0.75} \log k)} n^{\mathcal{O}(1)}$ algorithm on n-vertex unit disk graphs.*

Proof. Given k we set the value $c = k^\epsilon$ for a value of ϵ to be fixed later. The algorithm will pass down the value of c to recursive calls such that c remains fixed even though k changes. The algorithm proceeds as follows. Given an instance (G, k), it finds a maximum clique C of G. Recall that we can find a maximum clique in unit disk graphs in polynomial time [7,31]. If $|C| > k + 1$, then we return that G does not have vertex cover of size at most k. Next we check whether $|C| \leq c$.

If $|C| \leq c$, then the considered graph is in $\mathcal{G}_U^c$. Using Lemma 2 and the properties of a vertex cover, we can conclude that $\mathbf{tw}(G) \leq \mathcal{O}(k^{0.5+\epsilon})$. In this case we apply the known algorithm for VERTEX COVER that given a tree decomposition of width t of a graph G on n vertices, finds a minimum sized vertex cover in time $2^t n^{\mathcal{O}(1)}$ [8]. Hence, in this case the running time of our algorithm will be $2^{\mathcal{O}(k^{0.5+\epsilon} \log k)} n^{\mathcal{O}(1)}$.

In the case that $|C| > c$ we know that any vertex cover F of G contains almost all of the vertices in C, in particular, $|C \setminus F| \leq 1$. The algorithm branches on all $1 + |C|$ possibilities for $X = F \cap C$ and recursively solves the problem on $(G - X, k - |X|)$. If for some guess we have a yes answer, then we return yes, else, we return no. The running time of this step is guided by the following recurrence $T(k) \leq |C| \cdot T(k - (|C| - 1)) + T(k - |C|)$. Since $|C| \geq c$ a simple induction shows that $T(k) \leq (3c)^{4k/c}$ which again is upper bounded by $2^{\mathcal{O}(\frac{4k \log c}{c})} \leq 2^{\mathcal{O}(\frac{k \log k}{c})}$. Substituting k^ϵ for c this yields that the total number of branches explored by the algorithm is upper bounded by $2^{\mathcal{O}(k^{1-\epsilon} \log k)}$. Now we choose ϵ such the number of branches and the time spent in each branch is the same. Thus we choose an ϵ such that $2^{\mathcal{O}(k^{1-\epsilon} \log k)} = 2^{\mathcal{O}(k^{0.5+\epsilon})}$. This gives us that $\epsilon = 1/4$ is asymptotically best possible. Thus our algorithm runs in time $2^{\mathcal{O}(k^{0.75} \log k)} n^{\mathcal{O}(1)}$, concluding the proof. $\square$

Again, if we notice, the proof of Theorem 3 does not use any property of VERTEX COVER. Theorem 3 is applicable for all bidimensional problems for which there is a good branching strategy on cliques. However, for problems such as LONGEST PATH, given an undirected graph G and a non-negative integer k, does there exists a path of length at least k, the above approach does not work. In the next section we give our first decomposition theorem that allows us to do a dynamic programming algorithm with running time dependence subexponential on the parameter.

5   Clique-Grid Decomposition Theorem

In this section, we introduce a family of "grid-like" graphs, called clique-grid graphs, that is tailored to design subexponential algorithms on unit-disk graphs. Given a unit disk graph G, we extract the properties of G that we would like to exploit, and show that they can be captured by an appropriate clique-grid graph. Let us begin by giving the definition of a clique-grid graph. Roughly speaking, a graph G is a clique-grid graph if each of its vertices can be embedded into a single cell of a grid (where multiple vertices can be embedded into the same cell), ensuring that the subgraph induced by each cell is a clique, and that each cell can interact (via edges incident to its vertices) only with cells at "distance" at most 2. Formally,

Definition 2 ([20]). **[clique-grid graphs]** *A graph G is a clique-grid graph if there exists a function $f : V(G) \to [t] \times [t']$, for some $t, t' \in \mathbb{N}$, such that the following conditions are satisfied.*

1. *For all $(i, j) \in [t] \times [t']$, it holds that $f^{-1}(i, j)$ is a clique.*
2. *For all $\{u, v\} \in E(G)$, it holds that if $f(u) = (i, j)$ and $f(v) = (i', j')$ then*

$$|i - i'| \le 2 \text{ and } |j - j'| \le 2.$$

Such a function f is a representation of G.

We note that a notion similar to clique-grid graph was also used by Ito and Kadoshita [26]. One can show that unit-disk graphs are clique-grid graph.

Lemma 3 ([20]). *Let D be a set of points in the Euclidean plane, and let G be the unit disk graph of D. Then, a representation f of G can be computed in polynomial time.*

We conclude this section by introducing the definition of an ℓ-NCTD, which is useful for doing our dynamic programming algorithms.

Definition 3. *A tree decomposition $\mathcal{T} = (T, \beta)$ of a clique-grid graph G with representation f is a nice ℓ-clique tree decomposition, or simply an ℓ-NCTD, if for the root r of T, it holds that $\beta(r) = \emptyset$, and for each node $v \in V(T)$, it holds that*

- *There exist at most ℓ cells, $(i_1, j_1), \ldots, (i_\ell, j_\ell)$, such that $\beta(v) = \bigcup_{t=1}^{\ell} f^{-1}(i_t, j_t)$, and*
- *The node v is of one of the following types.*
  - **Leaf:** *v is a leaf in T and $\beta(v) = \emptyset$.*
  - **Forget:** *v has exactly one child u, and there exists a cell $(i, j) \in [t] \times [t']$ such that $f^{-1}(i, j) \subseteq \beta(u)$ and $\beta(v) = \beta(u) \setminus f^{-1}(i, j)$.*
  - **Introduce:** *v has exactly one child u, and there exists a cell $(i, j) \in [t] \times [t']$ such that $f^{-1}(i, j) \subseteq \beta(v)$ and $\beta(v) \setminus f^{-1}(i, j) = \beta(u)$.*
  - **Join:** *v has exactly two children, u and w, and $\beta(v) = \beta(u) = \beta(w)$.*

We have the following computation result about ℓ-NCTD.

Corollary 1 ([20]). *Given a clique-grid graph G that is a unit disk/square graph, a representation f of G and an integer $\ell \in \mathbb{N}$, in time $2^{\mathcal{O}(\ell)} \cdot n^{\mathcal{O}(1)}$, one can either correctly conclude that G contains an $\alpha\ell \times \alpha\ell$ grid as a minor, or compute a 5ℓ-NCTD of G, where $\alpha = \frac{1}{100 \cdot 599^3}$.*

5.1 Feedback Vertex Set

In this section, we show that FEEDBACK VERTEX SET admits a subexponential-time parameterized algorithm. More precisely, we prove the following.

Theorem 4 ([20]). FEEDBACK VERTEX SET *on unit-disk graphs can be solved in time $2^{\mathcal{O}(\sqrt{k}\log k)} \cdot n^{\mathcal{O}(1)}$.*

First, we prove that there is a $2^{\mathcal{O}(\sqrt{k}\log k)} \cdot n^{\mathcal{O}(1)}$ time algorithm which either concludes that there is no feedback vertex set of size k or outputs an $\mathcal{O}(\sqrt{k})$-NCTD of the input graph.

Lemma 4. *Let (G, O, k) be an instance of FEEDBACK VERTEX SET on unit disk graphs. Then, in time $2^{\mathcal{O}(\sqrt{k}\log k)} \cdot n^{\mathcal{O}(1)}$, one can either solve (G, O, k) or obtain an equivalent instance (G, f, k) of FEEDBACK VERTEX SET on clique-grid graphs together with an $\mathcal{O}(\sqrt{k})$-NCTD of G such that $|f^{-1}(L)| \leq k+2$ for any cell L.*

Proof. First, by using Lemmata 3, we obtain a representation f of G. Notice that if there is a cell L of f, such that $|f^{-1}(L)| \geq k+3$, then there is no feedback vertex set of size at most k in G, because $f^{-1}(L)$ is a clique of size at least $k+3$ in G. Now, by using Corollary 1 with $\ell = 200 \cdot 599^3 \cdot (\lceil\sqrt{k}\rceil + 1) = \mathcal{O}(\sqrt{k})$, we either correctly conclude that G contains a $2(\lceil\sqrt{k}\rceil + 1) \times 2(\lceil\sqrt{k}\rceil + 1)$ grid as a minor, or compute an $\mathcal{O}(\sqrt{k})$-NCTD of G. If there is a $2(\lceil\sqrt{k}\rceil+1) \times 2(\lceil\sqrt{k}\rceil+1)$ grid as a minor, then there are more than k vertex-disjoint cycles in G and hence (G, O, k) is a No-instance. □

Because of Lemma 4, to prove Theorem 4, we can focus on FEEDBACK VERTEX SET on clique-grid graphs, where the input also contains an $\mathcal{O}(\sqrt{k})$-NCTD. That is, the input of FEEDBACK VERTEX SET on clique-grid graphs is a tuple $(G, f, k, \mathcal{T})$ where G is a clique-grid graph with representation f, $|f^{-1}(L)| \leq k+2$ for all cells L of f and $\mathcal{T} = (T, \beta)$ is an $\mathcal{O}(\sqrt{k})$-NCTD of G. The following observation follows from the fact that $\mathcal{T} = (T, \beta)$ is an $\mathcal{O}(\sqrt{k})$-NCTD and $|f^{-1}(L)| \leq k+2$ for any cell L of f.

Observation 1. *For any $v \in V(T)$, $|\beta(v)| = \mathcal{O}(k^{1.5})$.*

Now, using a non-standard dynamic programming algorithm one can show the following which proves the main result of this section.

Lemma 5 ([20]). FEEDBACK VERTEX SET *on clique-grid graphs can be solved in time $2^{\mathcal{O}(\sqrt{k}\log k)} \cdot n^{\mathcal{O}(1)}$.*

6   Contraction Decomposition Theorem

In this section we give a Baker style contraction decomposition theorem (CDT) for UDG's. Notice that any clique of size n is a UDG. Thus, it is easy to see that we can have neither a vertex/edge form of Baker style decomposition theorems in UDGs. However, we obtain the following form of a CDT in UDGs.

Theorem 5. [Contraction Decomposition Theorem in UDGs, [19]] *Let G be an UDG and let $k \in \mathbb{N}$. Then, there exist fixed constants $\alpha, \beta \in \mathbb{N}$ (independent of G and k) and a family of subset of edges, $E_1, \ldots, E_{\alpha k}$, such that*

- *$\bigcup_{i=1}^{\alpha k} E_i = E(G)$,*
- *each edge $e \in E(G)$ belongs to at most β sets among $E_1, \ldots, E_{\alpha k}$, and*
- *contracting any E_i, $1 \leq i \leq \alpha k$, induces a graph of treewidth at most $\mathcal{O}(k^2)$.*

Moreover, such a family of subsets of edges, $E_1, \ldots, E_{\alpha k}$, together with tree decompositions of width at most $\mathcal{O}(k^2)$ of the respective graphs, can be found in polynomial time.

Notice that the decomposition theorem given in Theorem 5 is slightly weaker than a standard CDT in the following sense: an edge can participate in some constant number of sets rather than just one. In the literature, among other results, CDTs have been useful to design PTASes for problems that are contraction closed such as TSP [14], as well as FPT algorithms for cut problems such as MINIMUM BISECTION [5,8]. We remark that for this purpose, our slightly weaker form of a CDT suffices; that is, if the standard form of a CDT can be utilized, so is the weaker form.

In the realm of Parameterized Complexity, a direct application of Theorem 5 already brings us a $2^{\mathcal{O}(k^2)} n^{\mathcal{O}(1)}$-time algorithm for MINIMUM BISECTION; though faster than the known algorithm on general graphs [9], this running time is still not of the form $2^{\mathcal{O}(k)} n^{\mathcal{O}(1)}$. We obtain Theorem 5 as a corollary to a slightly more general form of a decomposition theorem. We use this decomposition theorem to design substantially faster FPT algorithm for well-studied cut problems such as MINIMUM BISECTION, STEINER CUT, s-WAY CUT, and EDGE MULTIWAY CUT-UNCUT on UDGs in [19].

The most central notion in our proof for Theorem 5 is that of a tree decomposition whose bags are "chunked into parts". After presenting its formal definition, we present an intuitive interpretation of this notion by relating it to our algorithmic applications.

Definition 4 ([19]). *Let G be a graph. A* chunked tree decomposition CTD *is a triple (T, β, ζ) that satisfies the following conditions.*

1. *(T, β) is a tree decomposition of G.*
2. *ζ is a function that assigns to each node $x \in V(T)$ a partition of $\beta(x)$.*
3. *For each node $x \in V(T)$ and for each part U of the partition $\zeta(x)$, the graph $G[U]$ is connected.*

The chunkiness *of* (T, β, ζ) *is the maximum number of parts in a partition assigned by* ζ, *that is,* $\max_{x \in V(T)} |\zeta(x)|$.

To understand the intuition behind this definition, we find it convenient to view a CTD (T, β, ζ) as follows. Suppose that each bag of the tree decomposition (T, β) was partitioned into chunks that are (not necessarily maximal) cliques. Such chunks are *(i)* hard to break (if the cliques are large enough) by the removal of only a few edges, but *(ii)* there can be many such chunks. The first property makes the resolution of a cut problem easy, while the second one makes it difficult. To circumvent this issue, we glue some chunks together, and think of a glued chunk as one unit that cannot be broken. Thus, although a glued chunk can potentially be easily broken (e.g., think of two large cliques connected by a single edge), we forbid this operation. Specifically, when the decomposition is utilized to construct a solution X for a cut problem, no edge having both endpoints inside a glued chunk should be inserted into X. This intuition gives rise to the following definition.

Definition 5 ([19]). *Let* (T, β, ζ) *be a CTD of a graph* G. *For any edge* $e \in E(G)$, *we say that* (T, β, ζ) *chunks* e *if it satisfies the following condition.*

– *For each node* $x \in V(T)$ *and for each part* U *of the partition* $\zeta(x)$, *if* $G[U]$ *is not a clique, then at most one endpoint of* e *belongs to* U.

More generally, for any $S \subseteq E(G)$, *we say that* (T, β, ζ) *chunks* S *if it chunks every edge in* S.

This definition bring us the so called parameteric version of our main decomposition theorem, formally stated as follows.

Theorem 6. [Linear-Chunkiness Decomposition Theorem in UDGs (Parameteric Version), [19]] *There exists a polynomial-time deterministic algorithm that, given a UDG* G *of a point set* D *and* $k \in \mathbb{N}$, *outputs a collection* $\mathcal{C}$ *of* $\mathcal{O}(k)$ *CTDs of* G *with the following properties.*

1. *Each CTD in the collection has chunkiness* $\mathcal{O}(k)$.
2. *For each subset of edges* $S \subseteq E(G)$ *of size at most* k, *there exists at least one CTD in* $\mathcal{C}$ *that chunks* S.

Using Theorem 6 and a non-trivial dynamic programming, we get the following algorithm for the MINIMUM BISECTION problem.

Theorem 7 ([19]). MINIMUM BISECTION *on UDGs is solvable in time* $2^{\mathcal{O}(k)} n^{\mathcal{O}(1)}$.

7   Concluding Remarks

In this survey article we described recent methods to design efficient parameterized algorithms for problems on unit-disk graphs. Most of these algorithms were based on graph decomposition theorems for unit-disk graphs. Extending these decomposition theorems to other geometric graph classes is a natural direction to pursue. Finding new applications to these decompositions theorems in designing approximation algorithms, exact algorithms and parameterized algorithms is another set of open problems.

References

1. Alber, J., Bodlaender, H.L., Fernau, H., Kloks, T., Niedermeier, R.: Fixed parameter algorithms for dominating set and related problems on planar graphs. Algorithmica **33**(4), 461–493 (2002)
2. Alber, J., Fellows, M.R., Niedermeier, R.: Polynomial-time data reduction for dominating set. J. ACM **51**(3), 363–384 (2004)
3. Baker, B.S.: Approximation algorithms for NP-complete problems on planar graphs. J. ACM **41**(1), 153–180 (1994)
4. Bodlaender, H.L., Fomin, F.V., Lokshtanov, D., Penninkx, E., Saurabh, S., Thilikos, D.M.: (meta) kernelization. J. ACM **63**(5), 44:1–44:69 (2016). http://dl.acm.org/citation.cfm?id=2973749
5. Bui, T.N., Peck, A.: Partitioning planar graphs. SIAM J. Comput. **21**(2), 203–215 (1992)
6. Chen, J., Fernau, H., Kanj, I.A., Xia, G.: Parametric duality and kernelization: lower bounds and upper bounds on kernel size. SIAM J. Comput. **37**(4), 1077–1106 (2007)
7. Clark, B.N., Colbourn, C.J., Johnson, D.S.: Unit disk graphs. Discret. Math. **86**(1–3), 165–177 (1990)
8. Cygan, M., et al.: Parameterized Algorithms. Springer, Cham (2015). https://doi.org/10.1007/978-3-319-21275-3
9. Cygan, M., Lokshtanov, D., Pilipczuk, M., Pilipczuk, M., Saurabh, S.: Minimum bisection is fixed parameter tractable. In: Symposium on Theory of Computing, STOC 2014, New York, NY, USA, 31 May–03 June 2014, pp. 323–332 (2014)
10. Dawar, A., Grohe, M., Kreutzer, S., Schweikardt, N.: Approximation schemes for first-order definable optimisation problems. In: LICS 2006, pp. 411–420 (2006)
11. Demaine, E.D., Fomin, F.V., Hajiaghayi, M., Thilikos, D.M.: Subexponential parameterized algorithms on graphs of bounded genus and H-minor-free graphs. J. ACM **52**(6), 866–893 (2005)
12. Demaine, E.D., Hajiaghayi, M.: Bidimensionality. In: Kao, M.Y. (ed.) Encyclopedia of Algorithms. Springer, Boston (2008). https://doi.org/10.1007/978-0-387-30162-4
13. Demaine, E.D., Hajiaghayi, M.: The bidimensionality theory and its algorithmic applications. Comput. J. **51**(3), 292–302 (2008)
14. Demaine, E.D., Hajiaghayi, M., Mohar, B.: Approximation algorithms via contraction decomposition. Combinatorica **30**(5), 533–552 (2010)
15. Diestel, R.: Graph Theory. Graduate Texts in Mathematics, vol. 173. Springer, Heidelberg (2000)
16. Dorn, F., Fomin, F.V., Thilikos, D.M.: Subexponential parameterized algorithms. Comput. Sci. Rev. **2**(1), 29–39 (2008)
17. Eisenstat, D., Klein, P., Mathieu, C.: An efficient polynomial-time approximation scheme for Steiner forest in planar graphs. In: Proceedings of the Twenty-Third Annual ACM-SIAM Symposium on Discrete Algorithms (SODA), pp. 626–638. SIAM (2012)

18. Eppstein, D.: Diameter and treewidth in minor-closed graph families. Algorithmica **27**, 275–291 (2000)
19. Fahad Panolan, S.S., Zehavi, M.: Contraction decomposition in unit disk graphs and algorithmic applications in parameterized complexity. In: Proceedings of the Twenty-Third Annual ACM-SIAM Symposium on Discrete Algorithms (SODA) (2019, to appear)
20. Fomin, F.V., Lokshtanov, D., Panolan, F., Saurabh, S., Zehavi, M.: Finding, hitting and packing cycles in subexponential time on unit disk graphs. In: 44th International Colloquium on Automata, Languages, and Programming, ICALP 2017. LIPIcs, 10–14 July 2017, Warsaw, Poland, vol. 80, pp. 65:1–65:15. Schloss Dagstuhl - Leibniz-Zentrum fuer Informatik (2017)
21. Fomin, F.V., Lokshtanov, D., Saurabh, S.: Excluded grid minors and efficient polynomial-time approximation schemes. J. ACM **65**(2), 10:1–10:44 (2018). https://doi.org/10.1145/3154833
22. Fomin, F.V., Thilikos, D.M.: New upper bounds on the decomposability of planar graphs. J. Graph Theory **51**(1), 53–81 (2006)
23. Gandhi, R., Khuller, S., Srinivasan, A.: Approximation algorithms for partial covering problems. J. Algorithms **53**(1), 55–84 (2004)
24. Grohe, M.: Local tree-width, excluded minors, and approximation algorithms. Combinatorica **23**(4), 613–632 (2003). https://doi.org/10.1007/s00493-003-0037-9
25. Gu, Q.P., Tamaki, H.: Improved bounds on the planar branchwidth with respect to the largest grid minor size. Algorithmica **64**(3), 416–453 (2012)
26. Ito, H., Kadoshita, M.: Tractability and intractability of problems on unit disk graphs parameterized by domain area. In: Proceedings of the 9th International Symposium on Operations Research and Its Applications (ISORA 2010), pp. 120–127 (2010)
27. Khanna, S., Motwani, R.: Towards a syntactic characterization of PTAS. In: STOC 1996, pp. 329–337. ACM (1996)
28. Klein, P.N., Marx, D.: A subexponential parameterized algorithm for subset TSP on planar graphs. In: Proceedings of the 25th Annual ACM-SIAM Symposium on Discrete Algorithms (SODA), pp. 1812–1830. SIAM (2014)
29. Lipton, R.J., Tarjan, R.E.: Applications of a planar separator theorem. SIAM J. Comput. **9**, 615–627 (1980)
30. Marx, D.: The square root phenomenon in planar graphs. In: Fomin, F.V., Freivalds, R., Kwiatkowska, M., Peleg, D. (eds.) ICALP 2013. LNCS, vol. 7966, p. 28. Springer, Heidelberg (2013). https://doi.org/10.1007/978-3-642-39212-2_4
31. Raghavan, V., Spinrad, J.: Robust algorithms for restricted domains. J. Algorithms **48**(1), 160–172 (2003)
32. Robertson, N., Seymour, P.D., Thomas, R.: Quickly excluding a planar graph. J. Comb. Theory Ser. B **62**(2), 323–348 (1994)

Facility Location Problem

r-Gatherings on a Star

Shareef Ahmed[1(✉)], Shin-ichi Nakano[2], and Md. Saidur Rahman[1]

[1] Graph Drawing and Information Visualization Laboratory,
Department of Computer Science and Engineering,
Bangladesh University of Engineering and Technology, Dhaka, Bangladesh
{shareefahmed,saidurrahman}@cse.buet.ac.bd
[2] Gunma University, Kiryu 376-8515, Japan
nakano@cs.gunma-u.ac.jp

Abstract. Let C be a set of n customers and F be a set of m facilities. An r-gather clustering of C is a partition of the points in clusters such that each cluster contains at least r points. The r-gather clustering problem asks to find an r-gather clustering which minimizes the maximum distance between any two points in a cluster. An r-gathering of C is an assignment of each customer $c \in C$ to a facility $f \in F$ such that each open facility has zero or at least r customers. The r-gathering problem asks to find an r-gathering that minimizes the maximum distance between a customer and its facility. In this work we consider the r-gather clustering and r-gathering problems when the customers and the facilities are lying on a "star". We show that the r-gather clustering problem and the r-gathering problem with points on a star with d rays can be solved in $O(rn + (r+1)^d dr)$ and $O(n + r^2m + d^2r^2(d + \log m) + (r+1)^d 2^d(r+d)d)$ time respectively.

Keywords: r-Gathering · Clustering · Facility location problem

1 Introduction

Let C be a set of n points. An r-gather clustering of C is a partition of the points of C in clusters such that each cluster contains at least r points. The cost of a cluster is the maximum distance between a pair of points in the cluster. The cost of an r-gather clustering is the maximum cost among the costs of the clusters. The r-gather clustering problem asks to find an r-gather clustering of C with minimum cost [2].

Let C be a set of n customers and F be a set of m facilities, $d(c, f)$ be the distance between $c \in C$ and $f \in F$. An r-gathering of C to F is an assignment A of C to F such that each facility has at least r or zero customers assigned to it. The cost of an r-gathering is $\max_{c \in C}\{d(c, A(c))\}$ which is the maximum distance between a customer and its facility. The r-gathering problem asks to find an assignment of C to F having the minimum cost [4]. This problem is also known as the min-max r-gathering problem. The other version of the problem is known as the min-sum r-gathering problem which asks to find an assignment

G. K. Das et al. (Eds.): WALCOM 2019, LNCS 11355, pp. 31–42, 2019.
https://doi.org/10.1007/978-3-030-10564-8_3

which minimizes $\sum_{c \in C} d(c, A(c))$ [7,9]. In this paper we consider the min-max r-gathering problem and we use the term r-gathering problem to refer the min-max version unless specified otherwise.

Both the r-gather clustering and r-gathering problems are NP-complete in general [2,4]. For r-gather clustering problem a 2-approximation algorithm is known [2]. For the r-gathering problem a 3-approximation algorithm is known and it is proved that the problem cannot be approximated within a factor less than 3 for $r > 3$ unless $P = NP$ [4]. Recently, both problems are considered in a setting where all the points are lying on a line. An $O(n \log n)$ time algorithm [3] based on the matrix search method [5], and an $O(rn)$ time algorithm [10] by reduction to the min-max path problem in a weighted directed graph [6] are known for the r-gather clustering problem when all the points are on a line. For the r-gathering problem an $O((n + m) \log(n + m))$ time algorithm [3] based on the matrix search method [1,5], an $O(n + m \log^2 r + m \log m)$ time algorithm [8], and an $O(n + r^2 m)$ time algorithm [10] by reduction to the min-max path problem in a weighted directed graph [6] are known when all the customers and facilities are on a line. Recently the r-gather clustering problem is studied on mobile setting and a 4-approximation distributed algorithm is known [11].

In this paper, we consider both the r-gathering clustering and r-gathering problem when the points are on a star. Consider a scenario where a number of streets meet in a junction, and residents live by the streets. We wish to set up emergency shelters on the streets so that each shelter can serve at least r residents. The distance between two points are measured along the lines. We also wish to locate shelters so that evacuation time span can be minimized. This scenario can be modeled by the r-gather clustering problem where all input points C are located on a star. In an r-gather clustering of C having the minimum cost, each emergency shelters is located at the center of each cluster. On the other hand, if the set F of possible locations of shelters on the star is also given with the set C of residents and we wish to find an assignment of C to F with minimizing the evacuation time so that each shelter serve at least r residents, then the scenario can be modeled by the r-gathering problem where the points of C and F are located on a star. In this case, an r-gathering corresponds to an assignment of residents to shelters such that each "open" shelter serves at least r residents and the r-gathering problem finds the r-gathering minimizing the evacuation time.

When the points are on a line, each cluster of an optimal r-gather clustering consists of consecutive points on the line [10]. However, when the points are on a star, some clusters may not consists of consecutive points in the optimal r-gather clustering. For example, see Fig. 1. We can observe that at least one cluster consists of non-consecutive points in any optimal solution. Figure 1 demonstrates an optimal solution for this scenario.

In this paper we give an $O(rn + (r + 1)^d dr)$ time algorithm for r-gather clustering problem on a star, and an $O(n + r^2 m + d^2 r^2 (d + \log m) + (r + 1)^d 2^d (r + d)d)$ time algorithm for the r-gathering problem on a star, where d is the number of rays that form the star.

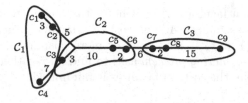

Fig. 1. An optimal 3-gather clustering on a star.

The rest of the paper is organized as follows. In Sect. 2 we define the problems and define terms used in the paper. In Sect. 3 we give an algorithm for the r-gather clustering problem on a star. In Sect. 4 we give an algorithm for the r-gathering problem on a star. Finally Sect. 5 is a conclusion.

2   Preliminaries

In this section we define two problems and some terms used in this paper.

Let $\mathcal{L} = \{l_1, l_2, \cdots, l_d\}$ be a set of d rays where all the rays of $\mathcal{L}$ share a common source point o. We call the set of rays $\mathcal{L}$ a *star* and the common source point o the *center* of the star. The *degree* of a star is the number d of rays which form the star. The Euclidean distance between two points p, q is denoted by $d_E(p, q)$. We denote by $d(p, q)$ the distance between two points p, q which is measured along the rays. If p and q are both located on the same ray, then $d(p, q) = d_E(p, q)$. On the other hand, if p and q are located on different rays, then $d(p, q) = d_E(p, o) + d_E(o, q)$. A cluster consists of points from two or more rays is a *multi-ray cluster*, otherwise a *single-ray cluster*. Two points p and q are the *end-points* of a cluster $\mathcal{C}$ if $d(p, q) = cost(\mathcal{C})$. A point p in a cluster $\mathcal{C}$ is a *far point* of $\mathcal{C}$, denoted by $e(C)$, if $d(o, p) \geq d(o, q)$ for each $q \in \mathcal{C}$.

We now define the first problem. Let $C = \{c_1, c_2, \cdots, c_n\}$ be n points located on a star. An *r-gather clustering* of C is a partition of the points of C into clusters such that each cluster contains at least r points. The cost of a cluster $\mathcal{C}$, denoted by $cost(\mathcal{C})$, is $\max_{p,q \in \mathcal{C}} d(p, q)$. The *r-gather clustering problem* asks to find an r-gather clustering such that the maximum cost among the costs of clusters is minimized, and such a clustering is called an optimal r-gather clustering. The following result is known [10]. Note that any cluster with $2r$ or more points can be divided into clusters so that each of which has at most $2r - 1$ points and at least r points.

Lemma 1 ([10]). *There is an optimal r-gather clustering in which each cluster has at most $2r - 1$ points.*

Let $C = \{c_1, c_2, \cdots, c_n\}$ be n customers and $F = \{f_1, f_2, \cdots, f_m\}$ be m possible locations for facilities located on a star. An *r-gathering* of C to F is an assignment A of C to F such that each facility has zero or at least r customers. A facility having one or more customers is called an *open facility*. We denote by F' the set of open facilities. $A(c)$ denotes the facility to which a customer c is assigned in an assignment A. The cost of a facility f, denoted by $cost(f)$, is $\max\{d(f, c_i) | A(c_i) = f\}$ if f has one or more customers, and is 0 if f has no customer. *The r-gathering problem* asks to find an r-gathering such that the maximum cost among the costs of facilities is minimized.

3 r-Gather Clustering on a Star

In this section we give an algorithm for r-gather clustering problem on a star. Let C be a set of points on a star $\mathcal{L} = \{l_1, l_2, \cdots, l_d\}$ of d rays with center o. We consider the set C as a union of d sets $C_1, C_2, \cdots, C_d$ where C_i is the set of customers on ray l_i. We have the following lemma.

Lemma 2. *There is an optimal r-gather clustering such that, for each C_i, the set of points in C_i assigned to multi-ray clusters is consecutive points on l_i including the nearest point to o.*

Proof. A pair c_m, c_s in C_i is called a reverse pair if c_m is assigned to a multi-ray cluster, c_s is assigned to a single-ray cluster, and $d(o, c_s) < d(o, c_m)$. Assume for a contradiction that A is an optimal r-gather clustering with the minimum number of reverse pairs but the number is not zero. Let c_s and c_m be a reverse pair in C_i with maximum $d(o, c_m)$. Let $\mathcal{C}_s$ and $\mathcal{C}_m$ be the clusters containing c_s and c_m, respectively. We have two cases.

Case 1: $\mathcal{C}_s$ has a point c in C_i with $d(o, c_m) < d(o, c)$ (Fig. 2).
Let c' be the nearest point to o in $\mathcal{C}_s$. Replacing $\mathcal{C}_s$ and $\mathcal{C}_m$ in the clustering by $\mathcal{C}_s \setminus \{c'\} \cup \{c_m\}$ and $\mathcal{C}_m \setminus \{c_m\} \cup \{c'\}$ generates a new r-gather clustering with less reverse pairs as illustrated in Fig. 2(a). A contradiction. Note that $cost(\mathcal{C}_s \setminus \{c'\} \cup \{c_m\}) \leq cost(\mathcal{C}_s)$ and $cost(\mathcal{C}_m \setminus \{c_m\} \cup \{c'\}) \leq cost(\mathcal{C}_m)$ hold.
Case 2: Otherwise. (Thus $d(o, c) < d(o, c_m)$ for every point c in $\mathcal{C}_s$.)
The same replacing results in a new r-gather clustering with less reverse pairs as illustrated in Fig. 2(b). A contradiction. Note that $cost(\mathcal{C}_s \setminus \{c'\} \cup \{c_m\}) \leq cost(\mathcal{C}_m)$ and $cost(\mathcal{C}_m \setminus \{c_m\} \cup \{c'\}) \leq cost(\mathcal{C}_m)$ hold.

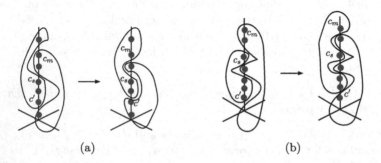

(a)   (b)

Fig. 2. (a) Illustration of Case 1 and (b) illustration of Case 2 of proof of Lemma 2.

Lemma 3. *If an optimal r-gather clustering has multi-ray clusters, then at most one multi-ray cluster contains more than r points.*   □

Proof. Assume for a contradiction that every optimal r-gather clustering has two or more multi-ray clusters having more than r points. Let A be an r-gather clustering with the minimum number of multi-ray clusters having more than r points. Let $\mathcal{C}_i$ and $\mathcal{C}_j$ be two multi-ray clusters having more than r points. Let s_i, t_i be the two endpoints of $\mathcal{C}_i$ and s_j, t_j be the two endpoints of $\mathcal{C}_j$. Without loss of generality, assume that t_j is the closest point to o among the four end-points. Let $\mathcal{C}'_j \subset \mathcal{C}_j$ be $\{c \in \mathcal{C}_j | d(o, c) > d(o, t_j)\}$. Any point $c \in \mathcal{C}'_j$ must be on the same ray as s_j, otherwise t_j would not be an end-point of $\mathcal{C}_j$. We have two cases.

Case 1: $|\mathcal{C}'_j| < r$.

Let $\mathcal{C}''_j$ be a set of $|\mathcal{C}_j| - r$ arbitrary points from $\mathcal{C}_j \setminus \mathcal{C}'_j$. We now derive a new r-gather clustering A' by replacing $\mathcal{C}_i$ and $\mathcal{C}_j$ by $\mathcal{C}_i \cup \mathcal{C}''_j$ and $\mathcal{C}_j \setminus \mathcal{C}''_j$. Since t_j is the closest point to o among the four end-points s_i, t_i, s_j, t_j and $d(o, c) \leq d(o, t_j)$ for any point $c \in \mathcal{C}''_j$, we have $d(o, c) \leq d(o, s_i)$ and $d(o, c) \leq d(o, t_i)$. Thus the cost of $\mathcal{C}_i \cup \mathcal{C}''_j$ does not exceed the cost of $\mathcal{C}_i$. Hence the cost of A' is not greater than the cost of A. Thus A' has less multi-ray clusters with more than r points, a contradiction.

Case 2: Otherwise. Thus $|\mathcal{C}'_j| \geq r$.

In this case we derive a new r-gather clustering A' by replacing $\mathcal{C}_i$ and $\mathcal{C}_j$ by $\mathcal{C}_i \cup (\mathcal{C}_j \setminus \mathcal{C}'_j)$ and $\mathcal{C}'_j$. In this case, $\mathcal{C}'_j$ is a single-ray cluster. By a similar argument of Case 1, the cost of A' does not exceed the cost of A. Thus A' has less multi-ray clusters having more than r points than A, a contradiction.

We now give the following lemma, which is used in the proof of Lemmas 5 and 6.

Lemma 4. *If $|C| \geq 2r$ and there is an optimal r-gather clustering consisting of only multi-ray clusters, then there is an optimal r-gather clustering with the multi-ray cluster consisting of the farthest point from o and its $r - 1$ nearest points.*

Proof. Let p be the farthest point from o and let N be the $r - 1$ nearest points of p. Assume for a contradiction that in every optimal solution $N \cup \{p\}$ is not a cluster. We first prove that $N \cup \{p\}$ is contained in the same cluster. Let A be an optimal solution with cluster $\mathcal{C}_p$ containing p having the maximum number of points in N. Let q be a point in N assigned to a cluster $\mathcal{C}_q \neq \mathcal{C}_p$. Since the number of points in $\mathcal{C}_p$ is at least r, there is a point $p' \in \mathcal{C}_p$ not in N. Let q' be the farthest point from o in $\mathcal{C}_q \setminus \{q\}$. We now derive a new r-gather clustering by replacing $\mathcal{C}_p$ and $\mathcal{C}_q$ by $\mathcal{C}_p \setminus \{p'\} \cup \{q\}$ and $\mathcal{C}_q \setminus \{q\} \cup \{p'\}$. Thus a contradiction. Note that, $cost(\mathcal{C}_p \setminus \{p'\} \cup \{q\}) \leq cost(\mathcal{C}_p)$ and $cost(\mathcal{C}_q \setminus \{q\} \cup \{p'\}) \leq \max\{cost(\mathcal{C}_p), cost(\mathcal{C}_q)\}$, since $d(o, p) \geq d(o, q')$.

We now prove that $N \cup \{p\}$ form a multi-ray cluster. Assume for a contradiction that in any optimal r-gather clustering $N \cup \{p\}$ is not a cluster. Let A' be an optimal r-gather clustering with cluster $\mathcal{C}_p$ containing p having the minimum number of points not in N. Let p'' be the farthest point in C_p not in the ray l_p containing p, and C_s be a cluster in A' other than C_p. Let s be the farthest point from o in C_s. We now derive a new r-gathering by replacing C_p and C_s with $C_p \setminus \{p''\}$ and $C_s \cup \{p''\}$ without increasing cost, a contradiction. Since $d(o, s) \leq d(o, p)$, we have $d(s, p'') \leq d(p, p'')$ and thus $cost(C_s \cup \{p''\}) \leq \max\{cost(C_p), cost(C_s)\}$.

We now have the following lemma.

Lemma 5. *If an optimal r-gather clustering consists of only multi-ray clusters, then there is an optimal r-gather clustering with at most $d-1$ multi-ray clusters.*

Proof. We give a proof by induction on the number d of rays in the star. Clearly, the claim holds for $d = 2$, since in such case only one multi-ray cluster can exist. Assume that the claim holds for any star with less than d rays. We now prove that the claim also holds for any star of d rays. Assume for a contradiction that every optimal solution has at least d multi-ray clusters. Let A be an optimal r-gather clustering with the minimum number of multi-ray clusters. Let p be the farthest point from o. By Lemma 4, there is an optimal r-gather clustering with the cluster C_p containing p and its $r - 1$ nearest points, denoted by N. Let l_p be the ray containing p. We have two cases.
Case 1: p and N are on ray l_p.
In this case there is an optimal r-gather clustering with a single ray cluster $N \cup \{p\}$, a contradiction.
Case 2: Otherwise. There is a point q in N which is not on l_p.
By Lemma 4 there is an optimal r-gathering with $\{p\} \cup N$, and since N consists of the $r - 1$ nearest neighbors of p, all the points on l_p are contained in C_p. Thus the points in $C \setminus C_p$ are lying on other $d - 1$ rays except l_p. By inductive hypothesis there is an optimal r-gather clustering of $C \setminus C_p$ with at most $d - 2$ multi-ray clusters. Thus the claim holds.   □

Corollary 1. *If an optimal r-gahter clustering consists of only multi-ray clusters, then C has at most $(d - 2)r + 2r - 1 = dr - 1$ points.*

We now give an outline of our algorithm which constructs an optimal r-gathering clustering on a star. We first choose every possible at most $dr - 1$ candidate points for multi-ray clusters. We find the optimal r-gather clustering consisting of only multi-ray clusters for each candidate points, by repeatedly searching for the farthest point from o and its $r - 1$ nearest point as a multi-ray cluster of the remaining set of points, by the algorithm **Multi-rayClusters**.
We now have the following lemma.

Lemma 6. *Let $A = \{C_1, C_2, C_3, \cdots, C_{|A|}\}$ be the clusters computed by Algorithm **Multi-rayClusters**. If A has only multi-ray clusters, then A is an optimal r-gather clustering of S.*

Proof. The proof of this lemma is immediate from Lemma 4.   □

Algorithm 1. Multi-rayClusters(C)

Input : A set C of points on a star
Output: An r-gather clustering with only multi-ray clusters
if $|C| < r$ **then**
| **return** $\emptyset$;
endif
$i \leftarrow 1$;
while $|C| \neq 0$ **do**
| **if** $|C| < 2r$ **then**
| | Create new cluster $\mathcal{C}_i = C$;
| **else**
| | $p \leftarrow$ farthest point from o in C;
| | $\mathcal{C}_i \leftarrow \{p, p_1, p_2, \cdots, p_{r-1}\}$ where p_i is the i-th nearest point of p in C;
| **endif**
| $C \leftarrow C \setminus \mathcal{C}_i$;
| $i \leftarrow i + 1$;
end
return $\{\mathcal{C}_1, \mathcal{C}_2, \mathcal{C}_3, \cdots, \mathcal{C}_{i-1}\}$

We now give an algorithm r**GatherClusteringOnStar** to construct an optimal r-gather clustering of C on a star. We have the following theorem.

Theorem 1. *The algorithm rGatherClusteringOnStar constructs an optimal r-gather clustering of C on star in $O(rn + (r + 1)^d dr)$ time.*

Proof. We first prove the correctness of the algorithm. By Lemma 2 multi-ray clusters in an optimal r-gathering are located near o, and by Corollary 1 the number of customers in the multi-ray clusters is at most $dr - 1$. The algorithm r**GatherClusteringOnStar** considers every possible choice of the set of points for multi-ray clusters having at most $dr - 1$ points. The algorithm considers the solution for each possible choice for multi-ray clusters with the solution obtained by 1-dimensional algorithm for remaining points on each ray, and choose the solution having minimum cost. Thus the algorithm produces an optimal r-gather clustering.

We now prove that the algorithm runs in linear time. We consider points in each ray are in sorted order according to the distance from o. The d nested loops iterates $\prod_{j=1}^{d}(n_j + 1)$ times. Thus the number of points involved in all calls to Multi-rayClusters is at most $(r + 1)^d dr$, since $\sum_{j=1}^{d} n_j = dr - 1$. Within each nested loop we repeatedly compute multi-ray clusters which takes linear time in total. We also compute single-ray clusters on each of the d rays. Rather than computing those single-ray clusters each time in the loop, we compute the r-gather clustering for points consisting of i farthest points from o, for each i, and for each ray in $O(rn)$ time total [10]. Thus to compute all the required cases for single-ray cluster we need total $O(rn)$ time. Thus the time complexity of the algorithm is $O(rn + (r + 1)^d dr)$. $\qquad\square$

If both r and d are constants then this is linear.

Algorithm 2. rGatherClusteringOnStar(C)

Input   : A set C of points on star $\mathcal{L} = \{l_1, l_2, l_3, \cdots, l_d\}$
Output: An optimal r-gather clustering of C
if $|C| < r$ **then**
  | **return** $\emptyset$;
endif
$Best \leftarrow \emptyset$;
Let $n_1, n_2, \cdots, n_d$ be the number of points of C in each ray $l_1, l_2, \cdots, l_d$;
for $i_1 \leftarrow 0$ **to** n_1 **do**
  for $i_2 \leftarrow 0$ **to** n_2 **do**
    for $i_3 \leftarrow 0$ **to** n_3 **do**
      $\cdots$;
      for $i_d \leftarrow 0$ **to** n_d **do**
        if $i_1 + i_2 + \cdots + i_d < dr$ **then**
          S be the set of points consisting of $i_1, i_2, \cdots, i_d$ closest
          points from o for ray $l_1, l_2, \cdots, l_d$;
          $R_m \leftarrow$ Multi-rayClusters(S);
          $R_i \leftarrow r$-gather clustering of remaining points of ray l_i by 1D
          algorithm;
          $R \leftarrow R_m \cup R_1 \cup R_2 \cup \cdots \cup R_d$;
          if R *is the best r-gather clustering so far* **then**
           | $Best \leftarrow R$;
          endif
        endif
      end
      $\cdots$;
    end
    if $i_1 + i_2 \geq dr$ **then**
    | break;
    endif
  end
  if $i_1 \geq dr$ **then**
  | break;
  endif
end
return $Best$;

---

4 r-Gathering on a Star

In this section we give an algorithm for the r-gathering problem on a star.

Let C be a set of customers and F be a set of facilities on a star $\mathcal{L} = \{l_1, l_2, \cdots, l_d\}$ of d rays with center o. We regard the set C as the union of d sets $C_1, C_2, \cdots, C_d$ where C_i is the set of customers on ray l_i. Similarly, F is the union of $F_1, F_2, \cdots, F_d$ where F_i is the set of facilities on ray l_i. In any optimal r-gathering each open facility serves at least r customers. However the number of customers assigned to an open facility can be more than $2r - 1$. In such case we regard the set of customers assigned to a facility as the union of clusters $C_1, C_2, \cdots, C_k$ sharing a facility and each of which satisfies $r \leq |C_i| < 2r$. Thus we can think of the r-gathering problem in a similar way to the r-gather clustering problem in Sect. 3, and Lemma 1 holds for the clusters of r-gathering. We denote by $A(C)$ the facility to which the customers in C is assigned in r-gathering A. We define the cost of a cluster C, denoted by $cost(C)$, in r-gathering A as $\max_{c \in C}\{d(c, A(c))\}$. It is easy to observe that Lemma 2 also holds for the clusters of r-gatherings. We now prove that Lemma 3 also holds for the r-gathering problem.

Lemma 7. *There is an optimal r-gathering including at most one multi-ray cluster having more than r customers.*   □

Proof. Omitted.

A customer on a ray $l \in \mathcal{L}$ is the *boundary customer* of l if it is the farthest customer on l from o. We now give the following lemma.

Lemma 8. *If $|C| \geq 2r$ and there is an optimal r-gathering A with only multi-ray clusters, then there is an optimal r-gathering with only multi-ray clusters satisfying the following (a) and (b). Let f be the farthest open facility from o in A and l be the ray containing f.*
(a) The boundary customer p of l and its $r - 1$ nearest customers form a multi-ray cluster, if l has a customer,
(b) All customers are assigned to f and the farthest boundary customer p from o and its $r - 1$ nearest customers form a multi-ray cluster, if l has no customer.

Proof. (a) We denote by N the set of the $r - 1$ nearest customers of p. We first prove that there is an optimal solution with the customers in $N \cup \{p\}$ are assigned to f. Assume for a contradiction that in any optimal solution $N \cup \{p\}$ are not assigned to f. Let A be an optimal solution with the maximum number of customers in $N \cup \{p\}$ are assigned to f. Let C_p be the multi-ray cluster assigned to f, and q be a customer in $N \cup \{p\}$ but $q \notin C_p$. Let q is assigned to f'. Since C_p has at least r customers, there is a customer $p' \in C_p$ not in $N \cup \{p\}$ and lying on a ray except l. We now derive a new r-gathering A' by reassigning q to f and p' to f'. Since $d(o, f') \leq d(o, f)$, we have $d(f', p') \leq d(o, f') + d(o, p') \leq d(o, f) + d(o, p') = d(f, p')$. Now if q is (1) not on l or (2) q is on l with $d(o, q) \leq d(o, f)$ then $d(f, q) \leq d(f, p')$. Otherwise, q is on l with $d(o, q) > d(o, f)$ holds, then we have $d(f, q) \leq d(f, p)$. Thus the cost of A' does not exceed the cost of A, and A' has more customers in $N \cup \{p\}$ assigned to f. A contradiction. Thus the customers in $N \cup \{p\}$ are contained in C_p.

We now prove that $N \cup \{p\}$ form a multi-ray cluster. Assume for a contradiction that in any optimal r-gathering $N \cup p$ is not a cluster. Let A' be an optimal r-gathering with the cluster C_p containing p having the minimum number of customers not in $N \cup \{p\}$. Since C_p is a multi-ray cluster, C_p has a customer s not in $N \cup \{p\}$ and lying on a ray except l. We can reassign s to some open facility $f' \neq f$ without increasing the cost, since $d(o, f') \leq d(o, f)$, $d(s, f')$ does not exceed $d(s, f)$. A contradiction.

(b) We first prove that all customers are assigned to f. Assume for a contradiction that there is an open facility $f' \neq f$ to which some customers are assigned. Since f is the farthest open facility from o and there is no customer on l, we can reassign all customers to f' without increasing the cost of the r-gathering. A contradiction.

The proof of the 2nd part of Lemma 8(b) is similar to the proof of Lemma 8(a).   □

We now prove that Lemma 5 also holds for r-gathering.

Lemma 9. *If an optimal r-gathering consists of only multi-ray clusters, then there is an optimal r-gathering consisting of at most $d - 1$ multi-ray clusters, where d is the number of rays containing a customer.*

Proof. Omitted.   □

We now give algorithm Multi-rayClusters2. If there is an optimal r-gathering with only multi-ray clusters, then the algorithm finds such an r-gathering, by repeatedly removing a cluster ensured by Lemma 8.

Lemma 10. *If there is an optimal r-gathering consisting of only multi-ray clusters, then Algorithm **Multi-rayClusters2** finds an optimal r-gathering. The running time of the algorithm is $O(2^d(r + d)d + (d + \log m)d)$.*

Proof. If there is an optimal r-gathering with only multi-ray clusters, then, by repeatedly removing a cluster ensured by Lemma 8, we can find a sequence $\mathcal{C}_1, \mathcal{C}_2, \cdots, \mathcal{C}_k$ of multi-ray clusters such that $\mathcal{C}_i$ consists of exactly r customers in $C \setminus (\mathcal{C}_1 \cup \mathcal{C}_2 \cup \cdots \mathcal{C}_{i-1})$ except the last cluster $\mathcal{C}_k$ with $r \leq |\mathcal{C}_k| \leq 2r - 1$. The algorithm checks every possible sequence of the rays containing the farthest open facility and chooses the best one as an optimal r-gathering. Note that if a cluster is a single-ray cluster, then the algorithm skips recursive call, since it try to find an r-gathering consisting of only multi-ray clusters.

We now estimate the running time of the algorithm.

By Lemma 9 the depth of the recursive calls is at most $d-1$. Thus, by the tree structure of the calls, the number of calls is at most $d!$. The algorithm repeatedly constructs a multi-ray cluster with exactly r customers by Lemma 8, which takes $O(r + d)$ time for each and $O(r + d)d$ time in total. The cluster is assigned to its best facility of the cluster. The best facility of a multi-ray cluster is the nearest facility to the mid-point of the farthest two customers on two different rays in the cluster. The best facility can be found in $O(d + \log m)$ time for each cluster. Thus the algorithm runs in $O(d!((r + d)d + (d + \log m))d)$ time.

We can improve the running time by modifying the algorithm to save the solution of each subproblem in a table. The number of distinct subproblems is the number of the combinations of the lines checked. Thus the number of distinct subproblem is $\sum_{j=1}^{d-1} \binom{d}{i} = O(2^d)$. Then the runtime is $O(2^d(r + d)d + (d + \log m)d)$.   □

Algorithm 3. Multi-rayClusters2(C, F)

Input : A set C of customers and a set of F of facilities on a star
Output: An r-gathering with only multi-ray clusters
if $|C| < r$ *or the number of rays containing customers is at most one* **then**
 | **return** $\emptyset$;
endif
if $|C| < 2r$ *or the number of rays containing customers is two* **then**
 | Assign C to its best facility; /* Lemma 8(b) */
 | **return** $\{C\}$;
endif
$Ans \leftarrow \emptyset$;
$Best \leftarrow \infty$;
for *each ray l_i containing a customer* **do**
 | $C_i \leftarrow p_i$ and its $r - 1$ nearest customers in C; /* Lemma 8(a) */
 | **if** C_i *is a multi-ray cluster* **then**
 | | Assign C_i to its best facility;
 | | $A \leftarrow \{C_i\} \cup$ Multi-rayClusters2$(C \setminus C_i, F)$;
 | | **if** $cost(A) < Best$ **then**
 | | | $Best \leftarrow cost(A)$;
 | | | $Ans \leftarrow A$;
 | | **endif**
 | **endif**
end
return Ans

Theorem 2. *An optimal r-gathering of C to F can be computed in $O(n + r^2 m + d^2 r^2 (d + \log m) + (r + 1)^d 2^d (r + d)d)$ time.*

Proof. Similar to Theorem 1 we can prove the number of possible choices of the customers for multi-ray clusters is at most $(r+1)^d dr$. For each choice we compute an r-gathering with Multi-rayClusters2 and compute r-gatherings of the remaining one-dimensional problems, then combine them to form an r-gathering of C to F. Then output the best one. This construction of multi-ray clusters needs $O(2^d (r + d)d + (d + \log m)d)$ for each. To eliminate redundant computation we precompute the best facilities of each pair in the dr customers which are candidate for the farthest two customers in possible multi-ray clusters. Such precomputation takes $O(d^2 r^2 (d + \log m))$ time. We can solve all possible one dimensional r-gathering problem in $O(n + r^2 m)$ time in total [10] and we store the solutions in a table. Note that when we solve one dimensional r-gathering problem of ray l, we may assign a cluster to the nearest facility to o located on other ray, however one can compute such f quickly. Thus the time complexity of finding an optimal r-gathering is $O(n + r^2 m + d^2 r^2 (d + \log m) + (r + 1)^d 2^d (r + d)d)$. $\square$

If both r and d are constant, then this is linear.

5   Conclusion

In this paper we presented an $O(rn + (r+1)^d dr)$ time algorithm to solve rhe r-gather clustering problem when all points are lying on a star with d rays. We also gave an $O(n + r^2 m + d^2 r^2 (d + \log m) + (r+1)^d 2^d (r + d)d)$ time algorithm to solve the r-gathering problem when all customers and facilities are lying on a star with d rays.

Can we solve the problems more efficiently or can we solve the problems on more general input like on a tree?

Acknowledgement. We thank CodeCrafters International and Investortools, Inc. for supporting the first author under the grant "CodeCrafters-Investortools Research Grant".

References

1. Agarwal, P.K., Sharir, M.: Efficient algorithms for geometric optimization. ACM Comput. Surv. **30**(4), 412–458 (1998)
2. Aggarwal, G., et al.: Achieving anonymity via clustering. ACM Trans. Algorithms **6**(3), 49:1–49:19 (2010)
3. Akagi, T., Nakano, S.: On r-gatherings on the line. In: Wang, J., Yap, C. (eds.) FAW 2015. LNCS, vol. 9130, pp. 25–32. Springer, Cham (2015). https://doi.org/10.1007/978-3-319-19647-3_3
4. Armon, A.: On min-max r-gatherings. Theor. Comput. Sci. **412**(7), 573–582 (2011)
5. Frederickson, G.N., Johnson, D.B.: Generalized selection and ranking: sorted matrices. SIAM J. Comput. **13**(1), 14–30 (1984)
6. Gabow, H.N., Tarjan, R.E.: Algorithms for two bottleneck optimization problems. J. Algorithms **9**(3), 411–417 (1988)
7. Guha, S., Meyerson, A., Munagala, K.: Hierarchical placement and network design problems. In: Proceedings 41st Annual Symposium on Foundations of Computer Science, pp. 603–612 (2000)
8. Han, Y., Nakano, S.: On r-gatherings on the line. In: Proceedings of the FCS 2016, pp. 99–104 (2016)
9. Karget, D.R., Minkoff, M.: Building steiner trees with incomplete global knowledge. In: Proceedings 41st Annual Symposium on Foundations of Computer Science, pp. 613–623 (2000)
10. Nakano, S.: A simple algorithm for r-gatherings on the line. In: Rahman, M.S., Sung, W.-K., Uehara, R. (eds.) WALCOM 2018. LNCS, vol. 10755, pp. 1–7. Springer, Cham (2018). https://doi.org/10.1007/978-3-319-75172-6_1
11. Zeng, J., et al.: Mobile r-gather: distributed and geographic clustering for location anonymity. In: Proceedings of the 18th ACM International Symposium on Mobile Ad Hoc Networking and Computing, Mobihoc 2017, pp. 7:1–7:10. ACM (2017)

Topological Stability of Kinetic k-centers

Ivor Hoog v.d.[1], Marc van Kreveld[1], Wouter Meulemans[2], Kevin Verbeek[2], and Jules Wulms[2(✉)]

[1] Department of Information and Computing Sciences,
Utrecht University, Utrecht, The Netherlands
{i.d.vanderhoog,m.j.vankreveld}@uu.nl
[2] Department of Mathematics and Computer Science,
TU Eindhoven, Eindhoven, The Netherlands
{w.meulemans,k.a.b.verbeek,j.j.h.m.wulms}@tue.nl

Abstract. We study the k-center problem in a kinetic setting: given a set of continuously moving points P in the plane, determine a set of k (moving) disks that cover P at every time step, such that the disks are as small as possible at any point in time. Whereas the optimal solution over time may exhibit discontinuous changes, many practical applications require the solution to be *stable*: the disks must move smoothly over time. Existing results on this problem require the disks to move with a bounded speed, but this model is very hard to work with. Hence, the results are limited and offer little theoretical insight. Instead, we study the *topological stability* of k-centers. Topological stability was recently introduced and simply requires the solution to change continuously, but may do so arbitrarily fast. We prove upper and lower bounds on the ratio between the radii of an optimal but unstable solution and the radii of a topologically stable solution—the topological stability ratio—considering various metrics and various optimization criteria. For $k = 2$ we provide tight bounds, and for small $k > 2$ we can obtain nontrivial lower and upper bounds. Finally, we provide an algorithm to compute the topological stability ratio in polynomial time for constant k.

Keywords: Stability analysis · Time-varying data · Facility location

1 Introduction

The *k-center problem* or *facility location problem* asks for a set of k disks that cover a given set of n points in the plane, such that the radii of the disks are as small as possible. The problem can be interpreted as placing a set of k facilities (e.g. stores) such that the distance from every point (e.g. client) to the closest facility is minimized. Since the introduction of the k-center problem by

W. Meulemans and J. Wulms are (partially) supported by the Netherlands eScience Center (NLeSC) under grant number 027.015.G02. K. Verbeek is supported by the Netherlands Organisation for Scientific Research (NWO) under project no. 639.021.541. A full version of the paper can be found at https://arxiv.org/abs/1810.00794.

G. K. Das et al. (Eds.): WALCOM 2019, LNCS 11355, pp. 43–55, 2019.
https://doi.org/10.1007/978-3-030-10564-8_4

Sylvester [20] in 1857, the problem has been widely studied and has found many applications in practice. Although the k-center problem is NP-hard if k is part of the input [15], efficient algorithms have been developed for small fixed k. Using rectilinear distance, the problem can be solved in $O(n)$ time [6,13,19] for $k = 2, 3$ and in $O(n \log n)$ time [17,18] for $k = 4, 5$. The problem becomes harder in Euclidean distance, and the currently best known algorithm for Euclidean 2-centers runs in $O(n \log^2 n (\log \log n)^2)$ time [3].

In recent decades there has been an increased interest, especially in the computational geometry community, to study problems for which the input points are moving, including the k-center problem. These problems are typically studied in the framework of *kinetic data structures* [1], where the goal is to efficiently maintain the (optimal) solution to the problem as the points are moving. The kinetic version of the k-center problem also finds a lot of practical applications in, for example, mobile networks and robotics.

A number of kinetic data structures have been developed for maintaining (approximate) k-centers [4,9–11], but in a kinetic setting another important aspect starts playing a role: *stability*. In many practical applications, e.g., if the disks are represented phys- 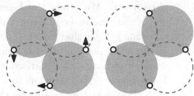 ically, or if the disks are used for visualization, the disks should move smoothly as the points are moving smoothly. As the optimal k-center may exhibit discontinuous changes as points move (see figure), we need to resort to approximations to guarantee stability.

The natural and most intuitive way to enforce stability is as follows. We assume that the points are moving at unit speed (at most), and bound the speed of the disks. Durocher and Kirkpatrick [8] consider this type of stability for Euclidean 2-centers and show that an approximation ratio of $8/\pi \approx 2.55$ can be maintained when the disks can move with speed $8/\pi + 1$. For k-centers with $k > 2$, no approximation factor can be guaranteed with disks of any bounded speed [7]. Similarly, in the black-box KDS model, de Berg et al. [2] show an approximation ratio of 2.29 for Euclidean 2-centers with maximum speed $4\sqrt{2}$.

However, this natural approach to stability is typically hard to work with and difficult to analyze. This is caused by the fact that several aspects are influencing what can be achieved with solutions that move with bounded speed:

1. How is the quality of the solution influenced by enforcing continuous motion?
2. How "far" apart are combinatorially different optimal (or approximate) solutions, that is, how long does it take to change one solution into another?
3. How often can optimal (or approximate) solutions change their combinatorial structure?

Ideally we would use a direct approach and design an algorithm that (roughly) keeps track of the optimal solution and tries to stay as close as possible while adhering to the speed constraints. However, especially the latter two aspects make this direct approach hard to analyze. It is therefore no surprise that most (if not all) approaches to stable solutions are indirect: defining a different structure that is stable in nature and that provides an approximation to what we really want to compute. Although interesting in their own right, such indirect approaches have several drawbacks: (1) techniques do not easily extend to other problems, (2) it is hard to perform better (or near-optimal) for instances where the optimal solution is already fairly stable, and (3) these approaches do not offer much theoretical insight in how optimal solutions (or, by extension, approximate solutions) behave as the points are moving. To gain a better theoretical insight in stability, we need to look at the aspects listed above, ideally in isolation.

Recently, Meulemans et al. [16] introduced a new framework for algorithm stability. This framework includes the natural approach to stability described above (called *Lipschitz stability* in [16]), but it also includes the definition of *topological stability*. An algorithm is topologically stable if its output behaves continuously as the input is changing. The topological stability ratio of a problem is then defined as the optimal approximation ratio of any algorithm that is topologically stable. A more formal definition is given below.

Due to the fact that it allows arbitrary speed, topological stability is mostly interesting from a theoretical point of view: it provides insight into the interplay between problem instances, solutions, and the optimization function; an insight that is invaluable for the development of stable algorithms. Nonetheless, topological stability has practical uses: an example of a very fast and stable change in a visualization can be found when opening a minimized application in most operating systems. The transition starts with the application having a very small size, even as small as a point. The application quickly grows to its intended size in a very smooth and fluid way, to help the user grasp what is happening.

k-Center Problem. An instance of the k-center problem arises from three choices to obtain variants of the problem: the number k of covering shapes, the geometry of the covering shapes and the criterion that measures solution quality. In this paper, we consider covering shapes in the *Euclidean* model, where the covering shapes are disks. The radius of a covering shape is the distance from its center to its boundary under L_2. Furthermore, the quality of a solution is the maximum radius of its covering shapes, the optimization criterion is to minimize this maximum radius. To refer to this problem, we use the notation k-EC-minmax.

Topological Stability. Let us now interpret topological stability, as proposed in [16], for the k-centers problem. Let $\mathcal{I}$ denote the input space of n (stationary) points in $\mathbb{R}^2$ and $\mathcal{S}^k$ the solution space of all configurations of k disks or squares of varying radii. Let Π denote the k-center problem with minmax criterion $f: \mathcal{I} \times \mathcal{S}^k \to \mathbb{R}$. We call a solution in $\mathcal{S}^k$ valid for an instance in $\mathcal{I}$ if it covers all points of the instance. An optimal algorithm OPT maps an instance of $\mathcal{I}$ to a solution in $\mathcal{S}^k$ that is valid and minimizes f.

To define instances on moving points and move towards stability, we capture the continuous motion of points in a topology $\mathcal{T}_{\mathcal{I}}$; an instance of moving points is then a path $\pi: [0,1] \to \mathcal{I}$ through $\mathcal{T}_{\mathcal{I}}$. Similarly, we capture the continuity of solutions in a topology $\mathcal{T}_{\mathcal{S}}^k$, of k disks or squares with continuously moving centers and radii. A *topologically stable* algorithm $\mathcal{A}$ maps a path π in $\mathcal{T}_{\mathcal{I}}$ to a path in $\mathcal{T}_{\mathcal{S}}^k$.[1] We use $\mathcal{A}(\pi, t)$ to denote the solution in $\mathcal{S}^k$ defined by $\mathcal{A}$ for the points at time t. The stability ratio of the problem Π is now the ratio between the best stable algorithm and the optimal (possibly nonstable) solution:

$$\rho_{\mathrm{TS}}(\Pi, \mathcal{T}_{\mathcal{I}}, \mathcal{T}_{\mathcal{S}}^k) = \inf_{\mathcal{A}} \sup_{\pi \in \mathcal{T}_{\mathcal{I}}} \sup_{t \in [0,1]} \frac{f(\pi(t), \mathcal{A}(\pi, t))}{f(\pi(t), \mathrm{OPT}(\pi(t)))}$$

where the infimum is taken over all topologically stable algorithms that give valid solutions. For a minimization problem ρ_{TS} is at least 1; lower values indicate better stability.

Contributions. In this paper we study the topological stability of the k-center problem. Although the obtained solutions are arguably not stable, since they can move with arbitrary speed, we believe that analysis of the topological stability ratio offers deeper insights into the kinetic k-center problem, and by extension, the quality of truly stable k-centers.

In Sect. 2, we prove various bounds on the topological stability for this problem. The ratio is $\sqrt{2}$ for $k = 2$; for arbitrary k, we prove an upper bound of 2 and a lower bound that converges to 2 as k tends to infinity. For small k, we show an upper bound strictly below 2 as well. In Sect. 3, we provide an algorithm to compute the cost of enforcing topological stability for an instance of moving points in polynomial time for constant k. Some proofs in the upcoming sections are sketched or omitted, while the details are described in the full version.

2   Bounds on Topological Stability

As illustrated above, some point sets have more than one optimal solution. If we can transform an optimal solution into another, by growing the covering disks or squares at most (or at least) a factor r, we immediately obtain an upper bound (or respectively a lower bound) of r on the topological stability. To analyze topological stability of k-center, we therefore start with an input instance for which there is more than one optimal solution, and continuously transform one optimal solution into another. This transformation allows the centers to move along a continuous path, while their radii can grow and shrink. At any point during this transformation, the intermediate solution should cover all points of the input. The maximum approximation ratio r that we need for such a transformation, gives a bound on the topological stability of k-center. We can simply consider the input to be static during the transformation, since for topological stability the solution can move arbitrarily fast. We start by introducing some tools to help us model and reason about these transformations.

---

[1] Whereas [16] assumes the black-box model, we allow omniscient algorithms, knowing the trajectories of the moving points beforehand. That is, the algorithm may use knowledge of future positions to improve on stability. This gives more power to stable algorithms, potentially decreasing the ratio. However, our bounds do not use this and thus are also bounds under the black-box model.

2-Colored Intersection Graphs. Consider a point set P and two sets of k convex shapes (disks, squares, ...), such that each set covers all points in P: we use R to denote the one set (red) and B to denote the other set (blue). We now define the *2-colored intersection* graph $G_{R,B} = (V,E)$: each vertex represents a shape $(V = R \cup B)$ and is either red or blue; E contains an edge for each pair of differently colored, intersecting shapes. A 2-colored intersection graph always contains equally many red nodes as blue nodes. Both colors in a 2-colored intersection graph must cover all points: there may be points only in the area of intersection between a blue and red shape. In the remainder, we use intersection graph to refer to 2-colored intersection graphs.

Lemma 1. *Consider two sets R and B of k convex translates each covering a point set P. If intersection graph $G_{R,B}$ is a forest, then R can morph onto B without increasing the shape size, while covering all points in P.*

Proof (sketch). We can always find a red leaf by a counting argument, which can then morph freely onto its blue neighbor. This removes these two nodes from $G_{R,B}$; repeating this argument gives a morph from R onto B. □

Without loss of generality, we assume here that the disks all have the same radius. We first need a few results on (static) intersection graphs, to argue later about topological stability.

Lemma 2. *Let R and B to be optimal solutions to a point set P for k-EC-minmax. Assume the intersection graph $G_{R,B}$ has a 4-cycle with a red degree-2 vertex. To transform R in such a way that $G_{R,B}$ misses one edge of the 4-cycle, while covering the area initially covered by both sets, it is sufficient to increase the disk radius of a red disk by a factor $\sqrt{2}$.*

Proof (sketch). To morph from R to B, a red disk r_1 has to grow to cover the intersection of an adjacent blue disk b with the other (red) neighbor r_2 of b. This allows r_2 to freely morph to the next adjacent blue disk, after which the intersection graph no longer has the 4-cycle.

Let D be the distance from point in an intersection of r_1 with b and the furthest point in an intersection of r_2 with an adjacent blue disk. One can conclude from the triangle inequality that for any pair of optimal solutions R and B that form a 4-cycle, D is at most $\sqrt{2}$ times the radius of a disk in R or B. □

Lemma 3. *Let R and B to be optimal solutions to a point set P for k-EC-minmax. Assume the intersection graph $G_{R,B}$ has only degree-2 vertices. To transform the disks of R onto B, while covering the area initially covered by both sets, it is sufficient to increase the disk radius by a factor $\left(1 + \sqrt{1 + 8\cos^2(\frac{\pi}{2k})}\right)/2$.*

Proof (sketch). To morph from R to B, a red disk r_1 has to grow to cover the intersection of an adjacent blue disk b with the other (red) neighbor r_2 of b (see dashed red disk in figure). We grow r_1 to fully cover its initial disk and the intersection between b and r_2. As a result, we now have to consider only r_1, b, r_2 without concerning ourselves with the other neighbor of r_1 or r_2.

We then assume that r_1 is the red disk that has to grow the least of all red disks in the instance. This allows us to make assumptions on the distances d_1 and d_2 between the center points of r_1 to b and from b to r_2 respectively and the angle α between them at b. In the worst case the instance is symmetric so that $d_1 = d_2 = d$ and $\alpha = \frac{\pi(k-1)}{k}$. Furthermore, for angle β we show that $\cos(\beta) = d/2$ and $\alpha + 2\beta = \pi$ in the worst case. We can finally find the radius of the dashed red disk by calculating x using the Law of Cosines: $x^2 = d^2 + 1^2 - 2d\cos(\alpha + \beta)$.   □

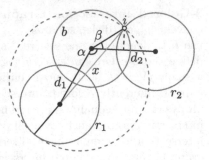

Lemma 4. *Let R and B to be optimal solutions to a point set P for k-EC-minmax. Assume the intersection graph $G_{R,B}$ has only degree-2 vertices. To transform the disks of R onto B, while covering the area initially covered by both sets, it may be necessary to increase the disk radius by a factor $2\sin(\frac{\pi(k-1)}{2k})$.*

Proof. Consider a point set of $2k$ points, positioned such that they are the corners of a regular $2k$-gon with unit radius, i.e., equidistantly spread along the boundary of a unit circle. There are now exactly two optimal solutions (see figure).

To morph from R to B, one of the red disks r_1 has to grow to cover the intersection of an adjacent blue disk b with the other (red) neighbor r_2 of b (see dashed red disk in the figure). Since the points are at equal distance from each other on a unit circle, they are the vertices of a regular $2k$-gon. The diameter of the disks in our optimal solution equals the length of a side of this regular $2k$-gon. This means that a red disk has to grow such that its diameter is equal to the distance between a vertex of the $2k$-gon and a second-order neighbor. Hence, r_1 has to grow to with a factor $2\sin(\frac{\pi(k-1)}{2k})$.   □

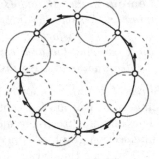

We are now ready to prove bounds on the topological stability of kinetic k-center. The upcoming sequence of lemmata establishes the following theorem.

Theorem 1. *For k-EC-minmax, we obtain the following bounds:*

- $\rho_{TS}(2\text{-}EC\text{-}minmax, \mathcal{T}_\mathcal{I}, \mathcal{T}_\mathcal{S}^2) = \sqrt{2}$

- $\sqrt{3} \le \rho_{TS}(3\text{-}EC\text{-}minmax, \mathcal{T}_\mathcal{I}, \mathcal{T}_\mathcal{S}^3) \le (1 + \sqrt{7})/2$

- $2\sin(\frac{\pi(k-1)}{2k}) \le \rho_{TS}(k\text{-}EC\text{-}minmax, \mathcal{T}_\mathcal{I}, \mathcal{T}_\mathcal{S}^k) \le 2$ *for $k > 3$.*

Lemma 5. $\rho_{TS}(k\text{-}EC\text{-}minmax, \mathcal{T}_\mathcal{I}, \mathcal{T}_\mathcal{S}^k) \le 2$ *for $k \ge 2$.*

Proof. Consider a point in time t where there are two optimal solutions; let R denote the solution that matches the optimal solution at $t - \varepsilon$ and B the solution at $t + \varepsilon$ for arbitrarily small $\varepsilon > 0$. Let C be the maximum radius of the disks in R and in B. Furthermore, let intersection graph $G_{R,B}$ describe the above situation. First we make a maximal matching between red and blue vertices that are adjacent in $G_{R,B}$, implying a matching between a number of red and blue disks. The intersection graph of the remaining red and blue disks has no edges, and we match these red and blue disks in any way.

We find a bound on the topological stability as follows. All the red disks that are matched to blue disks they already intersect grow to overlap their initial disk and the matched blue disk. Now the remaining red disks can safely move to the blue disks they are matched to, and adjust their radii to fully cover the blue disks. Finally, all red disks shrink to match the size of the blue disk they overlap to finish the morph (since each blue disk is now fully covered by the red disk that eventually morphs to be its equal). When all the red disks are overlapping blue disks, the maximum of their radii is at most $2C$, since the radius of each red disk grows by at most the radius of the blue disk it is matched to. $\square$

Lemma 6. $\rho_{TS}(k\text{-}EC\text{-}minmax, \mathcal{T_I}, \mathcal{T}_S^k) \geq 2\sin(\frac{\pi(k-1)}{2k})$ *for* $k \geq 2$.

Proof (sketch). The bound follows from Lemma 4, if we can show that a set of *moving* points that actually force this swap to happen. We let points moving on tangents of the circle defining the $2k$ points, to arrive at this situation at a time t, while ensuring that a swap before or after t would be more costly. $\square$

Lemma 7. $\rho_{TS}(2\text{-}EC\text{-}minmax, \mathcal{T_I}, \mathcal{T}_S^2) = \sqrt{2}$.

Proof. The lower bound follows directly from Lemma 6 by using $k = 2$. For the upper bound, consider a point in time t where there are two optimal solutions; let R denote the solution that matches the optimal solution at $t - \varepsilon$ and B the solution at $t + \varepsilon$ for arbitrarily small $\varepsilon > 0$. If $G_{R,B}$ is a forest, Lemma 1 applies and we do not need to increase the maximum radius during the morph. If $G_{R,B}$ contains a cycle, the entire graph must be a 4-cycle. Lemma 2 gives an upper bound of $\sqrt{2}$ for transforming the intersection graph $G_{R,B}$ to no longer have this 4-cycle, resulting in a tree. Finally, we can morph R into B without further increasing the maximum radius using Lemma 1. $\square$

Lemma 8. $\sqrt{3} \leq \rho_{TS}(3\text{-}EC\text{-}minmax, \mathcal{T_I}, \mathcal{T}_S^3) \leq (1 + \sqrt{7})/2$.

Proof (sketch). A case distinction can be made on how the intersection graph looks: If the intersection graph is a forest or there is a 6-cycle, we can respectively use Lemma 1 or Lemma 3 for the upper and Lemma 4 for the lower bound. However, in the remaining cases we carefully analyze how cycles can be broken until Lemma 1 can be applied. $\square$

The above proof shows the strengths and weaknesses of the earlier lemmata. While in many cases we can get close to tight bounds, dealing with high degree vertices in the intersection graph requires additional analysis. Furthermore, in general we cannot upper bound the approximation factor needed for stable solutions with bounded speed [7], but Theorem 1 can act as a lower bound for such bounded speed solutions.

3   Algorithms for k-Center on Moving Points

Topological stability captures the worst-case penalty that arises from making transitions in a solution continuous. In this section we are interested in the corresponding algorithmic problems that result in instance optimal penalties: how efficiently can we compute the (unstable) k-center for an instance with n moving points, and how efficiently can we compute the stable k-center? When we combine these two algorithms, we can determine for any instance how large the penalty is when we want to solve a given instance in a topologically stable way.

The second algorithm gives us a topologically stable solution to a particular instance of k-center. This solution can be used in a practical application requiring stability, for example as a stable visualization of k disks covering the moving points at all time. Since we are dealing with topological stability, the solution can sometimes move at arbitrary speeds. However, in many practical cases, we can alter the solution in a way that bounds the speed of the solution and makes the quality of the k-center only slightly worse.

3.1   An Unstable Euclidean k-Center Algorithm

Let P be a set of n points moving in the plane, each represented by a constant-degree algebraic function that maps time to the plane. We denote the point set at time t as $P(t)$ and we want to find the optimal set of k minimum covering disks that cover $P(t)$, denoted $\mathcal{B}^*(t)$. Observe that we can define $\mathcal{B}^*$ as the Cartesian product of k triples, pairs, and singletons of distinct points from the set $P(t)$. Not every triple is always relevant: if the circumcircle of the three points is not the boundary of the smallest covering disk, then the triple is irrelevant at that time. This formalization allows us to define what we call *candidate k-centers*.

Definition 1. *Any set of k disks $D_1, \ldots, D_k$ where each disk is the minimum covering disk of one, two or three points in $P(t)$ is called a **candidate k-center** and is denoted $\mathcal{B}(t)$. A candidate k-center is **valid** if the union of its disks cover all points of $P(t)$.*

This definition allows us to rephrase the goal of the algorithm: for each time t we want to compute the smallest value $C(t)$, such that there exists a valid candidate k-center $\mathcal{B}(t)$ where $C(t)$ is the cost in the minmax model. We can see these costs changing over time as curves in a graph that maps time to radii. There are $O(n^3)$ such curves. Using an analysis of the arrangement, lower envelope computation [12], and static k-center algorithms [5,14], we can show:

Theorem 2. *Given a set of n points whose positions in the plane are determined by constant-degree algebraic functions, the minmax Euclidean k-center problem can be solved in $O(n^{2k+5})$ or $n^{O(\sqrt{k})}$ time.*

3.2 A Stable Euclidean k-Center Algorithm

Intuitively, the unstable algorithm finds the lower envelope of all the *valid* radii by traversing the arrangement of all valid radii over time. At each time t a minimal enclosing disk D_1 (defined by a set of at most three points) in the set of optimal disks $\mathcal{B}^*(t)$ needs to be replaced with a new disk D_2, we "hop" from our previous curve to the curve corresponding to D_2. If we require that the algorithm is topologically stable these hops have a cost associated with them.

We first show how to model and compute the cost $C(t)$ of a topological transition between any two k-centers at a fixed time t. We then extend this approach to work over time. Let t be a moment in time where we want to go from one k-center $\mathcal{B}_1$ to another candidate k-center $\mathcal{B}_2$. The transition can happen at infinite speed but must be continuous. We denote the infinitesimal time frame around t in which we do the transition as $[0, T]$. We extend the concept of a k-center with a corresponding partition of P over the disks in the k-center:

Definition 2 (Disk set). *For each disk D_i of a candidate k-center $\mathcal{B}$ for $P(t)$ we define its **disk set** $P_i \subseteq P(t) \cap D_i$ as the subset of points assigned to D_i. A candidate k-center $\mathcal{B}$ with disk sets $P_1, \ldots, P_k$ is **valid** if the disk sets partition $P(t)$. We say $\mathcal{B}$ is **valid** if there exist disk sets $P_1, \ldots, P_k$ such that $\mathcal{B}$ with disk sets $P_1, \ldots, P_k$ is valid.*

k-centers with disk sets will change in the time interval $[0, T]$ while the points $P(t)$ do not move. In essence the time t is equivalent to the whole interval $[0, T]$. For ease of understanding we use t' to denote any time in the interval $[0, T]$. Observe that our definition for a topologically stable algorithm leads to an intuitive way of recognizing a stable transition:

Lemma 9. *A transition from one candidate k-center $\mathcal{B}_1(t)$ to another candidate k-center $\mathcal{B}_2(t)$ in the time interval $[0, T]$ is **topologically stable** if and only if the change of the disks' centers and radii is continuous over $[0, T]$ and at each time $t' \in [0, T]$, $\mathcal{B}(t')$ is **valid**.*

Proof. Note that by definition the disks must be transformed continuously and that all the points in $P(t)$ are covered in $[0, T]$ precisely when a valid candidate k-center exists. □

Now that we can recognize a topologically stable transition, we can reason about what such a transition looks like:

Lemma 10. *Any topologically stable transition from one k-center $\mathcal{B}_1(t)$ to another k-center $\mathcal{B}_2(t)$ in the timespan $[0, T]$ that minimizes $C(t)$ (the largest occurring minmax over $[0, T]$) can be obtained by a sequence of events where in each event, occurring at a time $t' \in [0, T]$, a disk $D_i \subset \mathcal{B}(t')$ adds a point to P_i and a disk $D_j \in \mathcal{B}(t')$ removes a point from P_j. We call this **transferring**.*

Proof. The proof is by construction. Assume that we have a transition from $\mathcal{B}_1(t)$ to $\mathcal{B}_2(t)$ and the transition that minimizes the maximum of all radii contains simultaneous continuous movement. Let this transition take place in $[0, T]$.

To determine $C(t)$ we only need to look at times $t' \in [0, T]$ where a disk $D_i \in \mathcal{B}$ adds a new point p to its disk set P_i and another disk D_j removes it from P_j. Only at t' must both disks contain p; before t' disk D_j may be smaller and after t' disk D_i may be smaller.

We claim that for any optimal simultaneous continuous movement of cost $C(t)$, we can discretize the movement into a sequence of events with cost no larger than $C(t)$. We do so recursively: If the movement is continuous then there exists a $t_0 \in [0, T]$ as the first time a disk $D_i \in \mathcal{B}$ adds a point to P_i. At t_0, D_i has to contain both P_i and p and must have a certain size d. All the other disks $D_j \in \mathcal{B}$ with $j \neq i$ only have to contain the points in P_j so they have optimal size if they have not moved from time 0. In other words, it is optimal to first let D_i obtain p in an event and to then continue the transition from $[t_0, T]$. This allows us to discretize the simultaneous movement into sequential events.   □

Corollary 1. *Any topologically stable transition from one k-center $\mathcal{B}_1(t)$ to another k-center $\mathcal{B}_2(t)$ in the timespan $[0, T]$ that minimizes $C(t)$ (the largest occurring* minmax *over $[0, T]$) can be obtained by a sequence of events where in each event the following happens:*

A disk $D_i \in \mathcal{B}_1(t)$ that was defined by one, two or three points in $P(t)$ is now defined by a new set of points in $P(t)$ where the two sets differ in only one element.

*With every event, P_i must be updated with a corresponding insert and/or delete. We call these events a **swap** because we intuitively swap out of the at most three defining elements.*

The Cost of a Single Stable Transition. Corollary 1 allows us to model a stable transition as a sequence of swaps but how do we find the optimal sequence of swaps? A single minimal covering disk at time t is defined by at most three unique elements from $P(t)$ so there are at most $O(n^3)$ subsets of $P(t)$ that could define one disk of a k-center. Let these $O(n^3)$ sets be the vertices in a graph G. We create an edge between two vertices v_i and v_j if we can transition from one disk to the other with a single swap and that transition is topologically stable. Each vertex is incident to only a constant number of edges (apart from degenerate cases) because during a swap the disk D_i corresponding to v_i can only add one element to P_i. Moreover, the radius of the disk is maximal on vertices in G and not on edges. The graph G has $O(n^3)$ complexity and takes $O(n^4)$ time to construct.

This graph provides a framework to trace the radius of the transition from a single disk to another disk. However, we want to transition from one k-center to another. We use the previous graph to construct a new graph G^k where each vertex w_i represents a set of k disks: a candidate k-center $\mathcal{B}_i$. We again create an edge between vertices w_i and w_j if we can go from the candidate k-center $\mathcal{B}_i$ to $\mathcal{B}_j$ in a single swap. With a similar argument as above, each vertex is only connected to $O(k)$ edges. The graph thus has $O(n^{3k})$ complexity and can be constructed in $O(n^{3k+1})$ time. Each vertex w_i gets assigned the cost of the k-center $\mathcal{B}_i$ where the cost is ∞ if $\mathcal{B}_i$ is invalid.

Any connected path in this graph from w_i to w_j without vertices with cost ∞ represents a stable transition from w_i to w_j by Corollary 1, where the cost of the path (transition) is the maximum value of the vertices on the path. We can now find the optimal sequence of swaps to transition from any vertex w_i to w_j by finding the cheapest path in this graph in $O(n^{3k} \log n)$ time, which is dominated by the $O(n^{3k+1})$ time it takes to construct the graph.

Maintaining the Cost of a Flip. For a single point in time we can now determine the cost of a topologically stable transition from a k-center $\mathcal{B}_i$ to $\mathcal{B}_j$ in $O(n^{3k+1})$ time. If we want to maintain the cost $C(t)$ for all times t, the costs of the vertices in the graph change over time. If we plot the changes of these costs over time, the graph consists of monotonously increasing or decreasing segments, separated by moments in time where two radii of disks are equal. These $O(n^{3k})$ events also contain all events where the structure of our graph G^k changes *and* all the moments where a vertex in our graph becomes *invalid* and thus gets cost ∞. The result of these observations is that we have a $O(n^{3k})$ size graph, with $O(n^{3k})$ relevant changes where with each change we need $O(n)$ time to restore the graph. This leads to an algorithm which can determine the cost of a topologically stable movement in $O(n^{6k+1})$ time.

Theorem 3. *Given a set of n points whose positions in the plane are determined by constant-degree algebraic functions, the stable minmax Euclidean k-center problem can be solved by an algorithm that runs in $O(n^{6k+1})$ time.*

If we run the unstable and stable algorithms on the moving points, we obtain two functions that map time to a cost. The maximum over time of the ratio of the cost is the stability ratio of the instance, obtained in the same running time.

4 Conclusion

We considered the topological stability of the kinetic k-center problem, in which solutions must change continuously but may do so arbitrarily fast. We proved nontrivial upper bounds for small values of k and presented a general lower bound tending towards 2 for large values of k. We also presented algorithms to compute topologically stable solutions together with the cost of stability for a set of moving points, that is, the growth factor that we need for that particular set of moving points at any point in time. A practical application of these algorithms would be to identify points in time where we could slow down the solution to explicitly show stable transitions between optimal solutions.

Future Work. It remains open whether a general upper bound strictly below 2 is achievable for k-EC-minmax. We conjecture that this bound is indeed smaller than 2 for any constant k. For this, we need more insight in how to resolve an intersection graph with more general structures. Our algorithms to compute the cost of stability for an instance have high (albeit polynomial) run-time complexity. Can the results for KDS (e.g. [2]) help us speed up these algorithms? Alternatively, can we approximate the cost of stability more efficiently?

Lipschitz stability requires a bound on the speed at which a solution may change [16]. This stability for $k > 2$ is unbounded, if centers have to move continuously [7]; A potentially interesting variant of the topology is one where a disk may shrink to radius 0, at which point it disappears and may reappear at another location. This alleviates the problem in the example; would it allow us to bound the Lipschitz stability?

References

1. Basch, J., Guibas, L.J., Hershberger, J.: Data structures for mobile data. J. Algorithms **31**(1), 1–28 (1999)
2. de Berg, M., Roeloffzen, M., Speckmann, B.: Kinetic 2-centers in the black-box model. In: Proceedings of 29th Symposium on Computational Geometry, pp. 145–154 (2013)
3. Chan, T.: More planar two-center algorithms. Comp. Geom. **13**(3), 189–198 (1999)
4. Degener, B., Gehweiler, J., Lammersen, C.: Kinetic facility location. Algorithmica **57**(3), 562–584 (2010)
5. Drezner, Z.: On a modified 1-center problem. Manag. Sci. **27**, 838–851 (1981)
6. Drezner, Z.: On the rectangular p-center problem. Nav. Res. Logist. **34**(2), 229–234 (1987)
7. Durocher, S.: Geometric Facility Location under Continuous Motion. Ph.D. thesis, University of British Columbia (2006)
8. Durocher, S., Kirkpatrick, D.: Bounded-velocity approximation of mobile Euclidean 2-centres. Int. J. Comput. Geom. Appl. **18**(03), 161–183 (2008)
9. Friedler, S., Mount, D.: Approximation algorithm for the kinetic robust k-center problem. Comput. Geom. **43**(6–7), 572–586 (2010)
10. Gao, J., Guibas, L., Hershberger, J., Zhang, L., Zhu, A.: Discrete mobile centers. Discret. Comput. Geom. **30**(1), 45–63 (2003)
11. Gao, J., Guibas, L., Nguyen, A.: Deformable spanners and applications. Comput. Geom. Theory Appl. **35**(1–2), 2–19 (2006)
12. Hershberger, J.: Finding the upper envelope of n line segments in $o(n \log n)$ time. Inf. Process. Lett. **33**(4), 169–174 (1989)
13. Hoffmann, M.: A simple linear algorithm for computing rectangular 3-centers. In: Proceedings of 11th Canadian Conference on Computational Geometry, pp. 72–75 (1999)
14. Hwang, R., Lee, R., Chang, R.: The slab dividing approach to solve the Euclidean p-center problem. Algorithmica **9**, 1–22 (1993)
15. Megiddo, N., Supowit, K.: On the complexity of some common geometric location problems. SIAM J. Comput. **13**(1), 182–196 (1984)
16. Meulemans, W., Speckmann, B., Verbeek, K., Wulms, J.: A framework for algorithm stability and its application to kinetic Euclidean MSTs. In: Bender, M.A., Farach-Colton, M., Mosteiro, M.A. (eds.) LATIN 2018. LNCS, vol. 10807, pp. 805–819. Springer, Cham (2018). https://doi.org/10.1007/978-3-319-77404-6_58
17. Nussbaum, D.: Rectilinear p-piercing problems. In: Proceedings of 1997 Symposium on Symbolic and Algebraic Computation, pp. 316–323 (1997)

18. Segal, M.: On piercing sets of axis-parallel rectangles and rings. In: Burkard, R., Woeginger, G. (eds.) ESA 1997. LNCS, vol. 1284, pp. 430–442. Springer, Heidelberg (1997). https://doi.org/10.1007/3-540-63397-9_33

19. Sharir, M., Welzl, E.: Rectilinear and polygonal p-piercing and p-center problems. In: Proceedings of 12th Symposium on Computational Geometry, pp. 122–132 (1996)

20. Sylvester, J.: A question in the geometry of situation. Q. J. Math. 1, 79 (1857)

A Linear Time Algorithm
for the r-Gathering Problem on the Line
(Extended Abstract)

Anik Sarker[1]([⊠]), Wing-kin Sung[2,3]([⊠]), and M. Sohel Rahman[1]([⊠])

[1] Department of CSE, BUET, ECE Building, West Palasi, Dhaka 1205, Bangladesh
aniksarkerbuet1997@gmail.com, msrahman@cse.buet.ac.bd
[2] School of Computing, National University of Singapore, Singapore, Singapore
ksung@comp.nus.edu.sg
[3] Genome Institute of Singapore, 60 Biopolis Street, Genome,
Singapore 138672, Singapore

Abstract. In this paper, we revisit the r-gathering problem. Given sets C and F of points on the plane and distance $d(c, f)$ for each $c \in C$ and $f \in F$, an r-gathering of C to F is an assignment A of C to open facilities $F' \subseteq F$ such that r or more members of C are assigned to each open facility. The cost of an r-gathering is $\max_{c \in C} d(c, A(c))$. The r-gathering problem computes the r-gathering minimizing the cost. In this paper we study the r-gathering problem when C and F are on a line and present a $O(|C|+|F|)$-time algorithm to solve the problem. Our solution is optimal since any algorithm needs to read C and F at least once.

1 Introduction

Facility location problems aim at finding an optimal placement of (some) facilities minimizing some cost function defined based on the actual application scenario. More formally, in the well-known basic facility location problem, we are given (1) a set C of customers, (2) a set F of facilities, (3) an opening cost $op(f)$ for each $f \in F$, and (4) a connecting cost $co(c, f)$ for each pair of $c \in C$ and $f \in F$, and then we need to open a subset $F' \subseteq F$ of facilities and find an assignment A from C to F' so that a designated cost is minimized. Facility location problems and variants thereof [4,5] have been studied heavily in both operations research and computational geometry literature.

In this article we study a relatively new but interesting variant which is referred to as the r-gathering problem. An r-gathering of customers C to facilities F is an assignment function $A : C \to F$ such that (1) $A(c) \in F$, (2) the set of open facilities is $F' = \{A(c) \mid c \in C\}$, (3) for each open facility $f \in F'$, at least r customers are assigned to it. The cost of an r-gathering is the maximum distance $d(c, f)$ between $c \in C$ and its assigned $A(c) \in F'$ among all customers $c \in C$, which is $\max_{c \in C} d(c, A(c))$. The r-gathering problem aims to find the r-gathering minimizing the cost. An example of the problem instance in a restricted setting

G. K. Das et al. (Eds.): WALCOM 2019, LNCS 11355, pp. 56–66, 2019.
https://doi.org/10.1007/978-3-030-10564-8_5

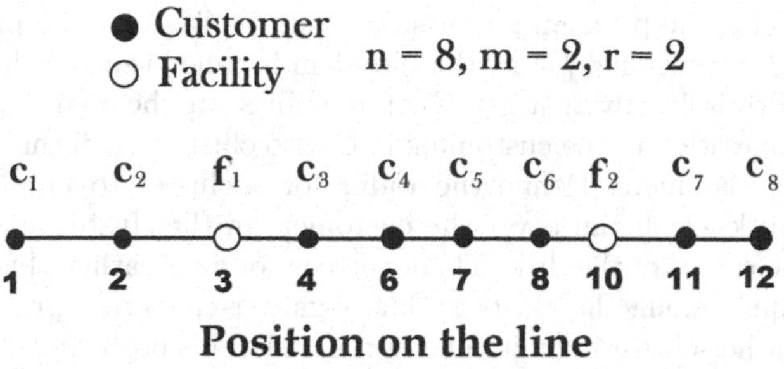

● **Customer**
○ **Facility** $n = 8, m = 2, r = 2$

c_1 c_2 f_1 c_3 c_4 c_5 c_6 f_2 c_7 c_8

●————●————○——●——●——●——●——○——●——●

1   2   3   4   6   7   8   10   11   12

Position on the line

Fig. 1. An example with 8 customers and 2 facilities.

where all the customers and the facilities are located on a line is presented in the example below.

Consider a horizontal line with 8 customers and 2 facilities (Fig. 1). Assume the set of positions of the customers is $C = \{c_1 = 1, c_2 = 2, c_3 = 4, c_4 = 6, c_5 = 7, c_6 = 8, c_7 = 11, c_8 = 12\}$ while the set of positions of the facilities is $F = \{f_1 = 3, f_2 = 10\}$. Suppose $r = 2$ (which means every open facility serves at least 2 customers). Then, the best solution is $A(c_1) = f_1, A(c_2) = f_1, A(c_3) = f_1, A(c_4) = f_1, A(c_5) = f_2, A(c_6) = f_2, A(c_7) = f_2, A(c_8) = f_2$. The cost can be calculated as follows:

$\max\{d(c_1, f_1), d(c_2, f_1), d(c_3, f_1), d(c_4, f_1), d(c_5, f_2), d(c_6, f_2), d(c_7, f_2), d(c_8, f_2)\}$
$= \max\{2, 1, 1, 3, 3, 2, 1, 2\}$
$= 3$

Besides being combinatorially interesting, like many other variants of the facility location problem, r-gathering problem has practical applications as well. The following interesting application has been borrowed from [7]:

Assume that we are planning an evacuation plan for the residents on a street, F is a set of possible locations for emergency shelters, and $d(c, f)$ is the time needed for a person $c \in C$ to reach a shelter $f \in F$. Then an r-gathering (when all C and F are on the line) corresponds to an evacuation assignment such that each open shelter serves r or more people, and the r-gathering problem finds an evacuation plan minimizing the evacuation time span.

We first give a brief literature review. In general, when C and F are placed on some points on a 2-dimensional plane, Armon [3] showed that this problem is NP-hard and gave a simple 3-approximation algorithm. They also proved that the problem cannot be approximated within a factor less than 3 for any $r \geq 3$ unless $P = NP$.

To the best of our knowledge, when all C and F are on the line, three works have been reported in the literature. Akagi et al. [2] presented an $O((|C| + |F|)\log(|C| + |F|))$-time algorithm. Subsequently, Han and Nakano [8] developed an $O(|C| + |F|\log^2 r + |F|\log|F|)$-time algorithm. Very recently, [7] proposed an $O(|C| + r^2|F|)$-time algorithm.

When we relax the problem and assume that facilities can be placed on any points on the 2-dimensional plane, this problem is called the r-gather-clustering problem [1]. Precisely, given a set C of n points on the plane, an r-gather-clustering is a partition of the customers in C into clusters such that each cluster has at least r customers. Define the radius of a cluster to be the minimum radius of the disk which can cover the customers in the cluster and the facility is placed at the center of the disk. Then, the cost of an r-gather-clustering is the maximum radius among the clusters. The r-gather-clustering problem aims to compute a r-gather-clustering minimizing the cost. This problem is NP-complete in general. A 2-approximation algorithm was presented in [1]. If C are on the line, Akagi et al. [2] presented an $O(|C|\log|C|)$-time algorithm. Very recently, Nakano [7] proposed an $O(r|C|)$ time algorithm to solve the problem by reducing the problem to the min-max path problem [6] in a weighted directed graph.

In this paper, we study the r-gathering problem when all C and F are on the line. We assume all points in C and F are sorted according to their position from left to right following previous literature ([2,7,8]). As stated above, three solutions have been proposed on this variant ([2,7,8]). However, none of those runs in linear time. We ask the question whether a linear time solution for this variant exists. We answer the question affirmatively by giving an $O(|C|+|F|)$-time algorithm. Our solution is optimal since any algorithm needs to read the list of customers and facilities at least once.

To develop our linear time algorithm, we present and prove a number of important lemmas and make use of a number of intermediate algorithms with inferior running times. We first present a simple $O(|C|^2\log|F|)$-time solution (Sect. 2). Then we present (and prove) some interesting lemmas and propose an improved algorithm that runs in $O(|C|\log|C|\log|F|)$ time (Sect. 3). Subsequently, we prove some more lemmas and improve the running time to $O(|F|+|C|\log|F|)$ (Sect. 4). Finally, we exploit an interesting relation to reach the desired $O(|C|+|F|)$-time solution (Sect. 5).

2   An $O(|F|+|C|^2\log|F|)$-time Solution

For the sake of notational ease, in what follows, we will use the following description of the problem. Suppose, $C=\{c_1,c_2,c_3,\ldots,c_n\}$ and $F=\{f_1,f_2,f_3,\ldots,f_m\}$ are points on the horizontal line and assume that they are sorted (according to their position) from left to right. The following lemma from [7] will be useful for us.

Lemma 1 ([7]). *There exists an optimal solution in which each facility is assigned to a cluster of customers which are consecutive in C.*   □

Let $c_a, c_{a+1}, \ldots, c_b$ be a sequence of consecutive points in C. Let $Cost(a,b)$ be the minimum cost to assign them to a single facility optimally.

Lemma 2. *Let $c_a, c_{a+1}, \ldots, c_b$ be a sequence of consecutive points in C. Let the center of a cluster be the midpoint of the two endpoints of the cluster (c_a and c_b).*

Algorithm 1. min-cost-r-gathering(C,F,r)

1: $PrefixCost(0) \leftarrow 0$
2: **for** each $i \in [1, r-1]$ **do**
3:      $PrefixCost(i) \leftarrow \infty$
4: **for** each $i \in [r, n]$ **do**
5:      $PrefixCost(i) \leftarrow \infty$
6:      **for** each $j \in [0, i-r]$ **do**
7:          Let F_{mid} be the nearest facility to the midpoint of [j+1,i] segment.
8:          $Cost(j+1, i) \leftarrow max(|C_i - F_{mid}|, |F_{mid} - C_{j+1}|)$.
9:          $CurrentCost \leftarrow max(PrefixCost(j), Cost(j+1, i))$
10:         $PrefixCost(i) \leftarrow min(PrefixCost(i), CurrentCost)$
11: **return** $PrefixCost(n)$

Then $Cost(a, b)$ will be minimized if the points in it are assigned to the nearest facility from the center of the cluster.

Proof. Let m be the centre of the cluster, i.e., $m = \frac{|c_b - c_a|}{2}$. Let f_x be the optimal facility. Since $Cost(a, b) = \max(|f_x - c_a|, |f_x - c_b|) = \frac{|c_a - c_b|}{2} + |f_x - m|$, this value will be minimum when f_x is the closest facility to m. The result follows.   □

Corollary 1. *$Cost(a, b)$ can be computed in $O(\log |F|)$ time.*

Proof. First, set $\mu = \frac{c_a + c_b}{2}$. By doing a binary search on $(f_1, f_2, \ldots, f_m)$, we can find f_x that is the closest to μ using $O(\log m) = O(\log |F|)$ time. Then, we report $Cost(a, b) = \max(|f_x - c_a|, |f_x - c_b|)$.   □

Now, we describe an $O(|F| + |C|^2 \log |F|)$-time dynamic programming solution. Let's define **PrefixCost(i)** as the minimum cost to serve customers in $\{c_1, c_2, c_3, \ldots, c_i\}$ such that each facility either serves 0 customer or at least r customers. By Lemma 1, we can assume the optimal solution partitions $\{c_1, \ldots, c_i\}$ into one or more consecutive clusters, where each cluster is served by one facility. Note that the ith point must be in the rightmost cluster of minimum size r and maximum size i. In other words, the rightmost cluster is $\{c_{j+1}, \ldots, c_i\}$ where j satisfies $0 \leq j \leq i - r$. Then, by definition of $PrefixCost$, we have $PrefixCost(i) = \max\{PrefixCost(j), Cost(j+1, i)\}$. Hence, we have the following recurrence:

$$PrefixCost(i) = \begin{cases} 0 & \text{if } i = 0 \\ \infty & \text{if } 1 \leq i < r \\ \min_{j=0}^{i-r} \max\{PrefixCost(j), Cost(j+1, i)\} & \text{otherwise} \end{cases} \tag{1}$$

Based on the recurrence, $PrefixCost(n)$ can be computed by Algorithm 1. In the algorithm, the facility F_{mid} can be found in $O(\log |F|)$ time using binary search (see Corollary 1). Hence, the algorithm runs in $O(|F| + |C|^2 \log |F|)$ time.

For the example in Fig. 1, when we run **min-cost-r-gathering**$(C, F, 2)$, the corresponding dynamic programming table $PrefixCost$ is shown in Fig. 2.

Index i	0	1	2	3	4	5	6	7	8
$PrefixCost(i)$	0	∞	2	2	3	4	3	3	3

Fig. 2. min-cost-r-gathering$(C, F, 2)$ for the example in Fig. 1.

3   An $O(|F| + |C| \log |C| \log |F|)$-time Solution

Previous section shows that $PrefixCost(i)$ can be computed by checking $PrefixCost(j)$ and $Cost(j+1, i)$ for $j \in [0, i - r]$. This approach requires us to check $O(i)$ entries. Can we do better? Indeed, here, we observe that $PrefixCost(i)$ can be computed by checking only a few of these entries. Below, we first define a *turning point sequence*, G_i for the ith point and state its different properties. Then, exploiting these properties, we present an improved algorithm. We start with the following useful lemma which basically says that the cost of a cluster will increase if you extend the cluster in either side.

Lemma 3. *For any index a, b such that $a > 1$ and $a \leq b$, we have:*
(i) $Cost(a, b) \leq Cost(a - 1, b)$.
(ii) $Cost(a, b) \leq Cost(a, b + 1)$,

Proof. (i) $Cost(a - 1, b)$ is the minimum cost to optimally assign consecutive points $c_{a-1}, c_a, \ldots, c_b$ to a single facility. Since, $c_a, c_{a+1}, \ldots, c_b$ is only a subset of the previous set of points, $Cost(a, b)$ cannot be greater than $Cost(a - 1, b)$.
(ii) Similarly as (i). □

Definition 1. *For a fixed point i, $i \geq r$, we define the turning point sequence G_i, which is a sequence of points $(a_0, \ldots, a_k)$ as follows:*
(1) The 0-th point is $a_0 = 0$ with $PrefixCost(a_0) = 0$;
(2) For $j \geq 1$, let a_j be the maximum index in $[a_{j-1} + 1, i - r]$ such that $PrefixCost(a_j)$ is minimum.

Remark 1. For the sake of ease in handling boundary conditions, we define $PrefixCost(a_{k+1}) = \infty$.

Lemma 4. *Let $G_i = (a_0, \ldots, a_k)$ be a turning point sequence for the ith point in C, $i \geq r$. We have:*

(i) $a_k = i - r$
(ii) $a_0 < a_1 < \ldots < a_k$
(iii) $PrefixCost(a_0) < PrefixCost(a_1) < \ldots < PrefixCost(a_k)$.

Proof. (i) Assume $a_k \neq i - r$. But then according to the definition, a_k cannot be the last point of the sequence and a_{k+1} must exist. So a_k must equal $i - r$.
(ii) For any $1 \leq j \leq k$, by definition, a_j is in the range $[a_{j-1} + 1, i - r]$. So, $a_{j-1} < a_j$ always holds.
(iii) By contrary, assume $PrefixCost(a_j) \leq PrefixCost(a_{j-1})$ for some $1 \leq j \leq k$. By definition, $PrefixCost(a_{j-1}) = \min_{\alpha = a_{j-2}+1}^{i-r} PrefixCost(\alpha)$ while $PrefixCost(a_j) = \min_{\alpha = a_{j-1}+1}^{i-r} PrefixCost(\alpha)$. Since $a_{j-1} > a_{j-2}$, we have $PrefixCost(a_j) \geq PrefixCost(a_{j-1})$. We arrive at a contradiction. □

Now note that, By Lemma 3, we have:

$$Cost(a_0 + 1, i) \geq Cost(a_1 + 1, i) \geq \ldots \geq Cost(a_k + 1, i).$$

And, by Lemma 4, we have:

$$PrefixCost(a_0) < PrefixCost(a_1) < \ldots < PrefixCost(a_k).$$

So, we have one sequence that is non-increasing and the other sequence that is strictly increasing. These properties will be helpful in our subsequent arguments.

Lemma 5. *Let $G_i, i \geq r$, be a turning point sequence. We have:*

$$PrefixCost(i) = \min_{j \in G_i} \max(PrefixCost(j), Cost(j + 1, i)).$$

Proof. The proof will be provided in the journal version. □

Lemma 6. *Suppose $G_i = (a_0, a_1, \ldots, a_k)$, $i \geq r$, is the turning point sequence for position i. There exists an index $p_i \in \{0, 1, \ldots, k\}$ such that $PrefixCost(a_j) \leq Cost(a_j + 1, i)$ for all $j \leq p_i$; and $PrefixCost(a_j) > Cost(a_j + 1, i)$ otherwise. We can find p_i using $O(\log |C| \log |F|)$ time.*

Proof. The proof will be provided in the journal version. □

Lemma 7. *Consider $G_i = (a_0, a_1, \ldots, a_k)$, $i > r$. Let p_i be the maximum index such that $PrefixCost(a_{p_i}) \leq Cost(a_{p_i} + 1, i)$. Then we have $PrefixCost(i) = \min(Cost(a_{p_i} + 1, i), PrefixCost(a_{p_i+1}))$.*

Proof. The proof will be provided in the journal version. □

Note that $p_i + 1$ in above Lemma may be point to an index which is out of bound of G_i ($p_i = k$). However, due to Remark 1, this does not affect the validity of the lemma. To apply the above lemmas to compute $PrefixCost(i)$, we need to obtain the turning point sequence G_i. Below lemma tells us how to construct G_i from G_{i-1}.

Lemma 8. *Suppose $G_{i-1} = (a_0, a_1, \ldots, a_k = i - 1 - r)$, $i - 1 \geq r$, is the turning point sequence for position $i - 1$. Now consider the computation of G_i. Let h be the maximum index (with $a_h \in G_{i-1}$) such that $PrefixCost(a_h) < PrefixCost(i - r)$. Then we have $G_i = (a_0, a_1, \ldots, a_h, i - r)$.*

Proof. The proof will be provided in the journal version. □

The above lemma implies that the turning point sequence can be maintained as a stack S so that $S[0] = a_0, \ldots, S[k] = a_k$. Suppose S stores the turning point sequence G_{i-1}. We can modify S to store G_i as follows.

1. While the stack is not empty and the point a_h on the top of the stack S satisfies $PrefixCost(a_h) \geq PrefixCost(i - r)$, pop the point a_h from the stack S;

Algorithm 2. min-cost-r-gathering1(C,F,r)

1: $PrefixCost(0) \leftarrow 0$
2: **for** $i = 1$ to $r - 1$ **do**
3:     $PrefixCost(i) \leftarrow \infty$
4: Let S be an empty stack
5: **for** $i = r$ to n **do**
6:     **while** S is not empty and $PrefixCost(Top(S)) \geq PrefixCost(i - r)$ **do**
7:         pop the element on top of S
8:     push $(i - r)$ at the top of S
9:     Find maximum index p such that $PrefixCost(S[p]) \leq Cost(S[p] + 1, i)$
10:     $PrefixCost(i) \leftarrow Cost(S[p] + 1, i)$
11:     **if** $p + 1 \leq Size(S)$ **then**
12:         $PrefixCost(i) \leftarrow min(PrefixCost(i), PrefixCost(S[p + 1]))$
13: **return** $PrefixCost(n)$

---

2. Push $(i - r)$ to the top of the stack S.

Since every point in C can be pushed and popped at most once, all turning point sequences can be constructed in $O(|C|)$ time. We can now describe the algorithm. We do the following -

- Set $PrefixCost(0) = 0$.
- Set $PrefixCost(1) = \ldots = PrefixCost(r - 1) = \infty$.
- Initialize an empty stack S (0 - indexed).
- For $i = r$ to $|C|$, iteratively build G_i by Lemma 8 and compute $PrefixCost(i)$ by Lemma 7.

The detail of the pseudocode is in Algorithm 2. The running time of the algorithm is $O(|F| + |C| \log |C| \log |F|)$. For the example in Fig. 1, when we run **min-cost-r-gathering1**$(C, F, 2)$, the corresponding dynamic programming table $PrefixCost$ is shown in Fig. 3.

index i	Stack S	p	Cost(S[p]+1, i)	PrefixCost(S[p+1])
0	-	-	-	0
1	-	-	-	∞
2	$\{0\}$	0	2	∞
3	$\{0, 1\}$	0	2	∞
4	$\{0, 2\}$	1	3	∞
5	$\{0, 3\}$	1	4	∞
6	$\{0, 3, 4\}$	2	3	∞
7	$\{0, 3, 4, 5\}$	2	3	4
8	$\{0, 3, 6\}$	1	4	3

Fig. 3. min-cost-r-gathering1$(C, F, 2)$ for the example in Fig. 1.

4   An $O(|F| + |C| \log |F|)$-time Solution

In this section we further improve the running time of our algorithm. In particular, we will show how we can avoid the binary search to compute p_i as reported in Lemma 6. Recall that p_i is the maximum index such that $PrefixCost(a_{p_i}) \leq Cost(a_{p_i} + 1, i)$. In what follows it will be convenient to define $x_{G_i} = a_{p_i}$ and $y_{G_i} = a_{p_i+1}$.

Lemma 9. *Suppose $G_{i-1} = (a_0, a_1, \ldots, i - r - 1)$, $i - 1 \geq r$ and following Lemma 8 we have computed, $G_i = (a_0, a_1, \ldots, a_h, i - r)$. Then the followings hold true:*

(i) *if $PrefixCost(x_{G_{i-1}}) < PrefixCost(i - r)$, then $x_{G_{i-1}} \in G_i$ and $x_{G_i} \geq x_{G_{i-1}}$.*

(ii) *if $PrefixCost(x_{G_{i-1}}) \geq PrefixCost(i - r)$, then $x_{G_{i-1}} \notin G_i$. And the followings hold true.*

    (a) *if $PrefixCost(i - r) \leq Cost(i - r + 1, i)$ then $x_{G_i} = i - r$*

    (b) *Otherwise, $x_{G_i} = a_h$.*

Proof. The proof will be provided in the journal version. □

We now need the following definitions.

Definition 2. *Suppose $G_i = (a_0, a_1, \ldots, a_h, i - r)$, $r \leq i \leq n$ is the turning sequence for position i and x_{G_i} and y_{G_i} are defined as before. We now define Opt_i as follows.*

$$Opt_i = \begin{cases} x_{G_i} & if \ Cost(x_{G_i} + 1, i) < PrefixCost(y_{G_i}) \\ y_{G_i} & otherwise \end{cases}$$

We also define $F_i(a_j) = max(PrefixCost(a_j), Cost(a_j + 1, i)), a_j \in G_i$.

Now we have the following lemma which will be the basis of further improvement of our algorithm.

Lemma 10. *Suppose $G_i = (a_0, a_1, \ldots, a_h, i - r)$, $r \leq i \leq n$ is the turning sequence for position i. Then the followings hold true.*

(i) *$PrefixCost(i) = F_i(Opt_i)$.*

(ii) *$F_i(a_{j-1}) \geq F_i(a_j)$ for $a_j \leq Opt_i$ and $F_i(a_{j+1}) > F_i(a_j)$ for $a_j \geq Opt_i$.*

(iii) *$Opt_i \geq Opt_{i-1}$.*

Proof. The proof will be provided in the journal version. □

We can now describe the new improved algorithm. We do not need any binary search to find Opt_i, instead we do the following:

- Set $PrefixCost(0) = 0$.
- Set $PrefixCost(1) = \ldots = PrefixCost(r - 1) = \infty$.
- Initialize an empty stack S (1 - indexed as opposed 0 - indexed).

Algorithm 3. min-cost-r-gathering2(C,F,r)

```
1:  PrefixCost(0) ← 0
2:  for i = 1 to r − 1 do PrefixCost(i) ← ∞
3:  OptIndx ← 1, NextOptIndx ← 2
4:  Let S be an empty stack
5:
6:  for i = r to n do
7:      while S not empty and PrefixCost(Top(S)) ≥ PrefixCost(i − r) do
8:          pop the element on top of S
9:      push (i − r) at the top of S
10:
11:     if OptIndx > Size(S) then
12:         OptIndx ← Size(S)
13:         NextOptIndx ← OptIndx + 1
14:
15:     while true do
16:         Opt ← S[OptIndx]
17:         F(Opt) ← max(PrefixCost(Opt), Cost(Opt + 1, i))
18:         if OptIndx = Size(S) then break
19:
20:         NextOpt ← S[NextOptIndx]
21:         F(NextOpt) ← max(PrefixCost(NextOpt), Cost(NextOpt + 1, i))
22:
23:         if F(Opt) < F(NextOpt) then break
24:         OptIndx ← NextOptIndx
25:         NextOptIndx ← NextOptIndx + 1
26:
27:     PrefixCost(i) ← F(Opt)
28: return PrefixCost(n)
```

– Initialize $OptIndx = 1, NextOptIndx = 2$. $OptIndx$ maintains index of Opt_i in S and $NextOptIndx = OptIndx + 1$.
– For $i = r$ to $|C|$:
  □ iteratively build G_i by Lemma 8
  □ if $(OptIndx > Size(S))$:
    $OptIndx = Size(S), NextOptIndx = OptIndx + 1$.
  □ While $OptIndx < Size(S)$ and $F_i(S[OptIndx]) \geq F_i(S[NextOptIndx])$:
    increment $OptIndx$ and $NextOptIndx$.
  □ $PrefixCost(i) = F_i(Opt)$.

The detail of the pseudocode is in Algorithm 3. We need $O(\log |F|)$ time to calculate $Cost(a, b)$ queries, so the running time of the algorithm is $O(|F| + |C| \log |F|)$.

5   An $O(|C| + |F|)$-time Solution

Previously we have argued that for a fixed i and an increasing j, optimal facility always remains the same or shifts to the right. Now we need a more general result along that line to achieve the final bit of improvement. So we present the following lemma.

Lemma 11. *Let f_g be the optimal facility for the consecutive sequence of customers $[s, \ldots, t]$ and f_h be the optimal facility for the consecutive sequence of customers $[u, \ldots, v]$. If $s \leq u$ and $t \leq v$, then we have $g \leq h$.*

Proof. The proof will be provided in the journal version.   □

The rest of the discussion can be understood in the context of Algorithm 3. We can now design an improved algorithm by answering $Cost(a, b)$ queries in constant time. This is done as follows using Lemma 11. In Algorithm 3, for each iteration of i (for loop in Line 6) we need to answer the query $Cost(a, i)$ for some specific values of a. So in the subsequent iteration (that is $i + 1$) the relation between the right limits in consecutive queries as stipulated in Lemma 11 (i.e. $s \leq v$) always holds. On the other hand, for a fixed i in the while loop of Line 15 Opt always increases. So similarly the relation between the left limits in consecutive queries as stipulated in Lemma 11 (i.e. $r \leq u$) always holds. Therefore clearly by Lemma 11 optimal facility in successive queries either remains the same or shifts to the right. So we can describe the improvement on Algorithm 3 as follows.

- Keep two variables: *OptFacility* for *OptIndx* and *NextOptFacility* for *NextOptIndx*. *OptFacility* maintains the optimal facility for cluster $(Opt + 1, i)$, while *NextOptFacility* maintains the optimal facility for cluster $(NextOpt + 1, i)$. Initialize *OptFacility* $= 1$ and *NextOptFacility* $= 1$.
- To calculate $F_i(Opt)$ at any iteration, we need to calculate $Cost(Opt+1, i)$. To do so, we can always shift *OptFacility* to the right as long as *OptFacility* $< m$ and the cost of the cluster $Opt + 1, \ldots, i$ decreases.
- Similarly to calculate $F_i(NextOpt)$ at any iteration, we can always shift *NextOptFacility* to the right as long as *NextOptFacility* $< m$ and cost of the cluster $NextOpt + 1, \ldots, i$ decreases.

Since, both *OptFacility* and *NextOptFacility* shift only to the right, we can answer all $Cost(a, b)$ queries in total $O(|F|)$ time. The overall running time of the algorithm is $O(|C| + |F|)$.

Acknowledgement. First and second authors are partially supported through a grant from Pubali Bank Ltd.

References

1. Aggarwal, G., et al.: Achieving anonymity via clustering. ACM Trans. Algorithms **6**(3), 49:1–49:19 (2010)

2. Akagi, T., Nakano, S.: On r-gatherings on the line. In: Wang, J., Yap, C. (eds.) FAW 2015. LNCS, vol. 9130, pp. 25–32. Springer, Cham (2015). https://doi.org/10.1007/978-3-319-19647-3_3
3. Armon, A.: On min-max r-gatherings. Theor. Comput. Sci. **412**(7), 573–582 (2011)
4. Drezner, Z.: Facility Location: A Survey of Applications and Methods. Springer, New York (1995)
5. Drezner, Z., Hamacher, H.W.: Facility Location: Applications and Theory. Springer, Heidelberg (2001)
6. Gabow, H.N., Tarjan, R.E.: Algorithms for two bottleneck optimization problems. J. Algorithms **9**(3), 411–417 (1988)
7. Nakano, S.: A simple algorithm for r-gatherings on the line. In: Rahman, M.S., Sung, W.-K., Uehara, R. (eds.) WALCOM 2018. LNCS, vol. 10755, pp. 1–7. Springer, Cham (2018). https://doi.org/10.1007/978-3-319-75172-6_1
8. Han, Y., Nakano, S.: On r-gatherings on the line. In: Proceedings of FCS 2016, pp. 99–104 (2016)

Computational Geometry

Maximum-Width Empty Square
and Rectangular Annulus

Sang Won Bae[1]([✉]), Arpita Baral[2], and Priya Ranjan Sinha Mahapatra[2]

[1] Division of Computer Science and Engineering,
Kyonggi University, Suwon, Republic of Korea
swbae@kgu.ac.kr
[2] Department of Computer Science and Engineering,
University of Kalyani, Kalyani, India
arpitabaral@gmail.com, priya@klyuniv.ac.in

Abstract. An annulus is, informally, a ring-shaped region, often described by two concentric circles. The maximum-width empty annulus problem asks to find an annulus of a certain shape with the maximum possible width that avoids a given set of n points in the plane. This problem can also be interpreted as the problem of finding an optimal location of a ring-shaped obnoxious facility among the input points. In this paper, we study square and rectangular variants of the maximum-width empty anuulus problem, and present first nontrivial algorithms. Specifically, our algorithms run in $O(n^3)$ and $O(n^2 \log n)$ time for computing a maximum-width empty axis-parallel square and rectangular annulus, respectively. Both algorithms use only $O(n)$ space.

1 Introduction

The problem of computing a minimum-size geometric object that encloses an input point set P is one of the central research problems in computational geometry. This type of problem has been extensively studied with direct applications to location of desirable facilities to customers P, for a variety of different geometric shapes including circles [18], rectangles [21], and annuli [1–3,5,12,17].

On the other hand, in some applications, the facility to be built among P is considered *obnoxious*, that is, every member in P wants to be as far away from it as possible. The problem of locating an obnoxious facility is often interpreted as the problem of finding a maximum-size empty geometric object among P. For examples, the center of a largest circle or square that is empty of P corresponds to an optimal location of a point obnoxious facility that maximizes the Euclidean or L_∞ distance, respectively, from its closest point in P. A largest empty circle or

S.W. Bae was supported by Basic Science Research Program through the National Research Foundation of Korea (NRF) funded by the Ministry of Education (2018R1D1A1B07042755). P.R.S. Mahapatra was supported by Research Project through Department of Atomic Energy (NBHM), Government of India with Ref. No. 2/48(19)/2014/R&D-II/1045.

© Springer Nature Switzerland AG 2019
G. K. Das et al. (Eds.): WALCOM 2019, LNCS 11355, pp. 69–81, 2019.
https://doi.org/10.1007/978-3-030-10564-8_6

square can be found in optimal $O(n \log n)$ time using the Voronoi diagram [20], and the best known algorithm that computes an empty axis-parallel rectangle of maximum area runs in $O(n \log^2 n)$ time by Aggarwal and Suri [4]. The widest empty corridor problem, in which one wants to find a widest empty strip among P of arbitrary orientation, is another interesting problem in this concept. After Houle and Maciel [14] presented an $O(n^2)$-time algorithm for this problem, a lot of variants and extensions have been addressed, including the widest L-shaped corridor problem [7], and the widest 1-corner corridor problem [9]. Note that these problems are equivalent to those of finding an optimal location of an obnoxious facility whose shape is of a line, a line segment, or a polygonal chain.

In this paper, along this line of research, we study the *maximum-width empty annulus problem*. Informally, an annulus is a ring-shaped region, often described by two concentric circles. Thus, the maximum-width empty annulus problem is to find an optimal location of a ring-shaped obnoxious facility among the input points P. Specifically, we discuss its square and rectangular variants, and present first nontrivial algorithms. Our algorithms run in $O(n^3)$ and $O(n^2 \log n)$ time for computing a maximum-width axis-parallel square and rectangular annulus, respectively, that is empty of a given set P of n points in the plane. Both algorithms use only $O(n)$ space.

There has been a little work on the maximum-width empty annulus problem. Díaz-Báñez et al. [10] first studied the problem for circular annulus, and proposed an $O(n^3 \log n)$-time and $O(n)$-space algorithm to solve it. To our best knowledge, there was no known correct algorithm in the literature for the maximum-width empty square or rectangular annulus problem. Mahapatra [16] considered the maximum-width empty rectangular annulus problem and claimed an incorrect $O(n^2)$-time algorithm. There is a missing argument in Observation 2 of [16], which incorrectly claimed that the total number of potential outer rectangles forming an empty rectangular annulus is $n - 1$.

Unlike the maximum-width empty annulus problem, the problem of finding a minimum-width annulus that encloses P has recently attained intensive interests from researchers. As a classical one, circular annuli have been studied earlier with applications to the roundness problem [11,19,22], and the currently best known algorithm runs in $O(n^{3/2+\epsilon})$ time [2,3]. Computing a minimum-width axis-parallel square or rectangular annulus that encloses n points P can be done in $O(n \log n)$ or $O(n)$ time, respectively [1,12]. Mukherjee et al. [17] considered the problem of identifying a rectangular annulus of minimum width that encloses P in arbitrary orientation, and presented an $O(n^2 \log n)$-time algorithm. Bae [5] studied a minimum-width square annulus in arbitrary orientation and showed that it can be solved in $O(n^3 \log n)$ time.

2   Problem Definition and Terminologies

Throughout the paper, we consider a Cartesian coordinate system of the plane $\mathbb{R}^2$ with the x- and y-axes. For any point p in the plane $\mathbb{R}^2$, we denote by $x(p)$ and

$y(p)$ its x- and y-coordinates. For an axis-parallel rectangle or square, its four sides are naturally identified by *top*, *bottom*, *left*, and *right* sides, respectively.

For an axis-parallel square, the intersection point of its two diagonals is called its *center*, and its *radius* is half its side length. An *axis-parallel square annulus* is the region between two concentric axis-parallel squares S and S', where $S' \subseteq S$. We call S and S' the *outer* and *inner* squares, respectively, of the annulus. The *width* of a square annulus is defined to be the difference of radii of its outer and inner squares. See Fig. 1(left) for an illustration.

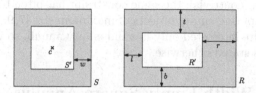

Fig. 1. A square annulus of width w with outer and inner squares S and S' having a common center c (left) and a rectangular annulus with outer and inner rectangles R and R' whose top-, bottom-, left-, right-widths are t, b, l, and r, respectively (right).

An *axis-parallel rectangular annulus* is the region obtained by subtracting the interior of an axis-parallel rectangle R' from another axis-parallel rectangle R such that $R' \subseteq R$. We call R and R' the *outer rectangle* and *inner rectangle* of the annulus, respectively. Consider a rectangular annulus A defined by its outer and inner rectangles, R and R'. By our definition, note that R and R' defining annulus A do not have to be concentric, so that A may not be a symmetric shape. The *top-width* of A is the vertical distance between the top sides of R and R', and the *bottom-width* of A is the vertical distance between their bottom sides. Analogously, the *left-width* and *right-width* of A are defined to be the horizontal distance between the left sides of R and R' and the right sides of R and R', respectively. Then, the *width* of A is defined to be the minimum of the four values: the top-width, bottom-width, left-width, and right-width of A. See Fig. 1(right) for an illustration.

In this paper, we only discuss squares, rectangles, square annuli, and rectangular annuli that are axis-parallel. Hence, we shall drop the term "axis-parallel", and any square, rectangle, or annulus we discuss is assumed to be axis-parallel.

Let P be a set of n points in $\mathbb{R}^2$. A square or rectangular annulus A is said to be *empty* of P, or just *empty* when there is no confusion, if the interior of A does not contain any point in P. Consider any empty square or rectangular annulus A. Then, A induces a partition of P into two subsets P_{out} and P_{in} such that P_{in} is the set of points in P lying in the interior or on the boundary of the inner square or rectangle of A, and $P_{\text{out}} = P \setminus P_{\text{in}}$. If both P_{out} and P_{in} are nonempty, then we say that A is *valid*. In this paper, we address the following problems:

MAXWIDTHEMPTYSQUAREANNULUS (MaxESA)
Input: A set of points P in $\mathbb{R}^2$
Output: A valid empty square annulus A of maximum width

MAXWIDTHEMPTYRECTANGULARANNULUS (MaxERA)
Input: A set of points P in $\mathbb{R}^2$
Output: A valid empty rectangular annulus A of maximum width

The constraint that the resulting empty annulus should be valid is essential to make the problem nontrivial; the same constraint has often been considered in the problem of computing empty objects of maximum size [7,9,10,15]. Throughout the paper, we are interested only in valid empty annuli, so we shall drop the term "valid" unless stated otherwise.

3   Maximum-Width Empty Square Annulus

In this section, we present an algorithm that computes a maximum-width valid empty square annulus for a given set P of n points.

Consider any empty square annulus A. Keeping the same partition of P by A, one can enlarge the outer square and shrink the inner square so that some points of P lie on the boundary of the outer and inner squares. This process implies the following observation. A side of a rectangle or a square is said to be *at infinity* if it is a translated copy of a line segment by a translation vector at infinity.

Observation 1. *There exists a maximum-width empty square annulus such that one side of its inner square contains a point of P and one of the following holds: (i) There are a pair of opposite sides of its outer square, each of which contains a point of P, (ii) there are two adjacent sides of its outer square, each of which contains a point of P, and the other two sides are at infinity, or (iii) One side of its outer square contains a point of P and the other three sides are at infinity.*

By Observation 1, we now have three different configurations of empty square annuli to search for. If this is case (iii), then observe that it corresponds to a maximum-width empty horizontal or vertical strip, which also can be reduced to the problem of finding the maximum gap in $\{x(p) \mid p \in P\}$ or in $\{y(p) \mid p \in P\}$. Hence, case (iii) can be handled in $O(n)$ time after sorting P.

On the other hand, if this is case (ii), then the resulting square annulus corresponds to a maximum-width empty "axis-parallel" L-shaped corridor. It is known that a maximum-width empty L-shaped corridor over all orientations can be computed in $O(n^3)$ time with $O(n^3)$ space by Cheng [7], while we are seeking only for axis-parallel ones. Here, we give a simple $O(n^2 \log n)$ time algorithm for this problem.

Theorem 1. *Given n points in the plane, one can compute a widest empty axis-parallel corridor in $O(n^2 \log n)$ time using $O(n)$ space.*

Now, we suppose that the solution falls in case (i) of Observation 1, so that both two opposite sides of its outer square contains a point of P. Without loss of generality, we assume that each of the top and bottom sides of the outer square of our target annulus contain a point of P. The other case can be handled in a symmetric way.

First, as preprocessing, we sort P in the decreasing order of their y-coordinates, so $P = \{p_1, p_2, \ldots, p_n\}$, where $y(p_1) \geq \cdots \geq y(p_n)$. We also maintain the list of points in P sorted in their x-coordinates. Our algorithm runs repeatedly for all pairs of indices (i, j) with $1 \leq i < j - 1 < n$, and finds a maximum-width empty square annulus such that the top and bottom sides of its outer square contain p_i and p_j, respectively.

From now on, we assume i and j are fixed. Let $P_{ij} := \{p_{i+1}, \ldots, p_{j-1}\}$, $r := (y(p_i) - y(p_j))/2$, and ℓ be the horizontal line with y-coordinate $y(\ell) = (y(p_i) + y(p_j))/2$. Provided that p_i lies on the top side and p_j lies on the bottom side of the outer square, the possible locations of its center is constrained to be on ℓ. For a possible center $c \in \ell$, let $S(c)$ be the square centered at c with radius r. Then, the corresponding inner square $S'(c)$ is determined by center c and the farthest point among those points in P_{ij} lying in the interior of $S(c)$. Here, the distance is measured by the L_∞ metric. More precisely, the radius of $S'(c)$ is exactly $\max_{p \in P_{ij} \cap S(c)} \|p - c\|_\infty$, where $\|\cdot\|_\infty$ denotes the L_∞ norm, and we want to minimize this over the relevant segment $C \subset \ell$ such that $S(c)$ for $c \in C$ contains p_i and p_j on its top and bottom sides. Note that the length of segment $C \subset \ell$ is exactly $2r - |x(p_i) - x(p_j)|$.

For the purpose, we define $f_p(c)$ for each $p \in P_{ij}$ and all $c \in \ell$ to be

$$f_p(c) = \begin{cases} \|p - c\|_\infty & \text{if } \|p - c\|_\infty < r \\ 0 & \text{otherwise} \end{cases},$$

and let $F(c) := \max_{p \in P_{ij}} f_p(c)$ be their upper envelope. Note that our goal is to minimize the upper envelope F over $C \subset \ell$.

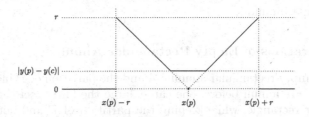

Fig. 2. Illustration of the graph of function f_p for $p \in P$.

Note that $x(p) - r < x(c) < x(p) + r$, or equivalently, $p \in S(c)$ if and only if $\|p - c\|_\infty < r$ for any $p \in P_{ij}$. As also observed in Bae [5], the function f_p is piecewise linear with at most three pieces over $c \in \ell$ such that $p \in S(c)$, and the three pieces have slopes -1, 0, and 1 in this order. Moreover, the height of the part of slope 0 is exactly $|y(p) - y(c)| = |y(p) - y(\ell)|$ and the extensions of the two pieces of slope -1 and 1 always cross at the point of x-coordinate $x(p)$ and height 0. See Fig. 2. These properties of f_p can be easily verified from the behavior of the L_∞ norm. By the above observations, one can explicitly compute F by computing the upper envelope of $O(n)$ line segments in $O(n \log n)$ time [13]. Applying this to all possible pairs (i, j) yields an $O(n^3 \log n)$ time algorithm.

In the following, we show how to improve this to $O(n^3)$ time by decomposing the function f_p into two functions g_p and h_p. For each $p \in P_{ij}$ and $c \in \ell$, define

$$g_p(c) = |x(p) - x(c)| \quad \text{and} \quad h_p(c) = |y(p) - y(c)| \quad \text{if } \|p - c\|_\infty < r,$$

and $g_p(c) = h_p(c) = 0$, otherwise. Also, let $G(c) := \max_{p \in P_{ij}} g_p(c)$ and $H(c) := \max_{p \in P_{ij}} h_p(c)$. As $\|p - c\|_\infty = \max\{|x(p) - x(c)|, |y(p) - y(c)|\}$, it is obvious that $f_p(c) = \max\{g_p(c), h_p(c)\}$, and hence $F(c) = \max\{G(c), H(c)\}$. We now show that the functions G and H can be explicitly computed in $O(n)$ time.

Lemma 1. *The functions G and H can be explicitly computed in $O(n)$ time.*

Since the function F is the upper envelope of G and H, we can compute F in $O(n)$ time using the explicit description of functions G and H. Note that the three functions F, G, and H are piecewise linear with $O(n)$ breakpoints. Consequently, we can compute F and find a lowest point of F over $C \subset \ell$ in $O(n)$ time, and hence a maximum-width empty square annulus of case (i) can be found in $O(n^3)$ time. Finally, we conclude the following theorem.

Theorem 2. *Given n points in the plane, a maximum-width empty square annulus can be computed in $O(n^3)$ time using $O(n)$ space.*

4   Maximum-Width Empty Rectangular Annulus

In this section, we present an algorithm computing a maximum-width empty rectangular annulus. First we give several basic observations on maximum-width empty rectangular annuli.

4.1   Configurations of Empty Rectangular Annuli

Consider any empty rectangular annulus A and the partition of P induced by A. As done for square annuli before, one can enlarge the outer rectangle of A and shrink its inner rectangle, while keeping the partition of P and not decreasing the width of A. This results in the following observation.

Observation 2. *There exists a maximum-width empty rectangular annulus such that each side of its outer rectangle either contains a point of P or lies at infinity, and every side of its inner rectangle contains a point of P.*

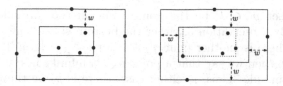

Fig. 3. Empty rectangular annuli of maximum width w: (left) Each side of the inner and outer rectangles contains a point. (right) A maximum-width empty rectangular annulus that is uniform and also top-anchored.

In Observation 2, note that each side of a rectangle is considered to include its endpoints. Thus, a point $p \in P$ can be contained in two adjacent sides of a rectangle if p is located at a corner. See Fig. 3(left).

A rectangular annulus A is said to be *width-uniform*, or simply *uniform*, if its top-width, bottom-width, left-width, and right-width are all equal to its width. In the following observation, we show that we can focus only on uniform rectangular annuli to solve our problem.

Observation 3. *There exists a maximum-width empty rectangular annulus A that is uniform such that the following property holds: each side of its outer rectangle either contains a point of P or lies at infinity, and at least one side of its inner rectangle contains a point of P.*

See Fig. 3(right) for an illustration of Observation 3. This observation suggests a specific configuration of annuli for us to solve the problem. First of all, we do not have to consider non-uniform annuli. Moreover, candidate outer rectangles are defined by at most four points in P. If we fix an outer rectangle R, then the inner rectangle that maximizes the width is also determined by searching points in $P \cap R$. This already yields an $O(n^5)$-time algorithm for our problem.

Let A be an empty rectangular annulus satisfying the condition described in Observation 2. We call A *top-anchored* (or, *bottom-anchored, left-anchored, right-anchored*) if both the top sides (or, bottom sides, left sides, right sides, resp.) of the outer and inner rectangles of A contain a point of P. For example, Fig. 3(right) shows an empty top-anchored rectangular annulus.

Observation 4. *There exists a maximum-width empty rectangular annulus A that satisfies the condition described in Observation 3 and is either top-anchored, bottom-anchored, left-anchored, or right-anchored.*

Our algorithm will find an empty anchored and uniform rectangular annulus of maximum width, which is the correct answer to our problem by Observation 4. In the following, we assume without loss of generality that there exists a maximum-width empty annulus that is uniform and top-anchored, and describe our algorithm for this case. The other three cases can be handled analogously.

Let $P = \{p_1, p_2, \ldots, p_n\}$ be the given set of points, sorted in the descending order of their y-coordinates, that is, $y(p_1) \geq y(p_2) \geq \cdots \geq y(p_n)$. Consider any empty top-anchored rectangular annulus A that satisfies the condition of

Observation 3. Let $p_i \in P$ be the point lying on the top side of the outer rectangle of A. By Observation 3, either the bottom side of the outer rectangle is at infinity or there is another point $p_j \in P$ for $i < j \leq n$ on it. If the bottom side is at infinity, then we say that a point p_∞ at infinity in the $(-y)$-direction lies on the bottom side. Thus, in either case, there is p_j on the bottom side of the outer rectangle for $i < j \leq n$ or $j = \infty$.

Since A is top-anchored, there is a third point $p_k \in P$ on the top side of the inner rectangle of A. Observe that the width of A is determined by the y-difference of p_i and p_k, that is, $y(p_i) - y(p_k)$. Thus, the maximum width for top-anchored empty rectangular annuli is one among $O(n^2)$ values $\{y(p_i) - y(p_k) \mid 1 \leq i \leq k \leq n\}$.

The problem becomes even simpler if we fix p_i on the top side of the outer rectangle, since the number of possible widths is reduced to n. An outlook of our algorithm that computes a maximum-width empty top-anchored rectangular annulus is as follows: (1) For each $p_i \in P$, find an empty annulus A_i^* with p_i lying on the top side of its outer rectangle whose width is the maximum among the set $\{y(p_i) - y(p_k) \mid i < k \leq n\}$ and then (2) output the one with maximum width among A_i^* for all $i \in \{1, \ldots, n\}$. In order to compute A_i^*, we try all possible points p_j that bound the bottom side of the outer rectangle.

In the following subsections, we first study the case where two points p_i and p_j on the top and bottom sides are fixed, and then move on to the case where only a point p_i on the top side is fixed. More precisely, we discuss a decision algorithm when two points on the top and bottom sides are fixed, and exploit it as a sub-procedure to solve the other case.

4.2 Decision When Two Points on Top and Bottom Are Fixed

Suppose that we are given p_i and p_j with $1 \leq i + 1 < j \leq n$ or $j = \infty$, and we consider only empty rectangular anuuli whose outer rectangle contains p_i and p_j on its top and bottom sides, respectively.

Here, we consider the following decision problem.

> *Given*: A positive real $w > 0$
> *Task*: Does there exist an empty rectangular annulus of width at least w whose outer rectangle contains p_i and p_j on its top and bottom sides, respectively?

Let $D_{ij}(w)$ denote the outcome of the above decision problem.

Observation 5. *If $D_{ij}(w)$ is TRUE, then $D_{ij}(w')$ is TRUE for any $w' \leq w$. On the other hand, if $D_{ij}(w)$ is FALSE, then $D_{ij}(w')$ is FALSE for any $w' \geq w$.*

Let $P_{ij} := \{p_{i+1}, \ldots, p_{j-1}\}$ for $i < j \leq n$, and $P_{i\infty} := \{p_{i+1}, \ldots, p_n\}$. In the following, we show that the decision problem for a given width $w > 0$ can be solved by a combination of certain operations on points P_{ij}, namely, the *y-range x-neighbor query* and the *range maximum-gap query*. Each of the two operations is described as follows (see Fig. 4):

Fig. 4. (left) the y-range x-neighbor query for (x, y_1, y_2) and its answer q_1 and q_2, (right) the range maximum-gap query for (x_1, x_2) and the maximum gap is g.

(i) The *y-range x-neighbor query*: Given three real numbers (x, y_1, y_2), this operation is to find two points q_1 and q_2 in P_{ij} such that q_1 is the rightmost one among points $P_{ij} \cap [-\infty, x] \times [y_1, y_2]$ and q_2 is the leftmost one among points $P_{ij} \cap [x, \infty] \times [y_1, y_2]$. Either q_1 and q_2 may be undefined if there is no point of $P_{i,j}$ in the corresponding range. If q_1 is undefined, then we return q_1 as a point at infinity such that $x(q_1) = -\infty$ and $y(q_1) = y_1$; if q_2 is undefined, then we return q_2 such that $x(q_2) = \infty$ and $y(q_2) = y_1$ (Fig. 4).

(ii) The *range maximum-gap query (in x-coordinates)*: Given two real numbers (x_1, x_2), find the *maximum gap* in the set of real numbers $\{x(p) \mid x_1 \leq x(p) \leq x_2, p \in P_{ij}\} \cup \{x_1, x_2\}$, where $x(p)$ denotes the x-coordinate of point p. The maximum gap in a set X of real numbers is the maximum difference between two consecutive elements when X is sorted. Notice that x_1 and x_2 are also included in the above set. Here, the output of the range maximum-gap query is to be the pair of values that define the maximum gap (Fig. 4).

We describe our algorithm for the decision problem as Algorithm 1.

Our decision algorithm, Algorithm 1, evaluates $D_{ij}(w)$ for a given w. As described in Algorithm 1, the decision is made by four calls of the y-range x-neighbor queries and the range maximum-gap queries. Thus, its running time depends on how efficiently we can handle these queries. If the algorithm decides that $D_{ij}(w)$ is TRUE, then it also returns a corresponding rectangular annulus, that is, an empty uniform annulus of width w with p_i and p_j on the top and bottom sides of the outer rectangle. This can be done by constructing its outer rectangle since its width w is fixed.

In the following, we show the correctness of our decision algorithm.

Lemma 2. *Algorithm 1 correctly computes $D_{ij}(w)$ for any given $w > 0$ in time $O(T)$, where T is an upper bound on time needed to perform a y-range x-neighbor query or a range maximum-gap query. Moreover, if $D_{ij}(w)$ is TRUE, then an empty rectangular annulus of width w such that p_i and p_j lie on the top and bottom side of its outer rectangle can be found in the same time bound.*

Note that the two operations can be easily done in linear time. We now show how to perform them in logarithmic time with an aid of the following data structures.

– Let $\mathcal{D}$ be the data structure on P described in Chazelle [6] that supports a segment dragging query for vertical line segments dragged by two horizontal rays. A segment dragging query is given by a segment and a direction along

---

Algorithm 1: Decision algorithm

Input: a width $w > 0$

Output: $D_{ij}(w)$, and an empty rectangular annulus of width w with p_i and p_j lying on the top and bottom sides of its outer rectangle, respectively, if $D_{ij}(w)$ is TRUE

1 **if** $y(p_i) - y(p_j) < 2w$ **then**
2 | Return FALSE.
3 Perform a y-range x-neighbor query for $(x(p_i), y(p_i) - w, y(p_i))$, and let q_l and q_r be the output with $x(q_l) \le x(p_i) \le x(q_r)$.
4 Perform a y-range x-neighbor query for $(x(p_j), y(p_j), y(p_j) + w)$, and let q'_l and q'_r be the output with $x(q'_l) \le x(p_j) \le x(q'_r)$.
5 Let p_l be the rightmost one in $\{q_l, q'_l\}$ and p_r be the leftmost one in $\{q_r, q'_r\}$.
6 **if** $\min\{x(p_i), x(p_j)\} < x(p_l) < \max\{x(p_i), x(p_j)\}$ *or* $\min\{x(p_i), x(p_j)\} < x(p_r) < \max\{x(p_i), x(p_j)\}$ **then**
7 | Return FALSE.
8 Perform a range maximum-gap query for $(x(p_l), \min\{x(p_i), x(p_j)\} + w)$, and let (l, l') be the output and $g_l := l' - l$ be the corresponding maximum gap.
9 Perform a range maximum-gap query for $(\max\{x(p_i), x(p_j)\} - w, x(p_r))$, and let (r', r) be the output and $g_r := r - r'$ be the corresponding maximum gap.
10 **if** $g_l \ge w$ *and* $g_r \ge w$ **then**
11 | Return TRUE, and the rectangular annulus of width w whose outer rectangle is defined by the top and bottom sides through p_i and p_j, respectively, and the left and right sides at $x = l$ and $x = r$, respectively.
12 **else**
13 | Return FALSE.

---

two rays and is to find the first point in P that is hit by the dragged segment. This structure can be constructed in $O(n \log n)$ time using $O(n)$ storage

– Let $\mathcal{X}_{ij}$ be a 1D range tree for the x-coordinates of points in P_{ij} with an additional field $maxgap(v)$ at each node v, where $maxgap(v)$ denotes the maximum gap in the canonical subset of v. Note that $maxgap(v) = 0$ if the canonical subset of v consists of only one element. The structure $\mathcal{X}_{ij}$ can be constructed using storage $O(|P_{ij}|)$ [8].

Now, suppose that we have already built these two structures $\mathcal{D}$ and $\mathcal{X}_{ij}$. Then, the two operations can be handled in $O(\log n)$ time as follows:

(i) For a y-range x-neighbor query for (x, y_1, y_2), we perform two segment dragging queries on $\mathcal{D}$ for a vertical line segment with endpoints (x, y_1) and (x, y_2) to both the left and the right directions. These two queries result in the rightmost point q_1 in the range $[-\infty, x] \times [y_1, y_2]$ and the leftmost point q_2 in the range $[x, \infty] \times [y_1, y_2]$.

(ii) For a range maximum-gap query for (x_1, x_2), we perform a 1D range search for the x-range $[x_1, x_2]$ on $\mathcal{X}_{ij}$ again to obtain a collection C of $O(\log n)$ nodes. The maximum gap in the x-coordinates of the points in P_{ij} in the range $[x_1, x_2]$ can be found by comparing $O(\log n)$ values: $maxgap(v)$ for all $v \in C$ and every gap between two consecutive canonical subset.

Algorithm 2: Computing A_i^*.

Input: A set $P = \{p_1, \ldots, p_n\}$ of points sorted by y-coordinates, and a point
    $p_i \in P$

Output: A_i^* with p_i and its width w_i^*

1 Set k to be $i + 1$, w to be 0, and A to be any annulus of width 0.
2 Build the data structure $\mathcal{D}$ for P and initialize $\mathcal{X}$ to be $\mathcal{X}_{i,i}$.
3 **for** *each* $j = i + 2, \ldots, n$ *and* $j = \infty$ **do**
4   Insert p_{j-1} if $j \leq n$, or p_n if $j = \infty$, into $\mathcal{X}$, so that now $\mathcal{X} = \mathcal{X}_{ij}$.
5   **while** $D_{ij}(y(p_i) - y(p_k))$ *is TRUE and* $k \leq j$ **do**
6   Set A to be the corresponding annulus of width $y(p_i) - y(p_k)$.
7   Set w to be $y(p_i) - y(p_k)$.
8   Increase k by 1.
9 Return the current A as A_i^* and the current w as w_i^*.

Therefore, we conclude the following:

Lemma 3. *Suppose that we already have two tree structures $\mathcal{D}$ and $\mathcal{X}_{ij}$. Then, Algorithm 1 correctly computes $D_{ij}(w)$ for any given $w > 0$ in time $O(\log n)$.*

4.3 Optimization When Only a Point on Top Is Fixed

Next, we describe how to find a maximum-width empty top-anchored rectangular annulus such that p_i lies on the top side of the outer rectangle.

Let w_i^* be the width of A_i^*. Observe that w_i^* lies in the set $W_i := \{y(p_i) - y(p_k) \mid i \leq k \leq n\}$. Instead of solving the optimization problem for each pair (p_i, p_j), we can rather solve the optimization problem when only a point p_i on top is fixed. Our algorithm that computes A_i^* and its width w_i^* is presented as in Algorithm 2.

Lemma 4. *Algorithm 2 can be implemented in $O(n \log n)$ time and $O(n)$ space for a fixed $p_i \in P$. Also, it correctly computes w_i^* and A_i^*.*

4.4 Putting It All Together

We are now ready to describe the overall algorithm to solve the MaxERA problem. Under the assumption that there exists a maximum-width empty rectangular annulus A^* that satisfies the condition of Observation 3 and is top-anchored, we execute Algorithm 2 for each $i = 1, \ldots n-1$ and choose the one with the maximum width as A^*. Its correctness is guaranteed by Lemma 4. The other three cases where there is a maximum width empty rectangular annulus that satisfies the condition of Observation 3 and is either bottom-anchored, left-anchored, or right-anchored, can be handled in a symmetric way. Thus, the overall algorithm runs for the four cases and outputs one with the maximum width.

Theorem 3. *Given a set P of n points in the plane, a maximum-width rectangular annulus that is empty with respect to P can be computed in $O(n^2 \log n)$ time and $O(n)$ space.*

References

1. Abellanas, M., Hurtado, F., Icking, C., Ma, L., Palop, B., Ramos, P.: Best fitting rectangles. In: Proceedings of European Workshop on Computational Geometry (EuroCG 2003) (2003)
2. Agarwal, P.K., Sharir, M.: Efficient algorithms for geometric optimization. ACM Comput. Surv. **30**(4), 412–458 (1998)
3. Agarwal, P.K., Sharir, M., Toledo, S.: Applications of parametric searching in geometric optimization. J. Algo. **17**(3), 292–318 (1994)
4. Aggarwal, A., Suri, S.: Fast algorithms for computing the largest empty rectangle. In: Proceedings of the Third Annual Symposium on Computational Geometry (SoCG 1987), pp. 278–290 (1987)
5. Bae, S.W.: Computing a minimum-width square annulus in arbitrary orientation. Theoret. Comput. Sci. **718**, 2–13 (2018)
6. Chazelle, B.: An algorithm for segment-dragging and its implementation. Algorithmica **3**(1), 205–221 (1988)
7. Cheng, S.W.: Widest empty L-shaped corridor. Inform. Proc. Lett. **58**(6), 277–283 (1996)
8. de Berg, M., Cheong, O., van Kreveld, M., Overmars, M.: Computational Geometry: Algorithms and Applications, 3rd edn. Springer-Verlag TELOS, Heidelberg (2008). https://doi.org/10.1007/978-3-540-77974-2
9. Díz-Báñez, J., López, M., Sellarès, J.: On finding a widest empty 1-corner corridor. Inform. Proc. Lett. **98**(5), 199–205 (2006)
10. Díz-Báñez, J.M., Hurtado, F., Meijer, H., Rappaport, D., Sellarès, J.A.: The largest empty annulus problem. Int. J. Comput. Geom. Appl. **13**(4), 317–325 (2003)
11. Ebara, H., Fukuyama, N., Nakano, H., Nakanishi, Y.: Roundness algorithms using the Voronoi diagrams. In: Abstract: 1st Canadian Conference on Computational Geometry (CCCG 1989), p. 41 (1989)
12. Gluchshenko, O.N., Hamacher, H.W., Tamir, A.: An optimal $O(n \log n)$ algorithm for finding an enclosing planar rectilinear annulus of minimum width. Oper. Res. Lett. **37**(3), 168–170 (2009)
13. Hershberger, J.: Finding the upper envelope of n line segments in $O(n \log n)$ time. Inform. Proc. Lett. **33**(4), 169–174 (1989)
14. Houle, M., Maciel, A.: Finding the widest empty corridor through aset of points. In: Toussaint, G. (ed.) Snapshots of Computational and Discrete Geometry, pp. 201–213, Department Computer Science, McGill University (1988)
15. Janardan, R., Preparata, F.P.: Widest-corridor problems. Nordic J. Comput. **1**, 231–245 (1994)
16. Mahapatra, P.R.S.: Largest empty axis-parallel rectangular annulus. J. Emerg. Trends Comput. Inf. Sci. **3**(6) (2012)
17. Mukherjee, J., Mahapatra, P.R.S., Karmakar, A., Das, S.: Minimum-width rectangular annulus. Theoret. Comput. Sci. **508**, 74–80 (2013)
18. Preparata, F.P., Shamos, M.I.: Computational Geometry: An Introduction. Springer, New York (1990). https://doi.org/10.1007/978-1-4612-1098-6
19. Roy, U., Zhang, X.: Establishment of a pair of concentric circles with the minimum radial separation for assessing roundness error. Comput. Aided Des. **24**(3), 161–168 (1992)
20. Toussaint, G.T.: Computing largest empty circles with location constraints. Int. J. Comput. Info. Sci. **12**(5), 347–358 (1983)

21. Toussaint, G.T.: Solving geometric problems with the rotating calipers. In: Proceedings of the IEEE MELECON 1983, pp. 1–4 (1983)
22. Wainstein, A.D.: A non-monotonous placement problem in the plane. In: Abstract: 9th All-Union Symposium USSR Software Systems for Solving Optimal Planning Problems, pp. 70–71 (1986)

Hard and Easy Instances
of L-Tromino Tilings

Javier T. Akagi[1], Carlos F. Gaona[1], Fabricio Mendoza[1], Manjil P. Saikia[2],
and Marcos Villagra[1(✉)]

[1] Universidad Nacional de Asunción,
NIDTEC, Campus Universitario, San Lorenzo 2619, Paraguay
mvillagra@pol.una.py
[2] Fakultät für Mathematik, Universität Wien,
Oskar-Morgenstern-Platz 1, 1090 Vienna, Austria

Abstract. In this work we study tilings of regions in the square lattice with L-shaped trominoes. Deciding the existence of a tiling with L-trominoes for an arbitrary region in general is NP-complete, nonetheless, we identify restrictions to the problem where it either remains NP-complete or has a polynomial time algorithm. First, we characterize the possibility of when an Aztec rectangle has an L-tromino tiling, and hence also an Aztec diamond; if an Aztec rectangle has an unknown number of defects or holes, however, the problem of deciding a tiling is NP-complete. Then, we study tilings of arbitrary regions where only 180° rotations of L-trominoes are available. For this particular case we show that deciding the existence of a tiling remains NP-complete; yet, if a region does not contain so-called "forbidden polyominoes" as subregions, then there exists a polynomial time algorithm for deciding a tiling.

Keywords: Polyomino tilings · Tromino · Efficient tilings
NP-completeness · Aztec rectangle · Aztec diamond · Claw-free graphs

1 Introduction

1.1 Background

A packing puzzle is a solitary game where a player tries to find a way to cover a given shape using polyominoes, where a polyomino is a set of squares joined together by their edges. The computational complexity of packing puzzles was studied by Demaine and Demaine [3] who showed that tiling a shape or region using polyominoes is NP-complete.

In this work we study tilings of regions in the square lattice with L-shaped trominoes (a polyomino of three cells) called an *L-Tromino* or simply tromino

---

M.P. Saikia is supported by the Austrian Science Foundation FWF, START grant Y463 and FWF SFB grant F50.

M. Villagra is supported by Conacyt research grant PINV15-208.

G. K. Das et al. (Eds.): WALCOM 2019, LNCS 11355, pp. 82–95, 2019.
https://doi.org/10.1007/978-3-030-10564-8_7

in this work. A cell in $\mathbb{Z}^2$ is a subset $[a, a+1] \times [b, b+1]$ and a region is any finite union of connected cells. At our disposal we have an infinite amount of trominoes and would like to know if a given region can be covered or tiled with trominoes.

The problem of tiling with trominoes was first studied by Conway and Lagarias [2] who presented an algebraic necessary condition for a region in order to have a tiling. Moore and Robson [8] showed that deciding if a region can be covered with trominoes is NP-complete. Later Horiyama *et al.* [5] presented another proof of NP-completeness by constructing an one-one reduction which implies that counting the number of tilings with trominoes is #P-complete. Counting the number of tilings with L-trominoes was also studied by Chin *et al.* [1] using generating functions.

1.2 Contributions

In this work we aim at identifying instances of the tiling problem with trominoes that either have efficient algorithms or it remains NP-complete. As a further generalization of the problem, we also consider regions with "defects" or holes, that is, we want to know if there is a tiling with trominoes without covering the defects. First we study the Aztec rectangle (and hence, also an Aztec diamond) [4,10] and show that any Aztec rectangle of side lengths a, b can be covered with trominoes if and only if $a(b+1) + b(a+1) \equiv 0 \pmod 3$ (Theorem 1), which implies the existence of a polynomial time algorithm for finding a tiling in an Aztec rectangle, and hence, an Aztec diamond. Then we show that for the cases when $a(b+1) + b(a+1) \equiv 0 \pmod 3$ does not hold, if an Aztec Rectangle has exactly one defect, then it can be covered with trominoes (Theorem 2). In general, however, deciding the tiling of an Aztec diamond with an unknown number of defects is NP-complete (Theorem 3).

In the second part of this paper we study a restricted case of the tiling problem where we only have 180° rotations of the trominoes available. Here we show that the problem remains NP-complete (Theorem 4) by slightly modifying the one-one reduction from the 1-in-3 Graph Orientation Problem of Horiyama *et al.* [5], whereas any Aztec rectangle has no tiling at all (Theorem 5). Nevertheless, we show that if a region does not contain any of the so-called "forbidden polyominoes" identified in this work, then that region has an efficient algorithm for deciding a tiling (Theorem 6). This latter result is proved by constructing a graph representation of the region, called an intersection graph, and identifying independent sets of certain size. If the intersection graph has a claw, then that claw will correspond to a forbidden polyomino, if the graph is claw-tree, however, we can use well-known efficient algorithms for finding independent sets, and hence, a tiling for the region.

Finally we close this paper in Sect. 5 where we study a relation between L-Trominoes and I-Trominoes. We introduce a technique for decomposing a region in simple parts that yields an efficient algorithm for finding L-Tromino covers. This tiling technique is a modification of the proof of Theorem 5 for tiling the Horiyama *et al.* [5] gadgets with I-Trominoes to tiling general regions with L-Trominoes.

2 Preliminaries

In this work we will use $\mathbb{Z}$ to denote the set of integers and $[a, b]$ to denote the discrete interval $\{a, a + 1, \ldots, b\}$.

A region R is a finite union of connected cells, where connected means that any two cells in R share one common edge (this convention is only restricted to the regions we study in this paper). If a cell is the set of points $[a, a+1] \times [b, b+1]$, we label such cell by (a, b) which we refer to as the *cell's coordinate*. Two cells are adjacent if the Manhattan distance, i.e., the L_1-norm, of their coordinates is 1; thus, two cells in diagonal to each other are not adjacent.

A *tromino* is a polyomino of 3 cells. In general there are two types of trominoes, the L-tromino and the I-tromino. An L-tromino is a polyomino of 3 cells with an L shape. An I-tromino is a polyomino of 3 straight cells with the form of an I. In this work we will mostly be dealing with L-Trominos and we will refer to them simply as trominoes; I-trominoes will appear later but we will make sure to clarify to which type of tromino we are referring to.

A *defect* is a cell that is "marked" in the sense that no tromino can be placed on top of that cell. A *cover* or *tiling* of a region R is a set of trominoes covering all cells of R that are not defects without overlapping and each tromino is packed inside R. The *size* of a cover is the number of tiles in it.

Definition 1. TROMINO *is the following problem:*
INPUT   : a region R with defects.
OUTPUT : "yes" if R has a cover and "no" otherwise.

Moore and Robson [8] proved that TROMINO is NP-complete and Horiyama *et al.* [5] proved that #TROMINO, the counting version of TROMINO, is #P-complete.

In this work we will also consider tilings where only trominoes with 180° rotations are used. More precisely, given a region R we want to find a cover where all trominoes are *right-oriented* as in Fig. 1(a) or *left-oriented* as in Fig. 1(b). We will refer to trominoes where only their 180° rotations are considered as *180-trominoes*. A *180-cover* of R is a cover with 180-trominoes.

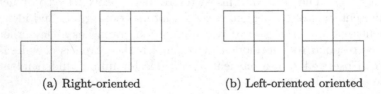

(a) Right-oriented   (b) Left-oriented oriented

Fig. 1. The 180-TROMINO problem either takes trominoes from the left figure or the right figure.

Definition 2. 180-TROMINO *is the following problem:*
INPUT   : a region R with defects.
OUTPUT : "yes" if R has a 180-cover and "no" otherwise.

3 Tiling of the Aztec Rectangle

The *Aztec Diamond of order n*, denoted AD(n), is the union of lattice squares $[a, a+1] \times [b, b+1]$, with $a, b \in \mathbb{Z}$, that lie completely inside the square $\{(x,y)\,|\,|x| + |y| \le n + 1\}$ [4]. Figure 2 shows the first four Aztec diamonds. Tilings of the Aztec diamond with dominoes was initially studied by Elkies *et al.* [4] and later by several other people.

The concept of an Aztec diamond can be very easily extended to that of an *Aztec rectangle*. We denote by $\mathcal{AR}_{a,b}$ the Aztec rectangle which has a unit squares on the southwestern side and b unit squares on the northwestern side; in the case when $a = b = n$ we get an Aztec diamond of order n. When dealing with Aztec rectangle, with no loss of generality, we always assume that $a < b$. As an example Fig. 3 shows $\mathcal{AR}_{4,10}$. Domino tilings of Aztec rectangles have been studied by various mathematicians starting with Mills *et al.* [6].

In the following subsections we study tilings of the Aztec rectangle using trominoes with and without defects, and then specialize them to Aztec diamonds.

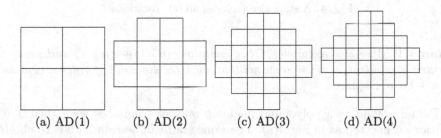

(a) AD(1)   (b) AD(2)   (c) AD(3)   (d) AD(4)

Fig. 2. Aztec diamonds of order 1, 2, 3 and 4.

3.1 Tilings with No Defects

For any Aztec rectangle $\mathcal{AR}_{a,b}$ with no defects, we can completely understand when there is a tiling. The following theorem gives a characterization.

Theorem 1. $\mathcal{AR}_{a,b}$ *has a cover if and only if* $a(b + 1) + b(a + 1) \equiv 0 \pmod{3}$.

As a corollary, we get the following for the Aztec diamond.

Corollary 1. AD(n) *has a cover if and only if* $n(n + 1) \equiv 0 \pmod{3}$.

To prove Theorem 1, first we present tilings of particular cases of the Aztec rectangle in Lemmas 2 and 3. The following lemma is trivial.

Fig. 3. Aztec rectangle $\mathcal{AR}_{4,10}$.

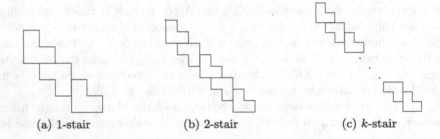

(a) 1-stair   (b) 2-stair   (c) k-stair

Fig. 4. A stair also includes all $90°$ rotations.

Lemma 1. *An Aztec rectangle, $\mathcal{AR}_{a,b}$ contains $a(b+1)+b(a+1)$ unit squares. Further, specializing $a = b = n$ we get that an Aztec diamond of order n contains $2n(n+1)$ unit squares.*

Define a *stair* as a polyomino made-up only of trominoes with their $180°$ rotations connected as in Fig. 4(a). The same stair can be rotated $90°$ to obtain another stair. A *k-stair* is a co-joined set of k stairs, where a stair is joined to another stair by matching their extremes; for example, in Fig. 4(b) we can see two stairs where the lowest extreme of the upper stair is matched with the upper extreme of the lower stair. This idea is easily extended to a set of k stairs thus giving a k-stair as in Fig. 4(c). A k-stair can also be rotated $90°$ to obtain another k-stair. The *height of a k-stair* is the number of steps in it. It is easy to see that the height of a k-stair is $3k + 2$. In addition, a single tromino would be a 0-stair.

Lemma 2. *If $3 \mid a,b$ and $\mathcal{AR}_{a,b}$ has a cover, then $\mathcal{AR}_{a+2,b+2}$ has a cover.*

Proof. If a,b are multiples of 3, then an $a/3$-stair and an $b/3$-stair can be used to tile around $\mathcal{AR}_{a,b}$ along the shorter and longer sides respectively, using the pattern of Fig. 5(a). This tiling increments the order of the Aztec rectangle by 2, thus obtaining a tiling for $\mathcal{AR}_{a+2,b+2}$.   □

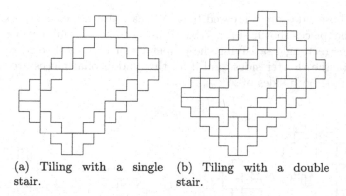

(a) Tiling with a single stair.   (b) Tiling with a double stair.

Fig. 5. Tilings of Lemmas 2 and 3.

Lemma 3. *If* $3 \mid a + 1, b + 1$ *and* $\mathcal{AR}_{a,b}$ *has a cover, then* $\mathcal{AR}_{a+4,b+4}$ *has a cover.*

Proof. To find a tiling for $\mathcal{AR}_{a+4,b+4}$ we use four copies of AD(2) added to the four corners of $\mathcal{AR}_{a,b}$. Then, to complete the tiling, we use two $(a - 2)/3$ and $(b - 2)/3$-stairs one on top of each other along the shorter and longer sides respectively, to complete the border. The entire construction follows the pattern of Fig. 5(b). This tiling increments the order of the Aztec rectangle by 4, thus obtaining a tiling for $\mathcal{AR}_{a+4,b+4}$. □

The above two Lemmas gives as easy corollaries the corresponding results for Aztec diamonds (in the spirit of Corollary 1.)

Now, let us prove Theorem 1.

Proof (Proof of Theorem 1). The values for which $a(b + 1) + b(a + 1) \equiv 0$ (mod 3) holds are $a, b = 3k$ and $a, b = 3k - 1$ for some $k \in \mathbb{Z}$.

Thus, the statement is equivalent to saying that for all positive integers k there is a tiling of $\mathcal{AR}_{a,b}$ where $3 \mid a, b$ or $3 \mid a + 1, b + 1$ and that there are no tilings for $\mathcal{AR}_{a,b}$ when $3 \mid a + 2, b + 2$.

We show the second part now, which is easy since if we have $\mathcal{AR}_{a,b}$ with a, b of the form $3k + 2$, then the number of lattice squares inside $\mathcal{AR}_{a,b}$ is not divisible by 3 and hence we cannot tile this region with trominoes.

We come to the first cases now. Using Lemmas 2 and 3, this part is clear if we can show the base induction case to be true.

The base case of Lemma 2 is shown in Fig. 6(a), which is $\mathcal{AR}_{3,6}$. Once we have a tiling of $\mathcal{AR}_{3,6}$, we can use Lemma 2 to create a tiling of an Aztec rectangle whose sides are increased by 2. We can also increase $\mathcal{AR}_{3,6}$ by using the additional pieces shown in Fig. 6(b,c) using them in combinations with any case of Aztec rectangle satisfying the properties of Lemma 2 to increase either the longer or the shorter sides, and if all three additional pieces are used then we can increase both sides of $\mathcal{AR}_{a,b}$.

Similarly, the base case of Lemma 3 is shown in Fig. 7(a), which is $\mathcal{AR}_{2,5}$. Once we have a tiling of $\mathcal{AR}_{2,5}$, we can use Lemma 3 to create a tiling of an Aztec rectangle whose sides are increased by 4. We can also increase $\mathcal{AR}_{a,b}$ by using the additional pieces shown in Fig. 7(b,c,d) using them in combinations with any case of Aztec rectangle satisfying the properties of Lemma 3 to increase either the longer or the shorter sides, and if all three additional pieces are used then we can increase both sides of $\mathcal{AR}_{a,b}$.   □

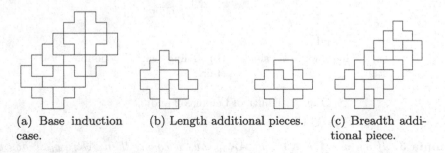

(a) Base induction case.   (b) Length additional pieces.   (c) Breadth additional piece.

Fig. 6. Base case of Lemma 2.

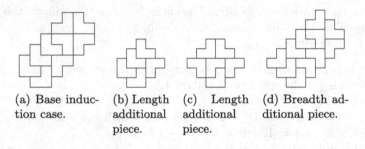

(a) Base induction case.   (b) Length additional piece.   (c) Length additional piece.   (d) Breadth additional piece.

Fig. 7. Base case of Lemma 3.

An $O(b^2)$ time algorithm is immediately obtained from the proof of Theorem 1, and also for Aztec diamonds (we omit the details due to lack of space).

3.2 Tiling with Defects

From Theorem 1 we know that for any positive integers a, b, the Aztec rectangles with no defects $\mathcal{AR}_{a,b}$ such that 3 divides a, b or 3 divides $a + 1, b + 1$ have a cover but if 3 divides $a + 2, b + 2$, then $\mathcal{AR}_{a,b}$ does not have a tiling. We show that if such an Aztec rectangle has exactly one defect, then it can be covered with trominoes.

Theorem 2. $\mathcal{AR}_{a,b}$ with a, b of the form $3k - 2$ with one defect has a cover.

Proof. To tile $\mathcal{AR}_{a,b}$ with one defect we use a construct which we call a *fringe* appearing in Fig. 8(a). It is easy to check that if a fringe has exactly one defect, then it can be covered with trominoes.

To construct a tiling for $\mathcal{AR}_{a,b}$ with one defect we place a fringe in a way that includes the defect and the left and right ends of the fringe touches the boundaries of the Aztec rectangle as in Fig. 8(b). Then we use the tiling pattern of Fig. 8(b) where we put stairs above and below the fringe.   □

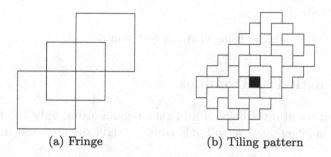

(a) Fringe   (b) Tiling pattern

Fig. 8. Tiling of $\mathcal{AR}_{a,b}$ with one defect. A *fringe* can be composed of any number of order 1 Aztec diamonds AD(1) joined by their upper right and lower left cells. An *reversed fringe* is obtained by joining order 1 Aztec diamonds by their upper left and lower right cells.

As an easy corollary, we obtain the corresponding result for Aztec diamonds.

Corollary 2. *For any positive integer k, the Aztec Diamond* AD$(3k - 2)$ *with one defect has a cover.*

We can consider many different classes of defects, and it is observed that some of these classes have easy tilings, as an example, we have in Fig. 9(a) an Aztec rectangle with four defects on its corners. A tiling of this region is shown in Fig. 9(b). In the combinatorics literature, tilings of regions with defects of several kinds for Aztec rectangle have been studied (see [10] for the most general class of boundary defects).

Remark 1. Similar defects can be studied for Aztec Diamonds as well. In fact, we can delete all cells in a fringe and obtain a tiling.

The proof of Theorem 2 gives an optimal $O(b^2)$ time algorithm for finding a cover for $\mathcal{AR}_{a,b}$ with one defect. In general, however, it is computationally hard to determine if $\mathcal{AR}_{a,b}$ with an unknown number of defects has a cover.

Theorem 3. *It is NP-complete to decide whether $\mathcal{AR}_{a,b}$ with an unbounded number of defects has a cover.*

Proof Sketch. The reduction is from tiling an arbitrary region R with defects. The idea is to embed R into $\mathcal{AR}_{a,b}$ for some sufficiently large n and insert defects in $\mathcal{AR}_{a,b}$ in a way that surrounds R.

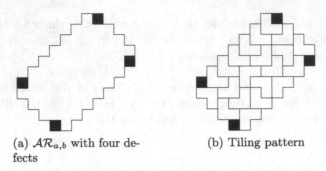

(a) $\mathcal{AR}_{a,b}$ with four defects

(b) Tiling pattern

Fig. 9. Tiling of $\mathcal{AR}_{a,b}$ with four defects.

4 Tiling with 180-Trominos

In this section we study tilings of arbitrary regions using only 180-trominoes. With no loss of generality, we will only consider right-oriented 180-trominoes.

4.1 Hardness

It is easy to see that even when restricted to 180-trominoes, deciding the existence of a tiling of an arbitrary region is still hard.

Theorem 4. 180-TROMINO *is NP-complete.*

Proof Sketch. The proof uses the same gadgets for the reduction for I-Trominoes from the *1-in-3 Graph Orientation Problem* of Horiyama *et al.* [5]. Take any gadget of Horiyama *et al.* [5] and partition each cell into 4 new cells. Thus, each I-tromino is transformed in a new 2×6 or 6×2 region (depending on the orientation of the I-tromino) which can be covered with four 180-trominoes as in Fig. 10. If a gadget is covered with I-trominoes, then the same gadget, after partitioning each cell into four new cells, can also be covered with 180-trominoes. To see the other direction of this implication, we exhaustively examined all possible ways to cover each 4-cell-divided gadget with L-trominoes, and observed that each gadget with its original cells can also be covered with I-trominoes (we omit the details here due to lack of space).

Theorem 4 also implies that the *Triangular Trihex Tiling Problem* of Conway and Lagarias [2] is NP-complete.

It is natural to think along these lines about tiling the Aztec rectangle (and hence, Aztec diamond) with 180-trominoes. However, we show that it is impossible.

Theorem 5. $\mathcal{AR}_{a,b}$ *does not have a 180-cover.*

Proof. Consider the southwestern side of any Aztec rectangle as in Fig. 11 and pick any one of the marked cells, say the cell at coordinate (c, d). There are only

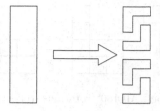

Fig. 10. I-Tromino to L-Tromino transformation using 180-Trominoes.

two ways to cover that cell with a right-oriented tromino. With one tromino we can cover the cells with coordinates $(c, d), (c, d + 1)$ and $(c + 1, d + 1)$, whereas with the other tromino we can cover the cells $(c, d), (c+1, d)$ and $(c+1, d+1)$. In either case the cells at (c, d) and $(c+1, d+1)$ are always covered, and depending on which tromino is chosen either the cell at $(c, d + 1)$ or $(c + 1, d)$ is covered. Therefore, if we cover the entire bottom-left side of an Aztec rectangle, there will always be a cell at $(c, d + 1)$ or $(c + 1, d)$ that cannot be covered. Note that any reversed fringe that is on top of the bottom-left side of any Aztec rectangle can be covered with 180-trominoes if it has one defect.

Corollary 3. AD(n) *does not have a 180-cover.*

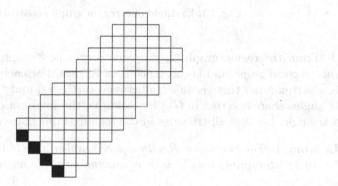

Fig. 11. Covering of an Aztec rectangle with right-oriented trominoes.

4.2   Efficient Tilings

In this section we identify a sufficient condition for a region to have an efficient algorithm that decides the existence of a 180-cover.

Theorem 6. *If a region R does not contain any of the forbidden polyominoes of Fig. 12 as a subregion, then there exists a polynomial-time algorithm that decides whether R has a 180-cover.*

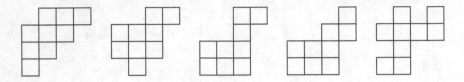

Fig. 12. Forbidden polyominoes. All 180° rotations, reflections and shear transformations are also forbidden polyominoes.

For the remaining of this section we present a proof of Theorem 6. Remember that, with no loss of generality, we only consider right-oriented trominoes. Given a region R we construct a graph G_R, which we call the *region graph* of R, as follows. For each cell (a, b) that is not a defect there is a vertex v_{ab}. There is an edge for each pair of adjacent cells and for each pair v_{ab} and $v_{(a+1)(b+1)}$. Note that this reduction is one-to-one. We present an example in Fig. 13.

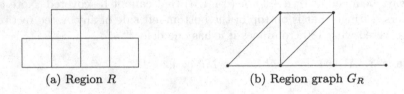

(a) Region R (b) Region graph G_R

Fig. 13. Example of a region graph construction.

From the region graph G_R we construct a new graph I_R which we call an *intersection graph* and is constructed as follows. For each triangle in G_R there is a vertex t and there is an edge between vertices t_i and t_j if the corresponding triangles share a vertex in G_R; for example, the intersection graph for Fig. 13 is a triangle, because all triangles in the region graph share at least one vertex.

Lemma 4. *For any region R with a fixed number of defects, the maximum number of 180-trominoes that fit in R equals the size of a maximum independent set in I_R.*

Proof. Let k be the maximum number of tiles that fit in R and let S be a maximum independent set in the intersection graph I_R. We claim that $|S| = k$.

Each triangle in the region graph G_R correspond to a position where a 180-tile can fit. If k is the maximum number tiles that can fit in R, then there exist k triangles in G_R, denoted T, that do not share any common vertex. Each triangle in T corresponds to a vertex in I_R and since none of the triangles in T share a common vertex, T defines an independent set in I_R and $k \leq |S|$.

To prove that $|S| = k$ suppose by contradiction that T is not a maximum independent set of I_R, that is, $k < |S|$. Since S is an independent set in I_R, there are $|S|$ triangles in G_R that do not share a common vertex. Thus, we can fit $|S|$ 180-trominos in R, which is a contradiction because $k < |S|$. $\square$

The idea for a proof of Theorem 6 is to construct a polynomial time algorithm that decides the existence of a 180-cover by deciding if a maximum independent set in I_R equals the number of cells of R divided by 3, which agrees with the number of trominoes covering R. Deciding the existence of a maximum independent set of a given size is a well-known NP-complete problem, nevertheless, it is known from the works of Minty [7], Sbihi [11] and Nakamura and Tamura [9] that for claw-free graphs[1] finding independent sets can be done in polynomial time. Hence, if I_R is claw-free, then we can use a polynomial time algorithm for finding independent sets to decide the existence of a 180-cover. If I_R has a claw, however, each claw will give one of the forbidden polyominoes.

In Lemma 5 below we show that 180-TROMINO is polynomial time reducible to deciding independent sets, which allow us to construct algorithms for 180-TROMINO using known algorithms for deciding independent sets. Then in Lemma 6 we show that if I_R has a claw, then that claw corresponds to a forbidden polyomino in the region R.

Lemma 5. *There is a many-one polynomial-time reduction from 180-TROMINO to the problem of deciding existence of an independent set of a given size.*

Proof. First the reduction constructs the region graph G_R and the intersection graph I_R. If the size of the largest independent set equals the number of cells of R divided by 3, then output "yes" because R has a 180-cover; otherwise output "no" because R does not have a 180-cover.

Suppose R has a 180-cover. If n is the number of cells in R, then the number of tiles in the 180-cover is $n/3$. By Lemma 4, the largest independent set in I_R equals $n/3$.

Now suppose R does not have a 180-cover. If n is the number of cells in R, then $n/3$ is not equal the maximum number of tiles that can fit in R. Thus, by Lemma 4, it holds that $n/3$ is not equal the size of the largest independent set in I_R. □

Lemma 6. *If I_R has a claw, then R has at least one forbidden polyomino.*

Proof Sketch. For any claw in I_R there is a vertex of degree 3 and three vertices of degree 1, and each vertex in I_R corresponds to a triangle in the region graph G_R. We refer to the triangle that corresponds to the degree 3 vertex as the *central triangle* and each degree 1 triangle is called an *adjacent triangle*. Thus, to obtain all forbidden polyominoes, we look at all posible ways to connect (by the vertices) each adjacent triangle to the central triangle in such a way that each adjacent triangle only connects to the central triangle in a single vertex and it is not connected to any other adjacent triangle; otherwise, if an adjacent triangle connects with two vertices of the central triangle or any two adjacent vertices connects with one another, then the induced graph does not corresponds to a claw. By exhaustively enumerating all possibilities, we can

---

[1] A graph is *claw-free* if it does not have $K_{1,3}$ (a claw) as an induced subgraph.

extract all polyominoes that correspond to claws in I_R. Then we partition this set of polyominoes in five equivalence classes, where two polyominoes are in the same class if and only if one can be obtained from the other by a 180° rotation, a reflection or shear transformation (we omit some details here due to lack of space).

Lemmas 5 and 6 complete the proof of Theorem 6.

5   I-Trominoes vs L-Trominoes

In Sect. 4 we saw that any gadget of Horiyama *et al.* [5] can be covered with I-trominoes if and only if the same gadget, after partitioning each cell into four new cells, can be covered with L-trominoes. In general, if R is any region and $R^{\boxplus}$ is the region R where each cell is partitioned into four cells, we have that if R can be covered with I-trominoes, then $R^{\boxplus}$ can be covered with L-trominoes. We do not know, however, if the other way of this implication holds in the general case. The following theorem partly answers this open problem.

Theorem 7. *Let R be a connected region of size n. The region $R^{\boxplus}$ has an L-Tromino cover if and only if 3 divides n.*

Proof Sketch. It is clear that if $R^{\boxplus}$ has an L-Tromino cover, then 3 divides n. Now suppose that 3 divides n. Say that a connected region R with n vertices is *detachable* if there exist a way to separate R in two connected subregions of sizes n_1 and n_2 such that 3 divides n_1 and 3 divides n_2. We can show that if R is not detachable, then $R^{\boxplus}$ can always be covered with L-Trominoes.

In order to construct a tiling for $R^{\boxplus}$ we first decompose R by recursively detaching it in connected subregions until all subregions obtained this way are not detachable. Since each subregion is not detachable, we construct an L-Tromino cover for each subregion and then join them to obtain a cover for $R^{\boxplus}$. We omit details due to the lack of space.

The proof of Theorem 7 gives an efficient algorithm to find covers for any $R^{\boxplus}$.

6   Concluding Remarks and Open Problems

In this work we studied the computational hardness of tiling arbitrary regions with L-trominoes. We showed restrictions to the problem that keeps it computationally intractable and identified concrete instances where an efficient tiling exists.

We conclude this paper with some open problems that we consider challenging and that we believe will fuel future research in the subject.

1. *Hardness of tiling the Aztec rectangle with a given number of defects.* In Sect. 3 we saw that an Aztec rectangle with 0 or 1 defects can be covered with L-trominoes in polynomial time, whereas in general the problem is NP-complete

when the Aztec rectangle has an unknown number of defects; with $2 + 3k$, for every k, an Aztec rectangle cannot be covered because the number of cells is not divisible by 3. It is open if there exist a polynomial time algorithm for deciding a tiling for an Aztec rectangle with a given number of defects.

2. *Tiling of orthogonally-convex regions.* In this work we showed several instances where a tiling can be found in polynomial time. In general, it is open if an orthogonally-convex region with no defects can be covered in polynomial time or if it is NP-complete to decide if a tiling exists.

References

1. Chin, P., Grimaldi, R., Heubach, S.: Tiling with L's and squares. J. Integer Sequences **10**(2), 3 (2007)
2. Conway, J.H., Lagarias, J.C.: Tiling with polyominoes and combinatorial group theory. J. Comb. Theor. Ser. A **53**(2), 183–208 (1990)
3. Demaine, E.D., Demaine, M.L.: Jigsaw puzzles, edge matching, and polyomino packing: connections and complexity. Graphs and Combinatorics **23**(1), 195–208 (2007)
4. Elkies, N., Kuperberg, G., Larsen, M., Propp, J.: Alternating-sign matrices and domino tilings. I. J. Algebraic Comb. **1**(2), 111–132 (1992)
5. Horiyama, T., Ito, T., Nakatsuka, K., Suzuki, A., Uehara, R.: Complexity of tiling a polygon with trominoes or bars. Discrete Comput. Geom. **58**(3), 686–704 (2017)
6. Mills, W.H., Robbins, D.P., Rumsey, H.: Alternating sign matrices and descending plane partitions. J. Comb. Theor. Ser. A **34**(3), 340–359 (1983)
7. Minty, G.J.: On maximal independent sets of vertices in claw-free graphs. J. Comb. Theor. Ser. B **28**(3), 284–304 (1980)
8. Moore, C., Robson, J.M.: Hard tiling problems with simple tiles. Discrete Comput. Geom. **26**(4), 573–590 (2001)
9. Nakamura, D., Tamura, A.: A revision of Minty's algorithm for finding a maximum weighted stable set of a claw-free graph. J. Oper. Res. Soc. Japan **44**(2), 194–204 (2001)
10. Saikia, M.P.: Enumeration of domino tilings of an aztec rectangle with boundary defects. Adv. Appl. Math. **89**, 41–66 (2017)
11. Sbihi, N.: Algorithme de recherche d'un stable de cardinalité maximum dans un graphe sans étoile. Discrete Math. **29**(1), 53–76 (1980)

The Prefix Fréchet Similarity

Christian Scheffer[✉] [iD]

Department of Computer Science, TU Braunschweig, 38106 Braunschweig, Germany
c.scheffer@tu-bs.de
http://www.Christian-Scheffer.de

Abstract. We present the *prefix Fréchet similarity* as a new measure
for similarity of curves which is e.g. motivated by evacuation analysis
and defined as follows. Given two (polygonal) curves T and T', we ask
for two *prefix curves* of T and T' which have a Fréchet distance no larger
than a given distance threshold $\delta \geq 0$ w.r.t. L_1 metric such that the sum
of the prefix curves is maximal. As parameterized Fréchet measures as,
e.g., the prefix Fréchet similarity are highly unstable w.r.t. to the value
of the distance threshold δ, we give an algorithm that computes exactly
the *profile* of the prefix Fréchet similarity, i.e., the complete functional
relation between δ and the prefix Fréchet similarity of T and T'. This is
the first efficient algorithm for computing exactly the whole profile of a
parametrized Fréchet distance.

While the running time of our algorithm for computing the profile of
the prefix Fréchet similarity is $\mathcal{O}\left(n^3 \log n\right)$, we provide a lower bound of
$\Omega(n^2)$ for the running time of each algorithm computing the profile of
the prefix Fréchet similarity, where n denotes the number of segments
on T and T'. This implies that our running time is at most a near linear
factor away from being optimal.

Keywords: Fréchet distance · Prefix curves · Curve matching

1 Introduction

The data which is generated by tracking moving objects has a long history and
is studied under a large range of various aspects [2,3,10,11]. A promising tool
to measure the similarity between curves is given via the *Fréchet distance*. The
Fréchet distance asks for a continuous matching between T and T' such that
the maximum distance between two matched points from T and T' is minimized
over all pairs of monotone walks.

Alt and Godau [1] gave an algorithm which computes the Fréchet distance
between T and T' in $\mathcal{O}\left(n^2 \log(n)\right)$ time, where n denotes the number of segments
from T and T'.

One application of measuring the similarity of curves is evacuation analysis.
During an evacuation, entities try to go away from the danger area as fast as
possible. The importance of the entities' movements decreases with increasing
distance to the danger area because the larger the distance to the danger area is,

© Springer Nature Switzerland AG 2019
G. K. Das et al. (Eds.): WALCOM 2019, LNCS 11355, pp. 96–107, 2019.
https://doi.org/10.1007/978-3-030-10564-8_8

the smaller is the influence of the danger area to the moving entity. This suggests to compare *prefix curves* of the entities' tracked movements, i.e., a subcurve Q of a tracked movement curve T such that Q and T have the same start point. In particular, we use L_2 to measure in the ambient space lengths of prefix curves which gives equal weight to both input curves T and T'. Furthermore, we use L_1 two measure distances in the ambient space between two points on different input curves which results in a polyhedral free space, see below for a definition.

Another motivation for considering prefix curves is the analysis how the infrastructure around a public hot spot, e.g., a main railway stations, is used and could be improved.

Furthermore, the analysis of football trajectories [9], in particular, the detection of similarities between different goal scenes provides an application for considering postfix curves which is the reversed problem and thus equivalent to considering prefix curves.

1.1   Related Work

In their pioneering work, Alt and Godau [1] give an algorithm that computes the Fréchet distance in $\mathcal{O}(n^2 \log n)$ time by applying parametric search. Buchin et al. [4] give a $\mathcal{O}(n^2 \log^2 n)$ algorithm computing the Fréchet distance while avoiding applying parametric search.

To decide if two prefix curves of T and T' are similar, we ask if they are within a Fréchet distance of an input threshold $\delta \geq 0$. Thus, our measure, abbreviated by $\mathcal{C}_\delta(T, T')$, can be seen as a specialization of the partial Fréchet similarity which was introduced by Buchin et al. [5]. In the man dog metaphor, the partial Fréchet similarity is defined as follows. We are searching for a pair of simultaneous walks on T and T' such that the sum of the lengths of subcurves from T and T' in which the needed leash is no larger than a given δ is maximized. This means that the prefix Fréchet similarity is upper-bounded by the partial Fréchet similarity. Buchin et al. [5] give an algorithm to compute the partial Fréchet similarity between two given polygonal curves in $\mathcal{O}\left(n^3 \log n\right)$ time. While Buchin et al. [5] measure distances between points w.r.t. the L_1 or L_∞ metric, De Carufel et al. [7] showed that the partial Fréchet similarity w.r.t. Euclidean distance is not computable exactly over rational numbers. Hence, they gave an approximation algorithm of time complexity $\mathcal{O}\left(n^3/\varepsilon \log \frac{n}{\varepsilon}\right)$, where ε is an approximation parameter verifying the additive approximation error. By setting $\delta := 0$ and assessing each leash length ℓ not by 0 or 1 but by $\ell - \delta = \ell$, we obtain the *integral Fréchet distance*, introduced by Buchin [6]. Maheshwari et al. give the first $(1 + \varepsilon)$-approximation algorithm with pseudo polynomial runtime depending on ε, the number of segments of the input curves, and the maximal ratio of the segments' lengths [12]. Furthermore, in a previous work [15], we provide a more flexible version of the partial Fréchet similarity, where leash lengths are measured w.r.t. an input vector set and an efficient algorithm for computing it.

Another problem setting which is similar to ours is the *partial curve matching problem* introduced by Maheshwari et al. [14]. In particular, this problem setting asks for a longest connected subcurve form T that is within a Fréchet distance of at most δ to the whole curve T'. Maheshwari et al. [14] gave an algorithm which solves the partial curve matching problem in $\mathcal{O}\left(n^2\right)$ time.

Furthermore, there are several other Fréchet measures that are parameterized via an input parameter δ, e.g., the *minimum backward Fréchet distance* [8] and Fréchet measures with bounded speed constraints [13].

1.2 Profiles of Parametrized Fréchet Distances

One drawback of many parametrized Fréchet distances is their instability regarding the choice of the input parameter δ. In particular, there are configurations of T and T' such that arbitrary small variations of δ result in arbitrary large changes of, e.g., the prefix Fréchet similarity or the partial Fréchet similarity, see Fig. 1. Motivated by that we provide an algorithm that computes the whole profile of the prefix Fréchet similarity.

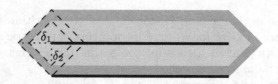

Fig. 1. Instability of the prefix Fréchet similarity regarding the choice of δ for $\delta_1 \approx \delta_2$ and $\delta_1 < \delta_2$: For $\delta = \delta_1$ the prefix Fréchet similarity is equal to 0 while $\delta = \delta_2$ implies that both curves are accepted completely, i.e., the prefix Fréchet similarity is the sum of the length of both curves. The above example also applies to the partial Fréchet similarity and the maximum walk Fréchet similarity.

1.3 Our Results

- We define a new Fréchet distance, the prefix Fréchet similarity, with applications in many practical areas, e.g., evacuation analysis, planning of public infrastructure, and the analysis of football trajectories.
- We provide a lower bound of $\Omega(n^2)$ for the running time of every algorithm computing the profile of the prefix Fréchet similarity, see Sect. 3.
- We give an algorithm that computes the profile of the prefix Fréchet similarity in $\mathcal{O}\left(n^3 \log n\right)$ time, see Sect. 4. This is the first efficient algorithm for computing exactly the whole profile of a parametrized Fréchet distance. Our lower bound of $\Omega(n^2)$ implies that our running time is at most a near linear factor away from being optimal.

2   Preliminaries

A *curve* is a continuous function $T : [0, \lambda] \to \mathbb{R}^2$ with a uniform parametrization. We denote by $|T|$ the *length* of T w.r.t. to the Euclidean norm. A *polygonal curve* T is a piecewise-linear curve, where the *complexity* n of T is defined as the number of segments of T. A *reparametrization* of $T : [0, \lambda] \to \mathbb{R}^2$ is defined as a continuous and monotone function $\alpha : [0, 1] \to [0, \lambda]$, with $\alpha(0) = 0$ and $\alpha(1) = \lambda$. A *matching* of two polygonal curves T and T' is defined as a pair (α, α') of reparametrizations for T and T'. The *value* of a matching (α, α') is defined as the maximum distance $d_1(T(\alpha(t)), T'(\alpha'(t)))$, for $t \in [0, 1]$, where d_p denotes the L_p metric. Finally, the *Fréchet distance* $\mathcal{F}(T, T')$ between T and T' is the infimum of all values of possible matchings.

Definition 1. *Let T and T' be two polygonal curves. A* prefix curve P *of* T : $[0, \lambda] \to \mathbb{R}^2$ *is the restriction of T to an interval $[0, \mu] \subseteq [0, \lambda]$, i.e., a curve $P : [0, \mu] \to \mathbb{R}^2$ such that $P(t) = T(t)$ for all $t \in [0, \mu] \subseteq [0, \lambda]$. Given a distance threshold $\delta \geq 0$, we define the* prefix Fréchet similarity $\mathcal{C}_\delta(T, T')$ *of T and T' and w.r.t. δ as the maximal sum of the lengths of two prefix curves T and T' such that P and P' are within a Fréchet distance of at most δ.*

The profile $\mathcal{P}(\cdot)$ *of the prefix Fréchet similarity is defined as the function mapping δ to $\mathcal{C}_\delta(T, T')$.*

This means that $|T| + |T'|$ is an upper bound for $\mathcal{C}_\delta(T, T')$ and that $\mathcal{C}_\delta(T, T') = |T| + |T'|$ is achieved if δ is chosen at least as large as the Fréchet distance between the entire curves T and T'. Thus, it suffices to compute $\mathcal{P}(\cdot)$ over the interval $[0, \mathcal{F}(T, T')]$.

We compute the profile of the prefix Fréchet similarity by considering the *free space diagrams* D_δ w.r.t. the distance thresholds $\delta \geq 0$ which are defined as follows. The *parameter space* $P \subset \mathbb{R}^2_{\geq 0}$ is defined as the axis aligned rectangle $[0, |T|] \times [0, |T'|] \subset \mathbb{R}^2_{\geq 0}$. Each parameter point $(\lambda, \lambda') \in P$ corresponds to the two points $T(\lambda)$ and $T'(\lambda')$. The *parameter grid* is an $(n + 1)$-by-$(n + 1)$ grid overlaying P such that the cell C in the ith column and jth row is the parameter cell of the ith segment of T' and the jth segment of T. This means, the height of the ith row corresponds to the length if the $i$the segment from T and the width of the jth column corresponds to the length of jth segment from T'.

We refer to the edges of the parameter grid by *parameter edges*. Given a distance threshold $\delta \geq 0$, the parameter space is refined by distinguishing for each point $p = (\lambda, \lambda') \in P$ whether p is *allowed*, i.e, $d_1(T(\lambda), T'(\lambda')) \leq \delta$ or *not allowed*, i.e., $d_1(T(\lambda), T'(\lambda')) > \delta$. We call the union of all allowed parameter points from P *free space* or *white space* and denote it by W_δ. Furthermore, we denote the closure of the union of all points that are not allowed by B_δ and call it the *black space*. The resulting decomposition of P is called the free space diagram D_δ of T and T'.

Furthermore, we call a continuous path $\pi : [0, 1] \to P$ *xy-monotone*, if and only, if $\pi(t) \leq_{xy} \pi(t')$ holds for all $t \leq t'$ from $[0, 1]$, where $a \leq_{xy} b$ abbreviates $a.x \leq b.x$ and $a.y \leq b.y$. A segment s is *monotone* if $a \leq_{xy} b$ or $b \leq_{xy} a$ hold for all $a, b \in s$. We call a point $p \in W_\delta$ *reachable with leash length* δ if there is an xy-monotone path $\pi \subset W_\delta$ with $\pi(0) = (0, 0)$ and $\pi(1) = p$.

Observation 1. $\mathcal{C}_\delta(T, T') = d_1((0, 0), p)$ *holds for a farthest reachable point p.*

Observation 1 implies that computing the profile of the prefix Fréchet similarity is equivalent to computing a continuous sequence of shortest path distances in a continuously deforming free space diagram for continuously varying $\delta \in [0, \mathcal{F}(T, T')]$.

3   A Quadratic Lower Bound for the Profile

In this section, we give a lower bound for the running time of an algorithm that computes the profile of the prefix Fréchet similarity of two polygonal input curves.

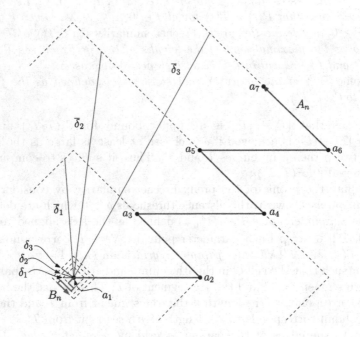

Fig. 2. The curves A_6 and B_6 used in the lower bound construction of Theorem 1. (Color figure online)

Theorem 1. *For each $n \in \mathbb{N}$, there is a pair of curves A_n, B_n such that the complexity of the profile of the prefix Fréchet similarity of A_n and B_n is $\Theta(n^2)$ where n is the total number of segments on A_n and B_n, see Fig. 2.*

Proof. Let $n \in \mathbb{N}$ be chosen arbitrarily. W.l.o.g. we assume that there is a $k \in \mathbb{N}$ such that $4k = n$. Otherwise we apply the below construction for $\lfloor \frac{n}{4} \rfloor 4$ and simply attach $n - \lfloor \frac{n}{4} \rfloor 4$ segments to A_n.

First, we give the construction of A_n and B_n.

We construct A_n as a sequence of $\frac{n}{2}$ segments whose lengths alternate between $\sqrt{2}$ and 1 and whose slopes alternate between 0 and -1, see the blue curve in Fig. 2.

Furthermore, we construct B_n as a sequence of $\frac{n}{2}$ segments whose slopes alternate between 1 and -1 such that A_n and B_n have the same starting point, see the red curve in Fig. 2. We start the construction of B_n with a diagonal segment pointing in south-western direction and with length $\frac{1}{n}$. Next, we attach a segment pointing in south-eastern direction and with length $\frac{1}{2n}$. Then we add a segment of length $\frac{1}{n}$ pointing in south-western direction and then a segment of length $\frac{1}{n}$. Next, we add a segment of length $\frac{1}{n}$ pointing in north-western direction and a segment of length $\frac{1}{n}$ and pointing in south-western direction. Finally, we repeat attaching the four last segments until B_n is made up of $\frac{n}{2}$ segments.

Next, we argue that the profile of the prefix Fréchet similarity of A_n and B_n has a complexity of $\frac{n^2}{4}$.

For $i \in \{1, \ldots, n/4\}$, let δ_i be the value such that the $2i$th segment of B_n lies on the boundary of the L_1 sphere with radius δ_i and center in the starting point of B_n, see Fig. 2. Furthermore, for $i \in \{1, \ldots, n/4\}$, let $\overline{\delta}_i$ be the value such that the $2i$th segment of A_n lies on the L_1 sphere with radius δ_i and center in the starting point of A_n. Furthermore, we denote the vertices of A_n and B_n by $a_1, \ldots, a_{n/2+1}$ and $b_1, \ldots, b_{n/2+1}$. For $i, j \in \{1, \ldots, \frac{n}{4}\}$, the prefix Fréchet similarity w.r.t. $\overline{\delta}_i + \delta_j$ is realized by the prefix curves between a_1 and a_{2i+1} and between b_1 and b_{2j+1} because the length of the entire curve B_n is smaller than the length of the segment $b_{2j}b_{2j+1}$. This means that the profile of the prefix Fréchet similarity has a discontinuity at $\overline{\delta}_i + \delta_j$ for each choice of $i, j \in \{1, \ldots, \frac{n}{2}\}$ resulting in $\frac{n^2}{4}$ discontinuities. This concludes the proof.

4   An Algorithm for Computing the Profile

In this section we show how to compute the profile of the prefix Fréchet similarity.

Theorem 2. *The profile of the prefix Fréchet Similarity of two given polygonal curves with n segments has a complexity of $\mathcal{O}\left(n^3\alpha(n^3)\right)$ and can be computed in $\mathcal{O}\left(n^3\log(n)\right)$ time.*

In the remainder of this section we give the proof for Theorem 2.

First, we define a discrete set of polynomial points, called *score points*, see Definition 2, such that for each $\delta > 0$ there is a score point which is a farthest reachable point, see Lemma 1.

Each score point p induces for $\delta \in [0, \mathcal{F}(T, T')]$ a function $p : A \to P$, called *score curve* that is piecewise continuous with $A \subseteq [0, \mathcal{F}(T, T')]$ and each score curve p induces a *score function* $\mathcal{S}_p : A \to \mathbb{R}_{>0}$ mapping δ onto its distance to $(0, 0)$ see Definition 3. We give different approaches for computing different versions of score curves and corresponding score functions, see Lemmata 2, 3 and 7. Finally, we compute the profile of the prefix Fréchet similarity, by computing the upper envelope of the reachable parts of all computed score functions, see Theorem 2.

We start by giving the definitions of different types of score points and approaches how to compute the corresponding score curves and score functions.

A key observation is that the allowed space $W_\delta \cap C$ of a single parameter cell $C \in P$ is formed by the intersection of C with a corresponding piecewise linear ellipse $\mathcal{E}_\delta$ [1], see Fig. 3. We often call these ellipses *free space ellipses*. All free space ellipses for a fixed parameter cell C but different values of the allowed leash length have two common axes G and H. As $\mathcal{E}_\delta$ and C are convex, it follows that $W_\delta \cap C = \mathcal{E}_\delta \cap C$ is convex and this, in turn, yields that the intersection of W_δ with each side of the boundary of C is convex.

We declare the following notations, see Fig. 3(a). For a parameter cell C we denote by L, R, U, and B its boundary's left, right, upper, and bottom side. Furthermore, we define $\mathcal{L}_\delta := \mathcal{E}_\delta \cap L$, $\mathcal{R}_\delta := \mathcal{E}_\delta \cap R$, $\mathcal{U}_\delta := \mathcal{E}_\delta \cap U$, and $\mathcal{B}_\delta := \mathcal{E}_\delta \cap B$. In addition to this, we define $\mathcal{G}_\delta := G \cap \mathcal{E}_\delta \cap C$ and $\mathcal{H}_\delta := H \cap \mathcal{E}_\delta \cap C$ if the segments corresponding to C do not lie parallel. Note, that each of these convex intersections could by empty.

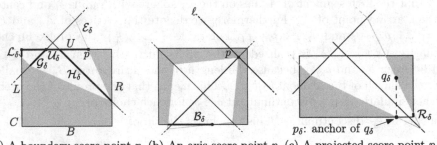

(a) A boundary score point p. (b) An axis score point p. (c) A projected score point p.

Fig. 3. Three different configurations of C and $\mathcal{E}$, distinguishing between the position of a farthest reachable point p.

Definition 2. *Let $\mathcal{L}_\delta$, $\mathcal{R}_\delta$, $\mathcal{U}_\delta$, $\mathcal{B}_\delta$, $\mathcal{G}_\delta$, and $\mathcal{H}_\delta$ be defined as above, see Fig. 3.*

- *Let $A \in \{L, R, U, B\}$. The end point point of the segment $\mathcal{A}_\delta := A \cap W_\delta$ which lies farther away from $(0,0)$ in L_1 norm is the* boundary score point *corresponding to A, see Fig. 3(b). The other end point of the segment $\mathcal{A}_\delta := A \cap W_\delta$ is the* anchor point *corresponding to A.*
- *Let $A \in \{G, H\}$. If $\mathcal{A}_\delta := A \cap W_\delta$ is monotone, the end point of the segment $\mathcal{A}_\delta$ which lies farther away from $(0,0)$ is the* axis score point *corresponding to A if A lies inside the parameter cell C, see Fig. 3(a).*
- *Let p_δ be an anchor point and q_δ its maximal horizontal or vertical projection in positive x- or y-axis direction onto the boundary of the black space such that $p_\delta q_\delta$ does not intersect the interior of the black space, see Fig. 3(c). We call q_δ the* projected score point *resulting from p_δ.*

The following Lemma is the key ingredient for computing the profile of the prefix Fréchet similarity.

Lemma 1. *The farthest reachable point $p \in D_\delta$ is a reachable score point.*

Proof. Let $\delta > 0$ be chosen arbitrarily but fixed. Let C be a parameter cell containing a farthest reachable point p and let $t \in T$, $t' \in T'$, and $\mathcal{E}_\delta$ be the segments and the ellipse corresponding to C. As p is the end of a longest xy-monotone path π starting from $(0,0)$ it follows that p lies on the boundary of the intersection of $\mathcal{E}_\delta$ and C. Otherwise, we could extend π to a longer path. This would be a contradiction to the construction of π as a longest xy-monotone path.

We distinguish between two cases:

- p lies on the boundary of C: This implies that there is a segment $v_1 v_2 \in \{\mathcal{L}, \mathcal{R}, \mathcal{U}, \mathcal{B}\}$ such that $p \in v_1 v_2$. In this case, p has to be an endpoint of the segment s. Otherwise, we could push p on s away from $(0,0)$ while maintaining reachability because p lies on the boundary of C.
- p lies not on the boundary of C: W.l.o.g. we assume that p does not lie on the axes u_ε or d_ε because otherwise p is an axis score point. This implies that the distance of p to $(0,0)$ can be increased by pushing p on the boundary of $\mathcal{E}_\delta$ away from $(0,0)$. For the sake of contradiction, we assume that p is not a projected score point. This implies that we do not loose reachability of p while moving p as described above. This is a contradiction and concludes the proof.

By Observation 1, we know that computing the profile of the prefix Fréchet similarity requires to consider a continuous deformation of the free space diagrams D_δ for $\delta \in [0, \mathcal{F}(T, T')]$. In particular, the allowed space is growing when the leash length increases. This causes that score points may be continuously shifted during the deformation of the free space which leads to the definition of score curves and functions.

Definition 3. *Let e be an arbitrarily chosen parameter edge or parameter axis and $\delta \geq 0$.*

If $\partial W_\delta \cap e$ holds a score point, we define p_δ as the unique score point from $\partial W_\delta \cap e$. The corresponding boundary or axis score curve $p : I \to P$ is defined as $p(\delta) := p_\delta$ where $I \subseteq [0, \mathcal{F}(T, T')]$.

If $\partial W_\delta \cap e$ holds an anchor point, we define q_δ as the unique anchor point from $\partial W_\delta \cap e$. The corresponding anchor curve $q(\delta) : I \to P$ is defined as $q(\delta) := q_\delta$ where $I \subseteq [0, \mathcal{F}(T, T')]$.

For each anchor point q_δ, let s_q be the corresponding projected score point. The corresponding projected score curve $q(\delta) : I \to P$ is defined as $s(\delta) := s_\delta$ where $I \subseteq [0, \mathcal{F}(T, T')]$.

The score function $\mathscr{S}_p : I \to \mathbb{R}_{\geq 0}$ corresponding to a boundary, axis, or projected score curve is defined as the function mapping $\delta \geq 0$ onto the distance between $q(\delta)$ and $(0,0)$ w.r.t. L_1.

By Lemmas 2 and 3 we provide algorithms for computing boundary, axis, and projected score functions.

Lemma 2. *We can compute in $\mathcal{O}(n^2)$ time all boundary and axis score functions. A boundary or axis score function has constant complexity.*

Lemma 3. *We can compute in $\mathcal{O}\left(n^3\right)$ time all projected score functions. Each projected score functions has a complexity of $\mathcal{O}\left(n\right)$.*

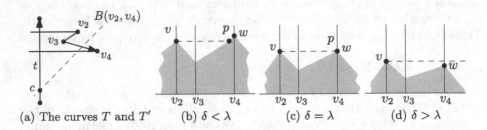

(a) The curves T and T' (b) $\delta < \lambda$ (c) $\delta = \lambda$ (d) $\delta > \lambda$

Fig. 4. Two curves T, T' causing a projected score curve that is not continuous in λ.

Proof. Projecting a boundary score point v to another segment of another free space ellipse results in a linear function, see Fig. 4. Thus, a projected score curve p is linear except for values for δ, called jump events, in which p jumps to another free space ellipse.

W.l.o.g. we assume that v lies on a vertical grid line. Let t be the segment corresponding to the row in which v lies, see Fig. 4(a)+(b). A jump event occurs when p has the same position as another boundary score point w that lies also on a vertical grid line, see Fig. 4(c)+(d). Let v_2 and v_4 be the vertices corresponding to the vertical grid lines on that v and w. Furthermore, let c be the intersection point between the segment t and $B(v_2, v_4)$ the bisector between v_2 and v_4. The situation that v and w lie horizontal is equivalent to the configuration that the intersection point c exists and that δ is equal to the distance λ between c and v_4 w.r.t. L_1.

Justified by the above observations, we compute a superset J of all jump events by computing all intersection points between the boundaries of bisectors and segments of the input curves. This can be done in $\mathcal{O}\left(n^3\right)$ time and leads to $\mathcal{O}\left(n^3\right)$ values which leads to an ordered sequence S of J in $\mathcal{O}\left(n^3 \log n\right)$ time.

By doing an increasing sweep through S we can compute all projected score curves simultaneously in $\mathcal{O}\left(n^3\right)$ time.

Next, we show how to compute these parts of score functions that are induced by reachable points of the corresponding score curves.

Definition 4. *Let p be an arbitrarily chosen score curve. We define the reachable part of the score function corresponding to p as the restriction of p to the union $A \subset [0, \mathcal{F}(T, T')]$ of maximal subsets of the domain of p such that $p(\delta)$ is reachable for all $\delta \in [\delta_1, \delta_2]$.*

Note that a maximal interval $[\delta, \mathcal{F}(T, T')] \subset [0, \mathcal{F}(T, T')]$ such that p is reachable inside W_δ is closed because a corresponding path in free space is also allowed to lie on the boundary of W_δ for each $\delta \geq 0$.

Lemma 4. *A boundary or axis score point p can only turn from non-reachable into reachable.*

Proof. Let $\delta_1 \in [0, \mathcal{F}(T, T')]$ be a point in time at which p turns from not-reachable into reachable and let $\delta_2 \geq \delta_1$.

As $\delta_1 \leq \delta_2$ it follows $W_{\delta_1} \subseteq W_{\delta_2}$. Furthermore, as $p(\delta_1)$ is reachable inside W_{δ_1}, there is an xy-monotone path $\pi \subset W_{\delta_1} \subseteq W_{\delta_2}$ between $(0,0)$ and $p(\delta_1)$. Let $\mathcal{E}_1$ and $\mathcal{E}_2$ be the free space ellipses of C corresponding to δ_1 and δ_2. Obviously, we have $\mathcal{E}_1 \subseteq \mathcal{E}_2$. Thus $p(\delta_1) \in \mathcal{E}_1$ implies $p(\delta_1) \in \mathcal{E}_2$. Since $\mathcal{E}_2$ is convex it follows $p(\delta_1)p(\delta_2) \subset \mathcal{E}_2$. As a witness for reachability of $p(\delta_2)$ we take the path $\pi \subset W_{\delta_2}$ concatenated with the segment $p(\delta_1)p(\delta_2) \subset W_{\delta_2}$. As π is xy-monotone and starts from $(0,0)$, it still remains to show $p(\delta_1) \leq_{xy} p(\delta_2)$. Assume this it not the case. Let $H \subset C$ be the segment such that $p(\delta_1), p(\delta_2) \in H$. As H is xy-monotone it follows that $p(\delta_2)$ lies closer to $(0,0)$ then $p(\delta_1)$ which in turn is element of $H \cap \mathcal{E}_2$. This is a contradiction because $p(\delta_2)$ is defined as the farthest point from $H \cap \mathcal{E}_2$ to $(0,0)$. By the above argument, it follows that δ_1 is unique. Furthermore, δ_1 is well defined because it is upper-bounded by the maximal distance between two points on T and T'.

Let e be a parameter edge and u an arbitrarily chosen point on e. We denote by δ_u the minimal leash length such that u is reachable by a leash length of δ_u. Furthermore, we denote by u_e the point from e such that $\min_e := \delta_{u_e}$ is minimal. We call $\min_e$ the *minimal needed leash length* of the parameter edge e.

Buchin et al. [4] gave an algorithm that computes simultaneously for all parameter edges e the point u_e and its corresponding minimal leash lengths $\min_e$ that is needed for u_e's reachability.

Lemma 5 ([4]). *We can compute in $\mathcal{O}(n^2)$ time for all parameter edges e the points $u_e \in e$ and the corresponding minimal needed leash lengths $\min_e$ simultaneously.*

The algorithm of Buchin et al. [4] implies directly how to compute the reachable parts of all boundary and axis score curves. In particular, let p be an arbitrary boundary or axis score curve and e the parameter edge contain p if p is a boundary score point. If e is an axis score point, let e be the parameter edge which is the left side of the cell containing p. We define the reachable part of p as $[\min_e, \mathcal{F}(T, T')]]$ where $\min_e$ is computed for each parameter edge e by a single application of the algorithm of Lemma 5.

Corollary 1. *We can compute simultaneously the reachable parts of all boundary and axis score curves in $\mathcal{O}(n^2)$ time.*

Finally, we provide an algorithm computing the reachable parts of projected score curves.

Lemma 6. *We can compute simultaneously the reachable parts of all projected score curves in $\mathcal{O}(n^3 \log n)$ time.*

Proof. First, we apply the algorithm of Lemma 5 as a single preprocessing step to compute all minimal leash lengths needed for all parameter edges.

We show how to compute simultaneously the reachable parts of all horizontal projected score curves lying in the same parameter row R in $\mathcal{O}\left(n^2 \log n\right)$ time. Applying this approach to all parameter rows and applying a symmetric approach to all parameter columns in parallel leads to the required approach.

We call a boundary score curve p or an anchor curve p *horizontal* if all its points lie on a horizontal parameter edge. Otherwise, p is *vertical*.

The reachability of a horizontal projected score point is equal to the reachability of the corresponding vertical anchor point. Thus, we compute the reachable parts of all vertical anchor curves in R by processing iteratively through all parameter cells in R from left to right as follows:

Let $C_1, \ldots, C_n$ be the parameter cells of R and $L_1, \ldots, L_n$ and $B_1, \ldots, B_n$ the left and bottom sides of $C_1, \ldots, C_n$. For each L_i with $i = 1, \ldots, n$, we consider the function $\lambda_i : L_i \to \mathbb{R}_{\geq 0}$ mapping a point $a \in L_i$ onto the smallest value $\delta \in [0, \mathcal{F}(T, T')]$ such that a is reachable with a leash length δ. Furthermore, we define $\lambda_i^\star(a) := \min_{L_i}$ for all $a \in L_i$ with $a \geq_{xy} u_{L_i}$ and $\lambda_i^\star(a) := \lambda_i(a)$ for all $a \in L_i$ with $a \leq_{xy} u_{L_i}$. Finally, we denote by a_i and p_i the vertical anchor and the vertical boundary score curves corresponding to L_i.

We compute the vertical anchor and vertical boundary score curves a_i and p_i of each vertical parameter edge of R by applying the same approach as used in Lemma 3. For $i = 1, \ldots, n$, we consider the minimal leash length $\min(B_i)$ of B_i.

We observe that the allowed space inside each parameter cell is convex because the L_1 metric is convex. Thus, we compute L_{i+1} as the minimum of $y = \min_{B_i}$ and the upper envelope of $\lambda_i^\star$, a_i, and p_i. Hence, the reachable part of a_i has a complexity of $\Theta(n)$ in the worst case. Furthermore, each iteration can be done in $\mathcal{O}\left(n \log n\right)$ time which results in a running time of $\mathcal{O}\left(n^2 \log n\right)$ for computing the reachable parts of all anchor curves in R.

Combining the approaches of Lemma 2, Lemma 3, Corollary 1 and Lemma 6 leads to an approach for computing the restrictions of all score functions to their reachable parts.

Lemma 7. *There is a $\mathcal{O}\left(n^3 \log n\right)$ runtime algorithm computing all score functions restricted to the reachable parts of the corresponding score curves.*

Combining Lemmata 1, 2, 3 and 7, yields that the profile of the prefix Fréchet similarity is the upper envelop of $\mathcal{O}\left(n^3\right)$ segments. Thus the profile of the prefix Fréchet similarity has a complexity of $\mathcal{O}\left(n^3 \alpha(n^3)\right)$ and can be computed in $\mathcal{O}\left(n^3 \log n\right)$ time. This concludes the proof of Theorem 2.

5   Conclusion

We introduced the prefix Fréchet similarity with applications in evacuation analysis, analysis of football trajectories, and planning of public infrastructure. We gave a $\mathcal{O}\left(n^3 \log n\right)$ runtime algorithm for computing the whole profile of the prefix Fréchet similarity which enables the choice of stable values for the distance threshold δ. Furthermore, we gave a family of pairs of curves whose profile of the prefix Fréchet similarity has a complexity of $\Theta(n^2)$. Thus the runtime of our algorithm is at most a near linear factor away from being optimal.

References

1. Alt, H., Godau, M.: Computing the Fréchet distance between two polygonal curves. Int. J. Comput. Geom. Appl. **5**, 75–91 (1995)
2. Brakatsoulas, S., Pfoser, D., Salas, R., Wenk, C.: On map-matching vehicle tracking data. In: Proceedings of the 31st International Conference on Very Large Data Bases, Trondheim, Norway, August 30–September 2 2005, pp. 853–864 (2005)
3. Buchin, K., Buchin, M., Gudmundsson, J.: Detecting single file movement. In: Proceedings of 16th ACM SIGSPATIAL International Symposium on Advances in Geographic Information Systems, ACM-GIS 2008, 5–7 November 2008, Irvine, California, USA, p. 33 (2008)
4. Buchin, K., Buchin, M., van Leusden, R., Meulemans, W., Mulzer, W.: Computing the Fréchet distance with a retractable leash. Discrete Comput. Geom. **56**(2), 315–336 (2016)
5. Buchin, K., Buchin, M., Wang, Y.: Exact algorithms for partial curve matching via the Fréchet distance. In: Proceedings of the Twentieth Annual ACM-SIAM Symposium on Discrete Algorithms, SODA 2009, New York, NY, USA, 4–6 January 2009, pp. 645–654 (2009)
6. Buchin, M.: On the computability of the Fréchet distance between triangulated surfaces. Ph.D. thesis, Department of Computer Science, Freie Universität, Berlin (2007)
7. Carufel, J.D., Gheibi, A., Maheshwari, A., Sack, J., Scheffer, C.: Similarity of polygonal curves in the presence of outliers. Comput. Geom. **47**(5), 625–641 (2014)
8. Gheibi, A., Maheshwari, A., Sack, J., Scheffer, C.: Minimum backward Fréchet distance. In: Proceedings of the 22nd ACM SIGSPATIAL International Conference on Advances in Geographic Information Systems, Dallas/Fort Worth, TX, USA, 4–7 November 2014, pp. 381–388 (2014)
9. Gudmundsson, J., Wolle, T.: Football analysis using spatio-temporal tools. Comput. Environ. Urban Syst. **47**, 16–27 (2014)
10. Keogh, E.J., Pazzani, M.J.: Scaling up dynamic time warping for datamining applications. In: Proceedings of the Sixth ACM SIGKDD International Conference on Knowledge Discovery and Data Mining, Boston, MA, USA, 20–23 August 2000, pp. 285–289 (2000)
11. Kim, M., Kim, S., Shin, M.: Optimization of subsequence matching under time warping in time-series databases. In: Proceedings of the 2005 ACM Symposium on Applied Computing (SAC), Santa Fe, New Mexico, USA, 13–17 March 2005, pp. 581–586 (2005)
12. Maheshwari, A., Sack, J., Scheffer, C.: Approximating the integral Fréchet distance. In: 15th Scandinavian Symposium and Workshops on Algorithm Theory, SWAT 2016, 22–24 June 2016, Reykjavik, Iceland, pp. 26:1–26:14 (2016)
13. Maheshwari, A., Sack, J., Shahbaz, K., Zarrabi-Zadeh, H.: Fréchet distance with speed limits. Comput. Geom. **44**(2), 110–120 (2011)
14. Maheshwari, A., Sack, J., Shahbaz, K., Zarrabi-Zadeh, H.: Improved algorithms for partial curve matching. Algorithmica **69**(3), 641–657 (2014)
15. Scheffer, C.: More flexible curve matching via the partial Fréchet similarity. Int. J. Comput. Geom. Appl. **26**(1), 33–52 (2016)

Probabilistic Analysis of Optimization Problems on Generalized Random Shortest Path Metrics

Stefan Klootwijk$^{(\boxtimes)}$, Bodo Manthey, and Sander K. Visser

Faculty of Electrical Engineering, Mathematics and Computer Science, University of Twente, P.O. Box 217, 7500 AE Enschede, The Netherlands
{s.klootwijk,b.manthey}@utwente.nl, s.k.visser@alumnus.utwente.nl

Abstract. Simple heuristics often show a remarkable performance in practice for optimization problems. Worst-case analysis often falls short of explaining this performance. Because of this, "beyond worst-case analysis" of algorithms has recently gained a lot of attention, including probabilistic analysis of algorithms.

The instances of many optimization problems are essentially a discrete metric space. Probabilistic analysis for such metric optimization problems has nevertheless mostly been conducted on instances drawn from Euclidean space, which provides a structure that is usually heavily exploited in the analysis. However, most instances from practice are not Euclidean. Little work has been done on metric instances drawn from other, more realistic, distributions. Some initial results have been obtained by Bringmann et al. (*Algorithmica*, 2013), who have used random shortest path metrics on complete graphs to analyze heuristics.

The goal of this paper is to generalize these findings to non-complete graphs, especially Erdős–Rényi random graphs. A random shortest path metric is constructed by drawing independent random edge weights for each edge in the graph and setting the distance between every pair of vertices to the length of a shortest path between them with respect to the drawn weights. For such instances, we prove that the greedy heuristic for the minimum distance maximum matching problem, the nearest neighbor and insertion heuristics for the traveling salesman problem, and a trivial heuristic for the k-median problem all achieve a constant expected approximation ratio. Additionally, we show a polynomial upper bound for the expected number of iterations of the 2-opt heuristic for the traveling salesman problem.

1 Introduction

Large-scale optimization problems, such as the traveling salesman problem (TSP), show up in many applications. These problems are often computationally intractable. However, in practice often ad-hoc heuristics are successfully used

---

A full version containing all proofs is available at https://arxiv.org/abs/1810.11232.

© Springer Nature Switzerland AG 2019
G. K. Das et al. (Eds.): WALCOM 2019, LNCS 11355, pp. 108–120, 2019.
https://doi.org/10.1007/978-3-030-10564-8_9

that provide solutions that come quite close to optimal solutions. In many cases these, often simple, heuristics show a remarkable performance, even though the theoretical results about those heuristics are way more pessimistic.

In order to explain this difference, probabilistic analysis has been widely used over the last decades. However, the challenge in probabilistic analysis is to come up with a good probabilistic model: it should reflect realistic instances, but also be sufficiently simple to make the analysis tractable.

So far, in almost all cases, either Euclidean space has been used to generate instances of metric optimization problems, or independent, identically distributed edge lengths have been used (e.g. [1,6]). However, both approaches have considerable shortcomings to explain the average-case performance of heuristics on general metric instances: the structure of Euclidean space is heavily used in the probabilistic analysis, but realistic instances are often not Euclidean. The independent, identically distributed edge lengths do not even yield a metric in the first place. In order to overcome these shortcomings, Bringmann et al. [3] have proposed and analyzed the following model to generate random metric spaces, which had already been proposed by Karp and Steele in 1985 [12]: given an undirected complete graph, start by drawing random edge weights for each edge independently and then define the distance between any two vertices as the total weight of the shortest path between them, measured with respect to the random weights.

1.1 Related Work

Bringmann et al. called the model described above *random shortest path metrics*. This model is also known as *first-passage percolation*, introduced by Hammersley and Welsh as a model for fluid flow through a (random) porous medium [7,9].

For first passage percolation in complete graphs, the expected distance between two fixed vertices is approximately $\ln(n)/n$ and the expected distance from a fixed vertex to the vertex that is most distant is approximately $2\ln(n)/n$ [3,10]. Furthermore, it is known that the expected diameter of the metric is approximately $3\ln(n)/n$ [8,10]. There are also some known structural properties of first passage percolation on the Erdős–Rényi random graph. Bhamidi et al. [2] have shown asymptotics for both the minimal weight of the path between uniformly chosen vertices in the giant component and for the hopcount, the number of edges, on this path. Bringmann et al. [3] used this model on the complete graph to analyze heuristics for matching, TSP, and k-median.

1.2 Our Results

As far as we know, no heuristics have been studied in this model for non-complete graphs yet. However, we believe that random shortest path metrics on non-complete graphs will bring us a step further in the direction of realistic input model.

This paper provides a probabilistic analysis of some simple heuristics in the model of random shortest path metrics on non-complete graphs. First, we provide some structural properties of generalized random shortest path metrics (Sect. 3), which can be seen as a generalization of the structural properties found by Bringmann et al. [3]. Although this generalization might seem straightforward at first sight, it brings up some new difficulties that need to be overcome. Most notably, since we do not restrict ourselves to the complete graph, we cannot make use anymore of its symmetry and regularity. This problem is partially solved by introducing two graph parameters, which we call the cut parameters of a graph (Definition 1).

Then, we use these structural insights to perform a probabilistic analysis for some simple heuristics for combinatorial optimization problems (Sect. 4), where the results are still depending on the cut parameters of a graph. Finally, we use these results, to show our main results, namely that these simple heuristics achieve constant expected approximation ratios for random shortest path metrics applied to Erdős–Rényi random graphs (Sect. 5).

2   Notation and Model

We use $X \sim P$ to denote that a random variable X is distributed using a probability distribution P. $\mathrm{Exp}(\lambda)$ is being used to denote the exponential distribution with parameter λ. In particular, we use $X \sim \sum_{i=1}^{n} \mathrm{Exp}(\lambda_i)$ to denote that X is the sum of n independent exponentially distributed random variables having parameters $\lambda_1, \ldots, \lambda_n$.

For $n \in \mathbb{N}$, we use $[n]$ as shorthand notation for $\{1, \ldots, n\}$. We denote the nth harmonic number by $H_n = \sum_{i=1}^{n} 1/i$. Sometimes we use exp to denote the exponential function. Finally, if a random variable X is stochastically dominated by a random variable Y, i.e., we have $F_X(x) \geq F_Y(x)$ for all x (where $X \sim F_X$ and $Y \sim F_Y$), we denote this by $X \precsim Y$.

Generalized Random Shortest Path Metrics. Given an undirected graph $G = (V, E)$ on n vertices, we construct the corresponding generalized random shortest path metric as follows. First, for each edge $e \in E$, we draw a random edge weight $w(e)$ independently from an exponential distribution[1] with parameter 1. Second, we define the distances $d : V \times V \to \mathbb{R}_{\geq 0} \cup \{\infty\}$ as follows: for every $u, v \in V$, $d(u, v)$ denotes the length of the shortest u, v-path with respect to the drawn edge weights. If no such path exists, we set $d(u, v) = \infty$. By doing so, the distance function d satisfies $d(v, v) = 0$ for all $v \in V$, $d(u, v) = d(v, u)$ for all $u, v \in V$, and $d(u, v) \leq d(u, s) + d(s, v)$ for all $u, s, v \in V$. We call the complete graph with distances d obtained from this process a generalized random shortest path metric. If $G = K_n$ (the complete graph on n vertices), then this generalized random shortest path metric is equivalent to the random shortest path metric as defined by Bringmann et al. [3]

---

[1] Exponential distributions are technically easiest to handle due to their memorylessness property. A (continuous, non-negative) probability distribution of a random variable X is said to be memoryless if and only if $\mathbb{P}(X > s + t \mid X > t) = \mathbb{P}(X > s)$ for all $s, t \geq 0$. [15, p. 294].

We use the following notation within generalized random shortest path metrics: $\Delta_{\max} := \max_{u,v} d(u,v)$ denotes the diameter of the graph. Note that $\Delta_{\max} < \infty$ if and only if G is connected. $B_\Delta(v) := \{u \in V \mid d(u,v) \leq \Delta\}$ denotes the 'ball' of radius Δ around v, i.e., the set containing all vertices at distance at most Δ from v. $\tau_k(v) := \min\{\Delta \mid |B_\Delta(v)| \geq k\}$ denotes the distance to the kth closest vertex from v (including v itself). Equivalently, one can also say that $\tau_k(v)$ is equal to the smallest Δ such that the ball of radius Δ around v contains at least k vertices.

Now, $B_{\tau_k(v)}(v)$ denotes the set of the k closest vertices to v. During our analysis, we make use of the size of the cut induced by this set, which we denote by $\chi_k(v) := |\delta(B_{\tau_k(v)}(v))|$, where $\delta(U)$ denotes the cut induced by U.

Erdős–Rényi Random Graphs. The main results of this work consider random shortest path metrics applied to Erdős–Rényi random graphs. An undirected graph $G(n,p) := G = (V,E)$ generated by this model has n vertices ($V = \{1,\ldots,n\}$) and between each pair of vertices an edge is included with probability p, independent of every other pair.

Working with the Erdős–Rényi random graph introduces an extra amount of stochasticity to the probabilistic analysis, since both the graph and the edge weights are random. In order to avoid this extra stochasticity as long as possible, in Sects. 3 and 4 we start our analysis using an arbitrary fixed (deterministic) graph G. Later on, in Sect. 5 we will consider Erdős–Rényi random graphs again.

3   Structural Properties

In order to analyze the structural properties of generalized random shortest path metrics, we first introduce the notion of what we call the cut parameters of a simple graph G.

Definition 1. *Let $G = (V,E)$ be a finite simple connected graph. Then we define the cut parameters of G by*

$$\alpha := \min_{\varnothing \neq U \subset V} \frac{|\delta(U)|}{\mu_U} \quad and \quad \beta := \max_{\varnothing \neq U \subset V} \frac{|\delta(U)|}{\mu_U},$$

where $\mu_U := |U| \cdot (|V| - |U|)$ is the maximum number of possible edges in the cut defined by U.

It follows immediately from this definition that $0 < \alpha \leq \beta \leq 1$ for any finite simple connected graph G. Moreover, for any such graph the following holds for all $\varnothing \neq U \subset V : \alpha \cdot \mu_U \leq |\delta(U)| \leq \beta \cdot \mu_U$. We observe that the cut parameters of the complete graph are given by $\alpha = \beta = 1$.

Distribution of $\tau_k(v)$. Now we have a look at the distribution of $\tau_k(v)$. For this purpose we use an arbitrary fixed undirected connected simple graph G (on n vertices) and let α and β denote its cut parameters.

The values of $\tau_k(v)$ are then generated by a birth process as follows. (Amongst others, a variant of this process for complete graphs has been analyzed by Davis and Prieditis [5] and Bringmann et al. [3].) For $k = 1$, we have $\tau_k(v) = 0$. For $k \geq 2$, we look at all edges (u, x) with $u \in B_{\tau_{k-1}(v)}(v)$ and $x \notin B_{\tau_{k-1}(v)}(v)$. By definition there are $\chi_{k-1}(v)$ such edges. Moreover the length of these edges is conditioned to be at least $\tau_{k-1}(v) - d(v, u)$. Using the memorylessness of the exponential distribution, we can now see that $\tau_k(v) - \tau_{k-1}(v)$ is the minimum of $\chi_{k-1}(v)$ (standard) exponential variables, or, equivalently, $\tau_k(v) - \tau_{k-1}(v) \sim \mathrm{Exp}(\chi_{k-1}(v))$. We use this result to find bounds for the distribution of $\tau_k(v)$.

Lemma 2. *For all $k \in [n]$ and $v \in V$ we have,*

$$\alpha k(n - k) \leq \chi_k(v) \leq \beta k(n - k).$$

Lemma 3. *For all $k \in [n]$ and $v \in V$ we have,*

$$\sum_{i=1}^{k-1} \mathrm{Exp}(\beta i(n - i)) \precsim \tau_k(v) \precsim \sum_{i=1}^{k-1} \mathrm{Exp}(\alpha i(n - i)).$$

Exploiting the linearity of expectation, the fact that the expected value of an exponentially distributed random variable with parameter λ is $1/\lambda$ and the fact that $\sum_{i=1}^{k-1} 1/(i(n-i)) = (H_{k-1} + H_{n-1} - H_{n-k})/n$, we obtain the following corollary.

Corollary 4. *For all $k \in [n]$ and $v \in V$ we have,*

$$\frac{H_{k-1} + H_{n-1} - H_{n-k}}{\beta n} \leq \mathbb{E}(\tau_k(v)) \leq \frac{H_{k-1} + H_{n-1} - H_{n-k}}{\alpha n}.$$

From this result, we can derive the following extensions of two known results. First of all, if we randomly pick two vertices $u, v \in V$, then averaging over k yields that the expected distance $\mathbb{E}[d(u, v)]$ between them is bounded between $\frac{H_{n-1}}{\beta(n-1)} \approx \ln(n)/\beta n$ and $\frac{H_{n-1}}{\alpha(n-1)} \approx \ln(n)/\alpha n$, which is in line with the known result for complete graphs, where we have $\mathbb{E}[d(u, v)] \approx \ln(n)/n$ [3,5,10]. Secondly, for any vertex v, the longest distance from it to another vertex is $\tau_n(v)$, which in expectation is bounded between $\frac{2H_{n-1}}{\beta n} \approx 2\ln(n)/\beta n$ and $\frac{2H_{n-1}}{\alpha n} \approx 2\ln(n)/\alpha n$, which also is in line with the known result for complete graphs, where we have an expected value of approximately $2\ln(n)/n$ [3,10].

It is also possible to find bounds for the cumulative distribution function of $\tau_k(v)$. To do so, we define $F_k(x) = \mathbb{P}(\tau_k(v) \leq x)$ for some fixed vertex $v \in V$.

Lemma 5. [3, Lemma 3.2] *Let $X \sim \sum_{i=1}^{n} \mathrm{Exp}(ci)$. Then, for any $a \geq 0$ we have $\mathbb{P}(X \leq a) = (1 - e^{-ca})^n$.*

Lemma 6. *For all $x \geq 0$ and $k \in [n]$ we have,*

$$(1 - \exp(-\alpha(n - k)x))^{k-1} \leq F_k(x) \leq (1 - \exp(-\beta nx))^{k-1}.$$

We can improve this result slightly.

Lemma 7. *For all $x \geq 0$ and $k \in [n]$ we have,*

$$F_k(x) \geq (1 - \exp(-\alpha n x/4))^n.$$

Using this improved bound for the cumulative distribution function of $\tau_k(v)$, we can derive the following tail bound for the diameter $\Delta_{\max}$.

Lemma 8. *Let $\Delta_{\max} = \max_{u,v \in V}\{d(u, v)\}$. For any fixed c we have $\mathbb{P}(\Delta_{\max} > c\ln(n)/\alpha n) \leq n^{2-c/4}$.*

Clustering. In this section we show that we can partition the vertices of generalized random shortest path metrics into a small number of clusters with a given maximum diameter. Before we prove this main result, we first provide a tail bound for $|B_\Delta(v)|$.

Lemma 9. *For $n \geq 5$ and for any fixed $\Delta \geq 0$ we have,*

$$\mathbb{P}\left(|B_\Delta(v)| < \min\left\{\exp(\alpha \Delta n/5), \frac{n+1}{2}\right\}\right) \leq \exp(-\alpha \Delta n/5).$$

We use the result of this lemma to prove our main structural property for generalized random shortest path metrics.

Theorem 10. *For any fixed $\Delta \geq 0$, if we partition the vertices into clusters, each of diameter at most 4Δ, then the expected number of clusters needed is bounded from above by $O(1 + n/\exp(\alpha \Delta n/5))$.*

4 Analysis of Heuristics

In this section we bound the expected approximation ratios of the greedy heuristic for minimum-distance perfect matching, the nearest neighbor and insertion heuristics for the traveling salesman problem, and a trivial heuristic for the k-median problem. For this purpose we still use an arbitrary fixed undirected connected simple graph G (on n vertices) and let α and β denote its cut parameters. The results in this section will depend on α and β.

Greedy Heuristic for Minimum-Distance Perfect Matching. The minimum-distance perfect matching problem has been widely analyzed throughout history. We do for instance know that the worst-case running-time for finding a minimum distance perfect matching is $O(n^3)$, which is high when considering a large number of vertices. Because of this, simple heuristics are often used, with the greedy heuristic probably being the simplest of them: at each step, add a pair of unmatched vertices to the matching such that the distance between the added pair of vertices is minimized. From now on, let GR denote the cost of the matching computed by this heuristic and let MM denote the value of an optimal matching.

The worst-case approximation ratio of this heuristic on metric instances is known to be $O(n^{\log_2(3/2)})$ [13]. Furthermore, for random shortest path metrics on complete graphs (for which the cut parameters are given by $\alpha = \beta = 1$) the heuristic has an expected approximation ratio of $O(1)$ [3]. We extend this last result to general values for α and β and show that the greedy matching heuristic has an expected approximation ratio of $O(\beta/\alpha)$.

Theorem 11. $\mathbb{E}[\mathsf{GR}] = O(1/\alpha)$.

Lemma 12. [11, Theorem 5.1(iii)] *Let* $X \sim \sum_{i=1}^{n} X_i$ *with* $X_i \sim \mathrm{Exp}(a_i)$ *independent. Let* $\mu = \mathbb{E}[X] = \sum_{i=1}^{n}(1/a_i)$ *and* $a_* = \min_i a_i$. *For any* $\lambda \leq 1$,

$$\mathbb{P}(X \leq \lambda\mu) \leq \exp(-a_*\mu(\lambda - 1 - \ln(\lambda))).$$

Lemma 13. *Let* S_m *denote the sum of the* m *lightest edge weights in* G. *For all* $\phi \leq (n-1)/n$ *and* $c \in [0, 2\phi^2]$ *we have*

$$\mathbb{P}\left(S_{\phi n} \leq \frac{c}{\beta}\right) \leq \exp\left(\phi n\left(1 + \ln\left(\frac{c}{2\phi^2}\right)\right)\right).$$

Furthermore, $\mathsf{TSP} \geq \mathsf{MM} \geq S_{n/2}$, *where* TSP *and* MM *are the total distance of a shortest TSP tour and a minimum-distance perfect matching, respectively.*

Theorem 14. *The greedy heuristic for minimum-distance perfect matching has an expected approximation ratio on generalized random shortest path metrics given by* $\mathbb{E}\left[\frac{\mathsf{GR}}{\mathsf{MM}}\right] = O(\beta/\alpha)$.

Nearest Neighbor Heuristic for TSP. The nearest-neighbor heuristic is a greedy approach for the TSP: start with some starting vertex v_0 as current vertex v; at every step, choose the nearest unvisited neighbor u of v as the next vertex in the tour and move to the next iteration with the new vertex u as current vertex v; go back to v_0 if all vertices are visited. From now on, let NN denote the cost of the TSP tour computed by this heuristic and let TSP denote the value of an optimal TSP tour.

The worst-case approximation ratio of this heuristic on metric instances is known to be $O(\ln(n))$ [14]. Furthermore, for random shortest path metrics on complete graphs (for which the cut parameters are given by $\alpha = \beta = 1$) the heuristic has an expected approximation ratio of $O(1)$ [3]. We extend this last result to general values for α and β and show that the nearest-neighbor heuristic has an expected approximation ratio of $O(\beta/\alpha)$.

Theorem 15. *For generalized random shortest path metrics, we have* $\mathbb{E}[\mathsf{NN}] = O(1/\alpha)$ *and* $\mathbb{E}\left[\frac{\mathsf{NN}}{\mathsf{TSP}}\right] = O(\beta/\alpha)$.

Insertion Heuristics for TSP. The insertion heuristics are another greedy approach for the TSP: start with an initial optimal tour on a few vertices chosen according to some predefined rule R; at every step, choose a vertex according to the same predefined rule R and insert this vertex in the current tour such that the total distance increases the least. From now on, let IN_R denote the cost of the TSP tour computed by this heuristic (with rule R) and let TSP still denote the value of an optimal TSP tour.

The worst-case approximation ratio of this heuristic for any rule R on metric instances is known to be $O(\ln(n))$ [14]. Furthermore, for random shortest path metrics on complete graphs (for which the cut parameters are given by $\alpha = \beta = 1$) the heuristic has an expected approximation ratio of $O(1)$ [3]. We extend this last result to general values for α and β and show that the insertion heuristic for any rule R has an expected approximation ratio of $O(\beta/\alpha)$.

Theorem 16. *For generalized random shortest path metrics, we have* $\mathbb{E}[\mathsf{IN}_R] = O(1/\alpha)$ *and* $\mathbb{E}\left[\frac{\mathsf{IN}_R}{\mathsf{TSP}}\right] = O(\beta/\alpha)$.

Running Time of 2-opt Heuristic for TSP. The 2-opt heuristic is an often used local search algorithm for the TSP: start with an initial tour on all vertices and improve the tour by 2-exchanges until no improvement can be made anymore. In a 2-exchange, the heuristic takes 'edges' $\{v_1, v_2\}$ and $\{v_3, v_4\}$, where v_1, v_2, v_3, v_4 are visited in this order in the tour, and replaces them by $\{v_1, v_3\}$ and $\{v_2, v_4\}$ to create a shorter tour.

We provide an upper bound for the expected number of iterations that 2-opt needs. In the worst-case scenario, this number is exponential. However, for random shortest path metrics on complete graphs (for which the cut parameters are given by $\alpha = \beta = 1$) an upper bound of $O(n^8 \ln^3(n))$ is known for the expected number of iterations [3]. We extend this result with a similar proof to general values for α and β and show an upper bound for the expected number of iterations of $O(n^8 \ln^3(n)\beta/\alpha)$.

We first define the improvement obtained from a 2-exchange. If $\{v_1, v_2\}$ and $\{v_3, v_4\}$ are replaced by $\{v_1, v_3\}$ and $\{v_2, v_4\}$, then the improvement made by the exchange equals the change in distance $\zeta = d(v_1, v_2) + d(v_3, v_4) - d(v_1, v_3) - d(v_2, v_4)$. These four distances correspond to four shortest paths (P_{12}, P_{34}, P_{13}, P_{24}) in the graph $G = (V, E)$. This implies that we can rewrite ζ as the sum of the weights on these paths. We obtain $\zeta = \sum_{e \in E} \gamma_e w(e)$, for some $\gamma_e \in \{-2, -1, 0, 1, 2\}$.

Since we are looking at the improvement obtained by a 2-exchange, we have $\zeta > 0$. This implies that there exists some $e = \{u, u'\} \in E$ such that $\gamma_e \neq 0$. Given this edge e, let $I \subseteq \{P_{12}, P_{34}, P_{13}, P_{24}\}$ be the set of all shortest paths of the 2-exchange that contain e. Then, for all combinations e and I, let $\zeta_{ij}^{e,I}$ be defined as follows:

- If $P_{ij} \notin I$, then $\zeta_{ij}^{e,I}$ is the length of the shortest path from v_i to v_j without using e.
- If $P_{ij} \in I$, then $\zeta_{ij}^{e,I}$ is the minimum of

- the length of a shortest path from v_i to u without using e plus the length of a shortest path from u' to v_j without using e and
- the length of a shortest path from v_i to u' without using e plus the length of a shortest path from u to v_j without using e.

Define $\zeta^{e,I} = \zeta_{12}^{e,I} + \zeta_{34}^{e,I} - \zeta_{13}^{e,I} - \zeta_{24}^{e,I}$.

Lemma 17. *For every outcome of the edge weights, there exists an edge e and a set I such that $\zeta = \zeta^{e,I} + \gamma w(e)$, where $\gamma \in \{-2, -1, 1, 2\}$ is determined by e and I.*

Lemma 18. *Let e and I be given with $\gamma = \gamma_e \neq 0$. Then $\mathbb{P}(\zeta^{e,I} + \gamma w(e) \in (0, x]) \leq x$. Moreover, $\mathbb{P}(\zeta \in (0, x]) = O(\beta n^2 x)$.*

Theorem 19. *The expected number of iterations of the 2-opt heuristic until a local optimum is found is bounded by $O(n^8 \ln^3(n)\beta/\alpha)$.*

Trivial Heuristic for k-Median. The goal of the (metric) k-median problem is to find a set $U \subseteq V$ of size k such that $\sum_{v \in V} \min_{u \in U} d(v, u)$ is minimized. The best known approximation algorithm for this problem achieves an approximation ratio of $2.675 + \varepsilon$ [4].

Here, we consider the k-median problem in the setting of generalized random shortest path metrics. We analyze a trivial heuristic for the k-median problem: simply pick k vertices independently of the metric space, e.g., $U = \{v_1, \ldots, v_k\}$. The worst-case approximation ratio of this heuristic is unbounded, even if we restrict ourselves to metric instances. However, for random shortest path metrics on complete graphs (for which the cut parameters are given by $\alpha = \beta = 1$) the expected approximation ratio has an upper bound of $O(1)$ and even $1 + o(1)$ for k sufficiently small [3]. We extend this result to general values for α and β and give an upper bound for the expected approximation ratio of $O(\beta/\alpha)$ for 'large' k and $\beta/\alpha + o(\beta/\alpha)$ for k sufficiently small.

For our analysis, let $U = \{v_1, \ldots, v_k\}$ be an arbitrary set of k vertices. Sort the remaining vertices $\{v_{k+1}, \ldots, v_n\}$ in increasing distance from U. For $k + 1 \leq i \leq n$, let $\rho_i = d(v_i, U)$ equal the distance from U to the $(i - k)$-th closest vertex to U. Let TR denote the cost of the solution generated by the trivial heuristic and let ME be the cost of an optimal solution to the k-median problem.

Observe that the random variables ρ_i are generated by a simple growth process analogously to the one described in Sect. 3 for $\tau_k(v)$. Using this observation, we can see that

$$\sum_{j=k}^{i-1} \mathrm{Exp}(\beta j(n - j)) \precsim \rho_i \precsim \sum_{j=k}^{i-1} \mathrm{Exp}(\alpha j(n - j)),$$

which in turn implies that $\mathsf{cost}(U) = \sum_{i=k+1}^{n} \rho_i$ is stochastically bounded by

$$\sum_{i=k}^{n-1} \mathrm{Exp}(\beta i) \precsim \mathsf{cost}(U) \precsim \sum_{i=k}^{n-1} \mathrm{Exp}(\alpha i).$$

From this, we can immediately derive bounds for the expected value of the k-median returned by the trivial heuristic.

Lemma 20. *Fix $U \subseteq V$ of size k. Then, we have $\mathbb{E}[\mathsf{TR}] = \mathbb{E}[\mathsf{cost}(U)]$ and*

$$\frac{1}{\beta} \left(\ln \left(\frac{n-1}{k-1} \right) - 1 \right) \leq \mathbb{E}[\mathsf{TR}] \leq \frac{1}{\alpha} \left(\ln \left(\frac{n-1}{k-1} \right) + 1 \right).$$

Before we provide our result for the expected approximation ratio of the trivial heuristic, we first provide some tail bounds for the distribution of the optimal k-median ME and the trivial solution TR.

Lemma 21. *Fix $U \subseteq V$ of size k. Then the probability density function f of $\sum_{i=k}^{n-1} \mathrm{Exp}(\beta i)$ is given by*

$$f(x) = \beta k \cdot \binom{n-1}{k} \cdot \exp(-\beta k x) \cdot (1 - \exp(-\beta x))^{n-k-1}.$$

Lemma 22. *Let $c > 0$ be sufficiently large and let $k \leq c'n$ for $c' = c'(c) > 0$ sufficiently small. Then we have*

$$\mathbb{P}\left(\mathsf{ME} \leq \left(\ln \left(\tfrac{n-1}{k} \right) - \ln \ln \left(\tfrac{n}{k} \right) - \ln(c) \right) / \beta \right) = n^{-\Omega(c)}.$$

Lemma 23. *Let $k \leq (1 - \varepsilon)n$ for some constant $\varepsilon > 0$. For every $c \in [0, 2\varepsilon^2)$, we have*

$$\mathbb{P}\left(\mathsf{ME} \leq c/\beta \right) \leq c^{\Omega(n)}.$$

Lemma 24. *For any $c \geq 4$ we have $\mathbb{P}\left(\mathsf{TR} > n^c \right) \leq \exp(-n^{c/4})$.*

Now we have obtained everything needed to provide an upper bound for the expected approximation ratio of the trivial heuristic.

Theorem 25. *Let $k \leq (1-\varepsilon)n$ for some constant $\varepsilon > 0$. For generalized random shortest path metrics, we have $\mathbb{E}\left[\frac{\mathsf{TR}}{\mathsf{ME}} \right] = O\left(\beta/\alpha \right)$. Moreover, if we have $k \leq c'n$ for some fixed $c' \in (0, 1)$ sufficiently small, then we have*

$$\mathbb{E}\left[\tfrac{\mathsf{TR}}{\mathsf{ME}} \right] = (\beta/\alpha) \cdot \left(1 + O\left(\tfrac{\ln \ln(n/k)}{\ln(n/k)} \right) \right).$$

5   Application to the Erdős-Rényi Random Graph Model

So far, we have analyzed random shortest path metrics applied to graphs based on their cut parameters (Definition 1). In this section, we first use a well-known result to show that instances of the Erdős–Rényi random graph model have 'nice' cut parameters with high probability. We then use this to prove our main results.

Lemma 26. *Let $G = (V, E)$ be an instance of the $G(n, p)$ model. For constant $\varepsilon \in (0, 1)$ and for any $p \geq c \ln(n)/n$ (as $n \to \infty$), in which $c > 9/\varepsilon^2$ is constant, the cut parameters of G are bounded by $(1 - \varepsilon)p \leq \alpha \leq \beta \leq (1 + \varepsilon)p$ with probability at least $1 - o\left(1/n^2 \right)$.*

Recall that from the result of Corollary 4 we could derive (approximate) bounds for the expected distance $\mathbb{E}[d(u,v)]$ between two arbitrary vertices in a random shortest path metric. Combining this with the result of the foregoing lemma, we can see that, for the case of the application to the Erdős–Rényi random graph model, w.h.p. over the random graph $\mathbb{E}[d(u,v)]$ is approximately bounded between $\ln(n)/((1+\varepsilon)np)$ and $\ln(n)/((1-\varepsilon)np)$ for any constant $\varepsilon \in (0,1)$. This is in line with the known result $\mathbb{E}[d(u,v)] \approx \ln(n)/np$ for p sufficiently large [2].

5.1 Performance of Heuristics

In this section, we provide the main results of this work. We use the results from Sect. 4 and Lemma 26 to analyze the performance of several heuristics in random shortest path metrics applied to Erdős–Rényi random graphs.

When a graph $G = (V, E)$ is created by the $G(n, p)$ model, there is a non-zero probability of G being disconnected. In a corresponding random shortest path metric this results in $d(u, v) = \infty$ for any two vertices $u, v \in V$ that are in different components of G. Observe that, if this is the case, then the identity of indiscernibles, symmetry and triangle inequality still hold. Thus we still have a metric and we can bound the expected approximation ratio for such graphs from above by the worst-case approximation ratio for metric instances.

Using this observation, we can prove the following results.

Theorem 27. *Let $\varepsilon \in (0,1)$ be constant. Let $G = (V, E)$ be a random instance of the $G(n, p)$ model, for p sufficiently large ($p \geq c\ln(n)/n$ as $n \to \infty$ for a constant $c > 9/\varepsilon^2$ satisfies), and consider the corresponding random shortest path metric. Then, we have*

$$\mathbb{E}\left[\frac{\text{GR}}{\text{MM}}\right] = O(1).$$

Theorem 28. *Let $\varepsilon \in (0,1)$ be constant. Let $G = (V, E)$ be a random instance of the $G(n, p)$ model, for p sufficiently large ($p \geq c\ln(n)/n$ as $n \to \infty$ for a constant $c > 9/\varepsilon^2$ satisfies), and consider the corresponding random shortest path metric. Then, we have*

$$\mathbb{E}\left[\frac{\text{NN}}{\text{TSP}}\right] = O(1) \quad and \quad \mathbb{E}\left[\frac{\text{IN}_R}{\text{TSP}}\right] = O(1).$$

For the last two results, we need the assumption that G is connected.

Theorem 29. *Let $\varepsilon \in (0,1)$ be constant. Let $G = (V, E)$ be a random instance of the $G(n, p)$ model, for p sufficiently large ($p \geq c\ln(n)/n$ as $n \to \infty$ for a constant $c > 9/\varepsilon^2$ satisfies), and consider the corresponding random shortest path metric. If G is connected, then the expected number of iterations of the 2-opt heuristic for TSP is bounded by $O(n^8 \ln^3(n))$.*

Theorem 30. *Let $\tilde{\varepsilon} \in (0,1)$ be constant. Let $G = (V, E)$ be a random instance of the $G(n,p)$ model, for p sufficiently large ($p \geq c\ln(n)/n$ as $n \to \infty$ for a constant $c > 9/\tilde{\varepsilon}^2$ satisfies), and consider the corresponding random shortest path metric. Let $\mathcal{E}'$ denotes the event that G is connected. Let $k \leq (1 - \varepsilon')n$ for some constant $\varepsilon' > 0$, then we have $\mathbb{E}\left[\frac{\text{TR}}{\text{ME}} \mid \mathcal{E}'\right] = O(1)$. Moreover, if we have $k \leq c'n$ for $c' \in (0,1)$ sufficiently small, then $\mathbb{E}\left[\frac{\text{TR}}{\text{ME}} \mid \mathcal{E}'\right] = 1 + \varepsilon + o(1)$.*

6   Concluding Remarks

We have analyzed heuristics for matching, TSP, and k-median on random shortest path metrics on Erdős–Rényi random graphs. However, in particular for constant values of p, these graphs are still dense. Although our results hold for decreasing $p = \Omega(\ln n/n)$, we obtain in this way metrics with unbounded doubling dimension. In order to get an even more realistic model for random metric spaces, it would be desirable to analyze heuristics on random shortest path metrics on sparse graphs. Hence, we raise the question to generalize our findings to sparse random graphs or sparse (deterministic) classes of graphs.

References

1. Ahn, S., Cooper, C., Cornuéjols, G., Frieze, A.: Probabilistic analysis of a relaxation for the k-median problem. Math. Oper. Res. **13**(1), 1–31 (1988). https://doi.org/10.1287/moor.13.1.1
2. Bhamidi, S., van der Hofstad, R., Hooghiemstra, G.: First passage percolation on the Erdős-Rényi random graph. Comb. Probab. Comput. **20**(5), 683–707 (2011). https://doi.org/10.1017/S096354831100023X
3. Bringmann, K., Engels, C., Manthey, B., Rao, B.V.R.: Random shortest paths: non-Euclidean instances for metric optimization problems. Algorithmica **73**(1), 42–62 (2015). https://doi.org/10.1007/s00453-014-9901-9
4. Byrka, J., Pensyl, T., Rybicki, B., Srinivasan, A., Trinh, K.: An improved approximation for k-median, and positive correlation in budgeted optimization. In: Indyk, P. (ed.) Proceedings of the Twenty-Sixth Annual ACM-SIAM Symposium on Discrete Algorithms (SODA 2015), pp. 737–756 (2015). https://doi.org/10.1137/1.9781611973730.50
5. Davis, R., Prieditis, A.: The expected length of a shortest path. Inf. Process. Lett. **46**(3), 135–141 (1993). https://doi.org/10.1016/0020-0190(93)90059-I
6. Frieze, A.M., Yukich, J.E.: Probabilistic analysis of the TSP (Chap. 7). In: Gutin, G., Punnen, A.P. (eds.) The Traveling Salesman Problem and Its Variations, pp. 257–307. Springer, Boston (2007). https://doi.org/10.1007/0-306-48213-4_7
7. Hammersley, J.M., Welsh, D.J.A.: First-passage percolation, subadditive processes, stochastic networks, and generalized renewal theory. In: Neyman, J., Le Cam, L.M. (eds.) Bernoulli 1713 Bayes 1763 Laplace 1813, pp. 61–110. Springer, Heidelberg (1965). https://doi.org/10.1007/978-3-642-49750-6_7
8. Hassin, R., Zemel, E.: On shortest paths in graphs with random weights. Math. Oper. Res. **10**(4), 557–564 (1985). https://doi.org/10.1287/moor.10.4.557
9. Howard, C.D.: Models of first-passage percolation. In: Kesten, H. (ed.) Probability on Discrete Structures, pp. 125–173. Springer, Heidelberg (2004). https://doi.org/10.1007/978-3-662-09444-0_3

10. Janson, S.: One, two and three times log n/n for paths in a complete graph with random weights. Comb. Probab. Comput. **8**(4), 347–361 (1999). https://doi.org/10.1017/S0963548399003892
11. Janson, S.: Tail bounds for sums of geometric and exponential variables. Stat. Probab. Lett. **135**, 1–6 (2018). https://doi.org/10.1016/j.spl.2017.11.017
12. Karp, R.M., Steele, J.M.: Probabilistic analysis of heuristics. In: Lawler, E.L., Lenstra, J.K., Rinnooy Kan, A.H.G., Shmoys, D.B. (eds.) The Traveling Salesman Problem: A Guided Tour of Combinatorial Optimization, pp. 181–205. Wiley, Hoboken (1985)
13. Reingold, E.M., Tarjan, R.E.: On a greedy heuristic for complete matching. SIAM J. Comput. **10**(4), 676–681 (1981). https://doi.org/10.1137/0210050
14. Rosenkrantz, D.J., Stearns, R.E., Lewis II, P.M.: An analysis of several heuristics for the traveling salesman problem. SIAM J. Comput. **6**(3), 563–581 (1977). https://doi.org/10.1137/0206041
15. Ross, S.M.: Introduction to Probability Models, 10th edn. Academic Press, Burlington (2010)

Optimal Partition of a Tree with Social Distance

Masahiro Okubo[1]([⊠]), Tesshu Hanaka[2]([⊠]), and Hirotaka Ono[1]([⊠])

[1] Graduate School of Informatics, Nagoya University,
Furo-cho, Chikusa-ku, Nagoya, Japan
okubo.masahiro@h.mbox.nagoya-u.ac.jp, ono@i.nagoya-u.ac.jp
[2] Department of Information and System Engineering, Chuo University,
1-13-27 Kasuga, Bunkyo-ku, Tokyo, Japan
hanaka.91t@g.chuo-u.ac.jp

Abstract. We study the problem to find a partition of a graph G with maximum social welfare based on social distance between vertices in G, called MaxSWP. This problem is known to be NP-hard in general. In this paper, we first give a complete characterization of optimal partitions of trees with small diameters. Then, by utilizing these results, we show that MaxSWP can be solved in linear time for trees. Moreover, we show that MaxSWP is NP-hard even for 4-regular graphs.

Keywords: Graph algorithm · Tree · Graph partition · Social distance

1 Introduction

With the development of Social Networking Services (SNS) such as Twitter, Facebook, Instagram and so on, it has become much easier than before to obtain graphs that represent human relationship, and there are many attempts to utilize such graphs for extracting useful information. Among them, grouping people according to the graph structures is focused and investigated from many standpoints. For example, if a community consisting of members with a common interest is found, advertising or promoting some products might be very effective for members of the community due to the strong interest.

Here, there are roughly two standpoints how we group communities. One is context based grouping, and the other is based on link structures. Previous work on community detection and grouping based on graph structure is summarized in [5,7,12], for example.

Basically, these studies formulate network structure identification (community detection, grouping, and partition) as an optimization problem (sometimes it is not explicitly conscious), and design a fast algorithm to (approximately) solve the optimization problem. Then, network structures to identify

This work is supported by JSPS KAKENHI Grant Numbers 17K19960, 17H01698, 18H06469.

G. K. Das et al. (Eds.): WALCOM 2019, LNCS 11355, pp. 121–132, 2019.
https://doi.org/10.1007/978-3-030-10564-8_10

are obtained as outputs of the proposed algorithm. Here, network structures to identify are already abstract, e.g., dense subgraphs; the proposed algorithm can be used not only for the original purpose but also other purposes. In fact, [13] is originally about boundary line detection in image data, but the proposed techniques are used for community detentions (e.g., [8]), and it is further used for the detection of industrial clusters in economic networks [6]. As above, the versatility of "optimization problems" is very useful. However, we may think that they do not best utilize the features or characteristics of the target network. For example, the criteria for group partition, image processing, and detecting industrial clusters in economic networks could be different. In other words, we might expect a better performance by considering an optimization problem specialized for community detection.

From these, we consider the problem for group partition (or simply say partition) in networks (graphs), taking into account the characteristics of SNS. In SNS, people communicate and exchange information with also a person who is not directly acquainted, i.e., followers. That is, in SNS, not only members with direct connections but also members without direct connections are loosely connected, which enables us to share information widely. Here, "looseness" is related to the degree of sharing information, and it is natural to define it as the distance (i.e., the length of a shortest path) between the persons on the network.

Based on such observation, Branzei et al. introduced a new grouping scale for human relations networks [4]. The definition of the utility in [4] is as follows: given a partition, the utility of an individual is defined as the sum of reciprocal distances to other people in the same coalition divided by the size of the coalition. Based on this, the social welfare of a partition is also defined as the sum of the utilities of all the members. Unfortunately, finding a partition with maximum social welfare (MaxSWP) is known to be NP-hard even on graphs with maximum degree 6 [3].

Also, the characterizations of optimal partitions are known only for trivial cases such as complete graphs and complete bipartite graphs [4]. Even for trees, it is not known whether MaxSWP can be solved in polynomial time. One of the reasons seems to be the objective function of MaxSWP. In a typical graph optimization problem, the objective function often forms a linear sum of weights, whereas the one of MaxSWP takes the form of a nonlinear function, which is the sum of the reciprocal distances.

1.1 Our Contribution

In this paper, we mainly study finding an optimal partition with social distance of a tree, which is one of the most basic and important structures in graph algorithm design. In the process of research, we first give a complete characterization of optimal partitions of paths. Although the argument is simple, it gives an insight about the hardness related to the nonlinearity of the utility and the social welfare. Next, we give a similar characterization of optimal partitions of trees. In the characterization, we find out sub-trees with small diameters appeared in optimal partitions of trees. By using the characterization, we design a linear-time algorithm for computing an optimal partition of a tree. Finally, we show that MaxSWP is NP-hard even for 4-regular graphs. This result strengthens the previous work for graphs with maximum degree 6 [3].

1.2 Related Work

Graph partition is one of the most basic and important problem in computer science and there are many studies about graph partition in various contexts, such as image processing and cluster analysis [5–7,12,13].

Graph partition with social distance has been studied in the context of coalition formation games [4]. In coalition formation games, each player has the utility based on the preference for other players in the same coalition. Intuitively, a player is happy if the utility is high, that is, there are many players he/she prefer in the same coalition. In the field of coalition formation games, many researchers study about desirable coalition formations, namely, partitions, in terms of maximum social welfare, stability, and core [1,11]. Furthermore, the price of anarchy (PoA) and the price of stability (PoS) are also well-studied for evaluating agents systems [2]. The PoA or PoS are more related to this paper because they are defined as the maximum and minimum ratio between a Nash stable solution and the best solution, respectively.

In coalition formation games on graphs, there are many utility functions for agents. For example, in [1,14], the utility of an agent is defined as the sum of edge-weights between him/her and other agents in the same coalition. The weight of an edge represents the strength of the relationship between agents. In social distance games, the utility is defined as the harmonic function of the distance between agents. This is based on the concept of the closeness centrality, which is one of classical measures for network analysis [3,4]. As mentioned above, finding the best partition, that is, a partition with maximum social welfare is NP-hard even on graphs with maximum degree 6 [3]. On the other hand, there is a 2-approximation algorithm for finding such a partition [4].

The organization of this paper is as follows. In Sect. 2, we give basic terminologies, notation, and definitions. In Sect. 3, we give a complete characterization of optimal partitions of paths. In Sect. 4, we propose a linear-time algorithm for MaxSWP on trees. Finally, we show that MaxSWP is NP-hard even on 4-regular graphs in Sect. 5. Due to the space limitation, we out the proofs of propositions lemmas and theorems. The detailed proofs can be found in [9].

2 Preliminaries

2.1 Terminologies

We use standard terminologies on graph theory. Let $G = (V(G), E(G))$ be a simple, connected, and undirected graph. For simplicity, we may denote $V(G)$ and $E(G)$ by V and E, respectively. We also denote the number of vertices and edges by n and m, respectively. A path from u to v of minimum length is called a *shortest path*, and the length is denoted by $dist_G(u,v)$. In a graph G, if there is no path from the vertex u to the vertex v, we define $dist_G(u,v) = \infty$. Let $G[C]$ be the subgraph induced by vertex set $C \subseteq V$. We sometimes denote $dist_{G[C]}(u,v)$ of $u,v \in C$ in $G[C]$ by $dist_C(u,v)$ for simplicity. For graph G, we denote the *diameter* of G by $diam(G) = \max_{u,v \in V, u \neq v} dist_G(u,v)$. For a vertex v, we denote the set of neighbors of v by $N(v) = \{u \mid (u,v) \in E\} \subseteq V$. We also define the degree of v as $d(v) = |N(v)|$. For $G = (V,E)$, we denote the maximum degree of G by $\Delta(G) = \max_{v \in V} d(v)$. For simplicity, we sometimes denote it by Δ. For a positive integer n, we define $[n] = \{1,\ldots,n\}$.

A graph $G = (V,E)$ is called a *path graph* denoted by P_n if $E = \{(v_i, v_{i+1}) \mid 1 \leq i < n\}$. We also sometimes simply call it a path. Moreover, if $E = \{(v_1, v_i) \mid 2 \leq i \leq n\}$, G is said to be a *star* and denoted by $K_{1,n-1}$. A graph G is a *tree* if G is connected and it has no cycle. We denote an n-vertex tree by T_n. Moreover, we denote a tree with the diameter d by T_n^d.

2.2   Coalition and Utility

The definitions here are based on [4]. Given a graph $G = (V,E)$ and $C \subseteq V$, we define the utility $U(v,C)$ of a vertex $v \in C$ as follows:

$$U(v,C) = \frac{1}{|C|} \sum_{u \in C \setminus \{v\}} \frac{1}{dist_{G[C]}(v,u)}.$$

By the definition, it satisfies that $0 \leq U(v,C) \leq 1$. In a graph $G = (V,E)$, a *partition* of G is defined as the family of sets of vertices $\mathcal{C} = \{C_1,\ldots,C_k\}$, where $C_1 \cup \cdots \cup C_k = V$ and $C_i \cap C_j = \emptyset$ for $i \neq j \in [k]$. Moreover, $C \in \mathcal{C}$ is called a *coalition* of partition $\mathcal{C}$. In particular, if $\mathcal{C} = \{V\}$, $\mathcal{C}$ is called the *grand* of G and V is called the *grand coalition*. If $\{v\} \in \mathcal{C}$ for a vertex $v \in V$, v is said to be an *isolated vertex* of partition $\mathcal{C}$. We define the utility of an isolated vertex as $U(v, \{v\}) = 0$. Next, we define the *social welfare* of a partition $\mathcal{C}$ in graph G as follows. We define the *social welfare* $\varphi(G, \mathcal{C})$ of partition $\mathcal{C}$ in $G = (V,E)$ as follows:

$$\varphi(G,\mathcal{C}) = \sum_{C \in \mathcal{C}} \sum_{v \in C} U(v,C).$$

If $\mathcal{C}$ is the grand of G, that is, $\mathcal{C} = \{V\}$, we simply denote $\varphi(G, \{V\})$ by $\varphi(G)$. We can observe that $\varphi(G,\mathcal{C})$ is bounded by $n-1$. Moreover, we define the *average social welfare* $\tilde{\varphi}(G,\mathcal{C})$ for partition $\mathcal{C}$ in G. The *average social welfare* of partition $\mathcal{C}$ in $G = (V,E)$ is defined as follows:

$$\tilde{\varphi}(G,\mathcal{C}) = \frac{\varphi(G,\mathcal{C})}{|V|}.$$

If $\mathcal{C}$ is the grand of G, that is, $\mathcal{C} = \{V\}$, we simply denote $\tilde{\varphi}(G, \{V\})$ by $\tilde{\varphi}(G)$. Finally, we define a partition $\mathcal{C}^*$ with *maximum* social welfare in graph G. A partition $\mathcal{C}^*$ is *maximum* if it satisfies that $\varphi(G, \mathcal{C}^*) \geq \varphi(G, \mathcal{C})$ for any partition $\mathcal{C}$ in G. We call the problem of finding a partition with maximum social welfare

MaxSWP. We also call an optimal solution of MaxSWP an *optimal partition*. In previous work, it is shown that MaxSWP is NP-hard even for graphs with maximum degree 6 [3]. On the other hand, it is known that the grand is the only optimal partition of MaxSWP on complete graphs and complete bipartite graphs [4].

Proposition 1 ([4].) *On complete graphs and complete bipartite graphs, the grand is the only optimal partition of* MaxSWP.

Branzei et al. showed that there exists a partition where the utility of each vertex v attains at least $1/2$ and a polynomial-time algorithm that finds such a partition for any graph [4].

Proposition 2 ([4].) *There is a polynomial-time algorithm that finds a partition such that each agent utility is at least $1/2$ for any graph.*

From Proposition 2, it can be easily seen that there exists a partition $\mathcal{C}$ that satisfies $\varphi(G, \mathcal{C}) \geq n/2$. Thus, the social welfare of an optimal partition for any graph is also at least $n/2$.

Corollary 1. *Any optimal partition $\mathcal{C}^*$ of graph G satisfies $\varphi(G, \mathcal{C}^*) \geq n/2$.*

Since for any G and $\mathcal{C}$, $\varphi(G, \mathcal{C})$ is bounded by $n - 1$, the algorithm proposed by Branzei et al. [4] is a 2-approximation algorithm. In the end of this section, we give another property of an optimal partition of MaxSWP.

Proposition 3. *For each coalition $C \in \mathcal{C}^*$ of optimal partition $\mathcal{C}^*$, $G[C]$ is connected.*

3   Optimal Partition of a Path

In this section, we characterize the optimal partition of a path P_n. In a path, the subgraph induced by a coalition is also a path by Proposition 3. By using this property and examining the average social welfare of P_n, we can identify the graph structures of coalitions in the optimal partitions of P_n. In the following, we first examine average social welfare of P_n. Then we give the optimal partition of P_n.

Let $h(k) = \sum_{i=1}^{k} 1/i$ be the harmonic function for some positive integer k. The social welfare and the average social welfare of P_n can be denoted by $\varphi(P_n) = (2 \sum_{k=1}^{n-1} h(k))/n$ and $\tilde{\varphi}(P_n) = (2 \sum_{k=1}^{n-1} h(k))/n^2$, respectively. Then, we obtain the following lemmas.

Lemma 1. *It holds that $\tilde{\varphi}(P_2) < \tilde{\varphi}(P_3)$, and $\tilde{\varphi}(P_n) > \tilde{\varphi}(P_{n+1})$ for $n \geq 3$.*

Lemma 2. *For an optimal partition $\mathcal{C}^*$ of a path P_n and a coalition $C \in \mathcal{C}^*$, $G[C]$ is either P_2, P_3 or P_4.*

Finally, we give the optimal partition of P_n.

Theorem 1. *The optimal partition of path* P_n *is*

1. $C^* = \{\{v_{3i-2}, v_{3i-1}, v_{3i}\} \mid 1 \leq i \leq n/3\}$ *if* $n \equiv 0$ (mod 3),
2. $C^* = \{\{v_1, v_2, v_3, v_4\}, \{v_{3i+2}, v_{3i+3}, v_{3i+4}\} \mid 1 \leq i \leq (n-4)/3\}$ *if* $n \equiv 1$ (mod 3), *and*
3. $C^* = \{\{v_1, v_2\}, \{v_{3i}, v_{3i+1}, v_{3i+2}\} \mid 1 \leq i \leq (n-2)/3\}$ *or* $\{\{v_1, v_2, v_3, v_4\}, \{v_5, v_6, v_7, v_8\}, \{v_{3i+6}, v_{3i+7}, v_{3i+8}\} \mid 1 \leq i \leq (n-8)/3\}$ *if* $n \equiv 2$ (mod 3).

4   Optimal Partition of a Tree

In Sect. 3, we identified the optimal partition of MaxSWP on a path. In this section, we consider MaxSWP on trees. Since a tree is more general and complicated than a path and the optimal structure of MaxSWP is quite different from typical graph optimization problems, MaxSWP is non-trivial even on trees.

To solve MaxSWP, we design an algorithm based on dynamic programming. However, we do not know which information we keep track of in dynamic programming since the optimal structure of MaxSWP is unknown. For this, we identify the small coalitions in the optimal partition. According to Corollary 1, if there is a coalition whose social welfare is less than $n/2$, it is not included in the optimal partition since the social welfare can be increased by dividing the coalition. Thus, we only keep track of the subgraph structures of coalitions of social welfare at least $n/2$. We can identify such coalitions by calculating the social welfare of each coalition. By using the subgraph structures of such coalitions, we design a linear-time algorithm for MaxSWP on a tree based on dynamic programming.

4.1   Social Welfare of Trees with Small Diameters

Since the utility of an agent is defined as the harmonic function with respect to the distance to all others in the same coalition, the diameter of the subgraph induced by each coalition affects the social welfare. Intuitively, the social welfare of a subgraph with large diameter is very low in a tree because a tree is quite sparse. Therefore, we characterize the subgraph structures of coalitions in the optimal partition in terms of small diameters.

We first consider trees $T_n^{\leq 2}$ with diameter at most 2. Such graphs are stars denoted by $K_{1,n-1}$. Since a star is a complete bipartite graph, it satisfies that $C^* = \{V\}$ by Property 1. Here, we investigate the number of vertices of a star that maximizes the average social welfare. This gives the upper bound of the social welfare of $T_n^{\leq 2} = K_{1,n-1}$.

Lemma 3. *For tree* $T_n^{\leq 2}$ *with diameter at most 2,* $n/2 \leq \varphi(T_n^{\leq 2}) \leq 9n/16$ *holds.*

Next, we consider trees with diameter 3. Any tree with diameter 3 can be represented as $T_n^3 = (V(T_n^3), E(T_n^3))$ where $V(T_n^3) = \{u_1, u_2, s_1, \ldots, s_k, t_1, \ldots, t_\ell\}$ and $E(T_n^3) = (u_1, u_2) \cup \{(u_1, s_i) \mid 1 \leq i \leq k\} \cup \{(u_2, t_j) \mid 1 \leq j \leq \ell\}$ for any $k, \ell \geq 1$ (see Fig. 1). Note that $n = |V(T_n^3)| = k + \ell + 2$. Then, the following lemma holds.

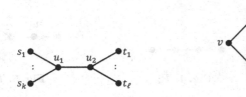

Fig. 1. Diameter 3 tree T_n^3 **Fig. 2.** Diameter 4 tree T_n^4

Lemma 4. *For tree T_n^3 with diameter 3, if $k = 2$ and $\ell \geq 7$, $k \geq 7$ and $\ell = 2$, $k > 3$ and $\ell \geq 3$, or $k \geq 3$ and $\ell > 3$, $\varphi(T_n^3) < n/2$ holds, and otherwise $\varphi(T_n^3) \geq n/2$.*

As with trees with diameter 3, we identify the types of trees with diameter 4 that satisfy $\varphi(T_n^4) \geq n/2$. Any tree with diameter 4 can be represented as $T_n^4 = (V(T_n^4), E(T_n^4))$ where $V(T_n^4) = \{v, u_1, \ldots, u_k, w_{1,1}, \ldots, w_{1,\ell_1}, \ldots, w_{k,1}, \ldots, w_{k,\ell_k}\}$ for $k \geq 2$ and $\ell_1, \ell_2 \geq 1$, and $E(T_n^4) = \{(v, u_i) \mid 1 \leq i \leq k\} \cup \{(u_i, w_{i,j}) \mid 1 \leq i \leq k, 1 \leq j \leq \ell_k\}$ (see Fig. 2). For each i, ℓ_i represents the number of leaves of u_i. We denote the total number of leaves of T_n^4 by $\alpha_k = \sum_{i=1}^{k} \ell_i$ and then the number of vertices of T_n^4 is represented as $n = |V(T_n^4)| = k + \alpha_k + 1$. Then, we obtain the following lemma.

Lemma 5. *For tree T_n^4 with diameter 4, $\varphi(T_n^4) \geq n/2$ holds if $(k, \alpha_k) = (2,2), (2,3), (3,2), (4,2)$, and otherwise $\varphi(T_n^4) < n/2$ holds.*

Finally, we show that the social welfare of the grand coalition of a tree with diameter at least 5 is less than $n/2$.

Lemma 6. *For tree T_n^μ with diameter $\mu \geq 5$, $\varphi(T_n^\mu) < n/2$ holds.*

By the above discussion, the optimal partition of a tree does not contain not only coalitions with large diameters but also particular coalitions with small diameters. In the following, we further refine the candidates for coalitions in the optimal partition of a tree.

First, we show that any optimal partition does not contain a coalition that consists of a tree with diameter 4. Thus, there is no coalition that consists of a tree with diameter at least 4 in the optimal solution by Lemma 6.

Lemma 7. *Any optimal partition of a tree T does not contain a coalition that consists of a tree $T^{\geq 4}$ with diameter at least 4.*

Moreover, we prove that the candidates for coalitions in the optimal partition are only three types.

Lemma 8. *Let C^* be the optimal partition of tree T_n. Then, the subgraph $G[C]$ induced by $C \in C^*$ is one of the following: (1) a star $K_{1,|C|-1}$ (see Fig. 3), (2) a path P_4 of length 3 (see Fig. 4), and (3) a tree T_5^3 of size five with diameter 3 (see Fig. 5).*

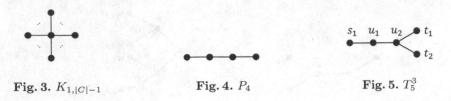

Fig. 3. $K_{1,|C|-1}$ Fig. 4. P_4 Fig. 5. T_5^3

4.2   Algorithm

In this section, we propose an algorithm that finds an optimal partition of a tree in linear time. This algorithm is based on dynamic programing with keeping track of the candidates identified by Lemma 8.

First, we introduce some notations to design our algorithm. Given a tree, we root it at arbitrary vertex r. We denote a subtree whose root is $v \in V$ by T_v and its partition by $\mathcal{C}_v$. For subtree T_v, we also denote the coalition including v by $C_v \in \mathcal{C}_v$.

By Lemma 8, the subgraph $G[C]$ induced by coalition $C \in \mathcal{C}^*$ is $K_{1,|C|-1}$, P_4 or T_5^3. The algorithm recursively computes a partition of T_v which attains the maximum social welfare for each v from the leaves of T.

Intuitively, our algorithm constructs coalitions in each step of dynamic programming. For example, a vertex u is added a coalition as an isolated vertex in T_u. In next step, vertex v must be added to the same coalition in T_v since the optimal solution does not contain an isolated vertex. Here, we keep track of not only coalition C_v, but also the position of v in the coalition since we compute a coalition with maximum social welfare by combining sub-coalitions in subtrees of T_v. For example, if v is positioned at a leaf of $K_{1,f}$, it is combined with a coalition that consists of $K_{1,f-1}$. On the other hand, if v is positioned at the center vertex of $K_{1,f}$, it is combined with f coalitions that consist of isolated vertices. Then we compute each coalition and sub-coalition with maximum social welfare including a new vertex in next step again. Since the optimal partition contains only coalitions of $K_{1,|C|-1}$, P_4 and T_5^3, we keep track of coalition C_v with the position of v that consists of them.

Let H be a subgraph induced by a coalition with the position of v. Then we consider the following types of H:

1. H is an isolated vertex of v, denoted by $H = (\{v\}, \emptyset)$,
2. H is a star $K_{1,f}$ and
   (a) v is the center of $K_{1,f}$, denoted by $H = K_{1,f}^{mid}$,
   (b) v is a leaf of $K_{1,f}$, denoted by $H = K_{1,f}^{leaf}$,
3. H is P_4 and
   (a) v is a leaf of P_4, denoted $H = P_4^{leaf}$,
   (b) root v is not a leaf of P_4, denoted by $H = P_4^{mid}$,

4. H is T_5^3 and
   (a) v is s_1, denoted by $H = T_5^3(s_1)$,
   (b) v is t_1 or t_2, denoted by $H = T_5^3(t)$,
   (c) v is u_1, denoted by $H = T_5^3(u_1)$,
   (d) v is u_2, denoted by $H = T_5^3(u_2)$.

We can observe that connected proper subgraphs of T_5^3 are subgraphs of P_4 and star $K_{1,3}$. Also connected proper subgraphs of P_4 are subgraphs of star $K_{1,2}$. Thus, by only keeping track of stars, we can treat P_4 and so T_5^3 as seen later.

Let $\mathcal{G}_v$ be the set of above subgraphs with the position of v in T_v. For T_v, we define the recursive formula $\rho(v, H) = \max_{C_v \ni C_v : G[C_v]=H} \varphi(T_v, C_v)$ as the maximum social welfare of the partition of T_v such that the subgraph induced by coalition C_v including v is $H \in \mathcal{G}_v$. We also define $\rho(v) = \max_{H \in \mathcal{G}_v} \rho(v, H)$ as the social welfare of the optimal partition of T_v. Then, the social welfare of the optimal partition of T with root r is denoted by $\rho(r) = \max_{H \in \mathcal{G}_r} \rho(r, H)$. Let w_j be the children of v where $1 \le j \le d(v) - 1$ in T_v. Then, we define the recursive formulas of $\rho(v, H)$ for $H \in \mathcal{G}_v$ to compute $\rho(r)$ as follows.

1. **H is an isolated vertex** $(\{v\}, \emptyset)$
   If $H = (\{v\}, \emptyset)$, $\varphi(H) = 0$. Since v separates trees T_{w_j}, $\rho(v, (\{v\}, \emptyset))$ is the sum of the social welfare of the optimal partition in T_{w_j}. Thus, $\rho(v, (\{v\}, \emptyset))$ is defined as $\rho(v, (\{v\}, \emptyset)) = \sum_{j=1}^{d(v)-1} \rho(w_j)$.
2. **H is a star $K_{1,f}$**
   (a) **$H = K_{1,f}^{mid}$ with center v**
       In this case, we include f children of v in coalition C_v. Note that f children are isolated vertices in subtrees of T_v since C_v forms $K_{1,f}$ in T_v. Let $\delta_j = \rho(w_j) - \rho(w_j, (\{w_j\}, \emptyset))$. Then δ_j means the difference between the maximum social welfare in T_{w_j} and the maximum social welfare of the partition such that w_j is an isolated vertex in T_{w_j}. In other words, δ_j is the cost to include w_j in C_v. Thus, choosing the smallest f children of δ_j maximizes $\rho(v, K_{1,f}^{mid}(v))$ since it consists of the social welfare of C_v, f optimal partitions of T_{w_j} such that w_j is an isolated vertex in T_{w_j}, and $d(v) - 1 - f$ optimal partitions of T_{w_j}. Let $w_1, w_2, \ldots, w_f$ be such children, where the indices are sorted in ascending order. Then, $\rho(v, K_{1,f}^{mid}(v))$ is defined as
       $$\rho(v, K_{1,f}^{mid}(v)) = \varphi(K_{1,f}) + \sum_{j=1}^{f} \rho(w_j, (\{w_j\}, \emptyset)) + \sum_{j=f+1}^{d(v)} \rho(w_j).$$
   (b) **$H = K_{1,f}^{leaf}$ with leaf v**
       Since C_v forms a star $K_{1,f}$ and v is a leaf of it in T_v, we include vertex v in a coalition of $K_{1,f-1}$ with center w_k that is a child of v in a subtree T_k. Thus, we need to choose such child w_k that maximize the social welfare of T_v. In this case, the maximum social welfare of T_v is the sum of the social welfare of the optimal partition of subtrees T_{w_j} except for T_{w_k}, $\varphi(K_{1,f})$, and the social welfare of the partition of T_{w_k} such that w_k is the center of $K_{1,f-1}$ of coalition C_{w_k} minus $\varphi(K_{1,f-1})$.

Thus, $\rho(v, K_{1,f}^{leaf})$ is defined as follows:

$$\rho(v, K_{1,f}^{leaf}) = \max_{k \in [d(v)]} \Big\{ \sum_{j \in [d(v)] \setminus \{k\}} \rho(w_j) + \varphi(K_{1,f}) + \rho(w_k, K_{1,f-1}^{mid}(w_k)) - \varphi(K_{1,f-1}) \Big\}$$

$$= \max_{k \in [d(v)]} \Big\{ \sum_{j=1}^{d(v)} \rho(w_j) - \rho(w_k) + \varphi(K_{1,f}) + \rho(w_k, K_{1,f-1}^{mid}) - \varphi(K_{1,f-1}) \Big\}$$

$$= \sum_{j=1}^{d(v)} \rho(w_j) + \varphi(K_{1,f}) - \varphi(K_{1,f-1}) + \max_{k \in [d(v)]} \big\{ \rho(w_k, K_{1,f-1}^{mid}) - \rho(w_k) \big\}.$$

We define $\rho(v, H)$ for the rest of $H \in \mathcal{G}_v$ in the same way.

3. **H is a path P_4**
   (a) **$H = P_4^{leaf}$ with leaf v**
   A path whose one of leaves is v consists of one $P_3 = K_{1,2}$ and v. Thus we choose one child of v whose coalition is $K_{1,2}$ and maximizes $\rho(v, P_4^{leaf})$.

$$\rho(v, P_4^{leaf}) = \varphi(P_4) - \varphi(P_3) + \sum_{j=1}^{d(v)} \rho(w_j) + \max_{k \in [d(v)]} \Big\{ \rho(w_k, K_{1,2}^{leaf}) - \rho(w_k) \Big\}.$$

   (b) **$H = P_4^{mid}$ with non-leaf v**
   A path whose one of non-leaf vertices is v consists of one $P_2 = K_{1,1}$, v, and one isolated vertex. Thus we choose two children of v such that each coalitions that includes them is $K_{1,1}$ and an isolated vertex, respectively, and they maximize $\rho(v, P_4^{mid})$.

$$\rho(v, P_4^{mid}) = \varphi(P_4) - \varphi(P_2) + \sum_{j=1}^{d(v)} \rho(w_j)$$

$$+ \max_{a,b \in [d(v)], a \neq b} \big\{ \rho(w_a, K_{1,1}^{leaf}) + \rho(w_b, (\{w_b\}, \emptyset)) - \rho(w_a) - \rho(w_b) \big\}.$$

4. **H is a tree T_5^3**
   (a) **$H = T_5^3(s_1)$ with $v = s_1$**
   Since v is s_1 of T_5^3 in T_v, we combine $K_{1,3}^{leaf}$ whose leaf is a child w_j of v with v. Thus, we choose such a child of v that maximizes $\rho(v, T_5^3(s_1))$. Then, $\rho(v, T_5^3(s_1))$ is defined as follows:

$$\rho(v, T_5^3(s_1)) = \varphi(T_5^3) - \varphi(K_{1,3}) + \sum_{j=1}^{d(v)} \rho(w_j) + \max_{k \in [d(v)]} \Big\{ \rho(w_k, K_{1,3}^{leaf}) - \rho(w_k) \Big\}.$$

   (b) **$H = T_5^3(t)$ with $v = t_1$ or t_2**
   Since v is t_1 or t_2 of T_5^3 in T_v, we combine P_4^{mid} in a subtree T_{w_j} with v. Thus, we choose such a child of v that maximizes $\rho(v, T_5^3(t))$. Then, $\rho(v, T_5^3(t))$ is defined as follows:

$$\rho(v, T_5^3(t)) = \varphi(T_5^3) - \varphi(P_4) + \sum_{j=1}^{d(v)} \rho(w_j) + \max_{k \in [d(v)]} \Big\{ \rho(w_k, P_4^{mid}) - \rho(w_k) \Big\}.$$

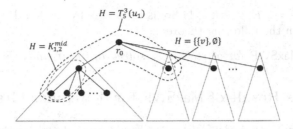

Fig. 6. Computing $\rho(r_0, H)$ for $H = T_5^3(u_1)$

(c) $H = T_5^3(u_1)$ **with** $v = u_1$

Since v is u_1 of T_5^3 in T_v, we combine one coalition of $K_{1,1}$ whose center is a child of v, one coalition of an isolated vertex, and v to construct coalition C_v. Thus, we choose two such children of v that maximizes $\rho(v, T_5^3(u_1))$. Then, $\rho(v, T_5^3(u_1))$ is defined as follows:

$$\rho(v, T_5^3(u_1)) = \varphi(T_5^3) - \varphi(P_3) + \sum_{j=1}^{d(v)} \rho(w_j)$$

$$+ \max_{a,b \in [d(v)], a \neq b} \{\rho(w_a, K_{1,2}^{mid}(w_a)) + \rho(w_b, (\{w_b\}, \emptyset)) - \rho(w_a) - \rho(w_b)\}.$$

(d) $H = T_5^3(u_2)$ **with** $v = u_2$

Since v is u_2 of T_5^3 in T_v, we combine one coalition of P_2 whose leaf is a child of v, two coalitions of isolated vertices, and v to construct coalition C_v. Note that such P_2 is a star $K_{1,1}^{leaf}$. Thus, we choose three such children of v that maximizes $\rho(v, T_5^3(u_2))$. Then $\rho(v, T_5^3(u_2))$ is defined as follows:

$$\rho(v, T_5^3(u_2)) = \varphi(T_5^3) - \varphi(K_{1,1}^{leaf}) + \sum_{j=1}^{d(v)} \rho(w_j)$$

$$+ \max_{a,b,c \in [d(v)], a \neq b \neq c} \{\rho(w_a, K_{1,1}^{leaf}) + \rho(w_b, (\{w_b\}, \emptyset)) + \rho(w_c, (\{w_c\}, \emptyset))$$

$$- \rho(w_a) - \rho(w_b) - \rho(w_c)\}.$$

Figure 6 shows an example of computing $\rho(r_0, H)$ where $H = T_5^3(u_1)$. To compute $\rho(r_0, H)$, we use the ρ's values of its subtrees. The pattern $H = T_5^3(u_1)$ contains one subtree with $H = K_{1,2}^{mid}$ and one with $H = \{\{v\}, \emptyset\}$. The best combination of these can be computed by the DP procedure (c) explained above.

Finally, we evaluate the running time of our algorithm. In Case 1, we can compute the recursive formula in time $O(d(v))$. In Case 2, for case (a), we need to compute largest $\rho(v, K_{1,f}^{mid}(v))$ among $f = 1, 2, \ldots, d(v) - 1$. This can be done by a binary search with SELECT, since δ_i is increasing and w_i's utility in $K_{1,f}$ is decreasing. We can find the optimal f in $O(d(v) + d(v)/2 + d(v)/4 \cdots) = O(d(v))$. Case (b) is also computable in the same running time. In Case 3, both cases can be computed in time $O(d(v))$ by memorizing the best score among its children. Finally, in Case 4, all the cases can be computed in $O(d(v))$ by a similar manner of Case 3. Thus the total running time of this algorithm is $\sum_{v \in V} O(d(v)) = O(n)$, since $\sum_{v \in V} d(v) = 2|E| = 2(n-1)$ holds for a tree by the handshaking lemma. Hence, we obtain the following theorem.

Theorem 2. MaxSWP *for a tree can be solved in linear time.*

5 Hardness Result of MaxSWP for 4-Regular Graphs

It is mentioned in [3] that MaxSWP is NP-hard for graphs with maximum degree 6, though the proof is omitted in the conference paper. Actually, we can show a stronger result, that is, MaxSWP is NP-hard even for 4-regular graphs. The proof is based on a reduction from a restricted variant of 3-SAT problem called M3XSAT(3L), which is shown to be NP-complete in [10, Lemma 5].

Theorem 3. MaxSWP *is NP-hard even for 4-regular graphs.*

References

1. Aziz, H., Brandt, F., Seedig, H.G.: Computing desirable partitions in additively separable hedonic games. Artif. Intell. **195**, 316–334 (2013)
2. Balliu, A., Flammini, M., Melideo, G., Olivetti, D.: Nash stability in social distance games. In: Proceedings of the Thirty-First AAAI Conference on Artificial Intelligence (AAAI 2017), pp. 342–348 (2017)
3. Balliu, A., Flammini, M., Olivetti, D.: On pareto optimality in social distance games. In: Proceedings of the Thirty-First AAAI Conference on Artificial Intelligence (AAAI 2017), pp. 349–355 (2017)
4. Brânzei, S., Larson, K.: Social distance games. In: Proceedings of the 22nd International Joint Conference on Artificial Intelligence (IJCAI 2011), pp. 91–96 (2011)
5. Fortunato, S.: Community detection in graphs. Phys. Rep. **486**(3), 75–174 (2010)
6. Kagawa, S., Okamoto, S., Suh, S., Kondo, Y., Nansai, K.: Finding environmentally important industry clusters: multiway cut approach using nonnegative matrix factorization. Soc. Netw. **35**(3), 423–438 (2013)
7. Newman, M.E.J.: Networks: An Introduction. Oxford University Press, Oxford (2010)
8. Newman, M.E.J.: Spectral methods for community detection and graph partitioning. Phys. Rev. E **88**, 042822 (2013)
9. Okubo, M., Hanaka, T., Ono, H.: Optimal partition of a tree with social distance. CoRR abs/1809.03392 (2018)
10. Porschen, S., Schmidt, T., Speckenmeyer, E., Wotzlaw, A.: XSAT and NAE-SAT of linear CNF classes. Discrete Appl. Math. **167**, 1–14 (2014)
11. Rahwan, T., Michalak, T.P., Wooldridge, M., Jennings, N.R.: Coalition structure generation: a survey. Artif. Intell. **229**, 139–174 (2015)
12. Schaeffer, S.E.: Graph clustering. Comput. Sci. Rev. **1**(1), 27–64 (2007)
13. Shi, J., Malik, J.: Normalized cuts and image segmentation. IEEE Trans. Pattern Anal. Mach. Intell. **22**(8), 888–905 (2000)
14. Sless, L., Hazon, N., Kraus, S., Wooldridge, M.: Forming k coalitions and facilitating relationships in social networks. Artif. Intell. **259**, 217–245 (2018)

Graph Drawing

Flat-Foldability for 1 × n Maps
with Square/Diagonal Grid Patterns

Yiyang Jia[⊠], Yoshihiro Kanamori, and Jun Mitani

University of Tsukuba, 1-1-1 Tennodai, Tsukuba, Ibaraki 305-8577, Japan
kaiyou@cgg.cs.tsukuba.ac.jp,
{kanamori,mitani}@cs.tsukuba.ac.jp

Abstract. In this paper, we propose three conclusions for $1 \times n$ maps with square/diagonal grid patterns. First, for a $1 \times n$ map consisting of all the vertical creases and all the diagonal creases as well as a mountain-valley assignment, if it obeys the local flat-foldability, then it can always be globally flat-folded and one of its flat-folded state can be reached in O(n) time. Second, for a $1 \times n$ map consisting of only square/diagonal grid pattern, it also can always be globally flat-folded and one of its flat-foldable state can be reached in O(n) time. We give theoretical proofs for both of them and propose corresponding algorithms. Then, we prove the NP-hardness of the problem of determining the global flat-foldability for a $1 \times n$ map consisting of a square/diagonal grid pattern and a specific mountain-valley assignment. Also, we show that given an order of the faces for an $m \times n$ map with all the vertical creases and all the diagonal creases assigned to be mountains or valleys, we can determine its validity in O(mn) time.

Keywords: Square/diagonal grid patterns · Flat-foldability · NP-hardness

1 Introduction

Given only a piece of paper with a given crease pattern and a mountain-valley assignment, determining whether this paper can be folded flat or not, is the most fundamental question in the studies of origami flat-foldability. Although the necessary and sufficient conditions for the local flat-foldability are precisely defined, a locally flat-foldable pattern can also be globally flat-unfoldable due to collisions of the parts of the paper during assembly. For the most general case, this problem was proven to be strongly NP-hard by Bern and Hayes [1]. Note that some errors were found in their paper and a correction was given by Akitaya et al. in 2015 [2]. In fact, even when putting limitations on the assignment of the crease pattern or the shape of the paper, this problem still seems to be unsolvable in polynomial time; for instance, the known results include the proof of NP-hardness of box-pleating patterns [2] and square paper shapes [3].

For flat-foldability, an attractive problem is the map folding problem. In the original form of this problem, the shape of the paper is a rectangle that can be partitioned into an $m \times n$ regular grid of squares, with the limitation that every non-boundary grid edge must be folded either as a mountain or as a valley. Unfortunately, finding a general

© Springer Nature Switzerland AG 2019
G. K. Das et al. (Eds.): WALCOM 2019, LNCS 11355, pp. 135–147, 2019.
https://doi.org/10.1007/978-3-030-10564-8_11

solution for this problem is inherently too hard. Therefore, this problem has been broken up into a series of sub-problems, which in turn brings about a series of transformations.

The most well-known results for these sub-problems and their transformations are as follows. Arkin et al. showed that for general $1 \times n$ maps with only vertical creases, their flat folded state can be computed in $O(n)$ time by applying 1D foldability to them; in this process, they defined two local operations: crimps and end folds [4]. Demaine et al. showed that for $m \times n$ maps, if this problem can be reduced to a 1D problem with a sequence of simple folds, NP boils down to P [4]. For $2 \times n$ maps, the folding problem can be solved in $O(n^9)$ time by Morgan's algorithm based on a "ray diagram structure" and three constraints for this structure [5]. In his method, a reduction to the hidden tree problem is used.

For the original case, some related problems popularly studied include the problems of deciding the validity of a given linear ordering of layers [6], identifying the classes of unfoldable mountain-valley patterns [7], counting the number of valid mountain-valley assignments [8], and so on.

Results have also been obtained for some more complex variations on the shapes of the map and creases. For instance, for an orthogonal polygon with horizontal and vertical creases, this problem was proven to be NP-complete by making a reduction from the PARTITION problem [3]. The NP-hardness of the flat-foldability of general maps with orthogonal creases has also been proven [3].

The maps with not only horizontal and vertical creases but also diagonal creases are defined as maps with square/diagonal grid patterns. An instance is shown in Fig. 1. For an $m \times n$ map with a square/diagonal grid pattern, Matsukawa et al. showed that when $m, n \geq 3$, there exist flat-unfoldable square/diagonal grid patterns without a specific mountain-valley assignment [9]. Taking this result as a reference, we aim to determine the flat-foldability of $1 \times n$ maps of square/diagonal grid patterns with or without mountain-valley patterns assigned. We propose a approach to prove that a $1 \times n$ map consisting of all assigned vertical and diagonal creases ($n - 1$ assigned vertical creases and $2n$ assigned diagonal creases) satisfying local flat-foldability can always be flat-folded (Fig. 2(a)). Also, we prove that a $1 \times n$ map consisting of only a square/diagonal grid pattern can always be flat-folded (Fig. 2(b)). However, determining the flat-foldability for a $1 \times n$ map with a square/diagonal grid pattern and a specific mountain-valley assignment is proven to be NP-complete (Fig. 2(c)). Some instances of the three different kinds of maps we concerned are illustrated in Fig. 2. In addition, we present an algorithm for determining the validity of a given linear ordering of faces of an $m \times n$ map consisting of all the vertical creases and all the diagonal creases in $O(mn)$ time.

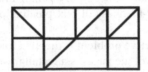

Fig. 1. A 2×4 map with a square/diagonal grid pattern.

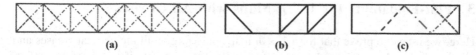

Fig. 2. Three different kinds of maps we concerned. (Color figure online)

2 Preliminary

We will use some of the definitions defined in [6]. A connected polygon in a plane is denoted as a paper, and assumed that its one side is white and the other side is colored. Suppose there is an embedded planar graph on this paper, which is denoted as a crease pattern, then an edge of the graph which is not on the boundary of the paper is denoted as a crease. The regions divided and bounded by some edges of the crease pattern are denoted as faces, i.e., a face is bounded by a set of creases and possibly by part of the boundary of the paper. Therefore, a crease is incident to exactly two faces. A crease can be labeled either as a mountain (the white sides of the two faces incident to the crease touch each other after the fold, denoted by "m") or as a valley (the colored sides of the two faces incident to the crease touch each other after the fold, denoted by "v"). If each crease in a crease pattern is assigned to be either mountain or valley, the assignment is called a mountain-valley assignment. An endpoint of a crease not on the boundary of the paper is denoted as a vertex. An $m \times n$ map is a rectangular paper that can be partitioned into an $m \times n$ grid of squares. A general map pattern is a crease pattern using a subset of the creases partitioning the map into regular unit squares, called a square grid pattern. Its extension, called a square/diagonal grid pattern, as introduced in Sect. 1, is a crease pattern using a subset of the creases partitioning the map into regular unit squares and partitioning each regular square into four regular isosceles right triangles in diagonal directions.

In our research, the folding process is defined as a sequence of functions $F = \{f_1, f_2, \ldots, f_a\}$, this is to take advantage of inverse functions when defining unfolding process. Conversely, the unfolding process is defined as a sequence of functions $U = \{u_1, u_2, \ldots, u_b\}$, and each u_j ($1 \leq j \leq b$) is an inverse function for some f_i ($1 \leq i \leq a$), i.e., $u_j = f_i^{-1}$. The initial state of paper P is defined as S_0 when it is not folded. The state of the paper after a folding or unfolding process s_k is denoted as S_k, i.e., each s_k names a function representing a folding or unfolding process from S_{k-1} to S_k, $s_k \in F$ or $s_k \in U$.

Lemma 1 [10]. The difference between the number of creases labeled mountains and the number of creases labeled valleys meeting at a non-boundary vertex is 2.

By Lemma 1, for a $1 \times n$ map consisting of all vertical and diagonal creases, if each non-boundary vertex has its creases labeled as three mountains and a valley or as three valleys and a mountain, then the map is locally flat-foldable; an instance is shown in Fig. 2(a). In our paper, we follow the common practice for illustrating mountain-valley assignments in a crease pattern, i.e., using the red line segments to denote mountains and using the blue line segments to denote valleys.

3 Flat-Foldability for 1 × n Maps with All Creases

Here, we intend to prove that for a 1 × n map consisting of all the vertical creases and all the diagonal creases and a mountain-valley assignment, if it obeys local flat-foldability, then it can always be flat-folded, one of its flat-folded state can be reached in O(n) time. We also propose an O(n) time algorithm for obtaining a flat-folded state of such a 1 × n map.

Theorem 1. A 1 × n map consisting of all the vertical creases and all the diagonal creases as well as a mountain-valley assignment can always be flat-folded if it obeys local flat-foldability. In this case, it means that each non-boundary vertex is surrounded with three mountains and a valley or a mountain and three valleys.

Proof. Although there might exist other possible ways to flatly fold a map, we prove this theorem by using only one folding method, as presented below.

First, let us consider the flat-folded state of only one square, i.e., a 1 × 1 map. There is only one vertex in its center and four labeled creases associated with the vertex (which divides the square into four triangles). Since the square has a colored side and a white side, there are eight kinds of valid mountain-valley assignments based on Lemma 1 (Fig. 3). Then we can tell that there are eight flat-folded states, i.e., eight different orders of four triangles in the flat-folded states.

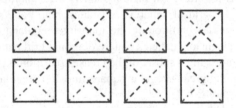

Fig. 3. Eight flat-folded states of one square.

Next, let us consider folding a 1 × n map. The folding process is as follows. Step 1, fold the map (with its longer side in the lateral direction) from the leftmost square, firstly fold its four diagonal creases, then fold the vertical crease on its right edge so as to get a strip with a triangular shape of four layers at the leftmost square. After this step, consider the folded layers and the triangle above them (the leftmost triangle in the second square) as an integrated part; therefore the map can be considered as being reduced to a 1 × (n − 1) map; Step 2, use the same operation to fold the adjacent square on the right side, also fold the diagonal creases; then fold the vertical crease on the right edge. Repeat the operation of Step 2 until all squares are folded.

With this method, in each step we fold the next unfolded adjacent square which is on the right side of the lastly folded square. For the already folded k (1 ≤ k ≤ n) squares, a triangular shape of 4 × k layers can be obtained.

In our folding, for every square, we consider both its own folded state and the crease located on its right edge. If it is labeled as a mountain, then the folded layers to its left are folded below the unfolded part of the strip all together, or otherwise they would be folded above the unfolded part of the strip all together. A demonstration for a 1 × 5 map is shown in Fig. 4.

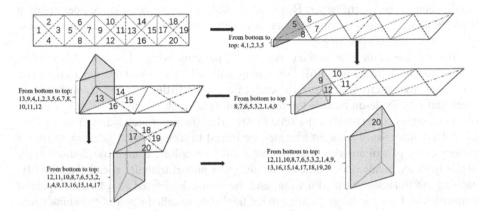

Fig. 4. A demonstration for folding a 1 × 5 map.

By following these folding steps, we can finally obtain the flat-folded state with each layer being a triangle. Thus, 1 × n maps consisting of all the vertical creases and all the diagonal creases and a mountain-valley assignment can definitely be flat-folded.

Since we only need to fold each part in a square once, the time complexity of this folding process is O(n). The algorithm to obtain the flat-folded state is shown below.

Algorithm 1

Input: Map M with a mountain-valley assignment.

Output: The order of the faces of the final flat-folded state of M.

1 Initialize a linked list L to Φ

2 Assign L by deciding the inner order of four faces in the first square

3 For every square in M
  Update L according to the crease left to the square
  Consider the folded layers and the leftmost triangle in this square as a whole, update L by deciding the order of the triangle layers according to the creases within this square

4 Return L

4 Algorithm for Determining the Validity of an Ordering in $m \times n$ Maps with All Creases

With other methods instead of the folding method described in Sect. 3, an $m \times n$ map consisting of all creases and a mountain-valley assignment can also be folded flat into a triangular shape, whose ordering of layers (in this case referring to all 1/4 square triangles) is specified differently in the final flat folded state. Therefore, we also want to consider different valid orderings of layers. From [11], we know that the postage-stamp folding problem is difficult; hence, a relevant compound problem of enumerating the number of ways in which an $m \times n$ map consisting of all creases can be folded up would seem to be much harder. However, another relevant problem, i.e., determining the validity of a given ordering of layers, can be solved in linear time with an extension of the method described in [6].

We will use definitions of the checkerboard patterns, wings, hinges, and butterflies similar to the definitions in [6]. For a map with all square and diagonal creases, a checkerboard pattern is defined for convenience to distinguish the colored-side-up faces and white-side-up faces in the final folded state, as shown in Fig. 6(a). The colors of the triangles correspond to the state of the grids (facing which side up) in the final flat folded state. Since adjacent triangles are forced to face different sides up, in such a pattern every pair of adjacent triangles have different colors. If in a folded state, a pair of triangles are incident to a same crease, then this pair of triangles is called a butterfly, each of the triangles is called a wing, and the crease is defined as a hinge. A pair of butterflies with a same hinge location in the final state is called a pair of twin butterflies. Figure 5 gives all possible states of a pair of twin butterflies; (a), (b) are called stacks, (c), (d) are called nests.

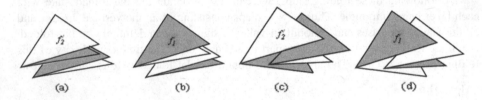

Fig. 5. Four states of a pair of butterflies.

Next, we define the directed network of the crease pattern, based on the fact that in our pattern, two triangles not adjacent in the initial map may also be adjacent in the final folded state if they have the property that the grids where they are located share the same point or edge. Thus, we have to extend the definition in [6] which only defined the relationship between adjacent squares in the initial map. Our directed network is defined as follows: first, suppose the leftmost triangle in the uppermost row is the uppermost layer in the final flat-folded state and facing white side up (Triangle 1 in Fig. 6(a)); next, for adjacent triangles, i.e., a pair of triangles sharing the same edge, label their order based on whether the crease between them is labeled a mountain or valley. For the simplicity of describing, we denote each Triangle c by t_c. For example, for Triangle 1 and Triangle 2 in Fig. 6(a), since the crease between them is labeled as a mountain and Triangle 1 is facing white side up, the order is defined as $t_1 < t_2$ (from top to bottom); similarly, for the four triangles in the grid located in the upper left corner in Fig. 6(a), the order is defined as $t_1 < t_2 < t_3 < t_4$ (from top to bottom: $\{t_1, t_2, t_3, t_4\}$). With this relationship, we can establish a directed network as in Fig. 6(b), which indicates the relationship between adjacent triangles; finally, we add the edges indicating the relationship between two triangles not adjacent but within adjacent grids. If their order can be determined by the relationship defined by the directed network we already have, then we add one directed edge between them; otherwise, we add a bidirectional edge between them. For instance, the newly defined network for Triangle 2 is shown in Fig. 6(c). The final directed network is obtained by adding the newly defined networks for every triangle to the original one.

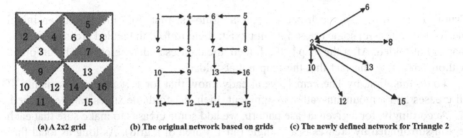

(a) A 2x2 grid   (b) The original network based on grids   (c) The newly defined network for Triangle 2

Fig. 6. A 2 × 2 grid, its original directed network, and the newly defined network for Triangle 2.

Theorem 2. Let P be an $m \times n$ map with a <u>mountain-valley assigned square/diagonal grid pattern</u> G. A linear ordering L of the faces of G is valid if and only if every pair of twin butterflies in L either stacks or nests, and <u>for the case of $m, n \geq 2$</u>, L must satisfy the directed network (a directed acyclic graph) of G.

Proof. The proof of Theorem 2 is similar to the proof in [6].

First, it is obvious that in a linear ordering, twin butterflies in L either stack or nest, or otherwise intersections would happen. When either m or n equals 1, it is easy to reach this conclusion by Theorem 1; for the case that $m, n \geq 2$, if the network is not a directed acyclic graph, then there would exist a layer not relating to other layers in the ordering. This shows the necessity.

Next, to prove the sufficiency, we only have to prove that it suffices to arrive at a final flat folded state of pattern G. The method of proof is the same as in [6]. First, decompose P into distinct triangles, where each triangle is a face of G with both a white side and a colored side. Then, stack these triangles on only one triangle t according to the linear ordering L. The checkerboard pattern of G decides either a face is facing the white side up or facing the colored side up. According to the location of hinges (matching with the edges of Triangle 1) of the butterflies in P, for each butterfly, join its two wings (along the hinge of it) such that its hinge lies above an edge of t. Since any pair of butterflies with the same hinge location either nests or stacks, there will be no intersection. This process leads to a folded state of P where L is the linear ordering of the faces of G in the folded state. Therefore, L is a valid linear ordering.

By Theorem 2, we can obtain the conclusion that the running time of this algorithm is $O(mn)$, which is identical to that of the algorithm presented in [6] since the difference happens only when checking the kind of butterflies and constructing the initial directed graph, both of which can be done by a simple traverse in linear time.

5 Flat-Foldability for $1 \times n$ Maps with a Square/Diagonal Grid Pattern

Here, we reach a conclusion for $1 \times n$ maps with only square/diagonal grid patterns (no specific mountain-valley assignment). An instance of such a map is given in Fig. 1.

Theorem 3. $1 \times n$ maps with only square/diagonal grid patterns can always be folded flat, and their flat-folded states can be reached in $O(n)$ time.

Proof. In brief, our approach is a process as follows: first, fold the map to an initial flat-folded state with some creases do not exist; then, unfold the added creases after the map is flat folded. After the map is flat folded, the creases added before can be unfolded without changing the fact that the map is flat folded.

To do this, first, by Theorem 1, we already know that for $1 \times n$ maps consisting of all creases and a mountain-valley assignment, their flat-foldable states can be reached.

Accordingly, for a given crease pattern, we add some creases to make sure that each region surrounded by creases is a triangle (an 1/4 area of a grid). As long as local flat-foldability is maintained, the kind of diagonal creases added has no effect on the final result. For simplicity, we fold this paper strip into an flat state with such an assignment that if we denote the creases by c_i ($i \in N^+$) following a left-to-right order, and for each four creases inside a grid, labeled in counterclockwise order, as shown in Fig. 7(b), the creases in the set $\{c_i \mid i = 10m, 10m + 4, 10m + 6, m \in N^+\}$ are labeled mountains, and the other creases are labeled valleys. The key is to keep four triangles in each grid adjacent to each other in the final flat-folded state (the ordering of the layers in the final flat-folded state is as follows: the triangles in the first grid, the triangles in the second grid, ..., the triangles in the nth grid). An instance is shown in Fig. 7, where the creases we added are indicated as the thick line segments.

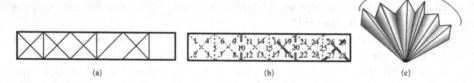

Fig. 7. A given crease pattern, its predefined mountain-valley assignment and the initial flat-folded state.

Next comes the unfolding process, i.e., unfolding the creases we added. First, we unfold the temporarily added vertical creases in left-to-right order (with the longer side of the map put in the lateral direction). For each crease c_i being unfolded, we perform the following process: Step 1. Unfold c_i; Step 2. Change the label for every vertical crease c_j ($j > i$) and diagonal crease $c_{2k+1}(2k + 1 > i, k \in N^+)$, which are in the adjacent grids of c_j on the same diagonal with the same labels. Note that each change during the unfolding process follows Lemma 1 (Maekawa's Theorem). An instance is shown in Fig. 8, we first unfold crease c_{10} while changing the labels of the vertical creases c_{15},

c_{20}, c_{25} and the diagonal creases c_{11}, c_{13}, c_{17}, c_{19}, c_{21}, c_{23}, c_{27}, and c_{29} and then unfold crease c_{20} while changing the labels of c_{25}, c_{21}, c_{23}, c_{27}, and c_{29}.

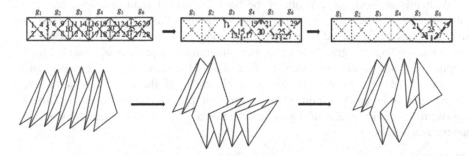

Fig. 8. Change of the state when unfolding the vertical creases c_{10} and c_{20}.

When the vertical crease between the kth grid and the $(k + 1)$th grid is unfolded, the change in the folded state is that the two triangles in the kth grid and the $(k + 1)$th grid adjacent to each other turn to be in a same layer, and the locations of the triangle layers from the $(k + 1)$th grid to the nth grid are changed. Intersections would not be produced because when the unfolding happens, we only have to move layers up or down the axis without influencing the layering order.

Finally, unfold the temporarily added diagonal creases. We intend to keep the global order unchanged while unfolding some creases. The process is like this: Step 1. Unfold the creases in the four corresponding layers of one grid; Step 2. If the order of the layers after (or before) the unfolded layers has been changed, then change the original mountain-valley assignment of the crease (only one) after (or before) this crease to keep the global order unchanged. For example, in Fig. 9, when we unfold creases c_{16} and c_{18} in g_4, we change the labels of c_{22}, c_{24}, c_{25}, c_{26}, and c_{28} so as to keep the global order; however, when we unfold the creases c_{26}, c_{27}, c_{28}, c_{29} in g_6, no other assignments have to be changed.

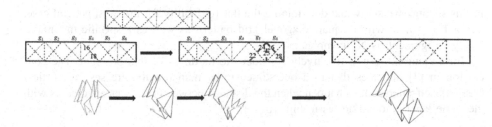

Fig. 9. Unfolding the diagonal creases.

When performing this process, we know that any unfolding operation will not change the order of the grids (from left to right). Compared with the initial flat-folded state (with the added creases actually do not exist), the unfolding only changes the layers of some grids from four layers into two layers or one layer. Whatever change happens within each grid, the order of grids is always maintained. Therefore, all $1 \times n$ square/diagonal grid patterns are flat-foldable.

With this method, we can obtain the final flat-folded state of a square/diagonal pattern in $O(n^2)$ time, since we have to determine the changes for other creases each time a crease is unfolded. The method introduced here can also be applied to $2 \times n$ square/diagonal grid patterns. In fact, the unfolding process is presented for the simplicity of our proof. To obtain the final flat-folded state, the real process can be simplified. We only have to decide the assignment of the creases between every adjacent pair of layers, based on which the time complexity can be reduced to $O(n)$. An algorithm for getting the final flat-folded state of a given $1 \times n$ square/diagonal grid pattern is given below.

Algorithm 2
Input: A $1 \times n$ square/diagonal grid pattern P
Output: The final flat-folded state S corresponding to P
1 Initialize S to ϕ
2 For each vertical crease v_i, do:
  Fold the creases in the grid between v_i and v_{i+1}. Whether they are labeled as mountains or valleys follows the principle of keeping the layering order of the grids consistent with the original order of the grids $(g_1, g_2, \cdots)$. Record the folded state of the strip between v_i and v_{i+1} as p_i;
  Fold v_{i+1} with an assignment which keeps the layering order of the layers to the right of v_{i+1} unchanged, reassign S by adding p_i to an appropriate location in S
3 Return S

6 Complexity of Determining the Flat-Foldability of a $1 \times n$ Map with a Square/Diagonal Grid Pattern and a Specific Mountain-Valley Assignment

In this section, we show that determining the flat-foldability for the most general case, i.e., a $1 \times n$ map with a square/diagonal grid pattern as well as a specific mountain-valley assignment, is NP-complete.

First though, we will intuitively consider the hardness of this problem. The conclusion in [4] indicates that in a one dimensional map (all the creases are vertical creases), self-intersections happen when the distance between two adjacent creases with the same label is too short, as in Fig. 10.

Fig. 10. A self-intersection in a one dimensional map.

A similar case could happen in a $1 \times n$ map with a square/diagonal grid pattern; however, sometimes diagonal creases influence the distances between the creases, causing self-intersections. This means that this problem is a combinatorial optimization problem. Hence, as a general case in combinatorial optimization problems, there is a great likelihood that this problem is NP-hard.

On the other hand, if a crease pattern and a mountain-valley assignment are given, we can easily determine the validity of a specific folding process (each s_k representing a folding operation from S_{k-1} to S_k by only folding one crease is given) by checking whether each s_k obeys the mountain-valley assignment and whether the folding operation is permitted according to the state of the layers in S_{k-1}. We only have to check each s_k once so as to determine the validity of a given folding process, which can be done in polynomial time. Thus, according to the definition of NP-completeness, this problem should not be considered as a problem harder than NP-complete problems. In the next paragraph, we reduce a well-known NP-complete problem, i.e., Partition Problem to this problem.

In [4], Arkin et al. have shown that determining whether a square crease pattern on an orthogonal piece of paper can be flat-folded or not is NP-complete. They used a reduction to Partition Problem by creating a pattern like in Fig. 11. We find that their pattern can also be used in our proof; hence, we only have to make corners with the crease shown in Fig. 12, so as to create a pattern as in Fig. 11 and it becomes easy show that the problem of determining the flat-foldability of a $1 \times n$ map with a square/diagonal grid pattern and a specific mountain-valley assignment to be NP-complete. Note that in our proof, L, W_1, W_2 mod $\varepsilon = 0$ (they are illustrated in Fig. 11 and their definitions can be found in [4]) since our operations are executed on a piece of paper formed by squares.

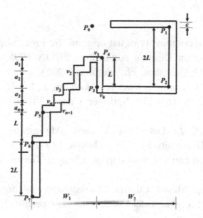

Fig. 11. The pattern for a reduction to Partition Problem (This figure was reproduced by the authors with reference to [4].)

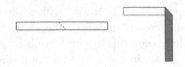

Fig. 12. Method of making a corner with a strip.

7 Conclusion

We reached three conclusions for $1 \times n$ maps with square/diagonal grid patterns: for a $1 \times n$ map with all vertical and diagonal creases as well as a mountain-valley assignment, if it obeys local flat-foldability, then it can always be folded flat and one of its flat-folded state can be reached in $O(n)$ time; for a $1 \times n$ map with only square/diagonal grid pattern, it also can always be folded flat and one of its flat-foldable states can be reached in $O(n)$ time. We gave theoretical proofs for both of them and proposed corresponding algorithms. Next, we proved that determining the flat-foldability of a $1 \times n$ map with a square/diagonal grid pattern and a specific mountain-valley assignment is NP-complete. Moreover, we showed that given an order of the faces of an $m \times n$ map with each vertical or diagonal crease assigned to be a mountain or a valley, we can determine the validity of the order in $O(mn)$ time.

An attractive problem remaining to be solved is determining the flat-foldability of $2 \times n$ maps. Also, related to this problem, it is an interesting question of defining flat-unfoldable classes for $1 \times n$ square/diagonal grid patterns with specific mountain-valley assignments.

References

1. Bern, M., Hayes, B.: The complexity of flat origami. In: Proceedings of the Seventh Annual ACM-SIAM Symposium on Discrete Algorithms SODA 1996, pp. 175–183. Society for Industrial and Applied Mathematics, Philadelphia (1996)
2. Akitaya, H.A., et al.: Box pleating is hard. In: Akiyama, J., Ito, H., Sakai, T. (eds.) JCDCGG 2015. LNCS, vol. 9943, pp. 167–179. Springer, Cham (2016). https://doi.org/10.1007/978-3-319-48532-4_15
3. Demaine, E., O'Rourke, J.: Geometric Folding Algorithms: Linkages, Origami, and Polyhedra. Cambridge University Press, Cambridge (2007)
4. Arkin, E.M., et al.: When can you fold a map? Comput. Geom. Theor. Appl. **29**(1), 23–46 (2004)
5. Morgan, T.: Map folding. Master's thesis, Massachusetts Institute of Technology (2012)
6. Nishat, R.I., Whitesides, S.: Canadian Conference on Computational Geometry, pp. 49–54 (2013)
7. Justin, J.: Aspects mathematiques du pliage de papier (mathematical aspects of paper fold). In: Huzita, H. (ed.) 1st International Meeting of Origami Science and Scientific Origami, pp. 263–277 (1989)
8. Hull, T.C.: Counting mountain-valley assignments for flat folds. arXiv preprint arXiv:1410.5022 (2014)

9. Matsukawa, Y., Yamamoto, Y., Mitani, J.: Enumeration of flat-foldable crease patterns in the square/diagonal grid and their folded shapes. J. Geom. Graph. **21**(2), 169–178 (2017)
10. Kasahara, K., Takahama, T.: Origami for the Connoisseur. Japan Publications Inc. (1978)
11. Koehler, J.E.: Folding a strip of stamps. J. Combinatorial Theor. **5**(2), 135–152 (1968)

(k, p)-Planarity: A Relaxation of Hybrid Planarity

Emilio Di Giacomo[1], William J. Lenhart[2], Giuseppe Liotta[1],
Timothy W. Randolph[3(✉)], and Alessandra Tappini[1]

[1] Università degli Studi di Perugia, Perugia, Italy
{emilio.digiacomo,giuseppe.liotta}@unipg.it,
alessandra.tappini@studenti.unipg.it
[2] Williams College, Williamstown, USA
wlenhart@williams.edu
[3] Columbia University, New York City, USA
t.randolph@columbia.edu

Abstract. We present a new model for hybrid planarity that relaxes existing hybrid representations. A graph $G = (V, E)$ is (k, p)-planar if V can be partitioned into clusters of size at most k such that G admits a drawing where: (i) each cluster is associated with a closed, bounded planar region, called a *cluster region*; (ii) cluster regions are pairwise disjoint, (iii) each vertex $v \in V$ is identified with at most p distinct points, called *ports*, on the boundary of its cluster region; (iv) each inter-cluster edge $(u, v) \in E$ is identified with a Jordan arc connecting a port of u to a port of v; (v) inter-cluster edges do not cross or intersect cluster regions except at their endpoints. We first tightly bound the number of edges in a (k, p)-planar graph with $p < k$. We then prove that $(4, 1)$-planarity testing and $(2, 2)$-planarity testing are NP-complete problems. Finally, we prove that neither the class of $(2, 2)$-planar graphs nor the class of 1-planar graphs contains the other, indicating that the (k, p)-planar graphs are a large and novel class.

Keywords: (k, p)-Planarity · Hybrid representations · Clustered graphs

1 Introduction

Visualization of non-planar graphs is one of the most studied graph-drawing problems in recent years. In this context, an emerging topic is hybrid representations (see, e.g., [1,2,5,7,10]). A hybrid representation simplifies the visual analysis of a non-planar graph by adopting different visualization paradigms for different portions of the graph. The graph is divided into (typically dense) subgraphs called *clusters* which are restricted to limited regions of the plane. Edges between vertices in the same cluster are called *intra-cluster* edges, and edges between vertices in different clusters are called *inter-cluster* edges. Inter-cluster edges are represented according to the classical node-link graph drawing

© Springer Nature Switzerland AG 2019
G. K. Das et al. (Eds.): WALCOM 2019, LNCS 11355, pp. 148–159, 2019.
https://doi.org/10.1007/978-3-030-10564-8_12

paradigm, while the clusters and their intra-cluster edges are represented by adopting alternative paradigms. A hybrid representation thus reduces the number of inter-cluster edges and the visual complexity of much of the drawing at the cost of creating cluster regions of high visual complexity. As a result, a hybrid representation provides an easy to read overview of the graph structure and it admits a "drill-down" approach when a more detailed analysis of some of its clusters is needed.

Different representation paradigms for clusters give rise to different types of hybrid representations. For example, Angelini et al. [1] introduce *intersection-link* representations, where clusters are represented as intersection graphs of sets of rectangles, while Henry et al. [10] introduce *NodeTrix* representations, where dense subgraphs are represented as adjacency matrices (see Fig. 1). Batagelj et al. employ hybrid representations in the (X, Y)-*clustering* model [2], where Y and X define the desired topological properties of the clusters and of the graph connecting the clusters, respectively. For instance, in a $(planar, k\text{-}clique)$-clustering of a graph each cluster is a k-clique and the graph obtained by contracting each cluster into a single node (called the *graph of clusters*) is planar. Given a graph G and a hybrid representation paradigm $\mathcal{P}$, the *hybrid planarity* problem asks whether G can be represented according to $\mathcal{P}$ with no inter-cluster edge crossings. Variants of the problem may or may not assume that the clustering is given as part of the input.

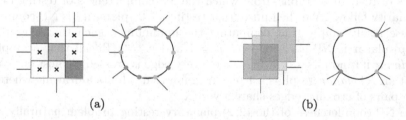

(a)   (b)

Fig. 1. (a) A NodeTrix representation of a 3-clique and a corresponding $(3, 4)$ representation. (b) An intersection-link representation of a 3-clique and a corresponding $(3, 2)$ representation.

In this paper, we present a general hybrid representation paradigm that relaxes the described hybrid paradigms. Given a graph $G = (V, E)$, a (k, p) *representation* Γ of G is a hybrid representation in which: (i) each cluster of G contains at most k vertices and is identified with a closed, bounded planar region; (ii) cluster regions are pairwise disjoint, (iii) each vertex $v \in V$ is represented by at most p distinct points, called *ports*, on the boundary of its cluster region; (iv) each inter-cluster edge $(u, v) \in E$ is represented by a Jordan arc connecting a port of u to a port of v. A (k, p) representation is (k, p)-*planar* if edge curves do not cross and do not intersect cluster regions except at their endpoints. We say that a graph G is (k, p)-*planar* if it can be clustered so that it admits a (k, p)-planar representation.

The definition of a (k, p) representation leaves the representation of clusters and intra-cluster edges intentionally unspecified. It is thus a relaxation of hybrid representation paradigms where the number of ports used by the inter-cluster edges depends on the geometry of the cluster regions. For example, in a NodeTrix representation, the squared boundary of each matrix allows four ports for every vertex except for the vertex in the first row/column of the matrix and the vertex in the last row/column of the matrix, which both have only three ports. Hence, a NodeTrix representation can be regarded as a constrained $(k, 4)$ representation (four ports for every vertex except for two, the vertices appear in the order imposed by the matrix); see Fig. 1(a). Similarly, a $(k, 2)$ representation relaxes an intersection-link representation with clusters represented as isothetic unit squares with their upper-left corners along a common line with slope 1; see Fig. 1(b). We also remark that the use of different ports to represent a vertex can be regarded as an example of *vertex splitting* [8,9]; however, while in the papers that use vertex splitting to remove crossings the multiple copies of each vertex can be placed anywhere in the drawing, in our model they are forced to lay within the boundary of the same cluster region.

The results of this paper are the following:

- In Sect. 2, we give an upper bound on the edge density of a (k, p)-planar graph and prove that this bound is tight for $p < k$.
- In Sect. 3, we observe that the class of $(4, 1)$-planar graphs coincides with the class of IC-planar graphs, from which the NP-completeness of testing $(4, 1)$-planarity follows. We then prove that testing $(2, 2)$-planarity is NP-complete. These results imply that computing the minimum k such that a graph is (k, p)-planar is NP-hard for both $p = 1$ and $p = 2$. Recall that a graph is *1-planar* if it admits a drawing where every edge is crossed at most once, and that an *IC-planar* graph is a 1-planar graph that admits a drawing where no two pairs of crossing edges share a vertex.
- The NP-completeness of the $(2, 2)$-planarity testing problem naturally suggests to further investigate the combinatorial properties of $(2, 2)$-planar graphs. In Sect. 4, we ask whether every 1-planar graph admits a $(2, 2)$-planar representation (see, e.g. Fig. 6). We prove the existence of 1-planar graphs that are not $(2, 2)$-planar and of $(2, 2)$-planar graphs that are not 1-planar. We also give a sufficient condition for 1-planar graphs to be $(2, 2)$-planar.

For reasons of space some proofs have been omitted or sketched, and can be found in the full version [6]. The corresponding statements are marked with [*].

2   Edge Density of (k, p)-Planar Graphs

In this section we give a tight bound on the number of edges of a (k, p)-planar graph when $p < k$. First, given a (k, p)-planar representation Γ, we define a *skeleton* of Γ to be a planar drawing Γ_S obtained by the following transformation. We first replace each port in Γ with a vertex. Each cluster region R_i of Γ is now an empty convex space surrounded by up to kp vertices. We connect these vertices in a cycle and triangulate the interior. For our purposes any triangulation is equivalent. The resulting representation is Γ_S. Figure 2(b) illustrates a skeleton of the $(2, 2)$-planar representation of Fig. 2(a).

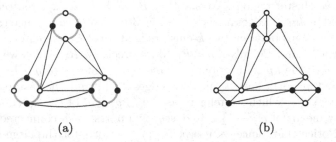

(a)   (b)

Fig. 2. (a) A $(2, 2)$-planar representation Γ of a graph G; (b) A skeleton Γ_S of Γ.

Theorem 1 [*]. *Let G be a (k, p)-planar graph with n vertices. G has $m \leq n(p + \frac{3}{k} + \frac{k}{2} - \frac{1}{2}) - 6$ edges. This bound is tight for any positive integers k, p and n such that $p < k$ and $n = N \cdot k$, where $N > 2$.*

Proof. Let Γ be a (k, p)-planar representation of G and let N be the number of clusters of G. As each cluster contains at most k vertices, G has at most $N \cdot \frac{k(k-1)}{2}$ intra-cluster edges.

Let R_i be a cluster region in Γ with p_i ports in total. Let Γ_S be a skeleton of Γ, and let n_S and m_S denote the number of vertices and the number of edges of Γ_S, respectively. When Γ_S is created, R_i is replaced with p_i vertices and $2p_i - 3$ edges if $p_i > 1$, or 0 edges if $p_i = 1$. Letting m_{inter} be the number of inter-cluster edges in G and s be the number of clusters in G containing a single vertex, we have,

$$m_S = m_{inter} + \sum_{i=1}^{N}(2p_i - 3) + s. \tag{1}$$

In other words, the total number of edges in Γ_S is equal to the number of inter-cluster edges in G plus the number of edges added for each cluster. Note that $m_S \leq 3n_S - 6$, as Γ_S is a planar drawing. As $\sum_{i=1}^{N} p_i = n_S$, rearranging generates $m_{inter} + 2n_S - 3N + s \leq 3n_S - 6$ and thus,

$$m_{inter} \leq n_S + 3N - 6 - s \leq N(kp + 3) - 6. \tag{2}$$

As m is equal to the sum of the number of inter-cluster and intra-cluster edges in G, we have

$$m \leq Nk(p + \frac{3}{k} + \frac{k}{2} - \frac{1}{2}) - 6. \tag{3}$$

If all clusters contain k vertices, then $N = \frac{n}{k}$ and Theorem 1 holds. Refer to the full version [6] for the proof that $m \leq n(p + \frac{3}{k} + \frac{k}{2} - \frac{1}{2}) - 6$ in the case where some clusters contain fewer than k vertices.

In order to show that the bound is tight for $p < k$, we describe a (k, p)-planar representation $\Gamma_{k,p}$ with $N = \frac{n}{k}$ clusters and $(kp + 3)N - 6$ inter-cluster edges. $\Gamma_{k,p}$ is possible for any pair of positive integers p and k such that $p < k$ and for any $N > 2$. $\Gamma_{k,p}$ has N clusters each with k vertices and thus kp ports. Let R_1 and R_2 be two cluster regions. We say that R_1 and R_2 are kp-*connected* if they are connected by $kp+1$ edges as shown in Fig. 3(a). (Note that, since the number of inter-cluster edges between two k-clusters is at most k^2, we can create $kp + 1$ edges between R_1 and R_2 only if $p < k$). More precisely, R_1, which we refer to as the *small end* of the kp-connection, is connected by means of $p + 1$ consecutive ports; the first p ports have k incident edges each, and the last port has an additional edge. R_2, which we refer to as the *large end* of the kp-connection, is connected by means of $p(k - 1) + 1$ consecutive ports, each connected to one or two edges. Notice that, since we use $p(k - 1) + 1$ ports for the large end, $p + 1$ for the small end and two ports can be shared by the two ends, each cluster region can be the small end of one kp-connection and the large end of another kp-connection. Thus, we can create a cycle with N clusters as shown in Fig. 3(b). In the resulting representation there are two faces of degree N: One is the outer face and the other one is inside the cycle. By triangulating these two faces with $N - 3$ edges for each face, we obtain the (k, p)-representation $\Gamma_{k,p}$. The number of inter-cluster edges of $\Gamma_{k,p}$ is thus $(kp + 1)N + 2N - 6 = (kp + 3)N - 6$.   □

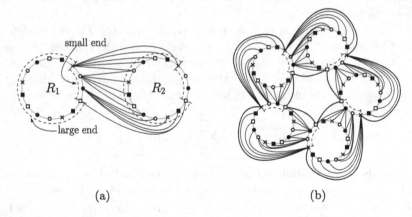

(a)   (b)

Fig. 3. (a) A kp-connection of two cluster regions R_1 and R_2 ($k = 5$, $p = 3$). (b) A cycle of $N = 5$ clusters; the bold edges highlight the two faces of degree N.

3   Recognition of (k, p)-Planar Graphs

This section considers the problem of testing (k, p)-planarity for the cases in which $p = 1$ and $p = 2$.

Theorem 2 [*]. $(k, 1)$-*planarity testing can be performed in linear time for* $k \leq$ *3, and it is NP-complete for* $k = 4$.

Proof. The first part of Theorem 2 follows from the fact that the class of $(k, 1)$-planar graphs coincides with the class of planar graphs for $k = 1, 2, 3$. The second part follows from the fact that the $(4, 1)$-planar graphs coincide with the IC-planar graphs [14]. Testing IC-planarity is known to be NP-complete [4]. $\square$

Corollary 1. *The problem of computing the minimum value of k such that a graph is $(k, 1)$-planar is NP-hard.*

We now focus on the $(2, 2)$-planarity testing problem, hereafter referred to as $(2, 2)$-PLANARITY. We show that $(2, 2)$-PLANARITY is NP-complete by a reduction from the NP-complete problem PLANAR MONOTONE 3-SAT [3]. We say that an instance of 3-SAT is *monotone* if every clause consists solely of positive literals (a *positive clause*) or solely of negative literals (a *negative clause*). A *rectilinear representation* of a 3-SAT instance is a planar drawing where each variable and clause is represented by a rectangle, all the variable rectangles are drawn along a horizontal line, and vertical segments connect clauses with their constituent variables. A rectilinear representation is *monotone* if it corresponds to a monotone instance of planar 3-SAT where positive clauses are drawn above the variables and negative clauses are drawn below the variables, as shown in Fig. 4(a). Given a monotone rectilinear representation Φ corresponding to a boolean formula F, the problem PLANAR MONOTONE 3-SAT asks if F has a satisfying assignment.

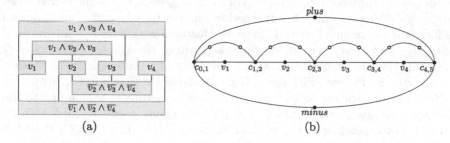

$$(a) \qquad\qquad (b)$$

Fig. 4. (a) A planar monotone representation of Φ_0. (b) The variable cycle of G_0 and false literal boundaries.

We denote by K_8^- the graph created by removing two adjacent edges from the complete graph K_8. In our reduction we make use of the following transformation. Let v be a vertex of G. we replace v with a copy of K_8^- by identifying v with the vertex of K_8^- with degree 5. After performing this operation we say that v is a K-vertex. The following lemma states a useful property of the K-vertices.

Lemma 1 [*]. *Let v be a K-vertex of a graph G and let G' be the K_8^- subgraph associated with v. In any $(2,2)$-planar representation of G, each vertex of G' is clustered with another vertex in G'.*

Theorem 3 [*]. $(2,2)$-PLANARITY *is NP-complete.*

Proof. $(2,2)$-PLANARITY is trivially in NP. We prove the NP-hardness of $(2,2)$-PLANARITY by reduction from PLANAR MONOTONE 3-SAT. Given an instance Φ of PLANAR MONOTONE 3-SAT, we construct a graph G that is a YES instance of $(2,2)$-PLANARITY if and only if Φ is a YES instance of PLANAR MONOTONE 3-SAT. For convenience, figures show the construction of the graph G_0 corresponding to the PLANAR MONOTONE 3-SAT instance Φ_0 in Fig. 4(a). In figures, we represent K-vertices and their associated K_8^- subgraphs with solid dots, while ordinary vertices are represented with hollow dots.

For each variable v_i of F (with $i = 1, \ldots, n$) create in G a K-vertex v_i and connect such K-vertices in a cycle, in the order implied by Φ (refer to Fig. 4(b)). Split each edge (v_i, v_{i+1}) of the cycle with a K-vertex $c_{i,i+1}$. Split the edge (v_1, v_n) with the vertices $c_{0,1}$ and $c_{n,n+1}$. Finally, duplicate the edge $(c_{0,1}, c_{n,n+1})$ and split the duplicated edges with the K-vertices *plus* and *minus*. We refer to this subgraph as the *variable cycle*. Given a variable v_i, let p_i be the number of positive clauses and q_i be the number of negative clauses of F in which v_i appears. For $1 \leq i \leq n$, connect $c_{i-1,i}$ to $c_{i,i+1}$ with a path of ordinary vertices of length equal to $max(p_i, q_i)$. We refer to these paths as *false literal boundaries*.

For each clause $C_j = (l_{j,1} \vee l_{j,2} \vee l_{j,3})$ in F, create a corresponding clause gadget in G. Create ordinary vertices $l_{j,1}, l_{j,2}, l_{j,3}$ and *open$_j$*, create a K-vertex *closed$_j$*, and add an edge between any pair of vertices, as in Fig. 5(a). Observe that in any $(2,2)$-planar representation of a clause gadget, two of the four vertices $l_{j,1}, l_{j,2}, l_{j,3}$ and *open$_j$* must be arranged in one cluster of size 2. This is due to the fact that by Lemma 1, *closed$_j$* must be clustered within its K_8^- subgraph. If $l_{j,1}, l_{j,2}, l_{j,3}$ and *open$_j$* were all clustered separately, the graph of clusters of G would contain a K_5 minor. Also, any 2-clustering of a clause gadget in which a literal vertex is clustered with *open$_j$* is $(2,2)$-planar, as shown in Fig. 5(b).

Now, connect the clause gadgets with a tree structure corresponding to the positions of clause rectangles in Φ. Let C_j be a clause rectangle in Φ with l_1, l_2, and l_3 corresponding to the vertical segments descending from C_j from left to right. If C_j is nested between vertical segments corresponding to literals m_1 and m_2 of another clause rectangle C_k, split the edges $(l_{j,1}, l_{j,3})$ and $(m_{k,1}, m_{k,2})$ with K-vertices and connect the new K-vertices with an edge. If C_j is nested under no other clause rectangle, split $(l_{j,1}, l_{j,3})$ with a K-vertex and connect the new vertex to *plus* if C_j corresponds to a positive clause and to *minus* otherwise. This procedure leads to a configuration consisting of two trees of clause gadgets connected as in Fig. 5(c). This concludes the construction of G. Refer to [6] for the proof that G is $(2,2)$-planar if and only if Φ has a satisfying assignment. □

Corollary 2. *The problem of computing the minimum value of k such that a graph is $(k,2)$-planar is NP-hard.*

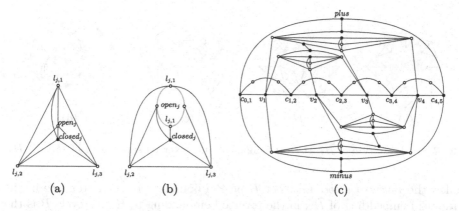

Fig. 5. (a) A clause gadget C_j. (b) A $(2,2)$-planar representation of the clause gadget C_j. (c) The graph G_0.

4 $(2,2)$-Planarity and 1-Planarity

The NP-completeness of $(2,2)$-PLANARITY suggests further investigation into the combinatorial properties of $(2,2)$-planar graphs. In this section, we study the relationship between $(2,2)$-planarity and 1-planarity. This is partly motivated by general interest in 1-planar graphs (see, e.g., [11]) and partly by the following observation.

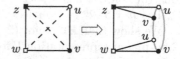

Fig. 6. Removal of a crossing in a $(2,2)$ representation.

Since a 1-planar graph admits a drawing where each edge is crossed by at most one other edge, it seems reasonable to remove each crossing of the drawing by clustering two of the vertices that are involved in the crossing as shown in Fig. 6. An n-vertex 1-planar graph has at most $4n-8$ edges [13]. By Theorem 1, a $(2,2)$-planar graph with n vertices has at most $4n-6$ edges, so it is not immediately clear that there are 1-planar graphs that are not $(2,2)$-planar.

As we are going to show, however, there is an infinite family of 1-planar graphs that are not $(2,p)$-planar for any value of $p \geq 1$. On the positive side, we demonstrate a large family of 1-planar graphs that are $(2,2)$-planar.

Theorem 4. *For every $h > 2$, there exists a 1-planar graph with $n = 5 \cdot 2^h - 8$ vertices and $m = 18 \cdot 2^h - 36$ edges that is not $(2,p)$-planar, for any $p \geq 1$.*

Proof. We define a recursive family of 1-plane graphs as follows. Graph $\overline{H}_1$ consists of a single *kite* K, which is a 1-plane graph isomorphic to K_4 drawn so that all the vertices are on the boundary of the outer face. Graph $\overline{H}_i$, for $i = 2, 3, \ldots$, has 2^i kites in addition to $\overline{H}_{i-1}$; these kites form a cycle in the outer face of $\overline{H}_{i-1}$, and each kite contains a vertex of the boundary of the outer face of $\overline{H}_{i-1}$ (note that $\overline{H}_{i-1}$ has 2^i vertices on the boundary of the outer face). See Fig. 7(a) for an example. The kites of $\overline{H}_i \setminus \overline{H}_{i-1}$ are called the *external kites of $\overline{H}_i$*. The embedding of $\overline{H}_i$ described in the definition will be called the *canonical embedding* of $\overline{H}_i$. We also consider another possible embedding,

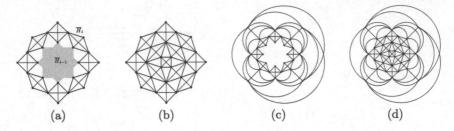

Fig. 7. (a) Definition of $\overline{H}_i$. (b)–(c) Canonical and reversed embedding of $\overline{H}_3$. (d) H_3.

called the *reversed embedding*. Let B be the boundary of the outer face in the canonical embedding of $\overline{H}_i$; in the reversed embedding of $\overline{H}_i$ the cycle B is the boundary of an inner face and all the rest of the graph is embedded outside B. See Fig. 7(b) and Fig. 7(c) for an example. For any $h > 2$, let $\overline{H}_h^c$ be a copy of $\overline{H}_h$ with a canonical embedding, and let $\overline{H}_h^r$ be a copy of $\overline{H}_h$ with a reversed embedding. The graph obtained by identifying the external kites of $\overline{H}_h^c$ with the external kites of $\overline{H}_h^r$ is denoted as H_h. Figure 7(d) shows the graph H_3. By construction, $\overline{H}_i$ has $n_i = 2^{i+1} - 4$ vertices and $m_i = 12 \cdot 2^i - 18$ edges. Hence, H_h has $n = 5 \cdot 2^h - 8$ vertices and $m = 18 \cdot 2^h - 36$ edges.

We show that H_h is not $(2,p)$-planar for any $p \geq 1$. Suppose that H_h has a $(2,p)$-planar representation Γ for some $p \geq 1$ and let G_C be the graph of clusters of H_h. Since Γ is planar, G_C must be planar. G_C can be obtained from H_h by contracting each pair of vertices that is assigned to each cluster region (and removing multiple edges). Contracting a pair of vertices u and v, the number of vertices reduces by one and the number of edges reduces by the number of paths of length at most 2 connecting u and v (for each path we remove one edge). In H_h, there are at most 4 such paths between any pair of vertices. Hence, if we contract q pairs of vertices, the number of vertices in G_C is $n' = n - q$, while the number of edges is $m' \geq m - 4q$. If G_C is planar, $m' \leq 3n' - 6$ and thus it must be $m - 4q \leq 3(n - q) - 6$, which gives $q \geq m - 3n + 6 = 3 \cdot 2^h - 6$, i.e. we must contract at least $3 \cdot 2^h - 6$ pairs of vertices. Since there are $5 \cdot 2^h - 8$ vertices, we can contract at most $\frac{5 \cdot 2^h - 8}{2}$ pairs. Thus, it must be $3 \cdot 2^h - 6 \leq 5 \cdot 2^{h-1} - 4$, i.e., $2^{h-1} \leq 2$, which can be satisfied only for $h \leq 2$.

Note that our argument is independent of the 1-planar embedding of H_h. This implies that the result holds for 1-planar graphs, not just for 1-plane graphs. $\square$

Theorem 4 motivates further investigation of the relationship between 1-planar and $(2,2)$-planar graphs. Note that there are infinitely many $(2,2)$-planar graphs that are not 1-planar. For example, observe that every graph obtained by connecting with an edge a planar graph and K_7 has such a property, because K_7 is not 1-planar (it has more than $4n - 8 = 20$ edges) but it is $(2,2)$-planar, as depicted in Fig. 8. In what follows, we describe a non-trivial family of 1-planar graphs that are also $(2,2)$-planar.

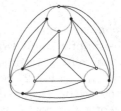

Fig. 8. A $(2,2)$-planar representation of K_7.

Let G be a 1-plane graph, and let $e_u = (u_1, u_2)$ and $e_v = (v_1, v_2)$ be a pair of crossing edges of G. Any pair $\langle u_i, v_j \rangle$, with $1 \le i, j \le 2$, is a *representative pair* of the edge crossing defined by e_u, e_v. An *independent set of distinct representatives* (*ISDR* for short) of G is a set of representative pairs such that there is exactly one representative pair per crossing and no two representative pairs in the set have a common vertex. Figure 9(b) shows an ISDR for the graph of Fig. 9(a).

We want to show that if a 1-plane graph G has an ISDR then it is $(2, 2)$-planar. The *crossing edges graph* of G, called *ce-graph* for short and denoted as $CE(G)$, is the subgraph of G induced by the crossing pairs of G. G is *pseudoforestal* if $CE(G)$ is a pseudoforest (i.e. it has at most one cycle in each connected component). For example, the 1-planar graph of Fig. 9(a) is pseudoforestal, as shown in Fig. 9(c). The pseudoforestal 1-planar graphs include non-trivial sub-families of 1-planar graphs, such as IC-planar graphs (whose *ce-graph* has maximum degree one), or the 1-planar graphs such that each vertex is shared by at most two crossing pairs (whose *ce-graph* has maximum degree two).

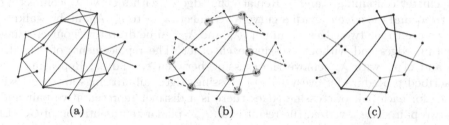

(a)   (b)   (c)

Fig. 9. (a) A 1-planar graph G. (b) An ISDR of G. For each pair of crossing edges the representative pair is indicated with a dashed line connecting the pair. Vertices shared by different crossing pairs are replicated in each pair. (c) The *ce-graph* $CE(G)$ of G.

Theorem 5. *A pseudoforestal 1-plane graph is $(2, 2)$-planar.*

Proof. We start by proving that a 1-plane graph G contains an ISDR if and only if G is pseudoforestal. It is known that a graph G can be oriented such that the maximum in-degree is k if and only if its pseudoarboricity is k (i.e. the edges of G can be partitioned into k pseudoforests) [12]. Thus, G is pseudoforestal if and only if $CE(G)$ can be oriented so that the maximum in-degree is one. We now show that this is a necessary and sufficient condition for the existence of an ISDR S in G. Assume that an ISDR exists. Let $e_u = (u_1, u_2)$ and $e_v = (v_1, v_2)$ be two crossing edges and let $\langle u_i, v_j \rangle$ $(1 \le i, j \le 2)$ be the representative pair of e_u and e_v. Direct e_u towards u_i and e_v towards v_j. Doing this for each pair of crossing edges defines an orientation for all edges of $CE(G)$. In this orientation each vertex of $CE(G)$ has in-degree at most 1, since no two pairs in S share a vertex. Now suppose that $CE(G)$ has an orientation such that each vertex has in-degree at most 1. For each pair of directed crossing edges $(u_1, u_2), (v_1, v_2)$ in

$CE(G)$, we add the pair $\langle u_2, v_2 \rangle$ to S. Since each vertex v in $CE(G)$ has in-degree at most 1, v is a vertex of at most one pair in S. Thus, the pairs selected for different crossing pairs are distinct and no two of them share a vertex.

We now describe how to use an ISDR S of G to construct a $(2,2)$-planar representation of G where each pair in S is represented as a 2-cluster that has 2 copies for each of its vertices. Let Γ be a 1-planar drawing of G that respects the 1-planar embedding of G. Consider any two crossing edges $e_u = (u_1, u_2)$ and $e_v = (v_1, v_2)$ and denote by c the point where they cross in Γ. Without loss of generality, assume that $\langle u_1, v_1 \rangle$ is the representative pair of e_u and e_v (see Fig. 10 for an illustration). Subdivide the edge e_u with a copy v_1' of v_1 placed between u_1 and c along e_u; analogously, subdivide the edge e_v with a copy u_1' of u_1. Add a curve λ_1 connecting u_1' to v_1' and a curve λ_2 connecting u_1 to v_1. By walking very close to the two edges e_u and e_v, these two curves can be drawn without crossing any existing edge and so that the closed curve λ formed by λ_1 and λ_2 together with the portion of e_u from u_1 to v_1' and the portion of e_v from v_1 to u_1' does not contain any vertex of Γ. Curve λ defines the cluster region for the cluster containing u and v. Replace the edge e_u with a curve λ_u connecting u_2 to u_1' and the edge e_v with a curve λ_v connecting v_2 to v_1'. Again, by walking very close to the two edges e_u and e_v, λ_u and λ_v can be drawn without crossing existing edges and without crossing each other. The replacements of e_u with λ_u and of e_v with λ_v remove the crossing between e_u and e_v. Repeating the described procedure for every pair of crossing edges, all crossings are removed. Since for each pair of crossing edges there is a distinct representative pair and no two pair share a vertex, the result is a $(2,2)$-planar representation of G. $\square$

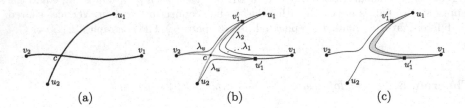

$$(a) \qquad\qquad (b) \qquad\qquad (c)$$

Fig. 10. (a) Two crossing edges e_u and e_v; (b) Construction of the cluster region and replacement of e_u and e_v; (c) The resulting drawing.

5   Conclusions and Open Problems

We introduced and studied (k, p)-planar graphs. We proved an upper bound on the number of edges of a (k, p)-planar graph, which is tight for $p < k$. This naturally motivates the problem of establishing a tight bound for $p \geq k$. We proved that (k, p)-planarity testing is NP-complete for $k = 4$ and $p = 1$ and for $k = 2$ and $p = 2$. It would be interesting to study the complexity of the problem for other values of k and p. Also, we investigated the relationship between 1-planar graphs and $(2,2)$-planar graphs. We showed that none of the two families is included in the other and described a subfamily that belongs to their intersection. An interesting research direction is to further investigate the relationship between 1-planar graphs and (k,p)-planar graphs for values of k larger than 2.

References

1. Angelini, P., Da Lozzo, G., Di Battista, G., Frati, F., Patrignani, M., Rutter, I.: Intersection-link representations of graphs. J. Graph Algorithms Appl. **21**(4), 731–755 (2017). https://doi.org/10.7155/jgaa.00437
2. Batagelj, V., Brandenburg, F., Didimo, W., Liotta, G., Palladino, P., Patrignani, M.: Visual analysis of large graphs using (X, Y)-clustering and hybrid visualizations. IEEE Trans. Vis. Comput. Graph. **17**(11), 1587–1598 (2011). https://doi.org/10.1109/TVCG.2010.265
3. de Berg, M., Khosravi, A.: Optimal binary space partitions for segments in the plane. Int. J. Comput. Geom. Appl. **22**(3), 187–206 (2012). http://www.worldscinet.com/doi/abs/10.1142/S0218195912500045
4. Brandenburg, F.J., Didimo, W., Evans, W.S., Kindermann, P., Liotta, G., Montecchiani, F.: Recognizing and drawing IC-planar graphs. Theor. Comput. Sci. **636**, 1–16 (2016). https://doi.org/10.1016/j.tcs.2016.04.026
5. Da Lozzo, G., Di Battista, G., Frati, F., Patrignani, M.: Computing NodeTrix representations of clustered graphs. In: Hu, Y., Nöllenburg, M. (eds.) GD 2016. LNCS, vol. 9801, pp. 107–120. Springer, Cham (2016). https://doi.org/10.1007/978-3-319-50106-2_9
6. Di Giacomo, E., Lenhart, W.J., Liotta, G., Randolph, T.W., Tappini, Λ.: (k, p)-planarity: A relaxation of hybrid planarity. CoRR abs/1806.11413 (2018)
7. Di Giacomo, E., Liotta, G., Patrignani, M., Tappini, A.: NodeTrix planarity testing with small clusters. In: Frati, F., Ma, K.-L. (eds.) GD 2017. LNCS, vol. 10692, pp. 479–491. Springer, Cham (2018). https://doi.org/10.1007/978-3-319-73915-1_37
8. Eades, P., de Mendonça N, C.F.X.: Vertex splitting and tension-free layout. In: Brandenburg, F.J. (ed.) GD 1995. LNCS, vol. 1027, pp. 202–211. Springer, Heidelberg (1996). https://doi.org/10.1007/BFb0021804
9. Eppstein, D., et al.: On the planar split thickness of graphs. Algorithmica **80**(3), 977–994 (2018). https://doi.org/10.1007/s00453-017-0328-y
10. Henry, N., Fekete, J., McGuffin, M.J.: NodeTrix: a hybrid visualization of social networks. IEEE Trans. Vis. Comput. Graph. **13**(6), 1302–1309 (2007). https://doi.org/10.1109/TVCG.2007.70582
11. Kobourov, S.G., Liotta, G., Montecchiani, F.: An annotated bibliography on 1-planarity. Comput. Sci. Rev. **25**, 49–67 (2017). https://doi.org/10.1016/j.cosrev.2017.06.002
12. Kowalik, Ł.: Approximation scheme for lowest outdegree orientation and graph density measures. In: Asano, T. (ed.) ISAAC 2006. LNCS, vol 4288, pp. 557 566. Springer, Heidelberg (2006). https://doi.org/10.1007/11940128_56
13. Pach, J., Tóth, G.: Graphs drawn with few crossings per edge. Combinatorica **17**(3), 427–439 (1997). https://doi.org/10.1007/BF01215922
14. Zhang, X., Liu, G.: The structure of plane graphs with independent crossings and its applications to coloring problems. Cent. Eur. J. Math. **11**(2), 308–321 (2013). https://doi.org/10.2478/s11533-012-0094-7

Drawing Clustered Graphs on Disk Arrangements

Tamara Mchedlidze[1], Marcel Radermacher[1(✉)], Ignaz Rutter[2],
and Nina Zimbel[1]

[1] Department of Computer Science, Karlsruhe Institute of Technology,
Karlsruhe, Germany
mched@iti.uka.de, radermacher@kit.edu
[2] Department of Computer Science and Mathematics,
University of Passau, Passau, Germany
rutter@fim.uni-passau.de

Abstract. Let $G = (V, E)$ be a planar graph and let $\mathcal{V}$ be a partition of V. We refer to the graphs induced by the vertex sets in $\mathcal{V}$ as *clusters*. Let $\mathcal{D}_\mathcal{C}$ be an arrangement of disks with a bijection between the disks and the clusters. Akitaya et al. [2] give an algorithm to test whether $(G, \mathcal{V})$ can be embedded onto $\mathcal{D}_\mathcal{C}$ with the additional constraint that edges are routed through a set of pipes between the disks. Based on such an embedding, we prove that every clustered graph and every disk arrangement without pipe-disk intersections has a planar straight-line drawing where every vertex is embedded in the disk corresponding to its cluster. This result can be seen as an extension of the result by Alam et al. [3] who solely consider biconnected clusters. Moreover, we prove that it is $\mathcal{NP}$-hard to decide whether a clustered graph has such a straight-line drawing, if we permit pipe-disk intersections.

1 Introduction

In practical applications, it often happens that a graph drawing produced by an algorithm has to be post processed by hand to comply with some particular requirements. Thus, the user moves vertices and modifies edges in order to fulfill these requirements. Interacting with large graphs is often time-consuming. It takes a lot of time to group and move the vertices or process them individually and to control the overall appearance of the produced drawing. The problem we study in this paper addresses this scenario. In particular, we assume that a user wants to modify a drawing of a large planar graph G. Instead we provide her with an abstraction of this graph. The user modifies the abstraction and thus providing some constraints on how the drawing of the initial graph should look like. Then our algorithm propagates the drawing of the abstraction to the initial graph so that the provided constraints are satisfied.

Work was partially supported by grant WA 654/21-1 of the German Research Foundation (DFG).

G. K. Das et al. (Eds.): WALCOM 2019, LNCS 11355, pp. 160–171, 2019.
https://doi.org/10.1007/978-3-030-10564-8_13

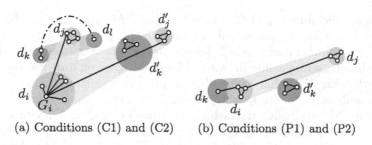

(a) Conditions (C1) and (C2)   (b) Conditions (P1) and (P2)

Fig. 1. (a) The blue disk arrangement satisfies the conditions (C1, C2) and (P1, P2). The disks d_k, d_l and d_i, d_j violate condition (C1). The disks d'_k and d_i, d'_j violate (C2). Note that the edge from disk d_i to d'_j has to cross the boundary of d'_k twice. (b) The disks d_k, d_i violate condition (P1) and d'_k and d_i, d_j violate condition (P2).

More formally, we model this scenario in terms of a *(flat) clustering* of a graph $G = (V, E)$, i.e., a partition $\mathcal{V} = \{V_1, \ldots, V_k\}$ of the vertex set V. We refer to the pair $\mathcal{C} = (G, \mathcal{V})$ as a *clustered graph* and the graphs G_i induced by V_i as *clusters*. The set of edges E_i of a cluster G_i are *intra-cluster edges* and the set of edges with endpoints in different clusters *inter-cluster edges*. A *disk arrangement* $\mathcal{D} = \{d_1, \ldots, d_k\}$ is a set of pairwise disjoint disks in the plane together with a bijective mapping $\mu(V_i) = d_i$ between the clusters $\mathcal{V}$ and the disks $\mathcal{D}$. We refer to a disk arrangement $\mathcal{D}$ with a bijective mapping μ as a *disk arrangement of* $\mathcal{C}$, denoted by $\mathcal{D_C}$. A $\mathcal{D_C}$-*framed drawing* of a clustered graph $\mathcal{C} = (G, \mathcal{V})$ is a planar drawing of G where each cluster G_i is drawn within its corresponding disk d_i. We study the following problem: given a clustered planar graph $\mathcal{C} = (G, \mathcal{V})$, an embedding ψ of G and a disk arrangement $\mathcal{D_C}$ of $\mathcal{C}$, does $\mathcal{C}$ admit a $\mathcal{D_C}$-framed straight-line drawing homeomorphic to ψ?

Related Work. Feng et al. [11] introduced the notion of *clustered graphs* and *c-planarity*. A graph G together with a recursive partitioning of the vertex set is considered to be a clustered graph. An embedding of G is *c-planar* if (i) each cluster c is drawn within a connected region R_c, (ii) two regions R_c, R_d intersect if and only if the cluster c contains the cluster d or vice versa, and (iii) every edge intersects the boundary of a region at most once. They prove that a c-planar embedding of a connected clustered graph can be computed in $O(n^2)$ time. It is an open question whether this result can be extended to disconnected clustered graphs. Many special cases of this problem have been considered [8].

Concerning drawings of c-planar clustered graphs, Eades et al. [10] prove that every c-planar graph has a c-planar straight-line drawing where each cluster is drawn in a convex region. Angelini et al. [5] strengthen this result by showing that every c-planar graph has a c-planar straight-line drawing in which every cluster is drawn in an axis-parallel rectangle. The result of Akitaya et al. [2] implies that in $O(n \log n)$ time one can decide whether an abstract graph with a flat clustering has an embedding where each vertex lies in a prescribed topological disk and every edge is routed through a prescribed topological pipe. In general they ask whether a simplicial map φ of G onto a 2-manifold M is a *weak embedding*, i.e., for every $\epsilon > 0$, φ can be perturbed into an embedding ψ_ϵ with $||\varphi - \psi_\epsilon|| < \epsilon$.

Godau [12] showed that it is $\mathcal{NP}$-hard to decide whether an embedded graph has a $\mathcal{D}_C$-framed straight-line drawing. The proof relies on a disk arrangement $\mathcal{D}_C$ of overlapping disks that have either radius zero or a large radius.

Banyassady et al. [6] study whether the intersection graph of unit disks has a straight-line drawing such that each vertex lies in its disk. They proved that this problem is $\mathcal{NP}$-hard regardless of whether the embedding of the intersection graph is prescribed or not. Angelini et al. [4] showed it is $\mathcal{NP}$-hard to decide whether an abstract graph G and an arrangement of unit disks have a $\mathcal{D}_C$-framed straight-line drawing. They leave the problem of finding a $\mathcal{D}_C$-framed straight-line drawing of G with a fixed embedding as an open question. Alam et al. [3] prove that it is $\mathcal{NP}$-hard to decide whether an embedded clustered graph has a c-planar straight-line drawing where every cluster is contained in a prescribed (thin) rectangle and edges have to pass through a defined part of the boundary of the rectangle. Further, they prove that all instances with biconnected clusters always admit a solution. Their result implies that graphs of this class have $\mathcal{D}_C$-framed straight-line drawings.

Ribó [14] shows that every embedded clustered graph where each cluster is a set of independent vertices has a straight-line drawing such that every cluster lies in a prescribed disk. In contrast to our setting Ribó allows an edge e to intersect a disk of a cluster G_i that does not contain an endpoint of e.

Contribution. A *pipe* p_{ij} of two clusters V_i, V_j is the *convex hull* of the disks d_i and d_j, i.e., the smallest convex set of points containing d_i and d_j; see Fig. 1. We refer to a topological planar drawing of G as an *embedding of G*. A $\mathcal{D}_C$-*framed embedding of G* is a $\mathcal{D}_C$-framed topological drawing of G with the additional requirement that (i) each intra-cluster edge entirely lies in its disk (ii) each inter-cluster edge uv intersects with a pipe p_{ij} if and only if u and v are vertices of the clusters G_i and G_j, respectively, and (iii) each edge crosses the boundary of a disk at most once. This concept is also known as *c-planarity with embedded pipes* [9]. An embedding ψ of G is *compatible with* $\mathcal{D}_C$ if ψ is homeomorphic to a $\mathcal{D}_C$-framed embedding of G. The result of Akitaya et al. can be used to decide whether an embedding ψ of G is compatible with $\mathcal{D}_C$.

The following two conditions are necessary, for $\mathcal{C}$ to have a $\mathcal{D}_C$-framed embedding: (C1) if $(V_i \times V_j) \cap E \neq \emptyset$ and $(V_k \times V_l) \cap E \neq \emptyset$ (i, j, k, l pairwise distinct), then the intersection of the pipes p_{ij} and p_{kl} is empty, and (C2) the set $p_{ij} \setminus d_k$ is connected. Thus, in the following we assume that $\mathcal{D}_C$ satisfies (C1) and (C2). A *planar* disk arrangement additionally satisfies the condition that (P1) the pairwise intersections of all disks are empty, and (P2) $(V_i \times V_j) \cap E \neq \emptyset$, the intersection of p_{ij} with all disks d_k (corresponding to V_k) is empty (i, j, k pairwise distinct). A planar disk arrangement can be seen as a thickening of a planar straight-line drawing of the graph obtained by contracting all clusters.

We prove that every clustered graph $(G, \mathcal{V})$ with planar disk arrangement $\mathcal{D}_C$ and an $\mathcal{D}_C$-framed embedding ψ has a $\mathcal{D}_C$-framed planar straight-line drawing homeomorphic to ψ. Taking the result of Akitaya et al. [2] into account, our

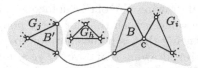

Fig. 2. A planar clustered graph C that is not simple.

result can be used to test whether an abstract clustered graph with connected clusters has a $\mathcal{D}_C$-framed straight-line drawing. Cluster G_i in Fig. 1 shows that in general clusters cannot be augmented to be biconnected, if the embedding is fixed. Hence, our result is generalization of the result of Alam et al. [3]. In Sect. 3 we show that the problem is $\mathcal{NP}$-hard in the case that the disk arrangements does not satisfy condition (P2). From now on we refer to a planar straight-line drawing of G simply as a drawing of G.

2   Drawing on Planar Disk Arrangements

In this Section we prove that every *simple* clustered graph with a planar disk arrangement $\mathcal{D}_C$ and $\mathcal{D}_C$-framed embedding has a $\mathcal{D}_C$-framed drawing. An embedded clustered graph C is *simple* if for every i, j, there is no cluster $G_h(i, j \neq h)$ embedded in the interior of the subgraph induced by $V_i \cup V_j$; see Fig. 2. Note that this is a necessary condition for the corresponding disk arrangement to be planar. A clustered graph $C = (G, \mathcal{V})$ is *connected* if each cluster G_i is connected.

We prove the statement by induction on the number of intra-cluster edges. In Lemma 1 we show that we can indeed reduce the number of intra-cluster edges by contracting intra-cluster edges. In Lemma 2, we prove that the statement is correct if the outer face is a triangle and C is connected. In Theorem 3 we extend this result to clustered graphs whose clusters are not connected.

Let $C = (G, \mathcal{V})$ with a disk arrangement $\mathcal{D}_C$ and a $\mathcal{D}_C$-framed embedding ψ. Let uv be an intra-cluster edge of G that is not an edge of a separating triangle. We obtain a *contracted clustered graph* C/e of C be removing v from G and connecting the neighbors of v to u. We obtain a corresponding embedding ψ/e from ψ by routing the edges $vw \in E, w \neq u$ close to uv.

Lemma 1. *Let $C = (G, \mathcal{V})$ be a connected simple clustered graph with a planar disk arrangement $\mathcal{D}_C$ and a $\mathcal{D}_C$-framed embedding ψ. Let e be an intra-cluster edge that is not an edge of a separating triangle. Then C has a $\mathcal{D}_C$-framed drawing that is homeomorphic to ψ if C/e has a $\mathcal{D}_C$-framed drawing that is homeomorphic to ψ/e.*

Proof. Let $e = uv$ and denote by $u_0, u_1, \ldots, u_k$ the neighbors of u and $v_0, v_1, \ldots, v_l$ the neighbors of v in C. Without loss of generality, we assume that $u_0 = v$ and $v_0 = u$. Since e is not an edge of a separating triangle the set $I := \{u_2, \ldots, u_{k-1}\} \cap \{v_2, \ldots, v_{l-1}\}$ is empty. Denote by u the vertex obtained by the contraction of e. Let G_i be the cluster of u and v, and let d_i be the corresponding disk in $\mathcal{D}_C$.

Consider a $\mathcal{D}_C$-framed drawing Γ/e of C/e homeomorphic to ψ/e. Then there is a small disk $d_u \subset d_i$ around u such that for every point p in d_u moving u to p yields a $\mathcal{D}_C$-framed drawing that is homeomorphic to ψ/e.

We obtain a straight-line drawing Γ of C from Γ/e as follows. First, we remove the edges uv_i from Γ/e. The edges u_1, u_k partitions d_u into two regions r_u, r_v such that the intersection of r_v with uu_i is empty for all $i \in \{2, \ldots, k-1\}$. We place v in r_v and connect it to u and the vertices $v_1, \ldots, v_l$. Since r_v is a subset of d_u and $I = \emptyset$, we have that the new drawing Γ is planar. Since v is placed in r_v, the edge uv is in between u_1 and u_k in the rotational order of edges around u. Hence, Γ is homeomorphic to ψ. Finally, Γ is a $\mathcal{D}_C$-framed drawing since, d_u is entirely contained in d_i and thus are u and v.   □

Lemma 2. *Let C be a connected simple clustered graph with a triangular outer face T, a planar disk arrangement $\mathcal{D}_C$, and a $\mathcal{D}_C$-framed embedding ψ. Moreover, let Γ_T be a $\mathcal{D}_C$-framed drawing of T. Then C has a $\mathcal{D}_C$-framed drawing that is homeomorphic to ψ with the outer face drawn as Γ_T.*

Proof. We prove the theorem by induction on the number of intra-cluster edges.

First, assume that every intra-cluster edge of C is an edge on the boundary of the outer face. Let Γ be the drawing obtained by placing every interior vertex on the center point of its corresponding disk and draw the outer face as prescribed by Γ_T. Since $\mathcal{D}_C$ is a planar disk arrangement and Γ_T is convex, the resulting drawing is planar and thus a $\mathcal{D}_C$-framed drawing of C that is homeomorphic to the embedding ψ.

Let S be a separating triangle of C that splits C into two subgraphs C_{in} and C_{out} so that $C_{in} \cap C_{out} = S$ and the outer face C_{out} and C coincide. Then by the induction hypothesis C_{out} has the $\mathcal{D}_C$-framed drawing Γ_{out} with the outer face drawn as Γ_T and C_{in} as a $\mathcal{D}_C$-framed drawing Γ_{in} with the outer face drawing as $\Gamma_{out}[S]$, where $\Gamma_{out}[S]$ is the drawing of S in Γ_{out}. Then we obtain a $\mathcal{D}_C$-framed drawing of C by merging Γ_{in} and Γ_{out}.

Consider an intra-cluster edge e that does not lie on the boundary of the outer face and is not an edge of a separating triangle. Then by the induction hypothesis, C/e has a $\mathcal{D}_C$-framed drawing with the outer face drawn as Γ_T. It follows by Lemma 1 that C has a $\mathcal{D}_C$-framed drawing homeomorphic to ψ.   □

Theorem 3. *Every simple clustered graph C with a $\mathcal{D}_C$-framed embedding ψ has a $\mathcal{D}_C$-framed drawing homeomorphic to ψ.*

Proof. We obtain a clustered graph C' from C by adding a new triangle T to the graph and assigning each vertex of T to is own cluster. Let Γ_T be a drawing of T that contains all disks in $\mathcal{D}_C$ in its interior. We obtain a new disk arrangement $\mathcal{D}'_C$ from $\mathcal{D}_C$ by adding a sufficiently small disk for each vertex of Γ_T. The embedding ψ together with Γ_T is a $\mathcal{D}'_C$-framed embedding ψ' of C'.

According to Feng et al. [11] there is a simple connected clustered graph C'' that contains C' as a subgraph whose embedding ψ'' is $\mathcal{D}_C$-framed and contains ψ'. By Lemma 2 there is a $\mathcal{D}_C$-framed drawing Γ'' of C'' homeomorphic to ψ'' with the outer face drawn as Γ_T. The drawing Γ'' contains a $\mathcal{D}_C$-framed drawing of C.   □

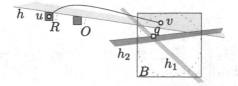

Fig. 3. Regulator

3 Drawing on General Disk Arrangements

We study the following problem referred to as $\mathcal{D}_{\mathcal{C}}$-FRAMED DRAWINGS OF NON-PLANAR ARRANGEMENTS. Given a planar clustered graph $\mathcal{C} = (G, \mathcal{V})$, a disk arrangement $\mathcal{D}_{\mathcal{C}}$ that is not planar, i.e., $\mathcal{D}_{\mathcal{C}}$ satisfies condition (C1) and (C2) but not (P1) and (P2), and a $\mathcal{D}_{\mathcal{C}}$-framed embedding ψ of G, is there a $\mathcal{D}_{\mathcal{C}}$-framed straight-line drawing Γ that is homeomorphic to ψ and $\mathcal{D}_{\mathcal{C}}$? Note that if the disks $\mathcal{D}_{\mathcal{C}}$ are allowed to overlap (condition (P1)) and G is the intersection graph of $\mathcal{D}_{\mathcal{C}}$, the problem is known to be $\mathcal{NP}$-hard [6]. Thus, in the following we require that the disks do not overlap, but there can be disk-pipe intersections, i.e, $\mathcal{D}_{\mathcal{C}}$ satisfies conditions (C1), (C1) and (P1) but not (P2). By Alam et al. [3] it follows that the problem restricted to thin touching rectangles instead of disks is $\mathcal{NP}$-hard. We strengthen this result and prove that in case that the rectangles are axis-aligned squares and are not allowed to touch the problem remains $\mathcal{NP}$-hard. Our illustrations contain blue dotted circles that indicate how the square in the proof can be replaced by disks.

To prove $\mathcal{NP}$-hardness we reduce from PLANAR MONOTONE 3-SAT [7]. For each literal and clause we construct a clustered graph $\mathcal{C}$ with an arrangement of squares $\mathcal{D}_{\mathcal{C}}$ of $\mathcal{C}$ such that each disk contains exactly one vertex. We refer to these instances as *literal* and *clause gadgets*. In order to transport information from the literals to the clauses, we construct a *copy* and *inverter gadget*. The design of the gadgets is inspired by Alam et al. [3], but due to the restriction to squares rather than rectangles, requires a more careful placement of the geometric objects. The green and red regions in the figures of the gadget correspond to *positive* and *negative* drawings of the literal gadget. The green and red line segments indicate that for each truth assignment of the variables our gadgets indeed have $\mathcal{D}_{\mathcal{C}}$-framed straight-line drawings. Negative versions of the literal and clause gadget are obtained by mirroring vertically. Hence, we assume that variables and clauses are positive. Each gadget covers a set of checkerboard cells. This simplifies the assembly of the gadgets for the reduction. The full version of the proof can be found on arXiv [13].

An *obstacle of a pipe* p_{ij} is a disk $d_k, i, j \neq k$, that intersects p_{ij}. The *obstacle number of a pipe* p_{ij} is the number of obstacles of p_{ij}. Let $P = \{p_{ij} \mid V_i \times V_j \cap E \neq \emptyset\}$. The *obstacle number of a disk arrangement* $\mathcal{D}_{\mathcal{C}}$ is maximum obstacle number of all pipes p_{ij} with $V_i \times V_j \cap E \neq \emptyset$.

Regulator. The *regulator gadget* restricts the feasible placements of a vertex v that lies in the interior of a square B; refer to Fig. 3. Let h_1, h_2 be two half planes

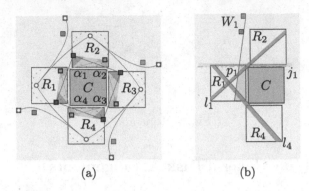

(a)   (b)

Fig. 4. Literal gadget

such that the intersection q of their supporting lines lies in B. In a $\mathcal{D}_C$-framed drawing of the regulator gadget the placement of v is restricted by a half plane h that excludes a placement of v in $h_1 \cap h_2$ but allows for a placement in $h_1 \cap B$ or $h_2 \cap B$. We refer to $h \cap B$ as the *regulated region of* B.

Literal Gadget. The *positive literal gadget* is depicted in Fig. 4. The *center block* is a unit square C with corners $\alpha_1, \alpha_2, \alpha_3, \alpha_4$ in clockwise order. For each corner α_i of C consider a line l_i that is tangent to C in α_i, i.e, $l_i \cap C = \{\alpha_i\}$. Let p_i be the intersection of lines l_{i-1} and l_i where $l_0 = l_4$; refer to Fig. 4b. Let $R_1, \ldots, R_4$ be four pairwise non-intersecting squares that are disjoint from C such that R_i contains p_i in its interior. We add a cycle $v_1 v_2 v_3 v_4 v_1$ such that $v_i \in R_i$. We refer to the vertex v_i as the *cycle vertex* of the *cycle block* R_i. For each i, let j_i be a half plane that contains R_{i+1} but does not intersect C. We place a regulator W_i of v_i with respect to h_{i-1} and h_i and position it such that it lies in j_i, where h_i is the half plane spanned by l_i with $C \not\subseteq h_i$.

We now describe the two combinatorially different realizations of the literal gadgets. Consider R_1 and its two adjacent squares R_2 and R_4. Let Q_i be the regulated region of R_i with respect to W_i. We refer to $\overline{h_2} \cap \overline{h_4} \cap Q_1$ as the *infeasible region of* R_1, where $\overline{h_i}$ denotes the complement of h_i. The intersection $h_1 \cap Q_1$ is the *positive region* P_1 of R_1. The region $\overline{h_4} \cap Q_1$ is the *negative region* N_1 of R_1. All these regions are by construction not empty. The positive, negative and infeasible region of $R_i, i \neq 1$ are defined analogously.

Property 4. If Γ is a $\mathcal{D}_C$-framed drawing of a positive (negative) literal gadget, then no cycle vertex v_i lies in the infeasible region of R_i. Moreover, either each cycle vertex v_i lies in the positive region P_i or each vertex v_i lies in the negative region N_i.

Property 5. The positive and negative placements induce a $\mathcal{D}_C$-framed drawing of the literal gadget, respectively.

Copy and Inverter Gadget. The copy gadget in Fig. 5 connects two positive literal gadgets X and Y such that a drawing of X is positive if and only if the

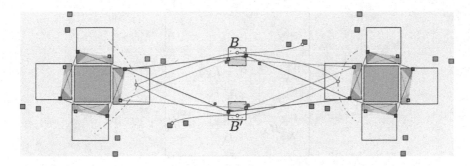

Fig. 5. Copy gadget

drawing of Y is positive. The inverter gadget connects a positive literal gadget X to a negative literal gadget Y such that the drawing of X is positive if and only if the drawing of Y is negative. The construction of the gadgets uses ideas similar to the construction of the literal gadget. In contrast to the literal gadget, we replace the center block by four squares.

Property 6. Let Γ be a $\mathcal{D}_C$-framed drawing of two positive (negative) literals gadgets X and Y connected by a copy gadget. Then the $\mathcal{D}_C$-framed of X in Γ is positive if and only if the $\mathcal{D}_C$-framed drawing of Y is positive.

Property 7. The positive (negative) placement of two literals gadgets X, Y induces a $\mathcal{D}_C$-framed drawing of a copy [inverter] gadget that connects X and Y.

Clause Gadget. We construct a *clause gadget* with respect to three positive literal gadgets X, Y, Z arranged as depicted in Fig. 7. The negative clause gadget, i.e., a clause with three negative literal gadgets, is obtained by mirroring vertically.

We construct the clause gadget in two steps. First, we place a *transition block* T_A close to each literal gadget $A \in \{X, Y, Z\}$. In the second step, we connect the transition block to a vertex k in a *clause block* K such that for every placement of k in K at least one drawing of the literal gadgets has to be positive.

Consider the literal gadget X and let R_X be the right-most cycle block of X. Let h_X be a negative half plane of R_X, i.e., h_X contains the positive region P_X but not the negative region N_X, refer to Fig. 6. We now place a transition block T_X such that the intersection $T_X \cap h_X$ has small area. Further, let p_X^+ and p_X^- be the positive and negative placements of X, respectively. Let q_X^- be a point in $T_X \cap h_X$. Let i be the intersection point of the supporting line l_X of h_X and the line segment $p_X^- q_X^-$. We place an obstacle O_X^1 such that l_X is tangent to O_X^1 in point i. Finally, we place a *transition vertex* t_X in the interior of T_X and route the edge $v_X t_X$ through $h_X \cup T_X \cup R_X$, where $v_X \in R_X$.

Consider a half plane h_X' such that $O_X^1 \not\subseteq h_X'$ and $N_X \not\subseteq h_X'$ and such that the supporting line l_X' of h_X' contains p_X^+ and is tangent to O_X^1. Let q_X^+ be a point $h_X' \cap R_X$. Observe that for q_X^+ and q_X^- there is a positive and negative drawing of X, respectively. Further, if X has a negative drawing, then t_X lies in the region $h_X \cap T_X$. In the following, we refer to $h_X \cap T_X$ as the *negative region*

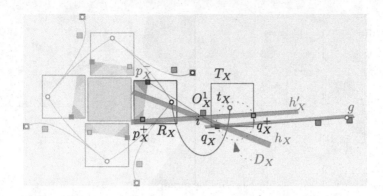

Fig. 6. Construction of the transition block.

of T_X. The transition blocks of Y and Z are constructed analogously with only minor changes. The transition block T_Z of Z is constructed with respect to the top-most cycle block. Note that we can choose the points $q_A^+, A \in \{X, Y, Z\}$ independent from each other as long as each of them induces a positive drawing the literal gadget A.

Denote by x-max S the maximum x-coordinate of a point in a bounded set $S \subset \mathbb{R}^2$. Note that x-max $D_X \cap h_X >$ x-max $T_X \cap h_X$, refer to Fig. 6. To ensure that our construction remains correct for disks we add a regulator R with a respect a half plane g such that x-max $D_X \cap h_x \cap g =$ x-max $T_X \cap h_x \cap g$ and g contains q_X^+, q_X^-.

Given the placement of the transition block T_X, T_Y and T_Z as depicted in Fig. 7, we construct the *clause block* K as follows. We choose a point $q_{X,Y}$. Let l_X^- and l_Y^- be the lines through the points $q_X^-, q_{X,Y}$, and $q_Y^-, q_{X,Y}$, respectively. Further, consider a line l_Z^- with $q_Z^- \in l_Z^-$ such that the intersection point $q_{A,Z} := l_Z^- \cap l_A^-, A \in \{X, Y\}$ lies in between q_A^- and $q_{X,Y}$. Further, let l_X^+ be the line through $q_X^+, q_{Y,Z}$, l_Y^+ the line through $q_Y^+, q_{X,Z}$, and let l_Z^+ be the line through q_Z^+ and $q_{X,Y}$. Let h_A be a half plane that does not contain the negative region N_A and whose supporting line contains the intersection i_A of l_A^- and l_A^+. We place obstacles O_A^2 such that $O_A^2 \not\subseteq h_A$ and the supporting line of h_A is tangent to O_A^2 in point i_A. We place the clause box K such that it contains $q_{X,Y}, q_{Y,Z}$, $q_{X,Z}$ and a new vertex k in its interior. We finish the construction by routing the edges kt_A through $K \cup h_A \cup T_A, A \in \{X, Y, Z\}$, where $t_A \in T_A$.

By construction we have that for each $y \in \{q_Y^-, q_Y^+\}$ and $z \in \{q_Z^-, q_Z^+\}$ the points y, z and $q_{Y,Z}$ induce a $\mathcal{D}_C$-framed drawing. The analog statement for the points $q_{X,Z}$ and $q_{X,Y}$ is also true. Further, if $h_X \cap h_Y \cap h_Z = \emptyset$, then there is no $\mathcal{D}_C$-framed drawing such that each vertex t_A lies on q_A^-. Figure 7 shows that there is an arrangement of the clause block and the obstacles such that $h_X \cap h_Y \cap h_Z$ indeed is empty.

Property 8. There is no $\mathcal{D}_C$-framed drawing of the clause gadget such that the drawing of each literal gadget is negative. For all other combinations of positive and negative drawings of the literal gadgets there is a $\mathcal{D}_C$-framed drawing of the clause gadget.

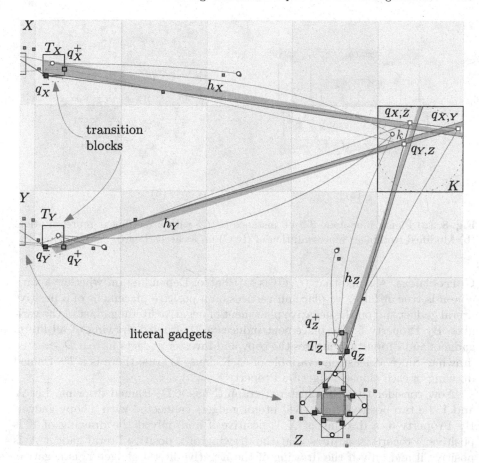

Fig. 7. Construction of the clause block.

Reduction. We reduce from a planar monotone 3-SAT instance (U, C); refer to Fig. 8. We modify its rectilinear representation such that each vertex and clause rectangle covers sufficiently many cells of a checkerboard and each edge covers the entire column between its two endpoints. We place positive literal gadgets in each blue cell of a rectangle corresponding to a variable. We place a clause gadget in each positive clause rectangle R_c such that it is aligned with the right-most edge of R_c. The literal gadget X of a variable x is connected to its corresponding literal gadget X' in R_c by a placing a literal gadget in each blue cell that is covered by the Γ-shape that connects X to X'; refer to Fig. 8b. Finally, we place a copy gadget in each orange cell between two literal gadgets of the same variable. The negative clauses are obtained by mirroring the modified rectilinear representation vertically and repeating the construction for the positive clauses. To negate the state of the variable we place the inverter gadget immediately below a variable (red cells in Fig. 8b).

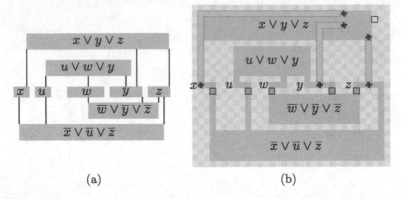

(a)   (b)

Fig. 8. (a) Planar monotone 3-SAT instance (U, C) with a rectilinear representation. (b) Modified rectilinear representation of (U, C) on a checkerboard.

Correctness. Assume that (U, C) is satisfiable. Depending on whether a variable u is true or false, we place all vertices on a positive placement of a positive literal gadget and on the negative placement of negative literal gadget of the variable. By Property 5, the placement induces a $\mathcal{D}_C$-framed drawing of all literal gadgets and Property 7 ensures the copy and inverter gadgets have a $\mathcal{D}_C$-framed drawing. Since at least one variable of each clause is true, there is a $\mathcal{D}_C$-framed drawing of each clause gadget by Property 8.

Now consider that the clustered graph C has a $\mathcal{D}_C$-framed drawing. Let X and Y be two positive (negative) literal gadgets connected with a copy gadget. By Property 6, a drawing of X is positive if and only if the drawing of Y is positive. Property 6 ensures that the drawing of a positive literal gadget X is positive if and only if the drawing of the negative literal gadget Y is negative, in case that both are joined with an inverter gadget. Further, Property 4 states that each cycle vertex lies either in a positive or a negative region. Thus, the truth value of a variable u can be consistently determined by any drawing of a literal gadget of u. By Property 8, the clause gadget K has no $\mathcal{D}_C$-framed drawing such that all drawings of the literal gadgets of K are negative. Thus, the truth assignment indeed satisfies C.

Theorem 9. *The problem $\mathcal{D}_C$-FRAMED DRAWINGS OF NON-PLANAR ARRANGEMENTS with axis-aligned squares is $\mathcal{NP}$-hard even when the clustered graph C is restricted to vertex degree 5 and the obstacle number of $\mathcal{D}_C$ is two.*

4   Conclusion

We proved that every clustered planar graph with a planar disk arrangement $\mathcal{D}_C$ and a $\mathcal{D}_C$-framed embedding ψ has a $\mathcal{D}_C$-framed straight-line drawing homeomorphic to ψ. If the requirement of the disk arrangement to satisfy condition (P2) is dropped, we proved that it is $\mathcal{NP}$-hard to decide whether C has a $\mathcal{D}_C$-framed straight-line drawing. We are not aware whether our problem is known to be in $\mathcal{NP}$. We ask whether techniques developed by Abrahamsen et al. [1] can be used to prove $\exists\mathbb{R}$-hardness of our problem.

Angelini et al. [4] showed that if $\mathcal{C}$ is not embedded and all squares have the same size, it is $\mathcal{NP}$-hard to decide whether $\mathcal{C}$ has a $\mathcal{D}_\mathcal{C}$-framed drawing. They posed as an open problem whether the same is true for embedded graphs. In our construction, the squares have constant number of different side lengths and the side length of the largest square is 32 time longer then the side length of the smallest rectangle. We conjecture that our construction can be modified to show that it is indeed $\mathcal{NP}$-hard to decide whether a clustered graph $\mathcal{C}$ with a non-planar arrangement of squares (disk) of unit size and a $\mathcal{D}_\mathcal{C}$-framed embedding ψ has a $\mathcal{D}_\mathcal{C}$-framed drawing that is homeomorphic to ψ. Further, we ask whether the obstacle number can be reduced to one.

References

1. Abrahamsen, M., Adamaszek, A., Miltzow, T.: The art gallery problem is $\exists\mathbb{R}$-complete. In: STOC 2018, pp. 65–73. ACM (2018)
2. Akitaya, H.A., Fulek, R., Tóth, C.D.: Recognizing weak embeddings of graphs. In: Czumaj, A. (ed.) SODA 2018, pp. 274–292. SIAM (2018)
3. Alam, M., Kaufmann, M., Kobourov, S.G., Mchedlidze, T.: Fitting planar graphs on planar maps. J. Graph Alg. Appl. **19**(1), 413–440 (2015)
4. Angelini, P., et al.: Anchored drawings of planar graphs. In: Duncan, C., Symvonis, A. (eds.) GD 2014. LNCS, vol. 8871, pp. 404–415. Springer, Heidelberg (2014). https://doi.org/10.1007/978-3-662-45803-7_34
5. Angelini, P., Frati, F., Kaufmann, M.: Straight-line rectangular drawings of clustered graphs. Disc. Comput. Geom. **45**(1), 88–140 (2011)
6. Banyassady, B., Hoffmann, M., Klemz, B., Löffler, M., Miltzow, T.: Obedient plane drawings for disk intersection graphs. In: Ellen, F., Kolokolova, A., Sack, J.R. (eds.) Algorithms and Data Structures. LNCS, vol. 10389, pp. 73–84. Springer, Cham (2017). https://doi.org/10.1007/978-3-319-62127-2_7
7. de Berg, M., Khosravi, A.: Optimal binary space partitions for segments in the plane. Int. J. Comput. Geom. Appl. **22**(3), 187–206 (2012)
8. Bläsius, T., Rutter, I.: A new perspective on clustered planarity as a combinatorial embedding problem. Theor. Comput. Sci. **609**(2), 306–315 (2016)
9. Cortese, P.F., Di Battista, G., Patrignani, M., Pizzonia, M.: On embedding a cycle in a plane graph. Disc. Math. **309**(7), 1856–1869 (2009)
10. Eades, P., Feng, Q., Lin, X., Nagamochi, H.: Straight-line drawing algorithms for hierarchical graphs and clustered graphs. Algorithmica **44**(1), 1–32 (2006)
11. Feng, Q.-W., Cohen, R.F., Eades, P.: Planarity for clustered graphs. In: Spirakis, P. (ed.) ESA 1995. LNCS, vol. 979, pp. 213–226. Springer, Heidelberg (1995). https://doi.org/10.1007/3-540-60313-1_145
12. Godau, M.: On the difficulty of embedding planar graphs with inaccuracies. In: Tamassia, R., Tollis, I.G. (eds.) GD 1994. LNCS, vol. 894, pp. 254–261. Springer, Heidelberg (1995). https://doi.org/10.1007/3-540-58950-3_377
13. Mchedlidze, T., Radermacher, M., Rutter, I., Zimbel, N.: Drawing Clustered Graphs on Disk Arrangements (2018). https://arxiv.org/abs/1811.00785
14. Ribó Mor, A.: Realization and Counting Problems for Planar Structures. Ph.D. thesis, FU Berlin (2006). https://refubium.fu-berlin.de/handle/fub188/10243

Graph Algorithms

Computing the Metric Dimension by Decomposing Graphs into Extended Biconnected Components

(Extended Abstract)

Duygu Vietz$^{(\boxtimes)}$ (ID), Stefan Hoffmann, and Egon Wanke

Heinrich-Heine-University Duesseldorf,
Universitaetsstr. 1, 40225 Duesseldorf, Germany
duygu.vietz@hhu.de

Abstract. A vertex set $U \subseteq V$ of an undirected graph $G = (V, E)$ is a *resolving set* for G, if for every two distinct vertices $u, v \in V$ there is a vertex $w \in U$ such that the distance between u and w and the distance between v and w are different. The *Metric Dimension* of G is the size of a smallest resolving set for G. Deciding whether a given graph G has Metric Dimension at most k for some integer k is well-known to be NP-complete. A lot of research has been done to understand the complexity of this problem on restricted graph classes. In this paper, we decompose a graph into its so called *extended biconnected components* and present an efficient algorithm for computing the metric dimension for a class of graphs having a minimum resolving set with a bounded number of vertices in every extended biconnected component. Furthermore, we show that the decision problem METRIC DIMENSION remains NP-complete when the above limitation is extended to usual biconnected components.

Keywords: Graph algorithm · Complexity · Metric dimension
Resolving set · Biconnected component

1 Introduction

An undirected graph $G = (V, E)$ has metric dimension at most k if there is a vertex set $U \subseteq V$ such that $|U| \leq k$ and $\forall u, v \in V$, $u \neq v$, there is a vertex $w \in U$ such that $d_G(w, u) \neq d_G(w, v)$, where $d_G(u, v)$ is the distance (the length of a shortest path in an unweighted graph) between u and v. The metric dimension of G is the smallest integer k such that G has metric dimension at most k. The metric dimension was independently introduced by Harary and Melter [10] and Slater [22]. If for three vertices u, v, w, we have $d_G(w, u) \neq d_G(w, v)$, then we say that u and v are *resolved* by vertex w. If every pair of vertices is resolved by at least one vertex of a vertex set U, then U is a *resolving set* for G. The *metric dimension* of G is the size of a minimum resolving set. Such a smallest resolving

© Springer Nature Switzerland AG 2019
G. K. Das et al. (Eds.): WALCOM 2019, LNCS 11355, pp. 175–187, 2019.
https://doi.org/10.1007/978-3-030-10564-8_14

set is also called a *resolving basis* for G. In certain applications, the vertices of a resolving set are also called *resolving vertices*, *landmark nodes* or *anchor nodes*. This is a common naming particularly in the theory of sensor networks.

Determining the metric dimension of a graph is a problem that has an impact on multiple research fields such as chemistry [2], robotics [18], combinatorial optimization [21] and sensor networks [15]. Deciding whether a given graph G has metric dimension at most k for a given integer k is known to be NP-complete for general graphs [9], planar graphs [3], even for those with maximum degree 6 and Gabriel unit disk graphs [15]. Epstein et al. showed the NP-completeness for split graphs, bipartite graphs, co-bipartite graphs and line graphs of bipartite graphs [4] and Foucaud et al. for permutation and interval graphs [7,8].

There are several algorithms for computing the metric dimension in polynomial time for special classes of graphs, as for example for trees [2,18], wheels [14], k-regular bipartite graphs [20], amalgamation of cycles [17], outerplanar graphs [3], cactus block graphs [16] and chain graphs [6]. The approximability of the metric dimension has been studied for bounded degree, dense, and general graphs in [12]. There are many variants of the Metric Dimension problem, see [1,4,5,7,8,11,13,19].

In this paper, we introduce a concept that allows us to compute the metric dimension based on a tree structure given by the decomposition of a graph G into components like *bridges*, *legs*, and so-called *extended biconnected components*. An *extended biconnected component* H of G is an induced subgraph of G formed by a biconnected component H' of G extended by paths attached to vertices of the biconnected component H'. Each vertex of H' has at most one path attached to it. Each vertex at which a path is attached is a separation vertex in G and not adjacent to any vertex outside of extended biconnected component H. The idea of such a decomposition leads to a polynomial time solution for the Metric Dimension problem restricted to graphs having a minimum resolving set with a bounded number of vertices in every extended biconnected component. This result is especially noteworthy, because we also show that the decision problem METRIC DIMENSION remains NP-complete if the above limitation is extended to usual biconnected components.

2 Definitions and Basic Terminology

We consider *graphs* $G = (V, E)$, where V is the set of *vertices* and E is the set of *edges*. We distinguish between *undirected graphs* with edge sets $E \subseteq \{\{u, v\} \mid u, v \in V, u \neq v\}$ and *directed graphs* with edge sets $E \subseteq V \times V$. Graph $G' = (V', E')$ is a *subgraph* of $G = (V, E)$ if $V' \subseteq V$ and $E' \subseteq E$. It is an *induced subgraph* of G, denoted by $G|_{V'}$, if $E' = E \cap \{\{u, v\} \mid u, v \in V'\}$ or $E' = E \cap (V' \times V')$, respectively. A sequence of $k+1$ vertices $(u_1, \ldots, u_{k+1})$, $k \geq 0$, $u_i \in V$ for $i = 1, \ldots, k+1$, is an *undirected path of length* k, if $\{u_i, u_{i+1}\} \in E$ for $i = 1, \ldots, k$. The vertices u_1 and u_{k+1} are the *end vertices* of undirected path p. The sequence $(u_1, \ldots, u_{k+1})$ is a *directed path of length* k, if $(u_i, u_{i+1}) \in E$ for $i = 1, \ldots, k$. Vertex u_1 is the start vertex and vertex u_{k+1} is the end vertex of the directed path p. A path p is a *simple path* if all vertices are mutually distinct.

An undirected graph G is *connected* if there is a path between every pair of vertices. The *distance* $d_G(u,v)$ between two vertices u,v in a connected undirected graph G is the smallest integer k such that there is a path of length k between u and v. A *connected component* of an undirected graph G is a connected induced subgraph $G' = (V', E')$ of G such that there is no connected induced subgraph $G'' = (V'', E'')$ of G with $V' \subseteq V''$ and $|V'| < |V''|$. A vertex $u \in V$ is a *separation vertex* of an undirected graph G if $G|_{V \setminus \{u\}}$ (the subgraph of G induced by $V \setminus \{u\}$) has more connected components than G. Two paths $p_1 = (u_1, \ldots, u_k)$ and $p_2 = (v_1, \ldots, v_l)$ are *vertex-disjoint* if $\{u_2, \ldots, u_{k-1}\} \cap \{v_2 \ldots, v_{l-1}\} = \emptyset$. A graph $G = (V, E)$ with at least three vertices is *biconnected*, if for every vertex pair $u, v \in V$, $u \neq v$, there are at least two vertex-disjoint paths between u and v. A *biconnected component* $G' = (V', E')$ of G is an induced biconnected subgraph of G such that there is no biconnected induced subgraph $G'' = (V'', E'')$ of G with $V' \subseteq V''$ and $|V'| < |V''|$.

Definition 1 (Resolving set). *Let $G = (V, E)$ be a connected undirected graph. A vertex set $R \subseteq V$ is a* resolving set *for G if for every vertex pair $u, v \in V$, $u \neq v$, there is a vertex $w \in R$ such that $d_G(u, w) \neq d_G(v, w)$. The set R is a* minimum resolving set *for G, if there is no resolving set $R' \subseteq V$ for G with $|R'| < |R|$. Graph $G = (V, E)$ has* metric dimension $k \in \mathbb{N}$ *if k is the smallest integer such that there is a resolving set for G of size k.*

Definition 2. *Let $G = (V, E)$ be a connected undirected graph.*

1. **(leg, root, leaf, hooked leg, ordinary leg)** *A path $p = (u_1, \ldots, u_k)$, $k \geq 2$, of G is a* leg, *if vertex u_1 has degree one, the vertices $u_2, \ldots, u_{k-1}$ have degree 2, and vertex u_k has degree ≥ 3 in G. Vertex u_k is called the* root *of p. Vertex u_1 is called the* leaf *of p. A leg is called a* hooked leg, *if the removal of its root separates G into exactly two connected components, i.e. the edges at root u_k without edge $\{u_{k-1}, u_k\}$ belong to exactly one biconnected component. A leg is called an* ordinary leg, *if it is not a hooked leg, i.e. if the removal of its root separates G into more than two connected components.*

2. **(bridge)** *An edge $e \in E$ is called a* bridge *if $(V, E \setminus \{e\})$ is not connected and if e is not an edge between two vertices of one and the same leg.*

3. **(extended biconnected component (EBC))** *A biconnected component $H = (V_H, E_H)$ of G extended by the subgraphs of G induced by the vertices of the hooked legs with roots in V_H is an* extended biconnected component (EBC) *of G.*

4. **(component)** *Every subgraph induced by the vertices of an ordinary leg, every subgraph induced by the two vertices of a bridge, and every EBC is called a* component *of G.*

5. **(amalgamation vertex)** *Separation vertices of G that belong to at least two components, i.e. separation vertices without the degree two vertices of the legs and roots of the hooked legs are called* amalgamation vertices.

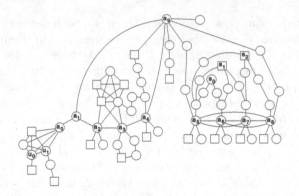

Fig. 1. Graph $G = (V, E)$ with ten amalgamation vertices $(a_0, \ldots, a_9)$, two hooked legs (at roots u_0 and u_1), 14 ordinary legs (two legs at each of the roots a_2, a_4, a_5, a_6, a_7, a_8 and a_9), one bridge ($\{a_0, a_1\}$) four EBCs and thus 19 components. The set of vertices that are drawn as squares is a minimum resolving set for G. See Fig. 2 for a DEBC-tree of G with root a_9.

Definition 3 (EBC-tree). *Let $G = (V, E)$ be a connected undirected graph. The EBC-tree $T = (V_T, E_T)$ for G is a tree with two types of nodes called c-nodes (nodes for the components of G) and a-nodes (nodes for the amalgamation vertices of G). T has a c-node for every component of G. The vertex set of the corresponding component of G represented by a c-node c is denoted by $\mathcal{V}(c)$.*

T has an a-node for every amalgamation vertex of G. The amalgamation vertex represented by a-node a is denoted by $\nu(a)$. Let V_c be the set of c-nodes and V_a be the set of a-nodes of T. Then $V_T = V_c \cup V_a$ and E_T is the set of all edges $\{c, a\}$ with $c \in V_c$, $a \in V_a$ and $\nu(a) \in \mathcal{V}(c)$.

Note that in an EBC-tree all leaves are c-nodes and there is no edge between two a-nodes and no edge between two c-nodes. All ordinary legs are represented by leaves, all bridges are represented by inner c-nodes, and all EBCs are represented by leaves or inner c-nodes.

Definition 4 (DEBC-tree). *Let $G = (V, E)$ be a connected undirected graph.*

1. *For the EBC-tree $T = (V_T, E_T)$ for G and a node $r \in V_T$ let $\overrightarrow{T} := (V_T, \overrightarrow{E}_T)$ be the directed EBC-tree (DEBC-tree) with root r that is defined as follows: $\overrightarrow{E}_T$ contains exactly one directed edge for every undirected edge of E_T such that for every node $u \in V_T$ there is a directed path to root r, i.e. all edges are directed from the leaves towards the root.*
2. *For a node $u \in V_T$, let $\overrightarrow{T}(u)$ be the subtree of $\overrightarrow{T}$ induced by all nodes v for which there is a directed path from v to u in $\overrightarrow{T}$. The root of $\overrightarrow{T}(u)$ is u.*
3. *For a subtree $\overrightarrow{T}(u)$ let V'_T be the set of c-nodes of $\overrightarrow{T}(u)$ and $\mathcal{V}(V'_T) := \bigcup_{v \in V'_T} \mathcal{V}(v)$. Then $G[u] := G|_{\mathcal{V}(V'_T)}$ is the subgraph of G induced by the vertices of $\mathcal{V}(V'_T)$.*

$G[u]$ is the induced subgraph of G represented by $\overrightarrow{T}(u)$. It is not necessary to refer to the a-nodes of $\overrightarrow{T}(u)$, because the vertices of G that are represented by the a-nodes are also represented by the c-nodes since for every a-node a there is a c-node c such that $\nu(a) \in \mathcal{V}(c)$. Note that the EBC- and DEBC-tree of G for an arbitrary root can be constructed in linear time with the help of any linear time algorithm for finding the biconnected components and bridges of G.

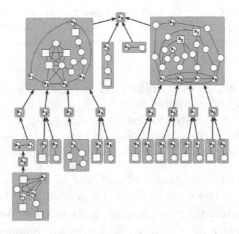

Fig. 2. A DEBC-tree $\overrightarrow{T} = (V_T, \overrightarrow{E}_T)$ at root a_9 for a the graph G from Fig. 1 with 10 amalgamation vertices $a_0, \ldots, a_9$, two hooked legs (at roots u_0 and u_1), 14 ordinary legs (two legs at each of the roots a_2, a_4, a_5, a_6, a_7, a_8 and a_9), one bridge ($\{a_0, a_1\}$) four EBCs and thus 19 components. The vertices that are drawn as squares build a minimum resolving set for G. The vertices of $\overrightarrow{T}$ (19 c-nodes and 10 a-nodes) are drawn as blue boxes, the directed edges as black arrows. For a c-node c the blue box for c contains the subgraph of G induced by $\mathcal{V}(c)$ and for an a-node a the blue box for a contains the vertex $\nu(a)$ of G. (Color figure online)

3 Computing the Metric Dimension Based on a Graph Decomposition

Without loss of generality we will use from now on the following assumptions:

1. $G = (V, E)$ is a connected undirected, but not biconnected graph.
2. $\overrightarrow{T} = (V_T, \overrightarrow{E}_T)$ is the DEBC-tree for G with root r.
3. V_a is the set of a-nodes of $\overrightarrow{T}$ and V_c is the set of c-nodes of $\overrightarrow{T}$.
4. Root $r \in V_a$ is an a-node.
5. Root r has at least two children (because G is not biconnected).

Property 1. For every subtree $\overrightarrow{T}(v)$, $v \in V_T$, of $\overrightarrow{T}$ we compute an information $h(v)$ satisfying the following properties:

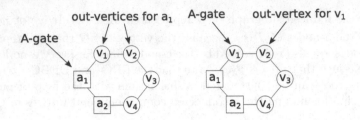

Fig. 3. The Figure shows graph $C_6 = (V, E)$ drawn two times. The set $A = \{a_1, a_2\} \subseteq V$ is a resolving set for C_6. The vertex a_1 is an A-gate with out-vertices v_1 and v_2 (left side). The vertex v_1 is an A-gate with out-vertex v_2 (right side).

1. For every a-node $a \in V_a$ with children $c_1, \ldots, c_k \in V_c$, $k \geq 1$, the information $h(a)$ can efficiently be computed from $h(c_1), \ldots, h(c_k)$.
2. For every c-node $c \in V_c$ with children $a_1, \ldots, a_k \in V_a$, $k \geq 0$, the information $h(c)$ can efficiently be computed from $h(a_1), \ldots, h(a_k)$ and $G|_{\mathcal{V}(c)}$.
3. The metric dimension of $G[r]$ can efficiently be computed from $h(r)$.

First we will describe the general idea of how to compute the metric dimension of G. The idea is based on dynamic programming. The properties above allow an efficient bottom-up processing of $\overrightarrow{T}$ as follows: We start by computing $h(c)$ for every leaf c of $\overrightarrow{T}$. Since the leaves are c-nodes without children we only need the subgraph $G|_{\mathcal{V}(c)}$ of G. For every inner a-node a with children $c_1, \ldots, c_k \in V_c$ we compute $h(a)$ from $h(c_1), \ldots, h(c_k)$. For this we don't need any information about G. For every inner c-node c with children $a_1, \ldots, a_k \in V_a$ we compute $h(c)$ from $h(a_1), \ldots, h(a_k)$ and additionally $G|_{\mathcal{V}(c)}$. Finally we compute the metric dimension of G from $h(r)$. Before we define $h(v)$ we need a few more definitions.

Definition 5 (Gate Vertex). *Let $A \subseteq V$ be a set of vertices. A vertex $v \in V$ is an A-gate of G, if there is a vertex $u \in V \setminus \{v\}$, such that for all $w \in A$ the equation $d_G(u, w) = d_G(u, v) + d_G(v, w)$ holds. Vertex u is called an out-vertex for A-gate v. See Fig. 3 for an example.*

Observation 1. *Let $A \subseteq V$ and $v \in V$ be an A-gate of G. Then there is an out-vertex $u \in V$ adjacent to v.*

Observation 2. *Let $A \subseteq V$, $v, u_1, u_2 \in V$ and v be an A-gate of G. If u_1 and u_2 are two out-vertices for A-gate v with the same distance to v, i.e. $d_G(u_1, v) = d_G(u_2, v)$, then both vertices u_1 and u_2 have the same distance to all vertices of A. In this case A is not a resolving set for G. Conversely, if A is a resolving set for G then all out-vertices have a different distance to A-gate v. A closer look shows that if A is a resolving set for G all out-vertices for A-gate v are on a shortest path between v and the out-vertex with longest distance to v, see Fig. 3.*

Definition 6 (v-resolving set, non-gate-v-resolving set). *Let $v \in V$.*

1. *A v-resolving set for G is a resolving set R for G with $v \in R$.*
2. *A minimum v-resolving set for G is a resolving set R for G with $v \in R$ such that there is no v-resolving set R' for G with $|R'| < |R|$.*
3. *A non-gate-v-resolving set for G is a v-resolving set R for G with $v \in R$ and v is not an R-gate in G.*
4. *A minimum non-gate-v-resolving set for G is a v-resolving set R for G with $v \in R$ and v is not an R-gate in G, such that there is no non-gate-v-resolving set R' for G with $|R'| < |R|$.*

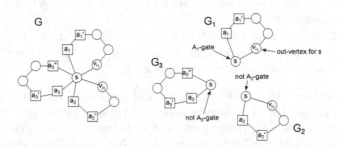

Fig. 4. A graph $G = (V, E)$ with a separation vertex $s \in V$ on the left side and the graphs G_1, G_2 and G_3 on the right side. The vertices of the resolving sets $A_1 = \{a_1, a_1'\}$ for G_1, $A_2 = \{a_2, a_2'\}$ for G_2 and $A_3 = \{a_3, a_3', a_3''\}$ for G_3 are drawn as squares. s is an A_1-gate in G_1 and an A_2-gate in G_2, therefore there are two vertices v_{i_1} and v_{j_1} that are not separated by $A := \bigcup_{i=1}^{3} A_i$ and A is not a resolving set for G.

Note that a minimum v-resolving set is not necessarily a minimum resolving set and a minimum non-gate-v-resolving set is not necessarily a minimum v-resolving set.

Lemma 1. *Let $v \in V$. Let $R_1 \subseteq V$ a minimum resolving set for G, $R_2 \subseteq V$ a minimum v-resolving set for G, and $R_3 \subseteq V$ a minimum non-gate-v-resolving set for G, then $|R_2| \leq |R_1| + 1$ and $|R_3| \leq |R_2| + 1$.*

Lemma 2. *Let $s \in V$ be a separation vertex and $V_1, \ldots, V_k$, $k > 1$, be the vertex sets of the connected components of $G|_{V \setminus \{s\}}$. Let R be a resolving set for G. Then there is at most one $i \in \{1, \ldots, k\}$ such that $V_i \cap R = \emptyset$.*

Lemma 3. *Let $s \in V$ be a separation vertex and $V_1, \ldots, V_k$, $k > 1$, be the vertex sets of the connected components of $G|_{V \setminus \{s\}}$ such that if $k = 2$ then every resolving set for G contains at least one vertex from V_1 and one vertex from V_2. Let $A \subseteq V$, $G_i := G|_{V_i \cup \{s\}}$, and $A_i := (A \cap V_i) \cup \{s\}$.*

1. *If A is a resolving set for G then $A \setminus \{s\}$ is resolving set for G.*

2. *A is a minimum resolving set for G iff for all $i \in \{1, \ldots, k\}$ A_i is a minimum s-resolving set for G_i and there is at most one $i \in \{1, \ldots, k\}$ such that s is an A_i-gate in G_i.*
3. *If A is a minimum resolving set for G, then $A' := A \cup \{s\}$ is a minimum s-resolving set for G.*

The proofs of the lemmas above can be found in the full version.

Now we define $h(v)$, $v \in V_T$, as introduced at the beginning of the section.

Definition 7.

1. *Let $a \in V_a$ be an a-node. We define $h(a) := (\alpha, \beta)$, where α is the size of a minimum non-gate-$\nu(a)$-resolving set for $G[a]$ and β is the size of a minimum $\nu(a)$-resolving set for $G[a]$.*
2. *Let $c \in V_c$ be a c-node with father $a \in V_a$ in $\overrightarrow{T}$. We define $h(c) := (\alpha, \beta)$, where α is the size of a minimum non-gate-$\nu(a)$-resolving set for $G[c]$ and β is the size of a minimum $\nu(a)$-resolving set for $G[c]$.*

To get familiar with this definition, we will investigate the smallest possible values for α and β. For an arbitrary node $v \in V_T$ with father $w \in V_T$ we have $h(v) \leq h(w)$, i.e. the i-th component of $h(v)$ is less than or equal to the i-th component of $h(w)$, since $G[v]$ is a subgraph of $G[w]$. Therefore we will first have a look at the leaves of $\overrightarrow{T}$, which are by definition c-nodes and afterwards at the fathers of the leaves, which are by definition a-nodes. For a leaf c with father a the graph $G[c]$ is either an EBC, or an ordinary leg. Let $G[c]$ be an ordinary leg. Then vertex $\nu(a) \in G[c]$ resolves all vertices in $G[c]$, so $\beta = 1$. Since $\nu(a)$ is a $\{\nu(a)\}$-gate in $G[c]$ every minimum non-gate-$\nu(a)$-resolving set contains another arbitrary vertex. Therefore $\alpha = 2$. Let $G[c]$ be an EBC, then every resolving set for $G[c]$ contains at least two vertices. Therefore $h(c) \geq (2, 2)$ (component-wise).

For an a-node a that has only leaves as children the graph $G[a]$ consists of EBCs and paths, that are connected by the separation vertex $\nu(a)$. Note that if a has exactly one child c the graph $G[c]$ is not an ordinary leg, since this contradicts the decomposition of G into EBCs, ordinary legs, and bridges. Thus every minimum $\nu(a)$-resolving set for $G[a]$ contains at least two vertices and the smallest values α and β for an a-node a are $h(a) = (2, 2)$, that leads to the following observation:

Observation 3. *Let S be a minimum resolving set for G. For any a-node $a \in V(\overrightarrow{T})$ the subgraph $G[a]$ contains at least one resolving node, i.e. $S \cap V(G[a]) \neq \emptyset$.*

We will now show that this definition satisfies Property 1.

Theorem 4. *For every a-node $a \in V_a$ with children $c_1, \ldots, c_k \in V_c$, $k \geq 1$, $h(a)$ can be computed from $h(c_1), \ldots, h(c_k)$.*

Proof. $k = 1$: If a has exactly one child c then $h(a) = h(c_1)$. Since $\nu(a) \in V(c_1)$ and c_1 is the only child of a, we can follow that $G[a] = G[c_1]$. Therefore a minimum non-gate-$\nu(a)$-resolving set for $G[c_1]$ is also a minimum non-gate-$\nu(a)$-resolving set for $G[a]$. The same holds for a minimum $\nu(a)$-resolving set.

$k \geq 2$: Let $h(c_i) = (\alpha_i, \beta_i)$, $i \in \{1, \ldots, k\}$. Then $h(a) = (\alpha, \beta)$ with $\alpha = (\sum_{i=1}^{k} \alpha_i) - (k-1)$ and $\beta = \begin{cases} \alpha, & \text{if } \beta_i = \alpha_i \; \forall i \\ \alpha - 1, & \text{else} \end{cases}$.

Let $A \subseteq V(G[a])$ and $A_i = A \cap V(G[c_i])$. The following conclusions are based on the facts of Lemmas 1 and 2. If every A_i, $1 \leq i \leq k$, is a minimum non-gate-$\nu(a)$-resolving set for $G[c_i]$ then $|A_i| = \alpha_i$ and A is a non-gate-$\nu(a)$-resolving set for $G[a]$, and thus $\alpha_1 + \cdots + \alpha_k - (k-1) \geq \alpha$. Conversely, if A is a minimum non-gate-$\nu(a)$-resolving set for $G[c_i]$ then $|A| = \alpha$ and every A_i is a non-gate-$\nu(a)$-resolving set for $G[c_i]$, and thus $\alpha \geq \alpha_1 + \cdots + \alpha_k - (k-1)$. Note that in both cases $\nu(a)$ is in every set A_i but only once in A. A is a $\nu(a)$-resolving set for $G[a]$ iff at most one of the A_i is a $\nu(a)$-resolving set for $G[c_i]$ and all other A_i are non-gate-$\nu(a)$-resolving sets. If two of sets A_{i_1}, A_{i_2} were $\nu(a)$-resolving sets but not non-gate-$\nu(a)$-resolving sets, then there would be two vertices $u_1 \in V(G[c_{i_1}]), u_2 \in V(G[c_{i_2}])$ such that for every $w \in A$ there is a shortest path to w via $\nu(a)$, i.e., A is not a resolving set for $G[a]$ (see Lemma 2). Therefore, if there is an index i such that $\beta_i < \alpha_i$ then $\beta = \alpha - 1$, otherwise $\beta = \alpha$.

Theorem 5. *For every c-node $c \in V_c$ with father $a \in V_a$ and children $a_1, \ldots, a_k \in V_a$, $k \geq 0$, $h(c)$ can be computed from $h(a_1), \ldots, h(a_k)$ and $G|_{\mathcal{V}(c)}$.*

To proof this theorem, we need the following lemma:

Lemma 4. *Let $c \in V_c$ be a c-node with father $a \in V_a$ and children $a_1, \ldots, a_k \in V_a$, $k \geq 0$. Let $R \subseteq V(G[c])$ with $\nu(a) \in R$. Let $R_i := (R \cap V(G[a_i])) \cup \{\nu(a_i)\}$, $i \in \{1, \ldots, k\}$, and $R^* := (R \cap \mathcal{V}(c))) \cup \{\nu(a_i) \mid 1 \leq i \leq k\}$. R is a $\nu(a)$-resolving set for $G[c]$ iff*

1. *R_i is a resolving set for $G[a_i]$ and*
2. *R^* is a $\nu(a)$-resolving set for $G|_{\mathcal{V}(c)}$ and*
3. *For every $i \in \{1, \ldots, k\}$ vertex $\nu(a_i)$ is neither an R_i-gate in $G[a_i]$ nor an R^*-gate in $G|_{\mathcal{V}(c)}$.*

Proof. See full version.

Proof. **of Theorem** 5 Graph $G[c]$ is composed by the graph $G|_{\mathcal{V}(c)}$ and the graphs $G[a_i]$, $i \in \{1, \ldots, k\}$. We compute $h(c) = (\alpha, \beta)$ by computing a minimum-non-gate-$\nu(a)$-resolving set A for $G[c]$ with $|A| = \alpha$ and a minimum $\nu(a)$-resolving set B for $G[c]$ with $|B| = \beta$ with the help of Lemma 4.

Let A_i be a minimum-non-gate-$\nu(a_i)$-resolving set for $G[a_i]$ and B_i be a minimum-$\nu(a_i)$-resolving set for $G[a_i]$, $i \in \{1, \ldots, k\}$. To compute sets A and B and thus α and β we can do the following:

For every subset $W \subseteq \mathcal{V}(c)$ that contains vertices $\nu(a), \nu(a_1), \ldots, \nu(a_k)$ and resolves all pairs $u, v \in \mathcal{V}(c)$ we determine a resolving set R_W for $G[c]$. R_W contains the vertices in W and for every $i \in \{1, \ldots, k\}$ either the vertices in A_i or in B_i. If vertex $\nu(a_i)$ is a W-gate in $G|_{\mathcal{V}(c)}$ then R_W contains the vertices in

B_i else the vertices in A_i. R_W is a $\nu(a)$-resolving set for $G[c]$ (Lemma 4) and by Lemma 3 we get that $R'_W := R_W \setminus \{\nu(a_1), \ldots, \nu(a_k)\}$ is a $\nu(a)$-resolving set for $G[c]$. Vertex set R'_W is a smallest $\nu(a)$-resolving set for $G[c]$ with the property $\mathcal{V}(c) \cap R'_W = W$ for a given W, that contains at least the vertices of W.

Then we have $B = \min\{R'_W \mid W \subseteq \mathcal{V}(c) \text{ is a resolving set for } G|_{\mathcal{V}(c)} \text{ with } \nu(a), \nu(a_1), \ldots, \nu(a_k) \in W\}$ with $\beta = |B|$ and $A = \min\{R'_W \mid W \subseteq \mathcal{V}(c) \text{ is a resolving set for } G|_{\mathcal{V}(c)} \text{ with } \nu(a), \nu(a_1), \ldots, \nu(a_k) \in W \text{ and } \nu(a) \text{ is not a } W\text{-gate in } G[c] \}$ with $\alpha = |A|$.

Theorem 6. *The metric dim. of $G[r]$ can efficiently be computed from $h(r)$.*

Proof. See full version.

4   Algorithm and Time Complexity

Let $G = (V, E)$ be a connected undirected graph with $|V| = n$ and $|E| = m$. To compute a resolving set for G we first compute the DEBC-tree $\overrightarrow{T} := (V_T, \overrightarrow{E}_T)$ for G. This can be done in $\mathcal{O}(n + m)$ with the help of any linear-time-algorithm for finding the biconnected components and bridges of G. Then we compute $h(c) = (\alpha_c, \beta_c)$ for every leaf c with father a in the DEBC-tree $\overrightarrow{T}$. We do this by checking for every subset $W \subseteq V(G[c])$ if $W' := W \cup \{\nu(a)\}$ is a resolving set for $G[c]$. We choose the size of the smallest set W' for β_c and the size of the smallest set W', such that a is not a W'-gate in $G[c]$ for α. This takes $\mathcal{O}(2^{n'} \cdot n' \cdot (n' + m'))$ time for $n' = |V(G[c])|$ and $m' = |E(G[c])|$, because we have $2^{n'}$ subset and for each subset we can test in time $n' \cdot (n' + m')$ whether it is a resolving set. Computing the h-values for the inner nodes of $\overrightarrow{T}$ can be done in $\mathcal{O}(2^n \cdot n \cdot (n + m))$, see Theorems 4 and 5. Thus the overall running time in $\mathcal{O}(2^n \cdot n \cdot (n + m))$.

Definition 8. *An undirected graph G is (minimum) k-EBC-bounded for some positive integer k, if there is a (minimum) resolving set R for G such that every EBC of G contains at most k vertices of R. R is called a (minimum) k-EBC-bounded-resolving set for G. Let $\mathcal{G}_k$ and $\mathcal{G}_k^{\min}$ be the class of graphs that are k-EBC-bounded and minimum-k-EBC-bounded, respectively. A set of graphs B is (minimum) EBC-bounded, if for every graph $G \in B$ there is a number k such that G is (minimum) k-EBC-bounded.*

Corollary 1. *The following problems can be solved in polynomial time for any fixed number k:*

1. *Given an undirected graph G. Is $G \in \mathcal{G}_k$?*
2. *Given a set of EBC-bounded graphs. Find the smallest number k' such that $G \in \mathcal{G}_{k'}$.*
3. *Given an undirected graph $G \in \mathcal{G}_k$. Compute a minimum k-EBC-bounded-resolving set for G.*
4. *Given an undirected graph $G \in \mathcal{G}_k^{\min}$. Compute a minimum resolving set for G and thus the metric dimension of G.*

To solve these problems we use our algorithm with slight modifications. Instead of checking every subset W' if it is resolving, we do the following: For the problems 1, 3 and 4 we only test those subsets with at most k vertices. For the problem 2 we run our algorithm for $k = 1$ and increase k successively by one until we get a resolving set. By doing so the running time of our algorithm can be bounded by $\mathcal{O}(n^k \cdot n \cdot (n + m))$. Obviously it holds that $\mathcal{G}'_k \subseteq \mathcal{G}_k$. Vice versa it holds that for all k there is a graph $G \in \mathcal{G}_2$ such that $G \notin \mathcal{G}_k^{\min}$, see Fig. 5. Moreover, the complexity of the following problems remains open:

1. Given an undirected graph G a fixed positive integer k. Is $G \in \mathcal{G}_k^{\min}$?
2. Given an undirected graph $G \in \mathcal{G}'$. Find the smallest integer k' such that $G \in \mathcal{G}_{k'}$.

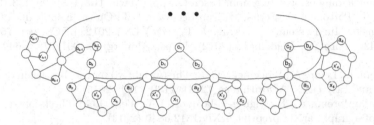

Fig. 5. Graph G_k with $k + 1$ EBCs. G_k is in $\mathcal{G}_2$ for all k. Every resolving set for G_k contains one of the vertices x_i, x'_i, $1 \le i \le k$, since there is no other vertex that can resolve them. The only vertex pairs that still need to be resolved are pairs a_i, b_i. It suffices to choose vertices a_i as resolving vertices. Then we get a 2-EBC-bounded-resolving-set with $2 \cdot k$ vertices and there is no other 2-EBC-bounded-resolving-set with less vertices. Nevertheless a minimum resolving set contains less vertices. By choosing vertices c_i instead of a_i one gets a minimum resolving set with $\frac{3}{2} \cdot k$ vertices. In this case one of the EBCs contains $\frac{1}{2} \cdot k$ resolving vertices and the others contain one vertex.

The following problem still remains NP-complete, see full version.

---

k-BOUNDED BC METRIC DIMENSION

Given: An undirected graph $G = (V, E)$ and a positive integer $r \in \mathbb{N}$ such that there is a minimum resolving set $R \subseteq V$ for G that contains at most k vertices from each biconnected component.

Question: Is the metric dimension of G at most r?

---

5   Conclusion

We have shown that METRIC DIMENSION can be solved in polynomial time on graphs having a minimum resolving set with a bounded number of resolving vertices in every EBC. Even more the algorithm can compute an according set in polynomial time. However, the problem remains NP-complete for graphs having a minimum resolving set with a bounded number of vertices in every biconnected component. This shows that the extended biconnected components cannot simply be downsized further. A next step can be to modify this algorithm to solve other variants of the Metric Dimension problem. The two open problems discussed at the end of Sect. 4 are also going to be investigated.

References

1. Belmonte, R., Fomin, F.V., Golovach, P.A., Ramanujan, M.: Metric dimension of bounded tree-length graphs. SIAM J. Discret. Math. **31**(2), 1217–1243 (2017)
2. Chartrand, G., Eroh, L., Johnson, M., Oellermann, O.: Resolvability in graphs and the metric dimension of a graph. Discret. Appl. Math. **105**(1–3), 99–113 (2000)
3. Díaz, J., Pottonen, O., Serna, M., van Leeuwen, E.J.: On the complexity of metric dimension. In: Epstein, L., Ferragina, P. (eds.) ESA 2012. LNCS, vol. 7501, pp. 419–430. Springer, Heidelberg (2012). https://doi.org/10.1007/978-3-642-33090-2_37
4. Epstein, L., Levin, A., Woeginger, G.J.: The (weighted) metric dimension of graphs: hard and easy cases. Algorithmica **72**(4), 1130–1171 (2015)
5. Estrada-Moreno, A., Rodríguez-Velázquez, J.A., Yero, I.G.: The k-metric dimension of a graph. arXiv preprint arXiv:1312.6840 (2013)
6. Fernau, H., Heggernes, P., van't Hof, P., Meister, D., Saei, R.: Computing the metric dimension for chain graphs. Inf. Process. Lett. **115**(9), 671–676 (2015)
7. Foucaud, F., Mertzios, G.B., Naserasr, R., Parreau, A., Valicov, P.: Algorithms and complexity for metric dimension and location-domination on interval and permutation graphs. In: Mayr, E.W. (ed.) WG 2015. LNCS, vol. 9224, pp. 456–471. Springer, Heidelberg (2016). https://doi.org/10.1007/978-3-662-53174-7_32
8. Foucaud, F., Mertzios, G.B., Naserasr, R., Parreau, A., Valicov, P.: Identification, location-domination and metric dimension on interval and permutation graphs. I. Bounds. Theor. Comput. Sci. **668**, 43–58 (2017)
9. Garey, M., Johnson, D.: Computers and Intractability: A Guide to the Theory of NP-Completeness. W.H. Freeman, New York (1979)
10. Harary, F., Melter, R.: On the metric dimension of a graph. Ars Combinatoria **2**, 191–195 (1976)
11. Hartung, S., Nichterlein, A.: On the parameterized and approximation hardness of metric dimension. In: 2013 IEEE Conference on Computational Complexity (CCC), pp. 266–276. IEEE (2013)
12. Hauptmann, M., Schmied, R., Viehmann, C.: Approximation complexity of metric dimension problem. J. Discret. Algorithms **14**, 214–222 (2012)
13. Hernando, C., Mora, M., Slater, P.J., Wood, D.R.: Fault-tolerant metric dimension of graphs. Convexity Discret. Struct. **5**, 81–85 (2008)
14. Hernando, M., Mora, M., Pelayo, I., Seara, C., Cáceres, J., Puertas, M.: On the metric dimension of some families of graphs. Electron. Notes Discret. Math. **22**, 129–133 (2005)
15. Hoffmann, S., Wanke, E.: METRIC DIMENSION for Gabriel unit disk graphs is NP-complete. In: Bar-Noy, A., Halldórsson, M.M. (eds.) ALGOSENSORS 2012. LNCS, vol. 7718, pp. 90–92. Springer, Heidelberg (2013). https://doi.org/10.1007/978-3-642-36092-3_10

16. Hoffmann, S., Elterman, A., Wanke, E.: A linear time algorithm for metric dimension of cactus block graphs. Theor. Comput. Sci. **630**, 43–62 (2016)
17. Iswadi, H., Baskoro, E., Salman, A., Simanjuntak, R.: The metric dimension of amalgamation of cycles. Far East J. Math. Sci. (FJMS) **41**(1), 19–31 (2010)
18. Khuller, S., Raghavachari, B., Rosenfeld, A.: Landmarks in graphs. Discret. Appl. Math. **70**, 217–229 (1996)
19. Oellermann, O.R., Peters-Fransen, J.: The strong metric dimension of graphs and digraphs. Discret. Appl. Math. **155**(3), 356–364 (2007)
20. Saputro, S., Baskoro, E., Salman, A., Suprijanto, D., Baca, A.: The metric dimension of regular bipartite graphs. arXiv/1101.3624 (2011). http://arxiv.org/abs/1101.3624
21. Sebö, A., Tannier, E.: On metric generators of graphs. Math. Oper. Res. **29**(2), 383–393 (2004)
22. Slater, P.: Leaves of trees. Congr. Numer. **14**, 549–559 (1975)

On the Algorithmic Complexity of Double Vertex-Edge Domination in Graphs

Y. B. Venkatakrishnan and H. Naresh Kumar[⊠]

Department of Mathematics, School of Humanities and Sciences,
SASTRA Deemed University, Thanjavur 613 401, India
venkatakrish2@maths.sastra.edu, nareshhari1403@gmail.com

Abstract. Let $G = (V, E)$ be a simple graph. A vertex $v \in V$ ve-dominates every edge uv incident to v, as well as every edge adjacent to these incident edges. A set $D \subseteq V$ is a double vertex-edge dominating set if every edge of E is ve-dominated by at least two vertices of D. The double vertex-edge dominating problem is to find a minimum double vertex-edge dominating set of G. In this paper, we show that minimum double vertex-edge dominating problem is NP-complete for chordal graphs. A linear time algorithm to find the minimum double vertex-edge dominating set for proper interval graphs is proposed. We also show that the minimum double vertex-edge domination problem cannot be approximated within $(1 - \varepsilon) \ln |V|$ for any $\varepsilon > 0$ unless NP$\subseteq$ DTIME($|V|^{O(\log \log |V|)}$). Finally, we prove that the minimum double vertex-edge domination problem is APX-complete for graphs with maximum degree 5.

Keywords: Double vertex-edge domination · Chordal graph
Proper interval graph · NP-complete · APX-complete

1 Introduction

Let $G = (V, E)$ be a simple connected graph. By an *open neighborhood* of a vertex v of G we mean the set $N_G(v) = \{u \in V(G) : uv \in E(G)\}$ and the *closed neighborhood*, $N_G[v] = N_G(v) \cup \{v\}$. The *degree* of a vertex v, denoted by $d_G(v)$, is the cardinality of its neighborhood. For a set $S \subseteq V$, the *subgraph* of G induced by S is defined as $G[S] = (S, E_S)$, where $E_S = \{xy : xy \in E(G), x, y \in S\}$. A set of vertices S is a *clique* in G if $G[S]$ is a maximal complete subgraph of G.

A graph G is a *chordal graph* if every cycle in G of length at least 4 has a chord. Let $\mathcal{F}$ be a nonempty family of sets. A graph $G = (V, E)$ is called an intersection graph for a finite family $\mathcal{F}$ of a nonempty set if there is a one-to-one correspondence between $\mathcal{F}$ and V such that two sets in $\mathcal{F}$ have nonempty intersection if and only if their corresponding vertices in V are adjacent.

Supported by National Board for Higher Mathematics, Mumbai, India. (Ref No.: NBHM/R.P.1/2015/Fresh/168).

G. K. Das et al. (Eds.): WALCOM 2019, LNCS 11355, pp. 188–198, 2019.
https://doi.org/10.1007/978-3-030-10564-8_15

We call $\mathcal{F}$ an intersection model of G. For an intersection model $\mathcal{F}$, we use $G(\mathcal{F})$ to denote the intersection graph for $\mathcal{F}$. If $\mathcal{F}$ is a family of intervals on a real line, then $G(\mathcal{F})$ is called an *interval graph* for $\mathcal{F}$ and $\mathcal{F}$ is called an interval model of G. If $\mathcal{F}$ is a family of intervals on a real line such that no interval in $\mathcal{F}$ properly contains another interval in $\mathcal{F}$, then $G(\mathcal{F})$ is called a *proper interval graph* for $\mathcal{F}$ and $\mathcal{F}$ is called a proper interval model of G.

A vertex $v \in V(G)$ is a *simplicial vertex* of G if $N_G[v]$ is a clique of G. An ordering $\alpha = (v_1, v_2, \ldots, v_n)$ is a *perfect elimination ordering* (PEO) of G if v_i is a simplicial vertex of $G_i = G[v_i, v_{i+1}, \ldots, v_n]$ for all i, $1 \leq i \leq n$. A PEO $\alpha = (v_1, v_2, \ldots, v_n)$ of a chordal graph is a *bi-compatible elimination ordering* (BCO) if $\alpha^{-1} = (v_n, v_{n-1}, \ldots, v_1)$ is also a PEO of G. This implies that v_i is simplicial in $G[v_1, v_2, \ldots, v_i]$ as well as in $G[v_i, v_{i+1}, \ldots, v_n]$. A graph G is chordal if and only if it has a PEO and proper interval graphs are characterized in terms of BCO, see [4].

A vertex $v \in V(G)$ dominates every vertex in its closed neighborhood. A set $S \subseteq V$ is a *double dominating set*, if each vertex in $V(G)$ is dominated by at least two vertices in S. The *double domination number* of a graph G, denoted by $\gamma_d(G)$, is the minimum cardinality of a double dominating set of G. The minimum double domination problem is to find a double dominating set of minimum cardinality. For more details on double domination, see [2,3].

A vertex $v \in V(G)$ vertex-edge dominates every edge uv incident to v, as well as every edge adjacent to these incident edges. A set $S \subseteq V$ is a *vertex-edge dominating set* (or simply, a *ve-dominating set*) if for every edge $e \in E$, there exists a vertex $v \in S$ such that v ve-dominates e. The *vertex-edge domination number* of a graph G, denoted by $\gamma_{ve}(G)$, is the minimum cardinality of a ve-dominating set of G. The minimum vertex-edge domination problem is to find a vertex-edge dominating set of minimum cardinality. The concept of vertex-edge domination was introduced by Peters [12] and studied further in [1,7–9].

A variant of vertex-edge domination, namely double vertex-edge domination was introduced in [6] and is defined as follows: a set $S \subseteq V$ is a *double vertex-edge dominating set* (or simply, a *double ve-dominating set*), abbreviated DVEDS, if every edge $e \in E$ is ve-dominated by at least two vertices of S. The *double vertex-edge domination number* of G, denoted by $\gamma_{dve}(G)$, is the minimum cardinality of a double ve-dominating set of G. The minimum double vertex-edge domination problem is to find a double vertex-edge dominating set of minimum cardinality. In [6], it is proved that determining the number $\gamma_{dve}(G)$ for bipartite graphs is NP-complete.

In this paper, it is proved that the double vertex-edge domination decision problem is NP complete for chordal graphs. A linear time algorithm for double vertex-edge domination for proper interval graphs is presented. We show that the MDVED problem cannot be approximated within $(1 - \varepsilon) \ln |V|$ for any $\varepsilon > 0$ unless NP $\subseteq$ DTIME $(|V|^{O(\log \log |V|)})$ and show that double vertex-edge domination problem is APX-complete for bounded-free graphs.

2   NP-Completeness Result

In this section, we show that the DVED problem is NP-complete for chordal graphs by proposing a polynomial reduction from a well known NP-complete problem, called Exact cover by 3-sets(X3C) which is defined below.

Exact cover by 3-sets (X3C)

INSTANCE: A finite set X with $|X| = 3q$ and a collection C of 3-element subsets of X.

QUESTION: Does C contain an exact cover for X, that is, a sub-collection $C' \subseteq C$ such that for every element in X belongs to exactly one member of C'?

We define the double vertex-edge domination decision problem.

Double Vertex-edge Domination Decision problem (DVEDD)

INSTANCE: A graph $G = (V, E)$ and a positive integer $k \leq |V|$.

QUESTION: Does there exist a DVED-set D in G such that $|D| \leq k$?

Theorem 1. *The double vertex-edge domination decision problem is NP-complete for chordal graphs.*

Proof. Double *ve*-domination is a member of NP, since we can check in polynomial time that a set of cardinality at most k is a double *ve*-dominating set. Now let us show how to transform any instance of X3C into an instance G of double *ve*-domination, so that one of them has a solution if and only if the other one has a solution. Let $X = \{x_1, x_2, \ldots, x_{3q}\}$ and $C = \{C_1, C_2, \ldots, C_t\}$ be an arbitrary instance of X3C.

Let $V(G) = \{x_i, y_i, z_i, p_i, t_i : 1 \leq i \leq 3q\} \cup \{c_i : 1 \leq i \leq t\}$, $E(G) = \{x_i c_j : x_i \in C_j, 1 \leq i \leq 3q, 1 \leq j \leq t\} \cup \{x_i y_i, y_i z_i, z_i p_i, p_i t_i : 1 \leq i \leq 3q\} \cup \{c_i c_j : 1 \leq i < j \leq t; j = 1, 2, \ldots, t\}$ and $k = 7q$.

The construction of the chordal graph $G = (V, E)$ associated with the instance of X3C, where $X = \{x_1, x_2, x_3, x_4, x_5, x_6\}$ and $C = \{C_1 = \{x_1, x_2, x_4\}, C_2 = \{x_2, x_3, x_6\}, C_3 = \{x_3, x_4, x_5\}, C_4 = \{x_3, x_5, x_6\}\}$ is shown in Figure 1.

Clearly, $G = (V, E)$ is a chordal graph as $\alpha = \{t_1, t_2, \ldots, t_{3q}, p_1, p_2, \ldots, p_{3q}, z_1, z_2, \ldots, z_{3q}, y_1, y_2, \ldots, y_{3q}, x_1, x_2, \ldots, x_{3q}, c_1, c_2, \ldots, c_m\}$ is a perfect elimination ordering of G. Define $X = \{x_1, x_2, \ldots, x_{3q}\}$, $Y = \{y_1, y_2, \ldots, y_{3q}\}$, $Z = \{z_1, z_2, \ldots, z_{3q}\}$, $P = \{p_1, p_2, \ldots, p_{3q}\}$, $T = \{t_1, t_2, \ldots, t_{3q}\}$ and $R = \{c_1, c_2, \ldots, c_t\}$.

Claim: C has an exact cover of size q if and only if G has a double vertex-edge dominating set of size at most $7q$.

Assume first that C has an exact cover C'. Then $\{c_j : C_j \in C'\} \cup Z \cup P$ is a double vertex-edge dominating set of cardinality $7q$.

Conversely, suppose that G has a double vertex-edge dominating set, say D, of cardinality at most $7q$. To dominate the edges $p_i t_i (1 \leq i \leq 3q)$ twice, the set of vertices $Z \cup P$ is contained in D. Let $D' = D \setminus (Z \cup P)$. Then $|D'| \leq q$. The vertex z_i dominates the edge $y_i x_i (1 \leq i \leq 3q)$ once. To dominate the edges $y_i x_i$, the set D' contains elements of R and X or R and Y. If D' contains the elements of R and Y, then for each $y \in Y$ remove $y \in D'$ and add its adjacent vertex x to D'. Without loss of generality we can assume that D' contains elements of R and X. Suppose D' contains a_1 vertices of the set X, a_2 vertices of the set R. Then $a_1 + a_2 \leq q$. The set of vertices in R should dominate $3q - a_1$ edges $y_i x_i$. The vertices in R is adjacent to at most three vertices of set X, $a_2 \geq q - \frac{a_1}{3}$. Therefore, $q \geq a_1 + a_2 \geq a_1 + q + \frac{a_1}{3} = q + \frac{2a_1}{3}$. This is possible only when $a_1 = 0$. This implies that $C' = \{C_j : c_j \in D'\}$ is an exact cover if and only if G has a double vertex-edge dominating set of cardinality at most $7q$.   □

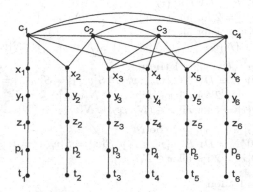

Fig. 1. Double vertex-edge domination for chordal graphs

3   Algorithm

We now present a linear time algorithm for a subclass of chordal graphs, namely the Proper Interval graph. Let G be a connected proper interval graph with a BCO $\sigma = (v_1, v_2, \ldots, v_n)$. Algorithm DVED-PROPER INTERVAL GRAPHS takes G as a input and returns a minimum double vertex-edge dominating set of G. Algorithm DVED-PROPER INTERVAL GRAPHS maintains two arrays D and S for selecting the vertices to the set DVE. If $D[v] = 0$, then there is an edge incident to the vertex v, which is not ve-dominated. If $D[v] = 1$, then there is an edge incident to the vertex v and is ve-dominated only once. If $D[v] = 2$, all the edges incident to v are ve-dominated at least twice. If $S[v] = 0$, then the vertex v is not in the set DVE so far constructed, otherwise $S[v] = 1$.

If the vertex x is assigned a value 0 in the array D, it is denoted by $D_0[x]$. Similarly $D_1[r]$ denotes the vertex x with assigned value 1 in the array D. Consider $v_{i_0} = v_i$ for any arbitrary i.

Algorithm 1. DVED-PROPER INTERVAL GRAPHS

Input: A connected proper interval graph G with BCO $\sigma = (v_1, v_2, \ldots, v_n)$ of
G and an array with $D[v_i] = 0$ for all v_i where $1 \leq i \leq n$.

Output: A minimum double vertex-edge dominating set of G.

1   $DVE \leftarrow \emptyset$;

2   **for** $i = 1$ *to* $n - 1$ **do**

3   **if** $D[v_i \neq 2]$ **then**

4   Let $N_{G_i}(v_i) = \{v_{i_1}, v_{i_2}, \ldots, v_{i_{r-1}}, v_{i_r}\}$, where $i_1 < i_2 < \ldots < i_{r-1} < i_r$;

5   **if** $D[v_i] = 0$ **then**

6   **if** $|N_{G_{i+1}}(v_{i+1})| \geq 2$ **then**

7           $DVE = DVE \cup \{v_{i+1_{r-1}}, v_{i+1_r}\}$, $S[v_{i+1_{r-1}}] = S[v_{i+1_r}] = 1$;

8           $D[x] = 2$ for all $x \in N[v_{i+1_{r-1}}] \cap N[v_{i+1_r}]$;

9           $D[x] = 1$ for all
           $x \in \left(N[v_{i+1_{r-1}}] \cup N[v_{i+1_r}]\right) \setminus \left(N[v_{i+1_{r-1}}] \cap N[v_{i+1_r}]\right)$;

10   **if** $|N_{G_{i+1}}(v_{i+1})| = 1$ **then**

11           $DVE = DVE \cup \{v_{i+1}, v_{i+2}\}, S[v_{i+1}] = S[v_{i+1}] = 1$;

12           $D[x] = 2$ for all $x \in N[v_{i+1}] \cap N[v_{i+2}]$;

13           $D[x] = 1$ for all $x \in (N[v_{i+1}] \cup N[v_{i+2}]) \setminus (N[v_{i+1}] \cap N[v_{i+2}])$;

14   **if** $|N_{G_{i+1}}(v_{i+1})| = 0$ **then**

15           $DVE = DVE \cup \{v_i, v_{i+1}\}$, $S[v_i] = S[v_{i+1}] = 1$;

16           $D[x] = 2$ for all $x \in N_{G_i}[v_i]$;

17   **if** $D[v_i] = 1$ **then**

18   **if** $|N_{G_{i+1}}(v_{i+1})| = 0$ **then**

19   **if** $S[v_{i+1}] = 1$ **then**

20             $DVE = DVE \cup \{v_i\}$, $D_0[x] = 1$ for all $x \in N[v_i]$;

21             $D_1[x] = 2$ for all $x \in N[v_i]$, $S[v_i] = 1$;

22   **else**

23             $DVE = DVE \cup \{v_{i+1}\}$, $D_0[x] = 1$ for all $x \in N[v_{i+1}]$;

24             $D_1[x] = 2$ for all $x \in N[v_{i+1}]$, $S[v_{i+1}] = 1$;

25   **else**

26   **if** $S[v_{i+1_r}] = 1$ **then**

27             $DVE = DVE \cup \{v_{i+1_{r-1}}\}$, $D_0[x] = 1$ for all $x \in N[v_{i+1_{r-1}}]$;

28             $D_1[x] = 2$ for all $x \in N[v_{i+1_{r-1}}]$, $S[v_{i+1_{r-1}}] = 1$;

29   **else**

30             $DVE = DVE \cup \{v_{i+1_r}\}$, $D_0[x] = 1$ for all $x \in N[v_{i+1_r}]$;

31             $D_1[x] = 2$ for all $x \in N[v_{i+1_r}]$, $S[v_{i+1_r}] = 1$;

32 **return** DVE

Theorem 2. *For $0 \leq i \leq n - 1$, the set DVE_i is contained in some minimum double vertex-edge dominating set of G.*

Proof. We prove the result by induction on the number of iterations i of the algorithm. The base case $i = 0$ is true as $DVE_0 = \emptyset$. Assume that the induction hypothesis is true for all positive integers less than or equal to $i-1$. Equivalently, the set DVE_{i-1} is contained in some minimum double vertex-edge dominating set, say D' of G. Notice that at the ith iteration of algorithm, the vertex v_i is being processed. If $D[v_i] = 2$, then the algorithm does not select any new vertex in to the set DVE_i. So $DVE_i = DVE_{i-1}$ and hence it is contained in D'. Now assume that $D[v_i] \neq 2$. Let $N_{G_{i+1}}(v_{i+1}) = \{v_{i+1_1}, v_{i+1_2} \ldots, v_{i+1_r}\}$ where $i + 1_1 \leq i + 1_2 \leq \ldots \leq i + 1_r$.

Case 1: $D[v_i] = 0$

Subcase 1: $|N_{G_{i+1}}(v_{i+1})| \geq 2$.

Since $D[v_i] = 0$, there exist an edge incident with v_i, which is not ve-dominated. Let $v_i v_{i_k}$ be such an edge. Let $v_\alpha, v_\beta \in D'$ be the vertices that double ve-dominate the edge $v_i v_{i_k}$. If $\{v_\alpha, v_\beta\} = \{v_{i+1_{r-1}}, v_{i+1_r}\}$, we are through, say $DVE_i \subset D'$. Let $\alpha, \beta \notin \{i+1_{r-1}, i+1_r\}$. To double ve-dominate the edge $v_i v_{i_k}$, v_α and v_β are adjacent to v_i or v_{i_k} or without loss of generality, we can assume v_α is adjacent to v_i and v_β is adjacent to v_{i_k}. Suppose v_α and v_β are adjacent to v_i. Since v_i is a simplicial vertex in $G[v_i, v_{i+1}, \ldots, v_n]$, v_α and v_β are adjacent to v_{i_k}. The vertex v_{i+1} is adjacent to v_i and hence the vertex v_{i+1} is adjacent to v_α and v_β. Since $v_{i+1_{r-1}}$ and v_{i+1_r} are adjacent to v_{i+1}, the vertices $v_{i+1_{r-1}}$ and v_{i+1_r} are adjacent to v_α and v_β. The set $D'' = (D' \setminus \{v_\alpha, v_\beta\}) \cup \{v_{i+1_{r-1}}, v_{i+1_r}\}$ is a minimum double vertex-edge dominating set of G and hence DVE_i is contained in a minimum double vertex-edge dominating set. Suppose the vertices v_α and v_β are adjacent to v_{i_k}. Since the vertex v_i is adjacent to v_{i+1} and v_{i_k}, we get v_{i+1} is adjacent to v_{i_k}. Now $v_{i+1_{r-1}}$ and v_{i+1_r} are adjacent to v_{i+1}. Hence $v_{i+1_{r-1}}$ and v_{i+1_r} are adjacent to v_{i_k}. The set $D'' = (D' \setminus \{v_\alpha, v_\beta\}) \cup \{v_{i+1_{r-1}}, v_{i+1_r}\}$ is a minimum double vertex-edge dominating set of G. Hence, DVE_i is contained in a minimum double vertex-edge dominating set of G.

Subcase 2: $|N_{G_{i+1}}(v_{i+1})| = 1$.

Since $D[v_i] = 0$ and degree of v_{i+1} is 2 in G_i, to double ve-dominate the edge $v_i v_{i+1}$, the minimum double vertex-edge dominating set should contain the vertices v_{i+1} and v_{i+2}. Hence $DVE_i = DVE_{i-1} \cup \{v_{i+1}, v_{i+2}\}$ and is contained in a minimum double vertex-edge dominating set of G.

Subcase 3: $|N_{G_{i+1}}(v_{i+1})| = 0$.

Since $D[v_i] = 0$, to double ve-dominate the edge $v_i v_{i+1}$, the minimum double vertex-edge dominating set should contain the vertices v_i and v_{i+1}. Hence $DVE_i = DVE_{i-1} \cup \{v_i, v_{i+1}\}$ is contained in a minimum double vertex-edge dominating set of G.

Case 2: $D[v_i] = 1$

Subcase 1: $|N_{G_{i+1}}(v_{i+1})| = 0$ and $S[v_{i+1}] = 1$.

Since $D[v_i] = 1$ and v_{i+1} belongs to DVE_{i-1} set, the edge $v_i v_{i+1}$ gets ve-dominated once. To double ve-dominate the edge $v_i v_{i+1}$, the double vertex-edge dominating set should contain the vertex v_i. Hence $DVE_i = DVE_{i-1} \cup \{v_i\}$ is contained in a minimum double vertex-edge dominating set of G.

Subcase 2: $|N_{G_{i+1}}(v_{i+1})| = 0$ and $S[v_{i+1}] = 0$.

Since $D[v_i] = 1$, the edge $v_i v_{i+1}$ gets ve-dominated once. To double ve-dominate the edge $v_i v_{i+1}$, the double vertex-edge dominating set should contain the vertex v_{i+1}. Hence $DVE_i = DVE_{i-1} \cup \{v_{i+1}\}$ is contained in a minimum double vertex-edge dominating set of G.

Subcase 3: $|N_{G_{i+1}}(v_{i+1})| \geq 1$ and $S[v_{i+1_r}] = 1$.

Since $D[v_i] = 1$, any edge incident with v_i is ve-dominated once, say $v_i v_k$. If $v_k = v_{i+1}$, then it is obvious that $DVE_{i-1} \cup \{v_{i+1_{r-1}}\}$ is contained in a minimum double vertex-edge dominating set of G. Suppose $v_k \neq v_{i+1}$. Since v_i is adjacent to v_{i+1} and v_k, v_{i+1} is adjacent to v_k. The vertex v_{i+1} is adjacent to $v_{i+1_{r-1}}$ and v_{i+1_r} and hence the vertices $v_{i+1_{r-1}}$ and v_{i+1_r} are adjacent to v_k. The vertex $v_{i+1_{r-1}}$ ve-dominates the edge $v_i v_k$. Hence $DVE_{i-1} \cup \{v_{i+1_{r-1}}\}$ is contained in a minimum double vertex-edge dominating set of G.

Subcase 4: $|N_{G_{i+1}}(v_{i+1})| \geq 1$ and $S[v_{i+1_r}] = 0$.

Since $D[v_i] = 1$, any edge incident with v_i is ve-dominated once, say $v_i v_k$. If $v_k = v_{i+1}$, then it is obvious that $DVE_{i-1} \cup \{v_{i+1_r}\}$ is contained in a minimum double vertex-edge dominating set of G. Suppose $v_k \neq v_{i+1}$. As given in previous subcase, the minimum double vertex-edge dominating set should contains the vertex v_{i+1_r}. Hence $DVE_{i-1} \cup \{v_{i+1_r}\}$ is contained in a minimum double vertex-edge dominating set of G. □

Next we show that the algorithm DVED-PROPER INTERVAL GRAPHS runs in linear time.

The BCO $\sigma = (v_1, v_2, \ldots, v_n)$ of a proper interval graph can be computed in linear time [10]. Each iteration of the algorithm DVED-PROPER INTERVAL GRAPHS checks the degree of the vertex v_{i+1} in the graph G_{i+1}. Thus the total time taken is $O(n + m)$.

Theorem 3. *For a given connected proper interval graph G with n vertices and m edges, the algorithm DVED-PROPER INTERVAL GRAPHS takes $O(n+m)$ time to compute a minimum double vertex-edge dominating set of G.*

4 Hardness Result

In this section, we show that there is no approximation algorithm for the minimum double vertex-edge domination problem with approximation factor better than $(1 - \varepsilon) \ln |V|$. To prove this, we require the following result by Lewis [8] on minimum vertex-edge domination problem.

Theorem 4. *[8] For a graph $G = (V, E)$, the minimum vertex-edge domination problem cannot be approximated within $(1 - \varepsilon) \ln |V|$ for any $\varepsilon > 0$ unless $NP \subseteq DTIME\, (|V|^{O(\log \log |V|)})$.*

Now we prove the approximation hardness for the double vertex-edge domination problem.

Theorem 5. *For a graph $G = (V, E)$, the minimum double vertex-edge domination problem cannot be approximated within $(1 - \varepsilon) \ln |V|$ for any $\varepsilon > 0$ unless $NP \subseteq DTIME\, (|V|^{O(\log \log |V|)})$.*

Proof. We present an approximation preserving reduction from the minimum vertex-edge domination problem to minimum double vertex-edge domination problem. This together with the inapproximability bound of the minimum double vertex-edge domination problem will provide the desired result.

Let $G = (V, E)$, where $V(G) = (v_1, v_2, \ldots, v_n)$ be the arbitrary instance of the minimum vertex-edge domination problem. Now, we construct the graph G', an instance of minimum double vertex-edge domination problem as follows: $V(G') = V(G) \cup \{x, y, z\}$ and $E(G') = E(G) \cup \{v_i x \; : \; 1 \leq i \leq n\} \cup \{xy, yz\}$.

Let S^* denote a minimum vertex-edge dominating set of G. By the construction of G', the vertices x, y with the set D double *ve*-dominates all the edges in G'. Thus the set $S^* \cup \{x, y\}$ is a double vertex-edge dominating set of G'. Hence the cardinality of a minimum double vertex-edge dominating set of G' must be less than or equal to $|S^*| + 2$.

Now, assume that the minimum double vertex-edge dominating set can be approximated within a ratio of α where $\alpha = (1 - \varepsilon)|V_{G'}|$ for some fixed $\varepsilon > 0$, by using some algorithm, say Algorithm DVED, that runs in polynomial time. For $k \geq 0$, be a fixed integer. Consider the following algorithm.

Algorithm 2. Algorithm VEDS

Input: A graph $G = (V, E)$.
Output: A minimum Vertex-edge dominating set S of graph G.
1   **if** *there exist a minimum vertex-edge dominating set S of G of cardinaliy $< k$* **then**
2 | **return** S;

    else
3 | Construct G';
4 | Compute a double vertex-edge dominating set D_{dve} in G' using the Algorithm DVED;
5 | Compute $S = D_{dve} \setminus \{x, y, z\}$;
6 | **return** S;

Note that Algorithm VEDS is a polynomial time algorithm and the Algorithm DVED is a polynomial time algorithm. If the cardinality of a minimum vertex-edge dominating set is at most k, then vertex-edge dominating set can be computed in polynomial time. Next we analyze the case when the minimum vertex-edge dominating set is greater than k.

Let S^* be the minimum vertex-edge dominating set of G and D^*_{dve} be the minimum double vertex-edge dominating set of G'. If S is a vertex-edge dominating set of G produced by Algorithm VEDS, then $|S| \leq |D_{dve}| \leq \alpha |D^*_{dve}| \leq \alpha(|S^*| + 2) < \alpha(1 + \frac{2}{k})|S^*|$. Therefore, Algorithm VEDS approximates the minimum vertex-edge dominating set with in the ratio $\alpha(1 + \frac{2}{k})$. Recall that $\alpha = (1 - \varepsilon)\ln|V_{G'}|$ for some fixed $\varepsilon > 0$. Choosing the integer $k > 0$ large enough so that $2/k < \varepsilon/2$, we note that

$$\alpha \left(1 + \frac{2}{k}\right) < (1 - \varepsilon)\left(1 + \frac{\varepsilon}{2}\right) \ln |V_{G'}| = (1 - \varepsilon') \ln |V_{G'}| \approx (1 - \varepsilon') \ln |V|.$$

where $\varepsilon' = \varepsilon/2 + \varepsilon^2/2$ and $|V_{G'}| = |V| + 3$. This proves that the minimum vertex-edge dominating set problem with in the ratio $(1 - \varepsilon') \ln |V|$ for some fixed $\varepsilon' > 0$. By Theorem 4, if the minimum vertex-edge domination problem can be approximated within $(1 - \varepsilon') \ln |V|$ for any $\varepsilon' > 0$, then NP $\subseteq$ DTIME $(|V|^{O(\log \log |V|)})$. Then the minimum double vertex-edge domination problem can be approximated within $(1 - \varepsilon) \ln |V_{G'}|$ for any $\varepsilon > 0$, then NP $\subseteq$ DTIME $(|V_{G'}|^{O(\log \log |V_{G'}|)})$. Hence, the minimum double vertex-edge domination problem cannot be approximated within $(1 - \varepsilon) \ln |V_{G'}|$ for any $\varepsilon > 0$ unless NP $\subseteq$ DTIME $(|V_{G'}|^{O(\log \log |V_{G'}|)})$. $\square$

5   APX-completeness

To prove the APX-completeness of minimum double vertex-edge domination problem, we use the concept called L-reduction, see [11]. Given two NP optimization problems π_1 and π_2 and a polynomial time transformation f from instances of π_1 to instances of π_2, we say that f is an L-reduction if there are positive constants α and β such that for every instance x of π_1 the following holds.

1. $opt_{\pi_2}(f(x)) \leq \alpha.opt_{\pi_1}(f(x))$.
2. for every feasible solution y of $f(x)$ with objective value $m_{\pi_2}(f(x), y) = c_2$ we can in polynomial time find a solution y' of x with $m_{\pi_1}(x, y') = c_1$ such that $|opt_{\pi_1}(x) - c_1| \leq \beta |opt_{\pi_2}(f(x)) - c_2|$.

To show the APX-completeness of the minimum double vertex-edge domination problem, we give an L-reduction from the minimum double domination problem. Let the minimum double domination problem and minimum double vertex-edge domination problem are defined as follows.

MIN DD SET-B
Instance: A graph $G = (V, E)$ with degree at most B.
Solution: A double dominating set of G, a subset $V' \subset V$ such that each vertex in $V(G)$ is dominated by at least two vertices of V'.
Measure: Cardinality of dominating set, $|V'|$.

MIN DVED SET-B
Instance: A graph $G = (V, E)$ with degree at most B.
Solution: A double vertex-edge dominating set of G, a subset $V' \subset V$ such that each edge $e \in E(G)$ gets ve-dominated by at least two vertices of V'.
Measure: Cardinality of vertex-edge dominating set, $|V'|$.

Theorem 6. [5] *MIN DD SET-4 is APX-complete.*

Theorem 7. *MIN DVED SET-5 is APX-complete.*

Proof. The minimum double domination problem is APX-complete for graphs with maximum degree 3. It is enough to establish an L-Reduction f from the instances of the minimum double domination problem for graphs with maximum degree 4 to the instances of the minimum double vertex-edge domination for graphs with maximum degree 5. Given a graph $G = (V, E)$, with degree at most 4, construct a graph G' with degree at most 5 by adding a vertex, say x_i, to every vertex v_i of G.

Claim: G has a double dominating set of cardinality at most k if and only if G' has a double ve-dominating set of cardinality k.

Proof of Claim: Let D be the minimum double dominating set of G, where $|D| \leq k$. It is easy to see that D is a double vertex-edge dominating set of G', since the edges of G' are double ve-dominated by the vertices of G. Thus, $\gamma_{dve}(G') \leq \gamma_d(G)$. Let D' be the minimum double vertex-edge dominating set of G', where $|D'| \leq k$. If $x_i \in D'$ and $v_i \notin D'$, then we can define $D'' = (D' \cup \{v_i\}) \setminus \{x_i\}$ is still a double ve-dominating set of G'. Suppose if, $x_i, v_i \in D$, then define $D'' = (D' \cup \{v_j\}) \setminus \{x_i\}$ is a double ve-dominating set of G', where $v_j \in N(v_i) \setminus \{x_i\}$. Thus, without loss of generality D' contains the vertices of v_j for some j. Clearly, D' is a double dominating set of G. Assume that there exist $v_i \in V(G)$ does not get double dominated by the vertices of D'. Since the edges in G' gets double ve-dominated by the vertices in D', to double ve-dominate the edge $v_i x_i$ either two neighbors of v_i should belongs to D' or $v_i \in D$ and a neighbor of v_i should belongs to D, a contradiction. Hence, $\gamma_d(G) \leq |D'| = \gamma_{dve}(G')$. This proves our claim.

Let D^* and S^* be a minimum double dominating set of G and a minimum double vertex-edge dominating set of G', respectively. By claim, we have $|S^*| = |D^*|$. Again $||D^*| - \gamma_d(G)| \leq ||S^*| - \gamma_{dve}(G')|$. Therefore, the reduction is an L-reduction with $\alpha = 1$ and $\beta = 1$. □

References

1. Bourtig, R., Chellali, M., Haynes, T.W., Hedetniemi, S.: Vertex-edge domination in graphs. Aequationes Math. **90,** 355–366 (2016)
2. Haynes, T., Hedetniemi, S., Slater, P.: Fundamentals of Domination in Graphs. Marcel Dekker, New York (1998)
3. Haynes, T., Hedetniemi, S., Slater, P. (eds.): Domination in Graphs: Advanced Topics. Marcel Dekker, New York (1998)
4. Jamison, R.E., Laskar, R.: Elimination orderings of chordal graphs. In: Proceedings of the Seminar on Combinatorics and Applications, Indian Statistical Institute, Calcutta, pp. 192–200 (1982)
5. Klasing, R., Laforest, C.: Hardness results and approximation algorithms of k-tuple domination in graphs. Inf. Process. Lett. **89**, 75–83 (2004)
6. Krishnakumari, B., Chellali, M., Venkatakrishnan, Y.B.: Double Vertex-edge Domination. In: Discrete Mathematics, Algorithms and Applictions (2017). https://doi.org/10.1142/S1793830917500458
7. Krishnakumari, B., Venkatakrishnan, Y.B., Krzywkowski, M.: Bounds on the vertex-edge domination number of a tree. Comptes Rendus Math. **352**, 363–366 (2014)

8. Lewis, J.R.: Vertex-edge and edge-vertex parameters in graphs, Ph.D. Thesis, Clemson University (2007)
9. Lewis, J., Hedetniemi, S., Haynes, T., Fricke, G.: Vertex-edge domination. Util. Math. **81**, 193–213 (2010)
10. Panda, B.S., Das, S.K.: A linear time recognition algorithm for proper interval graphs. Inform. Process. Lett. **87**(3), 153–161 (2003)
11. Papadimitriou, C.M., Yannakakis, M.: Optimization, approximation and complexity classes. J. Comput. System Sci. **43**, 425–440 (1991)
12. Peters, J., Theoretical and Algorithmic Results on Domination and Connectivity, Ph.D. Thesis, Clemson University (1986)

The Upper Bound on the Eulerian Recurrent Lengths of Complete Graphs Obtained by an IP Solver

Shuji Jimbo[1]($\boxtimes$)(iD) and Akira Maruoka[2]

[1] Okayama University, Okayama City, Okayama Prefecture, Japan
jimbo-s@okayama-u.ac.jp
[2] Ishinomaki Senshu University, Ishinomaki City, Miyagi Prefecture, Japan

Abstract. If the degree of every vertex of a connected graph is even, then the graph has a circuit that contains all of edges, namely an Eulerian circuit. If the length of a shortest subcycle of an Eulerian circuit of a given graph is the largest, then the length is called the Eulerian recurrent length of the graph. For an odd integer n greater than or equal to 3, $e(n)$ denotes the Eulerian recurrent length of K_n, the complete graph with n vertices. Values $e(n)$ for all odd integers n with $3 \leq n \leq 13$ have been found by verification experiments using computers. If n is 7, 9, 11, or 13, then $e(n) = n - 3$ holds, for example. On the other hand, it has been shown that $n - 4 \leq e(n) \leq n - 2$ holds for any odd integer n greater than or equal to 15 in previous researches. In this paper, it is proved that $e(n) \leq n - 3$ holds for every odd integer n greater than or equal to 15. In the core part of the proof of the main theorem, an IP (integer programming) solver is used as the amount of computation is too large to be solved by hand.

Keywords: Graph theory · Complete graphs · Eulerian circuits
Shortest subcycles · Computer experiments

1 Introduction

If every Eulerian circuit of an Eulerian graph G has a subcycle of length less than or equal to k, and there exists an Eulerian circuit that has no subcycle of length less than k, then k is called the *Eulerian recurrent length* of G. Here, a cycle of G, $C = v_0 \rightarrow v_1 \rightarrow v_2 \rightarrow \cdots \rightarrow v_k \rightarrow v_0$, is called a subcycle of a walk $W = w_0 \rightarrow w_1 \rightarrow w_2 \rightarrow \cdots \rightarrow w_l$, if C is a subsequence of W, that is, there is a nonnegative integer i such that $w_i = v_0, w_{i+1} = v_1, w_{i+2} = v_2, \cdots, w_{i+k} = v_k, w_{i+k+1} = v_0$. We use the book [8] for terminology and notation not defined in this paper. In other words, the Eulerian recurrent length of G is the maximum of the length of a shortest subcycle of an Eulerian circuit of G. In this paper, $e(G)$ denotes the Eulerian recurrent length of Eulerian graph G, and, for positive odd integer n greater than or equal to 3, $e(n)$ denotes the Eulerian recurrent length of the complete graph K_n consisting of n vertices, namely $e(K_n)$.

© Springer Nature Switzerland AG 2019
G. K. Das et al. (Eds.): WALCOM 2019, LNCS 11355, pp. 199–208, 2019.
https://doi.org/10.1007/978-3-030-10564-8_16

We will prove by contradiction that, $e(n) \leq n - 3$ holds for any odd integer $n \geq 13$. In the proof, a contradiction will be derived under the assumption that there is an Eulerian circuit of K_n such that the length of every subcycle of the Eulerian circuit is not less than $n-2$. For that purpose, given an Eulerian circuit C of K_n, we will define attributes of an edge of K_n described as positive and negative on C, and also define a notion called a position reversal on C. If we assume that there is an Eulerian circuit C of K_n such that the length of a shortest subcycle of C is not less than $n - 2$, then C must satisfy a rigid condition of occurrences of vertices around the position reversal (Theorem 1). On the other hand, it is readily follows that the total number of position reversals is not less than that of negative edges on any Eulerian circuit of any complete graph with odd number of vertices (Theorem 2). We will derive a contradiction from those results (Theorem 4). Using results of so large amount of computation that is impossible by hand, we will prove the core part of the proof (Theorem 3) of the main result. In this aspect, the proof is to some extent similar to the proof of the four color theorem established in the 1970s [2,3]. In recent works, Gonçalves et al. determined the domination number of all $n \times m$ grid graphs by a huge amount of computation [4]. In this paper, we have obtained an upper bound on the Eulerian recurrent lengths of complete graphs by solving a particular integer programming problem with computers. Our approach is thus similar to the one of Gonçalves et al. However, the optimum solution of the integer programming problem does not provide any Eulerian recurrent length directly. This part is different from the approach of Gonçalves et al.

It has been shown as an upper bound on the Eulerian recurrent length of complete graphs that $e(n) \leq n-2$ holds for any odd integer $n \geq 7$ [5]. The results of this paper improve the upper bound. On the other hand, it has been shown as a lower bound that $e(n) \geq n - 4$ holds for any odd integer $n \geq 7$ [5]. In the proof of the lower bound, a method to construct an Eulerian circuit of K_n that has no subcycle of length less than $n - 4$ was provided. In the method, decomposition of complete graph K_n with an odd number of vertices into Hamiltonian cycles is applied. Furthermore, it is clear that $e(3) = e(5) = 3$, and we have shown that $e(n) = n - 3$ holds for each $n \in \{7, 9, 11, 13\}$ by verification experiments using computers.

Other than the above, the following results related to the paper have been known. Complete bipartite graphs have high symmetry like complete graphs. We have provided the exact values of Eulerian recurrent lengths of complete bipartite graphs. This was shown by explicit construction of Eulerian circuits C of complete bipartite graphs $K_{m,n}$ with $m + n$ vertices such that the length of a shortest subcycle of C is identical to the trivial upper bound [5]. We have shown that the problem of computing the Eulerian recurrent length of a general undirected graph is NP-complete [6]. Furthermore, we have proved the results with respect to the hardness of approximating the Eulerian recurrent lengths under the assumption of P $\neq$ NP, including the following statement [7]. For any Eulerian graph G and any real number $\rho \geq 1$, there is no polynomial time algorithm that computes an approximation $e'(G)$ of $e(G)$, the Eulerian recurrent length of G, such that $e(G)/e'(G) \leq \rho$ always holds. In other words, there is no polynomial time approximation algorithm with bounded approximation ratio for solving the Eulerian recurrent length problem under the assumption of $P \neq NP$.

In Sect. 2, we will define terminology, notation, and notions proper to this paper. In Sect. 3, we will prove that, for any odd integer n greater than or equal to 13, there is no Eulerian circuit of K_n that has no subcycle of length less than $n - 2$, showing that $e(n) \leq n - 3$. In Sect. 4, we will state concluding remarks, including future work.

2   Preliminaries

The expression $|S|$ denotes the number of elements in a finite set S. For an integer i and a positive integer j, $i \bmod j$ denotes the unique nonnegative integer k such that $k < j$ and $k \equiv i \pmod{j}$.

Let $G = (V, E)$ be an Eulerian graph, and $C = v_0 \to v_1 \to v_2 \to \cdots \to v_{m-1} \to v_0$ an Eulerian circuit of G, where m equals $|E|$, the total number of edges of G. We will say that a vertex occurs at the *occurrence position*, or simply at the position, p on C, if the length of the subwalk of C from the initial vertex v_0 to the occurrence of the vertex is p. Notice that any integer q with $q \equiv p \pmod{m}$ indicates the identical position to p on C. Positions on C can, therefore, be regarded as elements of $\mathbb{Z}/m\mathbb{Z}$, the residue class ring modulo m. A position k is said to be from position i to position j on C, if there are integers i', j', and k' such that

$$i' \equiv i \pmod{m}, \quad j' \equiv j \pmod{m}, \quad k' \equiv k \pmod{m}, \quad \text{and} \quad i' \leq k' \leq j' < i' + m.$$

The length of the subwalk of C from position i to position j is, therefore, $(j - i) \bmod m$ by definition. Generally, if symbol C denotes an Eulerian circuit of G, then $C(j)$ denotes the vertex v_j of G at position j on C. Furthermore, for a position i on C, $N_C(i)$ denotes the nearest position to i other than i in the forward direction on C at which the vertex $C(i)$ occurs. Precisely, the position $N_C(i)$ is i if the vertex $C(i)$ occurs just once, and the only position that satisfies the following condition otherwise:

Both $C(i) = C(k)$ and $i < k < i + m$ hold, and, for any position j with $i < j < k$, $C(j) \neq C(i)$ holds.

We can regard function N_C as a bijection from and to residue class ring $\mathbb{Z}/m\mathbb{Z}$. The image of a position i on C under the inverse function for N, denoted by $N_C^{-1}(i)$, is the nearest position to i other than i in the backward direction on C at which the vertex $C(i)$ occurs. If n is an odd number greater than or equal to 7 and G is the complete graph K_n with n vertices, then, for any position i on any Eulerian circuit C of G, $N_C^{-1}(i)$, i, and $N_C(i)$ are three distinct positions on C.

Let e be an edge of an Eulerian graph G that connects the vertex $C(i)$ at position i and the vertex $C(i+1)$ at position $i+1$ on an Eulerian circuit C of G. Then, if either

$$N_C^{-1}(i+1) < N_C^{-1}(i) \text{ and } N_C(i) < N_C(i+1)$$

or

$$N_C^{-1}(i) < N_C^{-1}(i+1) \text{ and } N_C(i+1) < N_C(i)$$

holds, the edge e is said to be *negative* on C, and otherwise, e is said to be *positive* on C. Furthermore, if a quadruple (i, j, k, l) of positions on C satisfies the following condition, then (i, j, k, l) is said to be a *position reversal* on C:

Expressions $i + 1 \not\equiv j \pmod{m}$, $k + 1 \not\equiv l \pmod{m}$, $N_C(i) = l$, and $N_C(j) = k$ hold. And position j is from position i to position k on C, and position k is from position j to position l on C.

Let C be an Eulerian circuit of an Eulerian graph G, and a quadruple (i, j, k, l) of positions on C a position reversal. Moreover, let v and w be vertices of G. Then, if either $v = C(i), w = C(j)$ or $v = C(j), w = C(i)$ holds, then (i, j, k, l) is said to be a position reversal with respect to v and w on C. Furthermore, the first component i and the fourth component l of a position reversal (i, j, k, l) on C are said to be the *head* and *tail* of the position reversal, respectively. If a position on C is the head or tail of some position reversal on C, then the position is called a position reversal head or tail on C, respectively.

3   Improvement of the Upper Bound on the Eulerian Recurrent Lengths of Complete Graphs

In this section, we will prove by contradiction that, for any odd integer $n \geq 13$, the Eulerian recurrent length of K_n, the complete graph with n vertices, is not greater than $n - 3$, that is, $e(n) \leq n - 3$ holds. In the proof, a contradiction will be derived under the assumption that there is an Eulerian circuit of K_n such that the length of every subcycle of the Eulerian circuit is not less than $n - 2$. Let $C = C(0) \to C(1) \to C(2) \to \cdots \to C(m-1) \to C(0)$ be an Eulerian circuit of K_n, where m is $n(n-1)/2$, the total number of edges of K_n.

The following two conditions on an Eulerian circuit of K_n are equivalent:

Condition A: The length of every subcycle of C is not less than $n - 2$,

and

Condition B: For every integer i with $0 \leq i < m$, $N_C(i) - i \geq n - 2$.

First, we will prove two fundamental lemmas.

Lemma 1. *Let n be an integer greater than or equal to 6, and W a walk of the complete graph K_n with n vertices. Then, W satisfies one of the following conditions, if the length of W is $n + 3$:*

(a) *There is an edge of K_n that occurs twice on W.*

(b) *Walk W contains a subwalk W' such that the length of W' is $n-2$, and there is a vertex of K_n that occurs twice on W'.*

Proof. We describe W as $W = v_1 \rightarrow v_2 \rightarrow v_3 \rightarrow \cdots \rightarrow v_{n+4}$. Assume that, for any two integers i and j with $1 < i < j \leq n+4$, edges $v_{i-1}v_i$ and $v_{j-1}v_j$ are different. Furthermore, assume that, for any vertex v of K_n, v does not occur twice on any walk in the following subwalks of W, each of which is of length $n-2$:

$$W_1 = v_1 \rightarrow v_2 \rightarrow v_3 \rightarrow \cdots \rightarrow v_{n-1},$$
$$W_2 = v_2 \rightarrow v_3 \rightarrow v_4 \rightarrow \cdots \rightarrow v_n,$$
$$W_3 = v_3 \rightarrow v_4 \rightarrow v_5 \rightarrow \cdots \rightarrow v_{n+1},$$
$$W_4 = v_4 \rightarrow v_5 \rightarrow v_6 \rightarrow \cdots \rightarrow v_{n+2},$$
$$W_5 = v_5 \rightarrow v_6 \rightarrow v_7 \rightarrow \cdots \rightarrow v_{n+3}, \text{ and}$$
$$W_6 = v_6 \rightarrow v_7 \rightarrow v_8 \rightarrow \cdots \rightarrow v_{n+4}.$$

In what follows, we will deduce a contradiction under those assumptions.

Without loss of generality, we can regard $\{1, 2, 3, \ldots, n\}$ as the vertex set of K_n, and can assume that $v_1 = 1$, $v_2 = 2$, $v_3 = 3$, $\ldots$, and $v_{n-1} = n - 1$, since it is assumed that there is no vertex that occurs twice or more on W_1. If $\{6, 7, 8, \ldots, n - 1\} \cap \{v_n, v_{n+1}, v_{n+2}, v_{n+3}, v_{n+4}\} \neq \varnothing$, then there is a vertex that occurs twice on W_6. We therefore have $\{v_n, v_{n+1}, v_{n+2}, v_{n+3}, v_{n+4}\} \subseteq \{1, 2, 3, 4, 5, n\}$. Then, for any permutation $(v_n, v_{n+1}, v_{n+2}, v_{n+3}, v_{n+4})$ of five elements in $\{1, 2, 3, 4, 5, n\}$, we can verify that one of the following conditions holds:

$$\{\{1, 2\}, \{2, 3\}, \{3, 4\}, \{4, 5\}\}$$
$$\cap \{\{v_n, v_{n+1}\}, \{v_{n+1}, v_{n+2}\}, \{v_{n+2}, v_{n+3}\}, \{v_{n+3}, v_{n+4}\}\} \neq \varnothing,$$
$$5 \in \{v_n, v_{n+1}, v_{n+2}, v_{n+3}\},$$
$$4 \in \{v_n, v_{n+1}, v_{n+2}\},$$
$$3 \in \{v_n, v_{n+1}\}, \text{ and}$$
$$2 = v_n.$$

This contradicts the assumption at the beginning of the proof.   □

Lemma 2. *Let n be an odd integer greater than or equal to 7. If an Eulerian circuit C of the complete graph K_n with n vertices satisfies condition B, then for any integer i with $0 \leq i < m$, $n - 2 \leq N_C(i) - i \leq n + 3$ holds.*

Proof. Assume that an Eulerian circuit C of K_n satisfy condition B. By condition B, $n - 2 \leq N_C(i) - i$ clearly holds. In what follows, we will deduce $N_C(i) - i \leq n + 3$.

It follows from condition B that, for any subwalk $W = C(i+1) \rightarrow C(i+2) \rightarrow \cdots \rightarrow C(N_C(i) - 1)$ of $K_n = (V, E)$, there is no edge that occurs twice or more on W, and there is no vertex that occurs twice or more on some subwalk of W whose length is less than $n-2$. Furthermore, since W contains only vertices in $V-\{C(i)\}$, we can regard W as a walk of the complete graph K_{n-1} whose vertex set is $V-\{C(i)\}$. By Lemma 1, it follows that the length of W is $n-1+2=n+1$ or less. Hence, We have

$$((N_C(i)-1)-(i+1)) \bmod m = (N_C(i)-i-2) \bmod m \leqq n+1,$$

and conclude that

$$N_C(i)-i = (N_C(i)-i) \bmod m \leqq n+3$$

holds.   □

By Lemma 2, we can replace condition B on an Eulerian circuit C of a complete graph K_n with the following condition B$'$, since those conditions are equivalent.

Condition B$'$: For any integer i with $0 \leq i < m$, $n-2 \leqq N_C(i)-i \leqq n+3$.

The following theorem asserts that it is hard for a position reversal on an Eulerian circuit C of a complete graph K_n to exist, if C satisfies condition B$'$.

Theorem 1. *Let n be an integer greater than or equal to 7. Assume that there is an Eulerian circuit C of the complete graph $K_n = (V, E)$ with n vertices that satisfies condition B$'$. Let m denote $n(n-1)/2$, the total number of edges of K_n. Then, for any position reversal head i on C,*

$$N_C(i)-i = n+3, \tag{1}$$

$$N_C(i+1)-(i+1) = n-2, \tag{2}$$

and

$$N_C(i+2)-(i+2) = n-1 \tag{3}$$

all hold, and for any position reversal tail l on C,

$$l - N_C^{-1}(l) = n+3, \tag{4}$$

$$(l-1) - N_C^{-1}(l-1) = n-2, \tag{5}$$

and

$$(l-2) - N_C^{-1}(l-2) = n-1 \tag{6}$$

all hold.

Proof. If i is the head of a position reversal $(i, j, N_C(j), N_C(i))$ on C satisfying condition B$'$, then, by condition B$'$ and the definition of a position reversal, one of the following four conditions must hold.

(a) $N_C(i) \equiv i+(n+2)$, $j \equiv i+2$, and $N_C(j) \equiv i+n \pmod{m}$.
(b) $N_C(i) \equiv i+(n+3)$, $j \equiv i+2$, and $N_C(j) \equiv i+n \pmod{m}$.

(c) $N_C(i) \equiv i + (n+3)$, $j \equiv i+3$, and $N_C(j) \equiv i + (n+1)$ (mod m).
(d) $N_C(i) \equiv i + (n+3)$, $j \equiv i+2$, and $N_C(j) \equiv i + (n+1)$ (mod m).

In the case where (a) or (b) above holds, by condition B′ and the fact that there is no edge that occurs twice or more on C, we have $V = \{C(i), C(i+1), C(i+2), \ldots, C(i+n)\}$, where $C(i+2) = C(i+n)$. Then, it follows that whatever vertex in V $C(i+n+1)$ is, there is an edge that occurs twice or more on C or condition B′ does not hold. This is a contradiction. If there is an integer k such that $3 < k \le n$ and $C(i+n+1) = C(i+k)$ hold, then condition B′ does not hold. Since $C(i+2) = C(i+n)$ holds by (a) or (b), if $C(i+n+1) = C(i+2)$, then condition B′ does not hold. Since $C(i) = C(i+n+2)$ or $C(i) = C(i+n+3)$ holds by (a) or (b), if $C(i+n+1) = C(i)$, then condition B′ does not hold. Furthermore, since $C(i+2) = C(i+n)$, if $C(i+n+1) = C(i+1)$ or $C(i+n+1) = C(i+3)$, then the edge $C(i+n)C(i+n+1)$ is identical to the edge $C(i+1)C(i+2)$ or $C(i+2)C(i+3)$.

In the case where (c) above holds, by condition B′ and the fact that there is no edge in C that occurs twice or more on C, we have $V = \{C(i+3), C(i+4), C(i+5), \ldots, C(i+n+3)\}$, where $C(i+3) = C(i+n+1)$. Then, it follows that whatever vertex in V $C(i+2)$ is, there is an edge that occurs twice or more on C or condition B′ does not hold. This is a contradiction. If there is an integer k such that $3 \le k < n$ and $C(i+2) = C(i+k)$, then condition B′ does not hold. Since $C(i+3) = C(i+n+1)$ holds by (c), if $C(i+2) = C(i+n+1)$, then condition B′ does not hold. Since $C(i) = C(i+n+3)$ holds by (c), if $C(i+2) = C(i+n+3)$, then condition B′ does not hold. Furthermore, since $C(i+3) = C(i+n+1)$, if $C(i+2) = C(i+n)$ or $C(i+2) = C(i+n+2)$, then the edge $C(i+2)C(i+3)$ is identical to the edge $C(i+n)C(i+n+1)$ or $C(i+n+1)C(i+n+2)$.

Thus, condition (d) must hold. We therefore conclude that (1) and (3) hold. By condition B′, $C(i+1)$ must be identical to $C(i+n-1)$, $C(i+n)$, $C(i+n+1)$, $C(i+n+2)$, $C(i+n+3)$, or $C(i+n+4)$. Since $C(i) = C(i+n+3)$ and $C(i+2) = C(i+n+1)$, $C(i+1) \in \{C(i+n), C(i+n+1), C(i+n+2), C(i+n+3), C(i+n+4)\}$ is impossible. If either $C(i+1) = C(i+n)$ or $C(i+1) = C(i+n+2)$ holds, then the edge $C(i+1)C(i+2)$ is identical to the edge $C(i+n)C(i+n+1)$ or $C(i+n+1)C(i+n+2)$. If $C(i+1) = C(i+n+4)$ holds, then the edge $C(i)C(i+1)$ is identical to the edge $C(i+n+3)C(i+n+4)$. Thus, it follows that $C(i+1) = C(i+n-1)$, and we conclude that (2) holds.

It can be deduced in a similar manner that (4)–(6) all hold, if l is a position reversal tail on C. The detail is omitted. □

Theorem 2. *Let d be an integer with $d \ge 3$, and G an Eulerian graph. Then, if G is $2d$-regular, that is, every degree of G equals to $2d$, then, for any Eulerian circuit C of G, the total number of position reversals on C is not less than that of negative edges on C.*

Proof. Since G is $2d$-regular, every vertex of G occurs exactly d times on C. If there is a negative edge $e = vw = C(i)C(i+1)$ on C such that there is no position reversal with respect to v and w on C, then v and w occur alternatively and the same number of times on C except at positions i and $i+1$ as follows:

$$C(i+2) \to \cdots \to v \to \cdots \to w \to \cdots \to v \to \cdots \cdots w \to \cdots C(i-1).$$

This contradicts the assumption that e is a negative edge.   □

We use the following theorem for finite integer sequences to prove Theorem 4, the main theorem.

Theorem 3. *Let* $X = (x(0), x(1), x(2), \ldots, x(10))$ *and* $Y = (y(0), y(1), y(2), \ldots, y(10))$ *be nonnegative integer sequences of length 11 consisting of nonnegative integer components less than or equal to 5. Let* $M(X, Y)$ *denote the number of integers* $i \in \{2, 3, 4, 5, 6, 7, 8\}$ *such that*

$$\begin{array}{ll} either & x(i) < x(i+1) \quad and \quad y(i) < y(i+1), \\ or & x(i) > x(i+1) \quad and \quad y(i) > y(i+1). \end{array} \tag{7}$$

Let $M'(X, Y)$ *denote the number of integers* $i \in \{1, 2, 3, 4, 5, 6, 7\}$ *such that condition (7) holds. Let* $R(X, Y)$ *denote the number of integers* $i \in \{2, 3, 4, 5, 6, 7, 8\}$ *such that* $y(i) = 5$, $y(i+1) = 0$, *and* $y(i+2) = 1$. *Let* $R'(X, Y)$ *denote the number of integers* $i \in \{2, 3, 4, 5, 6, 7, 8\}$ *such that* $x(i) = 5$, $x(i-1) = 0$, *and* $x(i-2) = 1$.

Then, if X *and* Y *satisfy the following five conditions, (a), (b), (c), (d1), and (d2), then*

$$M(X, Y) + M'(X, Y) > R(X, Y) + R'(X, Y) \tag{8}$$

holds.

(a) *There is no pair of distinct integers* i *and* j *with* $0 \leq i \leq 10$ *and* $0 \leq j \leq 10$ *such that* $i - x(i) = j - x(j)$. *And, there is no pair of distinct integers* i *and* j *with* $0 \leq i \leq 10$ *and* $0 \leq j \leq 10$ *such that* $i + y(i) = j + y(j)$.

(b) *There is no integer* i *with* $0 \leq i \leq 9$ *such that* $|(i-x(i))-((i+1)-x(i+1))| = |x(i+1) - x(i) - 1| = 1$. *And, there is no integer* i *with* $0 \leq i \leq 9$ *such that* $|(i + y(i)) - ((i+1) + y(i+1))| = |y(i) - y(i+1) - 1| = 1$.

(c) *There is no pair of distinct integers* i *and* j *with* $0 \leq i \leq 10$ *and* $0 \leq j \leq 10$ *such that* $|(i - x(i)) - (j - x(j))| = |(i + y(i)) - (j + y(j))| = 1$.

(d1) *For each integer* i *with* $0 \leq i \leq 5$, *there is an integer* j *with* $0 \leq j \leq 5$ *such that* $x(i + j) = j$.

(d2) *For each integer* i *with* $0 \leq i \leq 5$, *there is an integer* j *with* $0 \leq j \leq 5$ *such that* $y(i + j) = 5 - j$.

Proof. By formulating conditions (a), (b), (c), (d1), and (d2), and conditions in the definitions of $M(X, Y)$, $M'(X, Y)$, $R(X, Y)$, and $R'(X, Y)$ into a set of linear inequalities, and setting the objective function as $M(X, Y) + M'(X, Y) - R(X, Y) - R'(X, Y)$, we have a minimization integer programming problem. By solving it with an IP solver, we obtain the inequality (8).   □

It seems to be impossible for a person to check that inequality (8) holds for any pair (X, Y) satisfying (a), (b), (c), (d1), and (d2). At the present time, by using SCIP [1], we can probably obtain an optimum solution of the integer programming problem in the proof of Theorem 3 within about ten seconds.

Theorem 4. *For any odd integer n greater than or equal to 15, there is no Eulerian circuit of the complete graph K_n with n vertices such that the length of a shortest subcycle of the Eulerian circuit equals $n - 2$.*

Proof. Assume that there exists an Eulerian circuit C of K_n with odd $n \geq 15$ satisfying Condition B'. Fix a position p on C. Let $x(i)$ denote the integer $(p + i) - N_C^{-1}(p + i) - (n - 2)$, and $y(i)$ the integer $N_C(p + i) - (p + i) - (n - 2)$, for $i \in \{0, 1, \ldots, 10\}$. Notice that $0 \leq x(i) \leq 5$ and $0 \leq y(i) \leq 5$ hold for any $i \in \{0, 2, \ldots, 10\}$. Then, $M(X, Y)$ is the number of positions i in $\{2, 3, \ldots, 8\}$ such that $C(p + i)C(p + i + 1)$ is a negative edge, $M'(X, Y)$ is the number of positions i in $\{2, 3, \ldots, 8\}$ such that $C(p + i)C(p + i - 1)$ is a negative edge, $R(X, Y)$ is the number of positions i in $\{2, 3, \ldots, 8\}$ such that $p + i$ is a position reversal head, and $R'(X, Y)$ is the number of positions i in $\{2, 3, \ldots, 8\}$ such that $p + i$ is a position reversal tail. Furthermore, the conditions (a), (b), (c), (d1), and (d2) in Theorem 3 must hold. Conditions (a), (b), and (c) follow from the definition of function $N_C(i)$. Conditions (d1) and (d2) follow from the following statements derived from condition B' and the definition of function $N_C(i)$:

If $j = N_C(i - (n - 2)) - i$, then $0 \leq j \leq 5$ and $x(i + j) = j$ hold. And, if $j = N_C^{-1}(i + (n + 3)) - i$, then $0 \leq j \leq 5$ and $y(i + j) = 5 - j$ hold.

Let $M(p)$, $M'(p)$, $R(p)$, and $R'(p)$ denote $M(X, Y)$, $M'(X, Y)$, $R(X, Y)$, and $R'(X, Y)$ for position p, respectively. It follows from inequality (8) that $M(p) + M'(p) > R(p) + R'(p)$ holds for any position p on C. Let M_C denote the total number of negative edges on C, and R_C the total number of position reversals on C. Since $\sum_p (M(p) + M'(p)) = 14M_C$ and $\sum_p (R(p) + R'(p)) = 14R_C$, we have $M_C > R_C$. This contradicts Theorem 2.   □

4   Concluding Remarks

We have proved that, for any odd integer n greater than or equal to 15, the Eulerian recurrent length of the complete graph with n vertices is not greater than $n - 3$, that is, $e(n) \leq n - 3$ holds as in Theorem 4, the main theorem. The proof of the main theorem depends on Theorem 3, and an IP (Integer Programming) solver is used for the proof of Theorem 3 as the amount of computation is too large to be solved by hand. It has been shown by verification experiments with computers that $e(n) = n - 3$ holds for each $n \in \{7, 9, 11, 13\}$. On the other hand, it follows from the result of this paper that $n - 4 \leq e(n) \leq n - 3$ holds for every odd integer $n \geq 15$.

Although we tried to find an Eulerian circuit of K_{15} whose shortest subcycle is of length 12 by performing a computer experiment for several days, we could not find such an Eulerian circuit. The computer experiment has not completed. We guess $e(15)$ to be 11. Furthermore, we suppose that the larger n is, the harder it is for $e(n) = n - 3$ to hold. We are now addressing the problem of proving that, for any odd integer n greater than or equal to 15, $e(n) = n - 4$ holds. We expect that, if we accomplish that, then outputs of a huge amount of computation will be used in the proof as in this paper.

Acknowledgments. This work was supported by JSPS KAKENHI Grant Number 15K00018.

References

1. Achterberg, T.: SCIP: solving constraint integer programs. Math. Programm. Comput. **1**(1), 1–41 (2009). http://mpc.zib.de/index.php/MPC/article/view/4
2. Appel, K., Haken, W., et al.: Every planar map is four colorable. Part I: discharging. Illinois J. Math. **21**(3), 429–490 (1977)
3. Appel, K., Haken, W., Koch, J., et al.: Every planar map is four colorable. Part II: reducibility. Illinois J. Math. **21**(3), 491–567 (1977)
4. Gonçalves, D., Pinlou, A., Rao, M., Thomassé, S.: The domination number of grids. SIAM J. Discrete Math. **25**(3), 1443–1453 (2011)
5. Jimbo, S.: On the Eulerian recurrent lengths of complete bipartite graphs and complete graphs. In: IOP Conference Series: Materials Science and Engineering, vol. 58, no. 1, p. 012019. IOP Publishing (2014)
6. Jimbo, S., Oshie, Y., Hashiguchi, K.: The NP-completeness of EULERIAN RECURRENT LENGTH. RIMS Kôkyûroku **1437**, 107–115 (2005)
7. Mori, K., Jimbo, S.: Proofs of impossibility of approximation algorithms for calculating Eulerian recurrent lengths. IPSJ SIG Notes, 2010-AL-128(11):1–4 (2010). (in Japanese)
8. Wilson, R.J.: Introduction to Graph Theory, 5th edn. Longman, Harlow (2010)

A Fast Algorithm for Unbounded Monotone Integer Linear Systems with Two Variables per Inequality via Graph Decomposition

Takuya Tamori[✉] and Kei Kimura

Department of Computer Science and Engineering,
Toyohashi University of Technology, 1-1 Tenpaku, Toyohashi,
Aichi 441-8122, Japan
t163353@edu.tut.ac.jp, kimura@cs.tut.ac.jp

Abstract. In this paper, we consider the feasibility problem of integer linear systems where each inequality has at most two variables. Although the problem is known to be weakly NP-complete by Lagarias, it has many applications and, importantly, a large subclass of it admits (pseudo-)polynomial algorithms. Indeed, the problem is shown pseudo-polynomially solvable if every variable has upper and lower bounds by Hochbaum, Megiddo, Naor, and Tamir. However, determining the complexity of the general case, pseudo-polynomially solvable or strongly NP-complete, is a longstanding open problem. In this paper, we reveal a new efficiently solvable subclass of the problem. Namely, for the *monotone* case, i.e., when two coefficients of the two variables in each inequality are opposite signs, we associate a directed graph to any instance, and present an algorithm that runs in $O(n \cdot s \cdot 2^{O(\ell \log \ell)} + n + m)$ time, where s is the length of the input and ℓ is the maximum number of the vertices in any strongly connected component of the graph. If ℓ is a constant, the algorithm runs in polynomial time. From the result, it can be observed that the hardness of the feasibility problem lies on large strongly connected components of the graph.

Keywords: Integer linear system · Integer programming
Two-variable-per-inequality system · Monotone system

1 Introduction

In this paper, we consider the feasibility problem of integer linear systems with *two variables per inequality* (TVPI). A TVPI system is formulated as follows.

$$a_i x_{j_i} + b_i x_{k_i} \geq c_i \quad (1 \leq i \leq m),$$

K. Kimura—Partially supported by JSPS KAKENHI Grant Number JP15H06286, Japan.

G. K. Das et al. (Eds.): WALCOM 2019, LNCS 11355, pp. 209–218, 2019.
https://doi.org/10.1007/978-3-030-10564-8_17

where a_i, b_i, c_i are integers and x_{j_i}, x_{k_i} are distinct variables for $i = 1, \ldots, m$. We are asked to find an integral vector x satisfying the inequalities. This problem has been studied extensively since 1980s. It can represent various problems, including many graph problems, scheduling problems and problems in artificial intelligence; see, e.g., [2,12,14]. Especially, the optimization problem (i.e., minimizing an objective function subject to a TVPI system) is a generalization of the *minimum weight vertex cover problem* and the *minimum weight 2-satisfiability problem*; see, e.g., [5].

Shostak [13] suggested that a TVPI system can be represented as a graph and proved that the fractional feasibility can be checked by examining paths and cycles in the graph. Here, the *fractional feasibility* is to determine if there exists a vector $x \in \mathbb{Q}^n$ satisfying the inequalities. Hochbaum and Naor [6] improved the running time to solve the problem using the Fourier-Motzkin elimination method. For general linear systems, the method does not run in polynomial time, however, they showed how to implement the method efficiently for TVPI systems. From these results, if the variables are not restricted to integers, then TVPI systems can be solved in strongly polynomial time. Whereas the integer linear system with only two variables is polynomially solvable as shown by Kannan [8] and Lenstra Jr. [11], the integral feasibility problem of TVPI systems is weakly NP-complete as shown by Lagarias [10].

A TVPI system is called *monotone*[1] if each inequality is of the form $a_i x_{j_i} - b_i x_{k_i} \geq c_i$, where a_i and b_i are nonnegative integers. It is observed by Veinott [15] that the set of all feasible integral solutions of a monotone TVPI system forms a distributive lattice. However, the feasibility problem of monotone TVPI systems is still weakly NP-complete as shown by Lagarias [10]. Hochbaum, Megiddo, Naor, and Tamir [5] showed that for monotone TVPI systems, a feasible solution can be computed in $O(m + n + \bar{u} + mU)$ time, i.e., pseudo-polynomial time if all variables are bounded, where n is the number of the variables, m is the number of the inequalities, $\bar{u}$ is the sum of the differences of the upper and lower bounds of the variables and U is the maximum value of the differences of the upper and lower bounds. Bar-Yehuda and Rawitz improve the time to $O(mU)$ [1]. However, the complexity of the problem without bounds on the variables is not known since the question is raised by Hochbaum and Naor [6] in 1994. Namely, while it is weakly NP-complete, it is open whether the problem is pseudo-polynomially solvable or strongly NP-complete.

The purpose of the paper is proposing a fast algorithm to a slightly restricted subclass of TVPI systems to tackle the open question. We mainly study monotone TVPI systems where each variable has only the nonnegativity constraint and does not necessarily have an upper bound. However, we note that the nonnegativity constraints can be easily removed. It is known that if a TVPI system is feasible, then there exists a feasible solution whose bit size is polynomially

---

[1] The word "monotone" is sometimes used to mean that the solution space is monotone, i.e., if a vector x is a solution of a problem, then a vector x' such that $x \leq x'$ is also a solution. However, we follow the standard notation in references for TVPI systems.

bounded by the length of the input (e.g., [9]). That is, letting B be a sufficiently large integer depending on the length of the input, we can assume that there exist upper and lower bounds such that $-B \leq x_i \leq B$ for every i without changing the feasibility of the system. Hence, if we consider a new variable $x_i' = x_i + B$ instead of original variable x_i for each i, then we obtain the nonnegativity constraint and an upper bound $0 \leq x_i' \leq 2B$ for every i.

For monotone TVPI systems with the nonnegativity constraints on the variables, we associate a directed graph to each instance and propose an algorithm that runs in $O(n \cdot s \cdot 2^{O(\ell \log \ell)} + n + m)$ time. Here, s is the length of the input of the problem and ℓ is the maximum number of the vertices in any strongly connected component of the associated graph. If ℓ is a constant, then the algorithm runs in polynomial time.

Compared with the $O(mU)$ time algorithm by Bar-Yehuda and Rawitz, our algorithm has an advantage that the time complexity of our algorithm is independent of U. In general, pseudo-polynomial time algorithms run in exponential time if the variables do not have bounds since the magnitude of a solution might increase to the exponential size of the length of the input. On the other hand, the exponential part of the time complexity of our algorithm only depends on ℓ. Therefore, our algorithm can solve the problem without bounds on the variables faster if ℓ is small.

From our result, it can be observed that the hardness of the feasibility problem lies on large strongly connected components of the graph. Therefore, to provide an efficient algorithm to all instances or to show strong NP-hardness, we have to deal with large strongly connected components in the graph.

In practice, for example some *scheduling problem* can be formulated as a TVPI system (see e.g., [4]). In such a formulation, the arcs in the associated graph correspond to the following: "some job should be done earlier than another job." In this context, we can observe that ℓ might be small depending on applications.

The rest of the paper is organized as follows. In Sect. 2, we prepare some terminologies and ideas for explaining our algorithm and state our main theorem. In Sect. 3, we explain our algorithm and analyze the running time of the algorithm. Then we prove our main theorem. In Sect. 4, we conclude the paper.

2   Preliminary

In this section, we describe our problem discussed in the paper. In Subsect. 2.1, we give a formulation of integer linear systems with two variables per inequality (TVPI). Then we introduce monotone TVPI systems and explain their nice characteristic used in our algorithm. In Subsect. 2.2, we refer the algorithms given by Lenstra Jr. and Kannan for integer linear systems with fixed number of variables. We use the Kannan's algorithm as a subroutine in our algorithm. In Subsect. 2.3, we reduce our problem to a problem without upper-bound inequalities (which are defined later). In Subsect. 2.4, we explain how to construct a directed graph from a monotone TVPI system. Our algorithm and its analysis rely on this directed graph.

2.1   Monotone TVPI Systems

We here formulate integer linear systems with two variables per inequality (TVPI). In the system, we are given n variables and m inequalities. For $i = 1, ..., m$, the i-th inequality is formulated as follows:

$$a_i x_{j_i} + b_i x_{k_i} \geq c_i,$$

where a_i, b_i, c_i are integers and x_{j_i}, x_{j_k} are distinct variables. The problem is to find a *nonnegative* integral vector x satisfying the inequalities.

A TVPI system is called *monotone* when the two coefficients of the two variables are opposite signs in every inequality. Without loss of generality, we can write down the i-th inequality of the monotone TVPI system as follows:

$$a_i x_{j_i} - b_i x_{k_i} \geq c_i,$$

where a_i and b_i are *nonnegative* integers, c_i is an integer. In the rest of the paper, we denote by M the set of inequalities of an instance of monotone TVPI systems. We also denote by M the instance itself. It was observed by Veinott [15] that the set of all feasible solutions of a monotone TVPI system forms a distributive lattice. This fact induces that monotone TVPI systems have the following nice characteristic. A solution x^* is called a *unique minimal solution* when for any other solution x, each component of x^* is less than or equal to the corresponding component of x. The following lemma is shown by Veinott [15] (see also [6]).

Lemma 1 ([6,15]).   *If a monotone TVPI system is feasible, then there exists a unique minimal solution.*

2.2   Polynomial Time Algorithm for Integer Linear Systems with Fixed Number of Variables

The feasibility problem of general integer linear systems is formulated as follows. Let n and m be positive integers, $A \in \mathbb{Z}^{m \times n}$, and $b \in \mathbb{Z}^m$. Then, the problem is to check whether there exists a vector $x \in \mathbb{Z}^n$ satisfying the linear system $Ax \geq b$. The vector is called *feasible solution*, and if there exists one, then the system is called *feasible*. In the optimization variant, we are asked to find a feasible vector $x \in \mathbb{Z}^n$ minimizing an objective function $c^T x$ for a given vector $c \in \mathbb{Z}^n$. The vector is called an *optimal solution*. Note that this problem is usually called an integer linear programming problem.

Lenstra Jr. [11] showed that if the number of the variables of the problem is a fixed constant, then the problem is solvable in polynomial time. That is, he showed the following theorem:

Theorem 1 ([11]).   *Given an instance of integer linear programming problem with n variables and m inequalities, if n is a fixed constant, then the problem can be solved in polynomial time.*

The algorithm of Lenstra Jr. runs in time $2^{O(n^2)}$ times a polynomial in the length of the input that contains binary encodings of the numbers, namely, it runs in weakly polynomial time for fixed n. Note that Kannan [7] improved the time to $O(s \cdot 2^{O(n \log n)})$, where s is the length of the input. In this paper, we will not touch the details of these algorithms but use the results in our algorithm. In particular, we set the vector c of an objective function as the vector whose every elements are all 1's. Thus, if we obtain an optimal solution for the objective function in a monotone TVPI system, then it is a unique minimal solution by Lemma 1.

2.3   Remove All Upper-Bound Inequalities

Let M be the set of m inequalities given by an instance of monotone TVPI systems. From the monotonicity, M has the following patterns of inequalities.

1. $a_i = b_i = 0$
2. $a_i > 0, b_i > 0$
3. $a_i > 0, b_i = 0$
4. $a_i = 0, b_i > 0$

For pattern 1, the inequality is satisfied if and only if $c_i \leq 0$. Thus, if there exists a inequality with pattern 1 such that $c_i > 0$, then M is infeasible.

Pattern 2 is the normal type of inequalities. This type of inequalities bounds a variable by the other one.

For pattern 3, the inequalities are one-variable inequalities. This type of inequalities gives lower bounds of variables.

For pattern 4, the inequalities are also one-variable inequalities. Note that this type of inequalities is transformed to $x_k \leq -\frac{c_i}{b_i}$. If $c_i \leq 0$, then M might be feasible. Otherwise, M is infeasible since $x_k \geq 0$. Thus, this pattern gives upper bounds of variables. We call an inequality of this type an *upper-bound inequality*.

Let $M_2 \subseteq M$ be the set of m_2 upper-bound inequalities and $M_1 = M \setminus M_2$ be the other m_1 inequalities. We rearrange the indices of M so that the index set I_1 of M_1 is $\{1, 2, ..., m_1\}$ and the index set I_2 of M_2 is $\{m_1 + 1, m_1 + 2, ..., m_1 + m_2 = m\}$. We use the following lemma which was observed by Chandrasekaran and Subramani [3]:

Lemma 2 ([3]). *Let M be the inequalities of a monotone TVPI system and x^* be a unique minimal solution of M_1. If x^* satisfies M_2, then M is feasible. Otherwise, M is infeasible.*

Proof. Since the former is easy, we only show the latter. Assume that the unique minimal solution x^* of M_1 does not satisfy M_2, that is, there exists an index $i \in I_2$ such that $-b_i x^*_{k_i} < c_i$. Moreover, assume that there exists a feasible solution x' of M. By definition, $x'_{k_i} \geq x^*_{k_i}$ holds. Therefore, we have $-b_i x'_{k_i} \leq -b_i x^*_{k_i} < c_i$. This is a contradiction since x' satisfies M_2 by the feasibility. Therefore, M is infeasible.   □

If we construct an algorithm for monotone TVPI systems which returns a unique minimal solution, then, by Lemma 2, we do not have to consider the whole inequalities M, but only smaller set of inequalities M_1. Hereafter, without loss of generality, we assume that there is no inequality which is pattern 1. It is possible to check which pattern each inequality has in $O(1)$ time.

2.4   Transform a Monotone TVPI System to a Directed Graph

Given a monotone TVPI system M with n variables, we transform it to a directed graph $G(M)$. The idea of the transformation is inspired by the work of Shostak [13] on the fractional feasibility. Let M_1 and M_2 be the subsets of M defined in previous subsection. For $i = 1, \ldots, m_1$, the i-th inequality (which has pattern 2 or 3) of M_1 can be written in the following form:

$$x_{j_i} \geq \frac{b_i}{a_i} x_{k_i} + \frac{c_i}{a_i}.$$

We denote by α_i and β_i the coefficient b_i/a_i and the intercept c_i/a_i, respectively. Thus, the inequality is written as

$$x_{j_i} \geq \alpha_i x_{k_i} + \beta_i.$$

We then construct a directed graph $G = (V, A)$ associated to M. Define $V := \{0, 1, \ldots, n\}$, where each vertex j corresponds to the variable x_j of the monotone TVPI system for $j = 1, \ldots, n$ and 0 is an additional vertex. Moreover, for any $i \in I_1$, we add to A an arc from k_i to j_i with weight (α_i, β_i). If α_i is zero, β_i becomes a lower bound of x_{j_i}. In this case, we add an arc which has weight $(1, \beta_i)$ from 0 to j_i.

Consider the strongly connected component decomposition of the graph G constructed above. Assume that each component has at most ℓ vertices. In the rest of the paper, we show the following theorem.

Theorem 2. *If each strongly connected component in the directed graph associated to a monotone TVPI system has at most ℓ vertices, then the feasibility problem is solvable in $O(n \cdot s \cdot 2^{O(\ell \log \ell)} + n + m)$ time, where s is the length of the input. Therefore, if ℓ is a fixed constant, the problem is solvable in polynomial time.*

3   Algorithm to Solve Monotone TVPI Systems

We are given an instance M of monotone TVPI systems such that for $i = 1, \ldots, m$, the i-th inequality is written as follows:

$$a_i x_{j_i} - b_i x_{k_i} \geq c_i,$$

where a_i and b_i are nonnegative integers, c_i is an integer. We have to find non-negative integral vector x satisfying the given inequalities. In this section, we

---

Algorithm 1. Solve monotone TVPI systems

1: Partition M into M_1 and M_2
2: Construct a directed graph $G(M_1)$
3: Decompose $G(M_1)$ to the strongly connected components
4: Order the strongly connected components $C_1, \ldots, C_p$ by topological sort and let $(C_1, \ldots, C_p)$ be the order
5: **for** $1 \leq k \leq p$ **do**
6:    Solve a minimization problem in C_k (using Kannan's algorithm)
7:    **if** obtain a unique minimal solution x_j^* for $j \in C_k$ **then**
8:       continue
9:    **else**
10:      **return** "infeasible"
11:    **end if**
12: **end for**
13: **if** x^* satisfies M_2 **then**
14:    **return** "feasible" and halt
15: **else**
16:    **return** "infeasible" and halt
17: **end if**

---

prove Theorem 2 by giving an algorithm that utilize the directed graph defined in Subsect. 2.4.

Our algorithm is shown in Algorithm 1. In what follows, we explain each line in detail.

In line 1, we partition the set M of m inequalities into the set M_2 of upper-bound inequalities and the set M_1 of the other inequalities as described in Subsect. 2.3. This line can be done in time $O(m)$, since the type of each inequality can be checked in time $O(1)$. In lines 2–12, we compute a unique minimal solution of M_1.

In line 2, we construct a directed graph $G(M_1)$ from M_1. Note that $G(M_1)$ has $n + 1$ vertices and m_1 arcs corresponding to the m_1 inequalities in M_1. As described in Subsect. 2.4, there is an arc from x_{k_i} to x_{j_i} with a weight (α_i, β_i) for each i, where we recall that the i-th inequality in M_1 can be written as follows:

$$x_{j_i} \geq \frac{b_i}{a_i} x_{k_i} + \frac{c_i}{a_i}$$
$$= \alpha_i x_{k_i} + \beta_i$$

since $a_i \neq 0$. (If $a_i = 0$, then such inequality is an upper-bound inequality.). This line can be done in $O(m_1 + n)$ time.

In line 3, we decompose $G(M_1)$ to the strongly connected components in time $O(m_1 + n)$. For instance, see Fig. 1. As assumed in statement of Theorem 2, each component has at most ℓ vertices.

In line 4, we order the strongly connected components by topological sort. This line can be done in $O(m_1 + n)$ time. Let the components be ordered as $(C_1, C_2, \ldots, C_p)$, where p is the number of the strongly connected components. For instance, see Fig. 2.

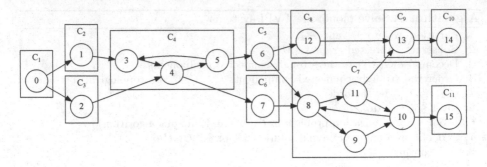

Fig. 1. A directed graph decomposed to strongly connected components

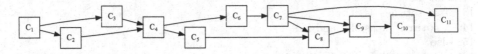

Fig. 2. The strongly connected components ordered by topological sort

In lines 5–12, we compute a unique minimal solution of M_1 inductively according to the order of the strongly connected components. Without loss of generally, we assume that $C_1 = \{0\}$. In the k-th round, we compute assignments to the variables in C_k in the unique minimal solution x^* for $k = 1, \ldots, p$. For $k = 1$, we set $x_0 = 0$. For $k > 1$, assume that the value of x_j^* is obtained for all $x_j \in C_r$ for $r \leq k - 1$. For C_k, there only exist incoming arcs from C_i with $i < k$ since $C_1, \ldots, C_p$ are sorted according to the strongly connected component. These arcs imply lower bounds on variables in C_k. Let L_k be the set of such lower bounds on C_k. Let $M^{(k)}$ be the set of the inequalities in M_1 such that both the vertices j_i and k_i corresponding to the variables x_{j_i} and x_{k_i} are contained in C_k. Then we minimize $\sum_{j \in C_k} x_j$ subject to L_k and $M^{(k)}$, using the algorithm of Kannan and substitute the optimal solution to x_j^* for $j \in C_k$.

To show the correctness of these lines, we prove the following Lemma.

Lemma 3. *Lines 5–12 correctly compute a unique minimal solution of M_1 in $O(n \cdot s \cdot 2^{O(\ell \log \ell)})$ time, where s is the length of the input.*

Proof. We prove this by induction on k in the algorithm. Note that the strongly connected components are sorted as $(C_1, C_2, \ldots, C_p)$.

(1) In the case of $k = 1$, we assign 0 to x_0. This is clearly minimal.

(2) In the case of $k > 1$, assume that until $(k-1)$-th strongly connected component, we have obtained a unique minimal solution $x_{j'}^*$ to $(\bigcup_{r=1}^{k-1} M^{(r)}) \cup (\bigcup_{r=1}^{k-1} L_r)$ for $j' \in \bigcup_{r=1}^{k-1} C_r$. Observe that for $j \in C_k$ each variable x_j to M_k has to satisfy the lower bounds L_k, since $x_{j'}^*$ so far is minimal for $j' \in \bigcup_{r=1}^{k-1} C_r$. Thus, $x_j \geq x_j^*$ must hold for $j \in C_k$, where x_j^* is a unique

minimal solution to $M^{(k)} \cup L_k$. Therefore, x_j^* for $j \in C_k$ is also a unique minimal solution to $(\bigcup_{r=1}^{k} M^{(r)}) \cup (\bigcup_{r=1}^{k} L_r)$. Hence, in each round, the variables are assigned some values that are less than or equal to the values in the unique minimal solution to M_1. Thus, if we obtain an assignment to all the variables in line 12, it is the unique minimal solution.

Now, we analyze the running time of the algorithm. Since we compute a unique minimal solution by minimizing $\sum_{j \in C_k} x_j$ subject to $M^{(k)} \cup L_k$ by Kannan's algorithm in time, each round takes $O(s \cdot 2^{O(\ell \log \ell)})$, where s is the length of the input. Note that k is at most n. Therefore, the overall running time of lines 5–12 is at most $O(n \cdot s \cdot 2^{O(\ell \log \ell)})$. This completes the proof. $\square$

In lines 13–17, we check whether the unique minimal solution to M_1 satisfies M_2 or not. This can be done in $O(m_2)$ time. From Lemma 2, if it satisfies M_2, then the problem is feasible, and, otherwise, the problem is infeasible.

The overall running time of the algorithm is $O(n \cdot s \cdot 2^{O(\ell \log \ell)} + n + m)$, where s is the length of the input. Therefore, we have shown Theorem 2.

4   Conclusion

In this paper, we present an algorithm for monotone TVPI integer linear systems without bounds on the variables that runs in polynomial time if the maximum number ℓ of the vertices in any strongly connected component of the associated graph is a fixed constant. Note that it is also a fixed-parameter algorithm parameterized by ℓ. Our algorithm utilize the one by Kannan [11] for solving integer linear programs with fixed number of variables. From our result, it can be observed that the hardness of monotone TVPI systems lies on large strongly connected components of the associated graph. Therefore, to provide an efficient algorithm to all instances or to show strong NP-hardness, we have to deal with large strongly connected components in the graph.

References

1. Bar-Yehuda, R., Rawitz, D.: Efficient algorithms for integer programs with two variables per constraint. Algorithmica **29**(4), 595–609 (2001)
2. Bordeaux, L., Katsirelos, G., Narodytska, N., Vardi, M.Y.: The complexity of integer bound propagation. J. Artif. Intell. Res. **40**, 657–676 (2011)
3. Chandrasekaran, R., Subramani, K.: A combinatorial algorithm for horn programs. Discrete Optim. **10**(2), 85–101 (2013)
4. Fügenschuh, A.: A set partitioning reformulation of a school bus scheduling problem. J. Sched. **14**(4), 307–318 (2011)
5. Hochbaum, D.S., Megiddo, N., Naor, J.S., Tamir, A.: Tight bounds and 2-approximation algorithms for integer programs with two variables per inequality. Math. Program. **62**(1–3), 69–83 (1993)
6. Hochbaum, D.S., Naor, J.S.: Simple and fast algorithms for linear and integer programs with two variables per inequality. SIAM J. Comput. **23**(6), 1179–1192 (1994)

7. Kannan, R.: Minkowski's convex body theorem and integer programming. Math. Oper. Res. **12**(3), 415–440 (1987)
8. Kannan, R.: A polynomial algorithm for the two-variable integer programming problem. J. Assoc. Comput. Mach. **27**(1), 118–122 (1980)
9. Korte, B., Vygen, J.: Combinatorial Optimization: Theory and Algorithms, 3rd edn. Springer, Heidelberg (2005). https://doi.org/10.1007/3-540-29297-7. Japanese translation from English
10. Lagarias, J.C.: The computational complexity of simultaneous diophantine approximation problems. SIAM J. Comput. **14**(1), 196–209 (1985)
11. Lenstra Jr., H.W.: Integer programming with a fixed number of variables. Math. Oper. Res. **8**(4), 538–548 (1983)
12. Schrijver, A.: Theory of Linear and Integer Programming. Wiley, New York (1986)
13. Shostak, R.: Deciding linear inequalities by computing loop residues. J. ACM **28**(4), 769–779 (1981)
14. Upadrasta, R., Cohen, A.: A case for strongly polynomial time sub-polyhedral scheduling using two-variable-per-inequality polyhedra. In: IMPACT 2012–2nd Workshop on Polyhedral Compilation Techniques (associated with HiPEAC), Paris, France (2012)
15. Veinott, A.F.: Representation of general and polyhedral subsemilattices and sublattices of product spaces. Linear Algebra Appl. **114–115**(1989), 681–704 (1989)

Multilevel Planarity

Lukas Barth, Guido Brückner, Paul Jungeblut, and Marcel Radermacher[✉]

Department of Computer Science, Karlsruhe Institute of Technology,
Karlsruhe, Germany
{lukas.barth,brueckner,radermacher}@kit.edu,
paul.jungeblut@student.kit.edu

Abstract. In this paper, we introduce and study the *multilevel-planarity testing* problem, which is a generalization of upward planarity and level planarity. Let $G = (V, E)$ be a directed graph and let $\ell : V \to \mathcal{P}(\mathbb{Z})$ be a function that assigns a finite set of integers to each vertex. A multilevel-planar drawing of G is a planar drawing of G such that the y-coordinate of each vertex $v \in V$ is $y(v) \in \ell(v)$, and each edge is drawn as a strictly y-monotone curve.

We present linear-time algorithms for testing multilevel planarity of embedded graphs with a single source and of oriented cycles. Complementing these algorithmic results, we show that multilevel-planarity testing is NP-complete even in very restricted cases.

1 Introduction

Testing a given graph for planarity, and, if it is planar, finding a planar embedding, are classic algorithmic problems. However, one is often not interested in just any planar embedding, but in one that has some additional properties. Examples of such properties include that a given existing partial drawing should be extended [3,17] or that some parts of the graph should appear clustered together [11,18].

There also exist notions of planarity specifically tailored to directed graphs. An *upward-planar drawing* is a planar drawing where each edge is drawn as a strictly y-monotone curve. While testing upward planarity of a graph is an NP-complete problem in general [15], efficient algorithms are known for single-source graphs and for embedded graphs [6,7]. One notable constrained version of upward planarity is that of level planarity. A *level graph* is a directed graph $G = (V, E)$ together with a level assignment $\gamma : V \to \mathbb{Z}$ that assigns an integer level to each vertex and satisfies $\gamma(u) < \gamma(v)$ for all $(u, v) \in E$. A drawing of G is level planar if it is upward planar, and we have $y(v) = \gamma(v)$ for the y-coordinate of each vertex $v \in V$. Level-planarity testing and embedding is feasible in linear time [19]. There exist further level-planarity variants on the cylinder and on the torus [1,4] and there has been considerable research on further-constrained versions of level planarity. Examples include ordering the vertices on each level according to so-called constraint trees [2,16], clustered level planarity [2,13], partial level planarity [8] and ordered level planarity [20].

© Springer Nature Switzerland AG 2019
G. K. Das et al. (Eds.): WALCOM 2019, LNCS 11355, pp. 219–231, 2019.
https://doi.org/10.1007/978-3-030-10564-8_18

Contribution and Outline. In this paper, we introduce and study the *multilevel-planarity testing* problem. Let $\mathcal{P}(\mathbb{Z})$ denote the power set of integers. The input of the multilevel-planarity testing problem consists of a directed graph $G = (V, E)$ together with a function $\ell : V \rightarrow \mathcal{P}(\mathbb{Z})$, called a multilevel assignment, which assigns admissible levels, represented as a set of integers, to each vertex. A multilevel-planar drawing of G is a planar drawing of G such that for the y-coordinate of each vertex $v \in V$ it holds that $y(v) \in \ell(v)$, and each edge is drawn as a strictly y-monotone curve. We start by discussing some preliminaries, including the relationship between multilevel planarity and existing planarity variants in Sect. 2. Then, we present linear-time algorithms that test multilevel planarity of embedded single-source graphs and of oriented cycles with multiple sources in Sects. 3 and 4, respectively. In Sect. 5, we complement these algorithmic results by showing that multilevel-planarity testing is NP-complete for abstract single-source graphs and for embedded multi-source graphs where it is $|\ell(v)| \leq 2$ for all $v \in V$. We finish with some concluding remarks in Sect. 6.

2   Preliminaries

This section consists of three parts. First, we introduce basic terminology and notation. Second, we discuss the relationship between multilevel planarity and existing planarity variants for directed graphs. Third, we define a normal form for multilevel assignments, which simplifies the arguments in Sects. 3 and 4.

Basic Terminology. Let $G = (V, E)$ be a directed graph. We use the terms *drawing, planar, (combinatorial) embedding* and *face* as defined by Di Battista et al. [10]. We say that two drawings are *homeomorphic* if they respect the same combinatorial embedding. A *multilevel assignment* $\ell : V \rightarrow \mathcal{P}(\mathbb{Z})$ assigns a finite set of possible integer levels to each vertex. An upward-planar drawing is *multilevel planar* if $y(v) \in \ell(v)$ for all $v \in V$. Note that any finite set of integers can be represented as a finite list of finite integer intervals. We choose this representation to be able to represent sets of integers that contain large intervals of numbers more efficiently.

A vertex of a directed graph with no incoming (outgoing) edges is a *source* (*sink*). A directed, acyclic and planar graph with a single source s is an *sT-graph*. An sT-graph with a single sink t and an edge (s, t) is an *st-graph*. In any upward-planar drawing of an st-graph, the unique source and sink are the lowest and highest vertices, respectively, and both are incident to the outer face. For a face f of a planar drawing, an incident vertex v is called *source switch* (*sink switch*) if all edges incident to f and v are outgoing (incoming). Note that a source switch or sink switch does not need to be a source or sink in G. We will frequently add incoming edges to sources and outgoing edges to sinks during later constructions, referring to this as *source canceling* and *sink canceling*, respectively. An *oriented path* of length k is a sequence of vertices $(v_1, v_2, \ldots, v_{k+1})$ such that for all $1 \leq i \leq k$ either the edge (v_i, v_{i+1}) or the edge (v_{i+1}, v_i) exists. A *directed path* of length k is a sequence of vertices $(v_1, v_2, \ldots, v_{k+1})$ such that for all $1 \leq i \leq k$

the edge (v_i, v_{i+1}) exists. Let $u, v \in V$ be two distinct vertices. Vertex u is a *descendant* of v in G, if there exists a directed path from v to u. A *topological ordering* is a function $\tau : V \rightarrow \mathbb{N}$ such that for every $v \in V$ and for each descendant u of v it is $\tau(v) < \tau(u)$.

Relationship to Existing Planarity Variants. Multilevel-planarity testing is a generalization of level planarity. To see this, let $G = (V, E)$ be a directed graph together with a level assignment $\gamma : V \rightarrow \mathbb{Z}$. Define $\ell(v) = \{\gamma(v)\}$ for all $v \in V$. It is readily observed that a drawing Γ of G is level planar with respect to γ if and only if Γ is multilevel planar with respect to ℓ. Therefore, level planarity reduces to multilevel planarity in linear time.

Multilevel-planarity testing is also a generalization of upward planarity. Without loss of generality, the vertices in an upward-planar drawing can be assigned integer y-coordinates, and there is at least one vertex on each level. Hence, upward planarity of G can be tested by setting $\ell(v) = [1, |V|]$ for all $v \in V$ and testing the multilevel planarity of G with respect to ℓ. Therefore, upward planarity reduces to multilevel planarity in linear time. By then restricting the multilevel assignment, multilevel planarity can also be seen as a constrained version of upward planarity. Garg and Tamassia [15] showed the NP-completeness of upward-planarity testing, which directly gives the following.

Theorem 1. *Multilevel-planarity testing is NP-complete.*

Multilevel Assignment Normal Form. A multilevel assignment ℓ has *normal form* if for all $(u, v) \in E$ it holds that $\min \ell(u) < \min \ell(v)$ and $\max \ell(u) < \max \ell(v)$. Some proofs are easier to follow for multilevel assignments in normal form. The following lemma justifies that we may assume without loss of generality that ℓ has normal form.

Lemma 1. *Let $G = (V, E)$ be a directed graph together with a multilevel assignment ℓ. Then there exists a multilevel assignment ℓ' in normal form such that any drawing of G is multilevel planar with respect to ℓ if and only if it is multilevel planar with respect to ℓ'. Moreover, ℓ' can be computed in linear time.*

Proof. The idea is to convert $\ell(v)$ into $\ell'(v) \subseteq \ell(v)$ for all $v \in V$ by finding a lower bound l_v and an upper bound u_v for the level of v, and then setting $\ell'(v) = \ell(v) \cap [l_v, u_v]$. To find the lower bound, iterate over the vertices in increasing order with respect to some topological ordering τ of G. Because all edges have to be drawn as strictly y-monotone curves, for each vertex $v \in V$ it must be $y(v) > \max_{(w, v) \in E} l_w$. So, set $l_v = \max(\min \ell(v), \max_{(w, v) \in E} l_w + 1)$. Analogously, to find the upper bound, iterate over V in decreasing order with respect to τ. Again, because edges are drawn as strictly y-monotone curves, for each vertex $v \in V$ it must hold true that $y(v) < \min_{(v, w) \in E} u_w$. Therefore, set $u_v = \min(\max \ell(v), \min_{(v, w) \in E} u_w - 1)$. This means that in any multilevel-planar drawing of G the y-coordinate of $v \in V$ is $y(v) \in \ell(v) \cap [l_v, u_v]$. So it follows that a drawing of G is multilevel planar with respect to ℓ if and only if it is multilevel planar with respect to ℓ'.

To see that the running time is linear, note that a topological ordering of G can be computed in linear time and every vertex and edge is handled at most twice during the procedure described above. Because every level candidate in ℓ is removed at most once, the total running time is $\mathcal{O}(n + \sum_{v \in V} |\ell(v)|)$, i.e., linear in the size of the input. $\square$

3   Embedded sT-Graphs

In this section, we characterize multilevel-planar sT-graphs as subgraphs of certain planar st-graphs. Similar characterizations exist for upward planarity and level planarity [12,21]. The idea behind our characterization is that for any given multilevel-planar drawing, we can find a set of edges that can be inserted without rendering the drawing invalid, and which make the underlying graph an st-graph. Thus, the graph must have been a subgraph of an st-graph. This technique is similar to the one found by Bertolazzi et al. [7], and in fact is built on top of it.

To use this characterization for multilevel-planarity testing, we cannot require a multilevel-planar drawing to be given. We show that if we choose the set of edges to be inserted carefully, the respective set of edges can be inserted into *any* multilevel-planar drawing for a fixed combinatorial embedding. An algorithm constructing such an edge set can therefore be used to test for multilevel planarity of embedded sT-graphs, resulting in Theorem 2. The algorithm is constructive in the sense that it finds a multilevel-planar drawing, if it exists. In Sect. 5, we show that testing multilevel planarity of sT-graphs without a fixed combinatorial embedding is NP-hard. Recall that every multilevel-planar drawing is upward planar. We now prove that the vertex with the largest y-coordinate on the boundary of each face is the same across all homeomorphic drawings.

Lemma 2. *Let $G = (V, E)$ be a biconnected sT-graph together with an upward-planar drawing Γ. For each inner face f of Γ and each vertex v incident to f, let $\angle_{\Gamma,f}(v)$ denote the angle defined by the two edges incident to v and f in Γ. Then the following properties hold:*

1. *There is exactly one sink switch t_f on the boundary of f with $\angle_{\Gamma,f}(t_f) \leq \pi$, namely the vertex with greatest y-coordinate among all vertices incident to f.*
2. *Let Γ' be an upward-planar drawing of G that is homeomorphic to Γ. Then the vertex t_f has the greatest y-coordinate of all vertices incident to f in Γ'.*

Proof. The first property was observed by Bertolazzi et al. [7, p. 138, fact 3]. To prove the second property, assume that there exists an upward-planar drawing Γ' of G and a face f such that in Γ', vertex t_f does not have the greatest y-coordinate of all vertices incident to f. Let $e_1 = (v_1, t_f)$ and $e_2 = (v_2, t_f)$ be the edges incident to f and t_f. Figure 1 shows exemplary drawings Γ and Γ'. Because G has a single source s, there exist directed paths p_1 and p_2 from s to v_1 and v_2, respectively. Then the left-to-right order of the edges e_1 and e_2 in Γ and Γ' is determined by the order of the outgoing edges at the last common vertex c on p_1 and p_2. Let $t' \neq t_f$ be the vertex with greatest y-coordinate of all vertices incident to f in Γ'. Then it holds that $\angle_{\Gamma',f}(t') \leq \pi$ and from the first property it follows that $\angle_{\Gamma',f}(t_f) > \pi$. Since Γ and Γ' have the same underlying combinatorial embedding, the clockwise cyclic walk around f is identical in both drawings. But because $\angle_{\Gamma,f}(t_f) \leq \pi$ and $\angle_{\Gamma',f}(t_f) > \pi$, the order of the outgoing edges of c is different in Γ and Γ'. Note that c either has an incoming edge or it is $s = c$, in which case the edge (s,t) lies to the left, i.e., the cyclic order of the edges around c is different in Γ and Γ'. Therefore, Γ and Γ' are not homeomorphic. □

Note that the result of Lemma 2 also holds for embedded sT-graphs that are not biconnected. Obviously it holds for any biconnected component. Any subgraph G' that does not belong to any biconnected component is an attached tree inside a face f given by the combinatorial embedding. If f is an inner face, the unique vertex t_f of that face with maximal y-coordinate must be higher than any vertex of G' in any upward planar drawing.

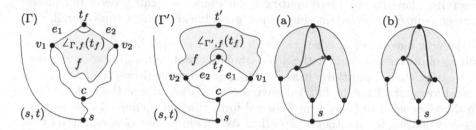

Fig. 1. Proof of Lemma 2.

Fig. 2. Not all edges are valid for the augmentation in Lemma 3.

Bertolazzi et al. showed that any sT-graph with an upward-planar embedding can be extended to an st-graph with an upward-planar embedding that extends the original embedding [6,7]. More formally, let $G = (V, E)$ be an sT-graph together with an upward-planar drawing Γ. Then there exists an st-graph $G_{st} = (V \dot\cup \{t\}, E \dot\cup E_{st})$ where t is the unique sink together with an upward-planar drawing Γ_{st} that extends Γ. Moreover, G_{st} and Γ_{st} can be computed in linear time. Note that in general it is possible for a given E_{st} to choose an upward-planar drawing Γ of G so that the additional edges in E_{st} cannot be added into Γ as y-monotone curves. For an example, see Fig. 2, where augmenting with the red and black edge works only for the drawing shown in (a), whereas augmenting with the blue and black edge works for both drawings. In Lemma 3 we therefore show that there is a set E_{st} that can be added into any drawing with the same combinatorial embedding as Γ. In a way, this is the most general set E_{st}.

Lemma 3. *Let $G = (V, E)$ be a directed sT-graph with a fixed combinatorial embedding. Then there exists an st-graph $G_{st} = (V \dot\cup \{t\}, E \dot\cup E_{st})$, where t is the unique sink, such that for any upward-planar drawing Γ of G there exists an* *upward-planar drawing Γ_{st} of G_{st} that extends Γ. Moreover, G_{st} can be computed in linear time.*

Proof. Start by finding an initial upward-planar drawing Γ_{init} of G in linear time using the algorithm due to Bertolazzi et al. [7]. The algorithm also outputs a matching st-graph G_{st} together with an upward-planar drawing Γ_{st} that extends Γ_{init}. Note that any edge $e \in E_{st}$ is drawn within some face of Γ_{init}. Because t is the only sink of G_{st}, it must have the highest y-coordinate among all vertices in every upward-planar drawing of G_{st}. Therefore, changing all edges in E_{st} drawn within the outer face to have endpoint t ensures that these edges can be drawn within the outer face of any upward-planar drawing Γ of G as y-monotone curves while preserving planarity. For any inner face f, Lemma 2 states that there is a unique t_f incident to f with greatest y-coordinate in every upward-planar drawing of G homeomorphic to Γ. So changing all edges in E_{st} that are drawn within f to have endpoint t_f ensures that these edges can be drawn within f in any upward-planar drawing Γ of G as y-monotone curves while preserving planarity. By precomputing t_f for every face, this procedure handles every edge in E_{st} in constant time, which gives linear running time overall.   □

We now have a set of edges that can be used to complete G into G_{st}. If a multilevel-planar drawing for the given combinatorial embedding of G respecting ℓ exists, then it must also exist for G_{st}. However, the property of ℓ being in normal form might not be fulfilled anymore in G_{st} because of the added edges. We therefore need to bring ℓ into normal form ℓ' again. Lemma 1 tells us that this does not impact multilevel planarity. We conclude that G is multilevel planar with respect to ℓ if and only if G_{st} is multilevel planar with respect to ℓ'. The final property we need is proved by Leipert [21, p. 117, Theorem 5.1], and described in an article by Jünger and Leipert [19].

Lemma 4. *Let G be an st-graph together with a level assignment γ. Then for any combinatorial embedding of G there exists a drawing of G with that embedding that is level planar with respect to γ.*

If ℓ' is in normal form, $\ell'(v) \neq \emptyset$ is a necessary and sufficient condition that there exists a level assignment $\gamma : V \to \mathbb{Z}$ with $\gamma(v) \in \ell'(v)$ for all $v \in V$. Setting $\gamma(v) = \min \ell'(v)$ is one possible such level assignment. Then G is level planar with respect to γ and therefore multilevel planar with respect to ℓ, resulting in the characterization of multilevel-planar st-graphs:

Corollary 1. *Let G be an st-graph together with a multilevel assignment ℓ in normal form. Then there exists a multilevel-planar drawing for any combinatorial embedding of G if and only if $\ell(v) \neq \emptyset$ for all v.*

For a constructive multilevel-planarity testing algorithm, we now first take the edge set computed by the algorithm by Bertolazzi et al. [7] and modify it using Lemma 3 to complete any sT-graph to an st-graph. Note that for this step, we need a fixed combinatorial embedding to be given, as is required by Point 2 of Lemma 2. Once arrived at an st-graph, we only need to check the premise of Corollary 1. This concludes the testing algorithm:

Theorem 2. *Let G be an embedded sT-graph with a multilevel assignment ℓ. Then it can be decided in linear time whether there exists a multilevel-planar drawing of G respecting that embedding. If so, such a drawing can be computed within the same running time.*

Our algorithm uses the fact that to augment sT-graphs to st-graphs, only edges connecting sinks to other vertices need to be inserted. For graphs with multiple sources and multiple sinks, further edges connecting sources to other vertices need to be inserted. The interactions that occur then are very complex: In Sect. 5, we show that deciding multilevel planarity is NP-complete for embedded multi-source graphs. In the next section, we identify oriented cycles as a class of multi-source graphs for which multilevel planarity can be efficiently decided.

4 Oriented Cycles

In this section, we present a constructive multilevel-planarity testing algorithm for oriented cycles, i.e., directed graphs whose underlying undirected graph is a simple cycle. We start by giving a condition for when an oriented cycle $G = (V, E)$ together with some level assignment γ admits a level-planar drawing. This condition yields an algorithm for the multilevel-planar setting.

In this section, γ is always level assignment and ℓ is always a multilevel assignment. Define $\max \gamma = \max\{\gamma(v) \mid v \in V\}$ and $\min \gamma = \min\{\gamma(v) \mid v \in V\}$. Further set $\max \ell = \max\{\max \ell(v) \mid v \in V\}$ and $\min \ell = \min\{\min \ell(v) \mid v \in V\}$. Let $S_{\min} \subset V$ be sources with minimal level, i.e., $S_{\min} = \{v \in V \mid \gamma(v) = \min \gamma\}$, and let $T_{\max} \subset V$ be the sinks with maximal level. We call sources in $S_{\min}$ *minimal sources*, sinks in $T_{\max}$ are *maximal sinks*. Two sinks $t_1, t_2 \in T_{\max}$ are *consecutive* if there is an oriented path between t_1 and t_2 that does not contain any vertex in $S_{\min}$. The set $T_{\max}$ is *consecutive* if all sinks in $T_{\max}$ are pairwise consecutive. We define consecutiveness for sources in $S_{\min}$ analogously. Because G is a cycle, consecutiveness of $T_{\max}$ also means that $S_{\min}$ is consecutive. If both $S_{\min}$ and $T_{\max}$ are consecutive, we say that γ is *separating*.

Lemma 5. *Let G be an oriented cycle with a level assignment γ. Then G is level planar with respect to γ if and only if $T_{\max}$ is consecutive.*

Proof. For the "if" part, augment G to a planar st-graph as follows. Let p_t be the oriented path of minimal length that contains all sinks in $T_{\max}$ and no vertex in $S_{\min}$, and let $t_1, t_2 \in T_{\max}$ denote its endpoints. Let p_s be the oriented path from t_2 to t_1 so that $p_s \cup p_t = G$ and $p_s \cap p_t = \{t_1, t_2\}$. Draw p_t from left to right; see Fig. 3. Below it, draw p_s from right to left. Fix some vertex $s_{\min} \in S_{\min}$ and add an edge from $s_{\min}$ to every source on the path p_t. Add a new vertex s to G, set $\gamma(s) = \min \gamma - 1$ and add an edge from s to every source on the path p_s. Thus, s is now the only source. Next, observe that any sink t_s on the path p_s is drawn to the left of $s_{\min}$ or to the right of $s_{\min}$. Add the edge (t_s, t_1) or the edge (t_s, t_2), respectively. Finally, add a new vertex t to G, set $\gamma(t) = \max \gamma + 1$ and add an edge from every sink on the path p_t to t. Thus, t is now the only sink.

All added edges (u, v) satisfy $\gamma(u) < \gamma(v)$. Hence, G is now an st-graph with a level assignment γ and so Lemma 4 gives that G is level planar with respect to γ.

For the "only if" part, assume that $T_{\max}$ is not consecutive. Then there are maximal sinks $t_1, t_2 \in T_{\max}$ and minimal sources $s_1, s_2 \in S_{\min}$ that appear in the order s_1, t_1, s_2, t_2 around the cycle underlying G. Because the chosen sinks and sources are highest and lowest vertices, respectively, the four edge-disjoint paths that connect them must intersect. □

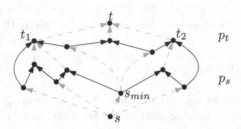

Fig. 3. An st-augmentation of an oriented cycle. The gray dashed edges are added for the st-augmentation. The blue edges belong to path p_t and the red edges to path p_s. (Color figure online)

Recall that any multilevel-planar drawing is a level-planar drawing with respect to some level assignment γ. Lemma 5 gives a necessary and sufficient condition for γ so that the drawing is level planar. Given a multilevel assignment ℓ, we therefore find an induced separating level assignment γ, or determine that no such level assignment exists. It must be $\ell(v) \neq \emptyset$ for all $v \in V$; otherwise, G admits no multilevel drawing. We find an induced level assignment γ that keeps the sets $S_{\min}$ and $T_{\max}$ as small as possible, because such a level assignment is, intuitively, most likely to be separating. To this end, let $S' \subset V$ denote the sources of G such that for $s' \in S'$ we have $\min \ell(s') = \min \ell$. Further, let $S'' \subseteq S'$ denote the sources of G such that for $s'' \in S''$ we have $\ell(s'') = \{\min \ell\}$. Likewise, let $T' \subset V$ denote the sinks of G such that for each $t' \in T'$ it holds that $\max \ell(t') = \max \ell$ let $T'' \subseteq T'$ denote the sinks of G such that for $t'' \in T''$ we have $\ell(t'') = \{\max \ell\}$.

Suppose $S'' \neq \emptyset$. Observe that due to the multilevel assignment, all sources in S'' have to be minimal sources. Therefore, set $S_{\min} = S''$. Otherwise, if $S'' = \emptyset$, pick any source $s' \in S'$ and set $S_{\min} = \{s'\}$. Proceed analogously to find $T_{\max}$. If $T'' \neq \emptyset$, set $T_{\max} = T''$. Otherwise, pick any sink $t' \in T'$ and set $T_{\max} = \{t'\}$. Note that if S'' or T'' are not empty there is no choice but to add all sources or sinks in them to $S_{\min}$ or $T_{\max}$. Otherwise $S_{\min}$ or $T_{\max}$ contains only one vertex, which guarantees that $T_{\max}$ is consecutive. Since ℓ is in normal form, any remaining vertex can be assigned greedily to its minimum possible level above all its ancestors. Hence, G is multilevel planar with respect to ℓ if and only if $T_{\max}$ is consecutive. We conclude the following.

Theorem 3. *Let G be an oriented cycle together with a multilevel assignment ℓ. Then it can be decided in linear time whether G admits a drawing that is multilevel planar with respect to ℓ. Furthermore, if such a drawing exists, it can be computed within the same time.*

5   Hardness Results

We now show that multilevel-planarity testing is NP-complete even in very restricted cases, namely for sT-graphs without a fixed embedding and for embedded multi-source graphs with at most two possible levels for each vertex.

5.1 sT-Graphs with Variable Embedding

In Sect. 3, we showed that testing multilevel planarity of embedded sT-graphs is feasible in linear time, because for every inner sink there is a unique sink switch to cancel it with. We now show that dropping the requirement that the embedding is fixed makes multilevel-planarity testing NP-hard. To this end, we reduce the SCHEDULING WITH RELEASE TIMES AND DEADLINES (SRTD) problem, which is strongly NP-complete [14], to multilevel-planarity testing. An instance of this scheduling problem consists of a set of tasks $T = \{t_1, \ldots, t_n\}$ with individual release times $r_1, \ldots, r_n \in \mathbb{N}$, deadlines $d_1, \ldots, d_n \in \mathbb{N}$ and processing times $p_1, \ldots, p_n \in \mathbb{N}$ for each task (we assume $0 \notin \mathbb{N}$), where $\sum_{i=1}^{n} p_i$ is bounded by a polynomial in n. See Fig. 4 (a) for an example. The question is whether there is a non-preemptive schedule $\sigma : T \to \mathbb{N}$, such that for each $i \in \{1, \ldots, n\}$ we get (1) $\sigma(t_i) \geq r_i$, i.e., no task starts before its release time, (2) $\sigma(t_i) + p_i \leq d_i$, i.e., each task finishes before its deadline, and (3) $\sigma(t_i) < \sigma(t_j) \implies \sigma(t_i) + p_i \leq \sigma(t_j)$ for any $j \in \{1, \ldots, n\} \setminus \{i\}$, i.e., no two tasks are executed at the same time.

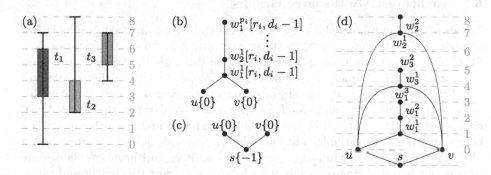

Fig. 4. A task gadget (b) for each task and one base gadget (c) that provides the single source are used to turn a SRTD instance (a) into a multilevel-planarity testing instance (d).

Create for every task $t_i \in T$ a *task gadget* $\mathcal{T}_i$ that consists of two vertices u, v together with a directed path $P_i = (w_i^1, w_i^2, \ldots, w_i^{p_i})$ of length $p_i - 1$; see Fig. 4(b).

For each vertex w_i^j on P_i set $\ell(w_i^j) = [r_i, d_i - 1]$, i.e., all possible points of time at which this task can be executed. Set $\ell(u) = \ell(v) = \{0\}$. Join all task gadgets with a *base gadget*. The base gadget consists of three vertices s, u, v and two edges $(s, u), (s, v)$, where u is placed to the left of v; see Fig. 4(c). Set $\ell(s) = \{-1\}$ and, again, set $\ell(u) = \ell(v) = \{0\}$. Merge all gadgets at their common vertices u and v; see Fig. 4(d). Because SRTD is strongly NP-complete, the size of the resulting graph is polynomial in the size of the input. Further, because the task gadgets may not intersect in a planar drawing and because they are merged at their common vertices u and v, they are stacked on top of each other, inducing a valid schedule of the associated tasks. Contrasting linear-time tests of upward planarity and level planarity for sT-graphs we conclude:

Theorem 4. *Let G be an sT-graph together with a multilevel assignment ℓ. Testing whether G is multilevel planar with respect to ℓ is NP-complete.*

Using a very similar reduction one can also show NP-completeness of multilevel-planarity testing for trees. Full proofs for Theorem 4 and for trees and can be found in [5].

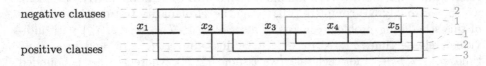

Fig. 5. A rectilinear embedding of the planar monotone 3-SAT instance $(x_1 \vee x_2 \vee x_5) \wedge (x_2 \vee x_3 \vee x_5) \wedge (x_3 \vee x_4 \vee x_5) \wedge (\neg x_1 \vee \neg x_2 \vee \neg x_5) \wedge (\neg x_3 \vee \neg x_4 \vee \neg x_5)$.

5.2   Embedded Multi-source Graphs

We show that multilevel-planarity testing for embedded directed graphs is NP-complete by reducing from PLANAR MONOTONE 3-SAT [9]. An instance $\mathcal{I} = (\mathcal{V}, \mathcal{C})$ of this problem is a 3-SAT instance with variables $\mathcal{V}$, clauses $\mathcal{C}$ and additional restrictions. Namely, each clause is *monotone*, i.e., it is either positive or negative, meaning that it consists of either only positive or only negative literals, respectively. The *variable-clause graph* of $\mathcal{I}$ consists of the nodes $\mathcal{V} \cup \mathcal{C}$ connected by an arc if one of the nodes is a variable and the other node is a clause that uses this variable. The variable-clause graph can be drawn such that all variables lie on a horizontal straight line, positive and negative clauses are drawn as horizontal line segments with integer y-coordinates below and above that line, respectively, and arcs connecting clauses and variables are drawn as non-intersecting vertical line segments; see Fig. 5. We call this a *planar rectilinear embedding* of $\mathcal{I}$.

Let $\Gamma_{\mathcal{I}}$ be a planar rectilinear embedding of $\mathcal{I}$. Transform this into a multilevel-planarity testing instance by replacing each positive or negative clause of $\mathcal{I}$ with a positive or negative clause gadget and merging them at common vertices. Figure 6(a) shows the gadget for the positive clause $(x_a \vee x_b \vee x_c)$. The vertices x_a, x_b and x_c are variables in $\mathcal{V}$. We call vertex p_i the *pendulum*. A variable $x \in \mathcal{V}$ is set to true (false) if it lies on level 1 (level 0). In a positive clause gadget p_i must lie on level 0, and so it forces one variable to lie on level 1, i.e., be set to true. The gadget for a negative clause $(\neg x_a \vee \neg x_b \vee \neg x_c)$ works symmetrically; its pendulum forces one variable to lie on level 0, i.e., be set to false; see Fig. 6 (b).

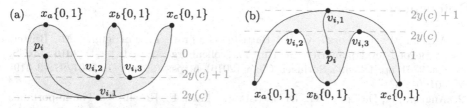

Fig. 6. Gadgets for the clauses $(x_a \vee x_b \vee x_c)$ (a) and $(\neg x_a \vee \neg x_b \vee \neg x_c)$ (b).

Theorem 5. *Let $G = (V, E)$ be an embedded directed graph together with a multilevel assignment ℓ. Testing whether G is multilevel planar is NP-complete, even if it is $|\ell(v)| \leq 2$ for all $v \in V$.*

A detailed proof for Theorem 5 and the graph that results from the instance in Fig. 5 can be found in [5].

6   Conclusion

In this paper we introduced and studied the multilevel-planarity testing problem. It is a generalization of both upward-planarity testing and level-planarity testing.

We started by giving a linear-time algorithm to decide multilevel planarity of embedded sT-graphs. The proof of correctness of this algorithm uses insights from both upward planarity and level planarity. In opposition to this result, we showed that deciding the multilevel planarity of sT-graphs without a fixed embedding is NP-complete. This also contrasts the situation for upward planarity and level planarity, both of which can be decided in linear time for such graphs.

We also gave a linear-time algorithm to decide multilevel planarity of oriented cycles, which is interesting because the existence of multiple sources makes many related problems NP-complete, e.g., testing upward planarity, partial level planarity or ordered level planarity. This positive result is contrasted by the fact that multilevel-planarity testing is NP-complete for oriented trees. Whether multilevel-planarity testing becomes tractable for trees with a given combinatorial embedding remains an open question. We also showed that deciding multilevel planarity remains NP-complete for embedded multi-source graphs where each vertex is assigned either to exactly one level, or to one of two adjacent levels. This result again contrasts the existence of efficient algorithms for testing upward planarity and level planarity of embedded multi-source graphs. The following table summarizes our results.

	Fixed combinatorial embedding			Not embedded		
	st-Graphs	sT-Graphs	Arbitrary	Cycles	sT-Graphs	Trees
Upward Planarity	$O(1)$ [6]	$O(n)$ [6]	P [6]	$O(n)$ [7]	$O(n)$ [7]	$O(1)$ [10]
Multilevel Planarity	$O(1)$ (Corollary 1)	$O(n)$ (Theorem 2)	NPC (Theorem 5)	$O(n)$ (Theorem 3)	NPC (Theorem 4)	NPC (Theorem 5)
Level Planarity	$O(1)$ [19]	$O(n)$ [19]	?	$O(n)$ [19]	$O(n)$ [19]	$O(n)$ [19]

References

1. Angelini, P., Da Lozzo, G., Di Battista, G., Frati, F., Patrignani, M., Rutter, I.: Beyond level planarity. In: Hu, Y., Nöllenburg, M. (eds.) GD 2016. LNCS, vol. 9801, pp. 482–495. Springer, Cham (2016). https://doi.org/10.1007/978-3-319-50106-2_37
2. Angelini, P., Da Lozzo, G., Di Battista, G., Frati, F., Roselli, V.: The importance of being proper (in clustered-level planarity and T-level planarity). Theoretical Comput. Sci. **571**, 1–9 (2015)
3. Angelini, P., et al.: Testing planarity of partially embedded graphs. ACM Trans. Alg. **11**(4), 32:1–32:42 (2015)
4. Bachmaier, C., Brandenburg, F.J., Forster, M.: Radial level planarity testing and embedding in linear time. J. Graph Alg. Appl. **9**(1), 53–97 (2005)
5. Barth, L., Brückner, G., Jungeblut, P., Radermacher, M.: Multilevel planarity (2018). https://arxiv.org/abs/1810.13297
6. Bertolazzi, P., Di Battista, G., Liotta, G., Mannino, C.: Upward drawings of triconnected digraphs. Algorithmica **12**(6), 476–497 (1994)
7. Bertolazzi, P., Di Battista, G., Mannino, C., Tamassia, R.: Optimal upward planarity testing of single-source digraphs. SIAM J. Comput. **27**(1), 132–169 (1998)
8. Brückner, G., Rutter, I.: Partial and constrained level planarity. In: Klein, P.N. (ed.) SODA 2017, pp. 2000–2011 (2017)
9. De Berg, M., Khosravi, A.: Optimal binary space partitions for segments in the plane. Int. J. Comput. Geom. Appl. **22**(3), 187–205 (2012)
10. Di Battista, G., Eades, P., Tamassia, R., Tollis, I.G.: Graph Drawing: Algorithms for the Visualization of Graphs, 1st edn. Prentice Hall PTR (1998)
11. Di Battista, G., Frati, F.: Efficient C-planarity testing for embedded flat clustered graphs with small faces. In: Hong, S.-H., Nishizeki, T., Quan, W. (eds.) GD 2007. LNCS, vol. 4875, pp. 291–302. Springer, Heidelberg (2008). https://doi.org/10.1007/978-3-540-77537-9_29
12. Di Battista, G., Tamassia, R.: Algorithms for plane representations of acyclic digraphs. Theoret. Comput. Sci. **61**(2), 175–198 (1988)
13. Forster, M., Bachmaier, C.: Clustered level planarity. In: Van Emde Boas, P., Pokorný, J., Bieliková, M., Štuller, J. (eds.) SOFSEM 2004. LNCS, vol. 2932, pp. 218–228. Springer, Heidelberg (2004). https://doi.org/10.1007/978-3-540-24618-3_18
14. Garey, M.R., Johnson, D.S.: Two-processor scheduling with start-times and deadlines. SIAM J. Comput. **6**(3), 416–426 (1977)
15. Garg, A., Tamassia, R.: On the computational complexity of upward and rectilinear planarity testing. SIAM J. Comput. **31**(2), 601–625 (2002)

16. Harrigan, M., Healy, P.: Practical level planarity testing and layout with embedding constraints. In: Hong, S.-H., Nishizeki, T., Quan, W. (eds.) GD 2007. LNCS, vol. 4875, pp. 62–68. Springer, Heidelberg (2008). https://doi.org/10.1007/978-3-540-77537-9_9

17. Jelínek, V., Kratochvíl, J., Rutter, I.: A Kuratowski-type theorem for planarity of partially embedded graphs. Comput. Geom. Theory Appl. **46**(4), 466–492 (2013)

18. Jelínková, E., Kára, J., Kratochvíl, J., Pergel, M., Suchý, O., Vyskočil, T.: Clustered planarity: small clusters in Cycles and Eulerian Graphs. J. Graph Alg. Appl. **13**(3), 379–422 (2009)

19. Jünger, M., Leipert, S.: Level planar embedding in linear time. In: Kratochvíyl, J. (ed.) GD 1999. LNCS, vol. 1731, pp. 72–81. Springer, Heidelberg (1999). https://doi.org/10.1007/3-540-46648-7_7

20. Klemz, B., Rote, G.: Ordered level planarity, geodesic planarity and bi-monotonicity. In: Frati, F., Ma, K.-L. (eds.) GD 2017. LNCS, vol. 10692, pp. 440–453. Springer, Cham (2018). https://doi.org/10.1007/978-3-319-73915-1_34

21. Leipert, S.: Level planarity testing and embedding in linear time. Ph.D. thesis, University of Cologne (1998)

Approximation Algorithms

Weighted Upper Edge Cover: Complexity and Approximability

Kaveh Khoshkhah[1($\boxtimes$)], Mehdi Khosravian Ghadikolaei[2], Jérôme Monnot[2], and Florian Sikora[2]

[1] Institute of Computer Science, Tartu University, Tartu, Estonia
khoshkhah@theory.cs.ut.ee
[2] Université Paris-Dauphine, PSL University, CNRS, LAMSADE, 75016 Paris, France
{mehdi.khosravian-ghadikolaei,
jerome.monnot,florian.sikora}@lamsade.dauphine.fr

Abstract. Optimization problems consist of either maximizing or minimizing an objective function. Instead of looking for a maximum solution (resp. minimum solution), one can find a minimum maximal solution (resp. maximum minimal solution). Such "flipping" of the objective function was done for many classical optimization problems. For example, MINIMUM VERTEX COVER becomes MAXIMUM MINIMAL VERTEX COVER, MAXIMUM INDEPENDENT SET becomes MINIMUM MAXIMAL INDEPENDENT SET and so on. In this paper, we propose to study the weighted version of *Maximum Minimal Edge Cover* called UPPER EDGE COVER, a problem having application in genomic sequence alignment. It is well-known that MINIMUM EDGE COVER is polynomial-time solvable and the "flipped" version is **NP**-hard, but constant approximable. We show that the weighted UPPER EDGE COVER is much more difficult than UPPER EDGE COVER because it is not $O(\frac{1}{n^{1/2-\varepsilon}})$ approximable, nor $O(\frac{1}{\Delta^{1-\varepsilon}})$ in edge-weighted graphs of size n and maximum degree Δ respectively. Indeed, we give some hardness of approximation results for some special restricted graph classes such as bipartite graphs, split graphs and k-trees. We counter-balance these negative results by giving some positive approximation results in specific graph classes.

Keywords: Maximum minimal edge cover
Graph optimization problem · Computational complexity
Approximability

1 Introduction

Considering a MaxMin or MinMax version of a problem by "flipping" the objective is not a new idea; in fact, such questions have been posed before for many

Supported by the Estonian Research Council, ETAG (Eesti Teadusagentuur), through PUT Exploratory Grant #620.

G. K. Das et al. (Eds.): WALCOM 2019, LNCS 11355, pp. 235–247, 2019.
https://doi.org/10.1007/978-3-030-10564-8_19

classical optimisation problems. Some of the most well-known examples include the MINIMUM MAXIMAL INDEPENDENT SET problem [7] (also known as MINIMUM INDEPENDENT DOMINATING SET), the MAXIMUM MINIMAL VERTEX COVER problem [6], the MINIMUM MAXIMAL MATCHING problem (also known as MINIMUM INDEPENDENT EDGE DOMINATING SET) [25], and the MAXIMUM MINIMAL DOMINATING SET problem (also called UPPER DOMINATING SET) [1]. However, to the best of our knowledge, weighted MaxMin and MinMax versions have not been considered so far, except for MINIMUM INDEPENDENT DOMINATING SET [10,20], and WEIGHTED UPPER DOMINATING SET problem [8]. MaxMin or MinMax versions of classical problems turn out to be much harder than the originals, especially when one considers complexity and approximation. For example, MAXIMUM MINIMAL VERTEX COVER does not admit any $n^{\frac{1}{2}-\epsilon}$ approximation [6], while VERTEX COVER admits a simple 2-approximation. MINIMUM MAXIMAL MATCHING is **NP**-hard (but 2-approximable) while MAXIMUM MATCHING is polynomial.

The focus of this paper is on *edge cover*. An *edge cover* of a graph $G = (V, E)$ is a subset of edges $S \subseteq E$ which covers all vertices of G. The *edge cover number* of $G = (V, E)$ is the minimum size of an *edge cover* of G. An optimal edge cover can be computed in polynomial time, even for the weighted version where a weight is given for each edge and one wants to minimize the sum of the weight of the edges in the solution (called here the *weighted edge cover number*). An edge cover $S \subseteq E$ is *minimal* (with respect to inclusion) if the deletion of any subset of edges from S destroys the covering property. Minimal edge cover is also known in the literature as an *enclaveless* set [24] or as a *nonblocker* set [14].

In this paper, we study the computational complexity of the *weighted upper edge cover number*, denoted here uec(G, w), that is the solution with maximum weight among all minimal edge covers. Formally, the associated optimization problem called the WEIGHTED UPPER EDGE COVER problem asks to find the largest weighted minimal edge cover of an edge-weighted graph.

WEIGHED UPPER EDGE COVER
Input: A weighted connected graph $G = (V, E, w)$, where $w(e) \geq 0$ for all $e \in E$.
Solution: Minimal edge cover $S \subseteq E$.
Output: Maximize $w(S) = \sum_{e \in S} w(e)$.

Hence, if S^* is an optimal solution of WEIGHED UPPER EDGE COVER on (G, w), then $w(S^*) = $ uec(G, w). The unweighted value of the optimal solution is $uec(G)$ (denoted *upper edge cover number*). To the best of our knowledge, the complexity of computing the weighted upper edge cover number has never been studied in the literature, while a lot of results appear for the unweighed case (corresponding to $w(e) = 1$ for all $e \in E$) [3,11,18,22]. The unweighted variant was firstly investigated in [21], where it is proven that the complexity of computing the upper edge cover number is equivalent to solve the dominating set problem because uec$(G) = |V| - \gamma(G)$ where $\gamma(G)$ is the size of minimum dominating set of graph G. We will consider the implications of this important remark afterwards in the paper.

We will now define a related problem useful in the following because it is proved in [21] that $S \subseteq E$ is a minimal edge cover of $G = (V, E)$ iff S is a spanning star forest of G *without trivial stars* (i.e. without stars consisting of a single vertex).

MAXIMUM   WEIGHTED   SPANNING   STAR   FOREST   PROBLEM   (MAXWSSF   in short)

Input: An edge-weighted graph (G, w) on n vertices where $G = (V, E)$ and $w(e) \geq 0$ for all $e \in E$.

Solution: Spanning star forest $\mathcal{S} = \{S_1, \ldots, S_p\} \subseteq E$.

Output: maximizing $w(\mathcal{S}) = \sum_{e \in \mathcal{S}} w(e) = \sum_{i=1}^{p} \sum_{e \in S_i} w(e)$.

Given an instance (G, w) of MAXWSSF, $opt_{MaxWSSF}(G, w)$ denotes the value of an *optimal spanning star forest*. Authors of [22] describe in details how to apply MAXWSSF model to alignment of multiple genomic sequence, a critical task in comparative genomics. They also show that this approach is promising with real data. In this model, taking weights into account is fundamental since it represents alignment score. Also, their model uses each edge of the spanning star forest to output the solution. Therefore, having trivial star is probably undesirable, which enforces the motivation of studying WEIGHED UPPER EDGE COVER.

The unweighted version (corresponding to the case $w(e) = 1$ for all edges e) is denoted by MAXSSF. In this case, the optimal value is $opt_{MaxSSF}(G)$. For unweighted graphs without isolated vertices, we have $uec(G) = opt_{MaxSSF}(G)$ since any spanning star forest (with possible trivial stars) can be (polynomially) converted into a star spanning forest without trivial stars (i.e. a minimal edge cover) with same size [21]. Hence, these two problems are completely equivalent even from an approximation point of view.

Concerning edge-weighted graphs, the relationship between WEIGHED UPPER EDGE COVER and MAXWSSF is less obvious. For instance, we only have the following inequality: $opt_{MaxWSSF}(G, w) \geq uec(G, w)$ because any minimal edge cover is a particular spanning star forest. However, the difference between these two values can be arbitrarily large as indicated in Fig. 1 (in the graph drawn in Fig. 1.(b), v_4 is an isolated vertex when ε goes to Infinity). This means that isolated vertices play an important role in feasible solutions. Given a spanning star forest $\mathcal{S} = \{S_1, \ldots, S_r\}$ of (G, w), we rename vertices such that there is some $p, 0 \leq p < r$ such that $S_i = \{v_i\}$ are trivial stars for all $1 \leq i \leq p$ (if $p = 0$, then there is no trivial stars), and S_j are non-trivial stars whose c_j is the center for all $j > p$ (if S_j is a single edge, both endpoints are considered as possible centers) We define Triv $= \{v_i \colon l \leq p\}$ as the set of isolated vertices of $(V, E(\mathcal{S}))$ where $E(\mathcal{S}) = \cup_{j > p}^{r} S_j$; moreover, V_l and V_c are respectively the set of leaves and the set of centers of stars in $V \setminus$ Triv. Finally, for $v \in V_l$, $e_v(\mathcal{S}) = c'v \in E(\mathcal{S})$ denotes the edge linking the center c' to the leaf v.

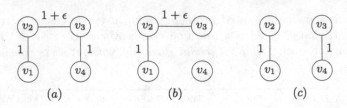

Fig. 1. (a) : The weighted graph $G = (V, E, w)$. (b) : Optimal solution of MaxWSSF(G, w). (c) : Optimal solution of Weighted Upper Edge Cover for G with value uec$(G, w) = 2$.

We mainly focus on specific solutions of MaxWSSF called *nice spanning star forests* defined as follows:

Definition 1. S *is a nice spanning star forest of* (G, w) *if* Triv $= \{v_i : i \leq p\}$ *is an independent set in* G *and all edges of* G *starting at* Triv *are linked to leaves of some ℓ-stars of* S *with* $\ell \geq 2$. *Moreover,* $w(uv) \leq w(e_v(S))$ *for* $u \in$ Triv, $v \in V_l$.

Property 2. Any spanning star forest of (G, w) can be polynomially converted into a nice one with at least the same weight.

It is well known that optimization problems are easier to approximate when the input is a complete weighted graphs satisfying the *triangle inequality*, like for example in the traveling salesman problem. Here, we introduce a generalization of this notion which works to any class of graphs.

Definition 3. *An edge weighted graph* (G, w) *where* $G = (V, E)$ *satisfies the* cycle inequality, *if for every cycle* C, *we have:*

$$\forall e \in C, \quad 2w(e) \leq w(C) = \sum_{e' \in C} w(e')$$

Clearly, for complete graphs, cycle and triangle inequality notions coincide. Definition 3 is interesting when focusing on classes of graphs like split graphs or k-trees. In this article, we are also interested in *bivaluate weights* (resp., *trivalued*) corresponding to the case $w(e) \in \{a, b\}$ with $0 \leq a < b$ (resp., $w(e) \in \{a, b, c\}$ where $0 \leq a < b < c$ are 3 reals). The particular case $a = 0$ and $b = 1$ (called here *binary weights*) is interesting by itself because MaxWSSF with binary weights exactly corresponds to MaxSSF and has been extensively studied in the literature. Moreover for instance, binary weighted Minimum Independent Dominating Set for chordal graphs has been studied in [15], where it is shown that this restriction is polynomial, but bivalued weighted Minimum Independent Dominating Set for chordal graphs with $a > 0$ is **NP**-hard [10].

Graph Terminology and Definitions: Throughout this paper, we consider edge-weighed undirected connected graphs $G = (V, E)$ on $n = |V|$ vertices and $m = |E|$ edges. Each edge $e = uv \in E$ between vertices u and v is weighted by a non-negative weight $w(e) \geq 0$; K_n denotes the *complete graph* on n vertices; a *bipartite graph* (resp., *split graph*) $G = (L \cup R, E)$ is a graph where the vertex set $L \cup R$ is decomposable into an independent set (resp., a clique) L and an independent set R. A *k-tree* is a graph which can be formed by starting from a k-clique and then repeatedly adding vertices in such a way that each added vertex has exactly k neighbors completely connected together (this neighborhood is a k-clique). For instance, 1-trees are trees and 2-trees are maximal series-parallel graphs. A graph is a *partial k-trees* (or equivalently with *treewidth* at most k) if it is a subgraph of a k-trees. The *degree* $d_G(v)$ of vertex $v \in V$ in G is the number of edges incident to v and $\Delta(G)$ is the *maximum degree* of the graph G. A *star* $S \subseteq E$ of a graph $G = (V, E)$ is a tree of G where at most one vertex has a degree greater than 1, or, equivalently, it is isomorphic to $K_{1,\ell}$ for some $\ell \geq 0$. The vertices of degree 1 (except the center when $\ell \leq 1$) are called *leaves* of the star while the remaining vertex is called *center* of the star. A ℓ-star is a star of ℓ leaves. If $\ell = 0$, the star is called *trivial* and it is reduced to a single vertex (the center); otherwise, the star is said *non-trivial*. A *spanning star forest* $\mathcal{S} = \{S_1, \ldots, S_p\} \subseteq E$ of G is a spanning forest into stars, that is, each S_i is a star (possibly trivial), $V(S_i) \cap V(S_j) = \emptyset$ and $\cup_{i=1}^{p} V(S_i) = V$. An *independent set* $S \subseteq V$ of a graph $G = (V, E)$ is a subset of vertices pairwise non-adjacent. The **NP**-hard problem MAXIS seeks an independent set of maximum size. The value of an optimal independent set of G is denoted $\alpha(G)$. A *matching* $M \subseteq E$ is a subset of pairwise non-adjacent edges. A matching M of G is *perfect* if all vertices of G are covered by M. A *dominating set* for a graph G is a subset D of V such that every vertex not in D is adjacent to at least one vertex of D. The *domination number* $\gamma(G)$ is the number of vertices in the smallest dominating set of G.

Related Work: UPPER EDGE COVER has been investigated intensively during the recent years for unweighed graphs, mainly using the terminologies of *spanning star forests* or *dominating sets*. The *minimum dominating set problem* (denoted MINDS) seeks the smallest dominating set of G of value $\gamma(G)$. As indicated before, we have $\text{uec}(G) = n - \gamma(G)$. Thus, using the complexity results known on MINDS, we deduce that UPPER EDGE COVER is **NP**-hard in planar graphs of maximum degree 3 [17], chordal graphs [5] (even in *undirected path graphs*, the class of vertex intersection graphs of a collection of paths in a tree), bipartite graphs, split graphs [4] and k-trees with arbitrary k [12], and it is *polynomial* in k-trees with fixed k, convex bipartite graphs [13], strongly chordal graphs [16]. Concerning the approximability, an **APX**-hardness proof with explicit inapproximability bound and a combinatorial 0.6-approximation algorithm is proposed in [22]. Better algorithms with approximation ratio 0.71 and 0.803 are given respectively in [3,11]. For any $\varepsilon > 0$, UPPER EDGE COVER is hard to approximate within a factor of $\frac{259}{260} + \varepsilon$ unless $\mathbf{P} = \mathbf{NP}$ [22]. It admits a **PTAS** in k-trees (with arbitrary k), although UPPER EDGE COVER remains **APX**-complete on *c-dense* graphs [18] (a graph is called *c-dense* if it contains at least $c\frac{n^2}{2}$ edges).

In contrast, for edge weighted graphs with non-negative weights, no result for WEIGHED UPPER EDGE COVER is known, although some results are given for

MAXIMUM WEIGHTED SPANNING STAR FOREST PROBLEM: a 0.5-approximation is given in [22] (which is the best ratio obtained so far) and polynomial-time algorithms for special classes of graphs such as trees and cactus graphs are presented in [22,23]. Negative approximation results are presented in [9,11,22]. In particular, MAXWSSF is **NP**-hard to approximate within $\frac{10}{11} + \varepsilon$ [9]. Two generalizations of WSSF, denoted MINEXTWSSF and MAXEXTWSSF, have been introduced very recently in [19] where the goal consists in *extending* some partial stars into spanning star forests. In this context, a partial feasible solution is given in advance and the goal is to extend this partial solution. Formally, the problem is defined as follow:

EXTENDED WEIGHTED SPANNING STAR FOREST PROBLEM (EXTWSSF in short)
Input: A weighted graph (G, w) and a packing of stars $\mathcal{U} = \{U_1, \ldots, U_r\}$ where $G = (V, E)$ and $w(e) \geq 0$ for $e \in E$.
Solution: Spanning star forest $\mathcal{S} = \{S_1, \ldots, S_p\} \subseteq E$ containing $\mathcal{U}$.
Output: $w(\mathcal{S}) = \sum_{e \in \mathcal{S}} w(e) = \sum_{i=1}^{p} \sum_{e \in S_i} w(e)$.

In [19], several results have been given for both *minimization* (MINEXTWSSF) and *maximization* (MAXEXTWSSF) versions of EXTWSSF (denoted MINEXTWSSF and MAXEXTWSSF respectively). Dealing with the minimization version for complete graphs: a dichotomy result of the computational complexity is presented depending on parameter c of the (extended) c-relaxed triangle inequality and an FPT algorithm is given. For the maximization version, a positive approximation of $1/2$ and a negative approximation result of $\frac{7}{8}$ (even for binary weights) are proposed.

A subset of vertices V' is called *non-blocking* if every vertex in V' has at least one neighbor in $V \setminus V'$. Actually, *non-blocking* is dual of dominating set and vice versa. For a given graph $G = (V, E)$ and a positive integer k, the NON-BLOCKER problem asks if there is a *non-blocking* set $V' \subseteq V$ with $|V'| \geq k$. Hence, for unweighted graphs, optimal value of *non-blocking* number equals the upper edge cover number. In [14] Dehne et al. propose a parameterized perspective of the NON-BLOCKER problem. They give a linear kernel and an **FPT** algorithm running in time $\mathcal{O}^*(2.5154^k)$. They also give faster algorithms for planar and bipartite graphs.

Contributions: The paper is organized in the following way. We first show in Sect. 2 that WEIGHTED UPPER EDGE COVER in complete graphs is equivalent for its approximation to MAXWSSF in general graphs. Then, we study WEIGHTED UPPER EDGE COVER for bipartite graphs, split graphs and k-trees respectively in Sects. 3, 4 and 5. Motivated by the above results mostly negative, we propose a constant approximation ratio algorithm in Sect. 6 for WEIGHTED UPPER EDGE COVER in bounded degree graphs.

Note that all results given in this paper are valid if G is isolated vertex free instead of connected.

2   Complete Graphs

In this section, we deal with edge-weighted complete graphs. This case seems to be the simplest one because the equivalence between UPPER EDGE COVER and MAXSSF for the unweighted case proven in [21] remains valid for the weighted case as proven in the following.

Theorem 4. MAXWSSF *in general graphs is equivalent to approximate* WEIGHTED UPPER EDGE COVER *in complete graphs.*

From Theorem 4 and from known results on MAXWSSF given in [9,22], we deduce the following:

Corollary 5. *In complete graphs,* WEIGHTED UPPER EDGE COVER *is 1/2-approximable but not approximable within* $\frac{10}{11} + \varepsilon$ *unless* $\boldsymbol{P = NP}$.

3   Bipartite Graphs

Let us now focus on bipartite graphs. We prove that, even in bipartite graphs with binary weights, WEIGHTED UPPER EDGE COVER is not $O(n^{\frac{1}{2}-\varepsilon})$ approximable unless $\mathbf{P = NP}$. Also, we show the problem is **APX**-complete even for bipartite graphs with fixed maximum degree Δ.

Theorem 6. WEIGHTED UPPER EDGE COVER *in bipartite graphs with binary weights and cycle inequality is as hard[1] as* MAXIS *in general graphs.*

Proof. We propose an approximation preserving **APX**-reduction from INDE-PENDENT SET (denoted MAXIS) to WEIGHTED UPPER EDGE COVER.

Given a connected graph $G = (V, E)$ with n vertices and m edges where $V = \{v_1, \ldots, v_n\}$, instance of MAXIS, we build a connected bipartite edge-weighted graph $H = (V_H, E_H, w)$ as follows (see also Fig. 2):

- For each $v_i \in V$, add a P_3 with edge set $\{v_i v_{i,1}, v_{i,1} v_{i,2}\}$.
- For each edge $e = v_i v_j \in E$ where $i < j$, add a middle vertex v_{ij} on edge e.
- $w(e) := \begin{cases} 1 & \text{if } e = v_i v_{i,1} \text{ for some } v_i \in V \\ 0 & \text{otherwise.} \end{cases}$

Clearly, H is a connected bipartite graph on $|V_H| = 3n + m$ vertices and $|E_H| = 2(m+n)$ edges. Moreover, weights are binary and instance satisfies cycle inequality.

Let S^* be a maximum independent set of G with size $\alpha(G)$. For each $e \in E$, let $v^e \in V \setminus S^*$ be a vertex which covers e; it is possible since $V \setminus S^*$ is a *vertex cover* of G. Moreover, $\{v^e : e \in E\} = V \setminus S^*$ since S^* is a maximum independent set of G. Clearly, $S' = \{v_{xy} v^e : e = xy \in E\} \cup \{v_{i,1} v_{i,2} : v_i \in V\} \cup \{v_i v_{i,1} : v_i \in S^*\}$

---

[1] The reduction is actually a Strict-reduction and it is a particular A-reduction which preserves constant approximation.

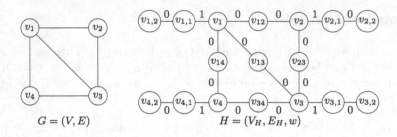

Fig. 2. Construction of H from G. The weights are indicated on edges.

covers all vertices of H and since it doesn't include any P_3, then S' is a minimal edge cover of H. By construction, $w(S') = |S^*| = \alpha(G)$. Hence, we deduce:

$$\mathrm{uec}(H, w) \geq \alpha(G) \tag{1}$$

Conversely, suppose S' is a minimal edge cover of H with weight $w(S')$. Let us make some simple observations of every minimal edge cover of H. Clearly, $\{v_{i1}v_{i2} : v_i \in V\}$ is part of every feasible solution because v_{i2} for $v_i \in V$ are leaves of H. Moreover, for each $e = v_iv_j \in E$ with $i < j$, at least one edge between v_iv_{ij} or v_jv_{ij} belongs to any minimal edge cover of H. If $v_iv_{ij} \notin S'$, it implies that $v_jv_{j,1} \notin S'$ is not a part of the feasible solution because of minimality of S'. Hence, $S = \{v_i : v_iv_{i,1} \in S'\}$ is an independent set of G with size $|S| = w(S')$. We deduce:

$$\alpha(G) \geq \mathrm{uec}(H, w) \tag{2}$$

Using inequalities (1) and (2) we deduce:

$$\alpha(G) = \mathrm{uec}(H, w) \tag{3}$$

In conclusion, for each minimal edge cover S' on H, there is an independent set S of G (computed in polynomial-time) such that $|S| \geq w(S')$.

From Theorem 6, we immediately deduce that WEIGHTED UPPER EDGE COVER in bipartite graphs is not in **APX** unless **P = NP**. However, using several results [2,17] concerning the **APX**-completeness of MAXIS in connected graph G with constant maximum degree $\Delta(G) \geq 3$ or **NP**-completeness of MAXIS in planar graphs, we obtain:

Corollary 7. WEIGHTED UPPER EDGE COVER *in bipartite (resp., planar bipartite) graphs of maximum degree Δ for any fixed $\Delta \geq 4$ and binary weights is **APX**-complete (resp. **NP**-complete).*

Using the strong inapproximation result for MAXIS given in [26], and because the reduction given in previous theorem is indeed a gap-reduction, we also deduce:

Corollary 8. *For any $\varepsilon > 0$, WEIGHTED UPPER EDGE COVER in bipartite graphs of n vertices is not $O(n^{\frac{1}{2} - \varepsilon})$ approximable unless $P = NP$, even for binary weights and cycle inequality.*

We also deduce one inapproximability result depending on the maximum degree.

Corollary 9. *For any constant $\varepsilon > 0$, unless $NP \subseteq ZPTIME(n^{\text{poly}\log n})$, it is hard to approximate* WEIGHTED UPPER EDGE COVER *on bipartite graphs of maximum degree Δ within a factor of $\Theta\left(\frac{1}{\Delta^{1-\varepsilon}}\right)$.*

4 Split Graphs

We will now focus on split graphs. Recall that a graph $G = (L \cup R, E)$ is a split graph if the subgraph induced by L and R is a maximum clique and an independent set respectively.

Theorem 10. WEIGHTED UPPER EDGE COVER *in split graphs with binary weights and cycle inequality is as hard[2] as* MAXIS *in general graphs.*

Corollary 11. WEIGHTED UPPER EDGE COVER *in split 3-subregular graphs is* **APX**-*complete and for any $\varepsilon > 0$,* WEIGHTED UPPER EDGE COVER *in split graphs of n vertices is not $O(n^{\frac{1}{2}-\varepsilon})$ approximable unless* $P = NP$.

5 k-trees

Recall that a k-tree is a graph which results from the following inductive definition: A K_{k+1} is a k-tree. If a graph G is a k-tree, then the addition of a new vertex which has exactly k neighbors in G such that these $k+1$ vertices induce a K_{k+1} forms a k-tree. As a main result in this section we prove WEIGHTED UPPER EDGE COVER is **APX**-complete in k-trees even for trivalued weights.

5.1 Negative Approximation Result

From Corollary 5, we already know that WEIGHTED UPPER EDGE COVER is **NP**-hard to approximate within a ratio strictly better than $\frac{10}{11}$ because the class of all k-trees contains the class of complete graphs. However, this lower bound needs a non-constant number of distinct values [9]. Here, we strengthen the result by proving the existence of lower bounds even for 3 distinct weights. On the other hand, WEIGHTED UPPER EDGE COVER in weighted complete graphs and k-trees with binary weights is not strictly approximable within ratio better than $\frac{259}{260} \approx 0.9961$ because it is proved in [22, Theorem 3.6] a lower bound of $\frac{259}{260} + \varepsilon$ for MAXSSF. Here, we slightly improve this latter bound to $\frac{179}{190} \approx 0.9421$ of WEIGHTED UPPER EDGE COVER with trivalued weights for k-trees.

---

[2] The reduction is actually a Strict-reduction and it is a particular A-reduction which preserves constant approximation.

Theorem 12. WEIGHTED UPPER EDGE COVER *is **APX**-hard in the class of k-trees, even for trivalued weights.*

Corollary 13. WEIGHTED UPPER EDGE COVER *is not approximable within $\frac{179}{190} + \varepsilon$ for every $\varepsilon > 0$ unless $\boldsymbol{P} = \boldsymbol{NP}$ in the class of weighted k-trees, even if there are only three distinct weights.*

5.2   Positive Approximation Result

Theorem 14. *In k-trees,* WEIGHTED UPPER EDGE COVER *is $\frac{k-1}{2(k+1)}$-approximable.*

6   Approximation for Bounded Degree Graphs

In this section, we propose some positive approximation results for graphs of bounded degree in complement to those given in Corollary 9.

Theorem 15. *In general graphs with maximum degree Δ, there is an approximation preserving reduction from* WEIGHTED UPPER EDGE COVER *to* MAX-EXTWSSF *with expansion $c(\rho) = \frac{1}{\Delta} \cdot \rho$.*

Proof. Consider an edge-weighted graph (G, w) of maximum degree $\Delta(G)$ bounded by Δ as an instance of WEIGHTED UPPER EDGE COVER. We make an instance (G, w, U) of MAXEXTWSSF by putting all pendant edges of G in the forced edge set U. Property 2 also works in this context since U is the set of pendant edges. In particular, we deduce $opt_{ExtWSSF}(G, w, U) \geq uec(G, w)$ because U belongs to any minimal edge cover. Let $\mathcal{S} = \{S_1, \ldots, S_r\} \subseteq E$ be a nice spanning star forest of (G, w) containing U satisfying:

$$w(\mathcal{S}) \geq \rho \cdot opt_{ExtWSSF}(G, w, U) \geq \rho \cdot uec(G, w) \tag{4}$$

For each $t \in \text{Triv}$, we choose two edges incident to it with maximum weights $e_1^t = tx_t$ and $e_2^t = ty_t$ in $E \setminus E(\mathcal{S})$ (since by construction $d_G(v) \geq 2$), i.e., $w(e_1^t) \geq w(e_2^t) \geq w(tv)$ for all possible v; let $W = \sum_{t \in \text{Triv}} (w(e_1^t) + w(e_2^t))$ be this global quantity. Also, recall that V_c and V_l are the set of vertices labeled by centers and leaves respectively according to $\mathcal{S}$. We build a new vertex weighted graph $G(\mathcal{S}) = G' = (V', E', w')$ with maximum degree $\Delta(G') \leq \Delta(G) - 1$ as follows:

- $V' = V_l$.
- $uv \in E'$ iff there exists $t \in \text{Triv}$ with $tx_t = tu$ and $ty_t = tv$.
- For $v \in V'$, we set $w'(v) = w(e_v(\mathcal{S}))^3$.

---

[3] We recall $e_v(\mathcal{S})$ is the edge of $\mathcal{S}$ linking leaf v to its center.

Clearly, G' is a graph with bounded degree $\Delta - 1$. We mainly prove that from any independent set $I \subseteq V'$ we can polynomially build an upper edge cover S_I of G satisfying:

$$w(S_I) \geq w'(I) + \left(W - \sum_{t \in \text{Triv}} w(e_1^t) \right) \geq w'(I) \qquad (5)$$

Let $I \subseteq V'$ be maximal independent set of G'. This implies $V' \setminus I$ is a vertex cover of G'. By construction of G', for every $t \in \text{Triv}$, at least one vertex x_t or y_t is not in I (say x_t in the worst case). Recall $e_{x_t}(S)$ is the edge of spanning star forest incident to x_t (since $x_t \in V_l$). We will iteratively apply the following procedure for all $t \in \text{Triv}$ to build S_I:

if the current ℓ-star S_r of S containing $e_{x_t}(S)$ satisfies $\ell \geq 2$ (it is true initially by hypothesis), then delete edge $e_{x_t}(S)$ from S, add edge e_1^t and update spanning star forest S. Otherwise, $\ell = 1$ and only add e_1^t. At the end of the procedure, we get a minimal edge cover S_I of G satisfying inequality (5).

Now, apply as solution of I the greedy algorithm of MaxIS for G' taking, at each step, one vertex with maximum weight w' and by removing all the remaining neighbors of it. It is well known that we have:

$$w'(I) \geq \frac{w'(V')}{\Delta(G') + 1} \geq \frac{w(S)}{\Delta(G)} \qquad (6)$$

Hence, using inequalities (4), (5) and (6), we get the expected result.

Using the 0.5-approximation of MaxExtWSSF given in [19], we deduce:

Corollary 16. Weighted Upper Edge Cover *is $\frac{1}{2\Delta}$-approximable in graphs with bounded degree Δ.*

7   Conclusion

In this article we gave positive and negative approximability aspects of Weighted Upper Edge Cover for special classes of graphs. We considered different types of weight function w for edges of input graph. Hardness of approximation on complete graphs when w satisfies cycle inequality remains open. Also for graphs with bounded degree Δ, we have shown that our problem is $\frac{1}{2\Delta}$-approximable while we proved it can not be better than $\Theta\left(\frac{1}{\Delta}\right)$. Finding a tighter approximation algorithm depending on Δ or on the average degree can be interesting.

References

1. AbouEisha, H., Hussain, S., Lozin, V., Monnot, J., Ries, B., Zamaraev, V.: A boundary property for upper domination. In: Mäkinen, V., Puglisi, S.J., Salmela, L. (eds.) IWOCA 2016. LNCS, vol. 9843, pp. 229–240. Springer, Cham (2016). https://doi.org/10.1007/978-3-319-44543-4_18
2. Alimonti, P., Kann, V.: Some APX-completeness results for cubic graphs. Theor. Comput. Sci. **237**(1–2), 123–134 (2000)
3. Athanassopoulos, S., Caragiannis, I., Kaklamanis, C., Kyropoulou, M.: An improved approximation bound for spanning star forest and color saving. In: Královič, R., Niwiński, D. (eds.) MFCS 2009. LNCS, vol. 5734, pp. 90–101. Springer, Heidelberg (2009). https://doi.org/10.1007/978-3-642-03816-7_9
4. Bertossi, A.A.: Dominating sets for split and bipartite graphs. Inf. Process. Lett. **19**(1), 37–40 (1984)
5. Booth, K.S., Johnson, J.H.: Dominating sets in chordal graphs. SIAM J. Comput. **11**(1), 191–199 (1982)
6. Boria, N., Croce, F.D., Paschos, V.T.: On the max min vertex cover problem. Discrete Appl. Math. **196**, 62–71 (2015)
7. Bourgeois, N., Croce, F.D., Escoffier, B., Paschos, V.T.: Fast algorithms for min independent dominating set. Discrete Appl. Math. **161**(4–5), 558–572 (2013)
8. Boyaci, A., Monnot, J.: Weighted upper domination number. Electr. Notes Discrete Math. **62**, 171–176 (2017)
9. Chakrabarty, D., Goel, G.: On the approximability of budgeted allocations and improved lower bounds for submodular welfare maximization and GAP. SIAM J. Comput. **39**(6), 2189–2211 (2010)
10. Chang, G.J.: The weighted independent domination problem is NP-complete for chordal graphs. Discrete Appl. Math. **143**(1–3), 351–352 (2004)
11. Chen, N., Engelberg, R., Nguyen, C.T., Raghavendra, P., Rudra, A., Singh, G.: Improved approximation algorithms for the spanning star forest problem. Algorithmica **65**(3), 498–516 (2013)
12. Corneil, D.G., Keil, J.M.: A dynamic programming approach to the dominating set problem on k-trees. SIAM J. Algebraic Discrete Methods **8**(4), 535–543 (1987)
13. Damaschke, P., Müller, H., Kratsch, D.: Domination in convex and chordal bipartite graphs. Inf. Process. Lett. **36**(5), 231–236 (1990)
14. Dehne, F., Fellows, M., Fernau, H., Prieto, E., Rosamond, F.: NONBLOCKER: parameterized algorithmics for MINIMUM DOMINATING SET. In: Wiedermann, J., Tel, G., Pokorný, J., Bieliková, M., Štuller, J. (eds.) SOFSEM 2006. LNCS, vol. 3831, pp. 237–245. Springer, Heidelberg (2006). https://doi.org/10.1007/11611257_21
15. Farber, M.: Independent domination in chordal graphs. Oper. Res. Lett. **4**(1), 134–138 (1982)
16. Farber, M.: Domination, independent domination and duality in strongly chordal graphs. Discrete Appl. Math. **7**, 115–130 (1984)
17. Garey, M.R., Johnson, D.S.: Computers and Intractability: A Guide to the Theory of NP-Completeness. W. H. Freeman & Co., New York (1979)
18. He, J., Liang, H.: Improved approximation for spanning star forest in dense graphs. J. Comb. Optim. **25**(2), 255–264 (2013)
19. Khoshkhah, K., Khosravian Ghadikolaei, M., Monnot, J., Theis, D.O.: Extended spanning star forest problems. In: Gao, X., Du, H., Han, M. (eds.) COCOA 2017. LNCS, vol. 10627, pp. 195–209. Springer, Cham (2017). https://doi.org/10.1007/978-3-319-71150-8_18

20. Lozin, V.V., Malyshev, D.S., Mosca, R., Zamaraev, V.: More results on weighted independent domination. Theor. Comput. Sci. **700**, 63–74 (2017)
21. Manlove, D.F.: On the algorithmic complexity of twelve covering and independence parameters of graphs. Discrete Appl. Math. **91**(1–3), 155–175 (1999)
22. Nguyen, C.T., Shen, J., Hou, M., Sheng, L., Miller, W., Zhang, L.: Approximating the spanning star forest problem and its application to genomic sequence alignment. SIAM J. Comput. **38**(3), 946–962 (2008)
23. Nguyen, V.H.: The maximum weight spanning star forest problem on cactus graphs. Discrete Math. Alg. Appl. **7**(2) (2015)
24. Slater, P.J.: Enclaveless sets and mk-systems. J. Res. Nat. Bur. Stand. **82**(3), 197–202 (1977)
25. Yannakakis, M., Gavril, F.: Edge dominating sets in graphs. SIAM J. Appl. Math. **38**(3), 364–372 (1980)
26. Zuckerman, D.: Linear degree extractors and the inapproximability of max clique and chromatic number. Theory Comput. **3**(1), 103–128 (2007)

Linear Pseudo-Polynomial Factor Algorithm for Automaton Constrained Tree Knapsack Problem

Soh Kumabe[1,2], Takanori Maehara[2(✉)], and Ryoma Sin'ya[3]

[1] The University of Tokyo, Tokyo, Japan
sou.kumabe@riken.jp
[2] RIKEN Center for Advanced Intelligence Project, Tokyo, Japan
takanori.maehara@riken.jp
[3] Akita University, Akita, Japan
ryoma@math.akita-u.ac.jp

Abstract. The *automaton constrained tree knapsack problem* is a variant of the knapsack problem in which the items are associated with the vertices of the tree, and we can select a subset of items that is accepted by a tree automaton. If the capacities or the profits of items are integers, it can be solved in pseudo-polynomial time by the dynamic programming algorithm. However, this algorithm has a quadratic pseudo-polynomial factor in its complexity because of the max-plus convolution. In this study, we propose a new dynamic programming technique, called *heavy-light recursive dynamic programming*, to obtain algorithms having linear pseudo-polynomial factors in the complexity. Such algorithms can be used for solving the problems with polynomially small capacities/profits efficiently, and used for deriving efficient fully polynomial-time approximation schemes. We also consider the k-subtree version problem that finds k disjoint subtrees and a solution in each subtree that maximizes total profit under a budget constraint. We show that this problem can be solved in almost the same complexity as the original problem.

Keywords: Knapsack problem · Dynamic programming
Tree automaton

1 Introduction

1.1 Background and Motivation

The knapsack problem seeks a set of items that maximizes total profit under a budget constraint. The problem is one of the most fundamental combinatorial optimization problems [12] and has many real-world applications such as scheduling [9], network design [15], and natural language processing [7]. The problem is NP-hard; however, if the profits or the weights of items are integers, the problem can be solved using the dynamic programming (DP) that runs in

© Springer Nature Switzerland AG 2019
G. K. Das et al. (Eds.): WALCOM 2019, LNCS 11355, pp. 248–260, 2019.
https://doi.org/10.1007/978-3-030-10564-8_20

pseudo-polynomial time. This algorithm is the basis for the fully-polynomial time approximation scheme (FPTAS) of the knapsack problem [9,13].

Here, we consider the *automaton constrained tree knapsack problem*, which is defined as follows. Let $T = (V(T), E(T))$ be a rooted tree where $V(T)$ is the set of vertices and $E(T)$ is the set of edges, $\mathcal{F}(\mathcal{A}) \subseteq 2^{V(T)}$ be a feasible domain represented by a top-down tree automaton (see Sect. 2.1 for details). We denote by $n = |V(T)|$ the number of vertices in T. Each $u \in V(T)$ has profit $p(u) \in \mathbb{R}_{\geq 0}$ and weight $w(u) \in \mathbb{R}_{\geq 0}$. For a vertex subset $X \subseteq V(T)$, we define $p(X) = \sum_{u \in X} p(u)$ and $w(X) = \sum_{u \in X} w(u)$. Let $C \in \mathbb{R}_{\geq 0}$ be the capacity. Then, the task is to solve the following optimization problem:

$$\text{maximize } p(X) \text{ subject to } w(X) \leq C, \ X \in \mathcal{F}(\mathcal{A}), \tag{1}$$

This is a quite general problem since any constraint on a tree specified by a monadic second-order logic formula is represented by a tree automaton [18]. For example, the precedence constrained problem [14], the connectivity constrained problem [8], and the independent set constrained problem [16] are particular cases of this problem (See Examples 1, 2, and 3).

As in the case of the standard knapsack problem, the automaton constrained tree knapsack problem can be solved by DP. If the tree automaton has a *polynomially bounded diversity of transitions* (see Sect. 2.1), the complexity of the algorithm is $O(\text{poly}(n)C^2)$ time if the weights are integers, and $O(\text{poly}(n)P^2)$ time if the profits are integers, where P is an upper bound of the optimal value (see Sect. 2.2). Several existing studies have considered particular cases of the problem and derived the corresponding realization of this algorithm [8,14,16].

In this study, we focus on the *pseudo-polynomial factors* C or P in the complexity. The quadratic pseudo-polynomial factors of the standard DP come from merging solutions to the subtrees, which is implemented by the max-plus (or min-plus) convolution, whose current best complexity is $O(N^2 \log \log N / \log^2 N)$, where N is the length of the arrays [2]. It is conjectured that the max-plus convolution requires $\Omega(N^{2-\delta})$ time for any $\delta > 0$ [1,2,6]. However, quadratic pseudo-polynomial factors are sometimes unacceptable. For example, in practice, we often encounter the case that C is polynomially greater than n (e.g., $n = 100$ and $C = 100,000$). In this case, quadratic pseudo-polynomial factors are not desirable. For another example, when we derive a FPTAS from the DP, we take $P \propto 1/\epsilon$; thus, a smaller degree in P implies a faster algorithm with the same accuracy. The purpose of this study is to derive algorithms for the problem that run in $O(\text{poly}(n)C)$ or $O(\text{poly}(n)P)$ time.

Thus far, the only studies that have addressed this issue are those on the precedence constrained knapsack problem. Johnson and Niemi [11] proposed a technique, called left-right DP, which runs in $O(nC)$ time. Cho and Shaw [4] proposed a variant of the left-right DP, called depth-first DP, which also runs in $O(nC)$ time. However, we do not know what kinds of constraints (other than the precedence constraint) admit algorithms with complexity that is linear in pseudo-polynomial factors.

1.2   Our Contribution

In this study, we introduce a new DP technique, called *heavy-light recursive dynamic programming (HLRecDP)*. This technique is motivated by Chekuri and Pal's recursive greedy algorithm for the *s-t* path constrained monotone submodular maximization problem [3] and its generalization to the logic constrained monotone submodular maximization problem [10]. It also generalizes the left-right DP and depth-first DP for precedence constrained problem to the automaton constrained problem. Formally, by using this technique, we obtain the following theorem. From now on, we denote the logarithm of base two by log.

Theorem 1. *Let $T = (V(T), E(T))$ be a tree with n vertices and $\mathcal{A}$ be a non-deterministic top-down tree automaton with the diversity of transitions $\delta(n)$. Let $p \in \mathbb{R}^V_{\geq 0}$, $w \in \mathbb{Z}^V_{\geq 0}$, and $C \in \mathbb{Z}_{\geq 0}$. Then, there is an algorithm for problem (1) that runs in $O(n^{\log(1+\delta(n))}C)$ time. In particular, if $\delta(n) = O(1)$, the algorithm runs in $O(poly(n)C)$ time.*[1]

This theorem gives a sufficient condition for admitting (pseudo-)polynomial time algorithms with linear pseudo-polynomial factors. By applying this theorem to the precedence constrained problem, we obtain $O(nC)$ time algorithm that is equivalent to the existing left-right DP [11] and depth-first DP [4] (Example 2).

   We then consider the *k-subtree* version problem. Let $k = O(1)$ be an integer. Then, the problem is to find k disjoint subtrees of the given tree and a feasible solution in each subtree such that the total profit is maximized under the total budget constraint. For example, the k connected component constrained problem is the k-subtree version of the precedence constrained problem. By using the property of the algorithm of Theorem 1 and divide-and-conquer techniques, we show that this problem can be solved in almost the same time complexity as the original problem.

Theorem 2. *Suppose that $\mathcal{A}$ is a prefix-closed top-down tree automaton with the bounded diversity of transitions, and the automaton constrained tree knapsack problem with $\mathcal{A}$ can be solved in $f(n)$ time by Algorithm 1. Let $k = O(1)$. Then, there exists an algorithm for the corresponding k-subtree version problem that runs in the following complexity:*

$k = 1$. *$O(f(n)\log n)$ if $f(n) = O(nC)$, and $O(f(n))$ time if $f(n) = O(n^e C)$ for some $e > 1$.*

$k \geq 2$. *$O(f(n)(\log n)^{\log k})$ time if $f(n) = O(n^e C)$ for some $e > 1$; the hidden constant is a polynomial in k.*

This theorem gives an $O(n \log nC)$ time algorithm for the connectivity constrained problem, and an $O(n^e C)$ time algorithm for any $e > 1$ for the k connected component constrained tree knapsack problem.

---

[1] For simplicity, we only consider the case in which the weights are integers. The same result is obtained when the profits are integers.

Organization of the Paper

The paper is organized as follows. In Sect. 2.1, we introduce top-down tree automata. In Sect. 2.2, we introduce the standard DP using a top-down tree automaton. In Sect. 3, we prove Theorem 1 by introducing the HLRecDP. In Sect. 4, we prove Theorem 2 using the divide-conquer technique with HLRecDP.

2 Preliminaries

2.1 Tree Automaton

A *non-deterministic top-down tree automaton ("automaton" for short)* [5] is a tuple $\mathcal{A} = (Q, \Sigma, Q_{\text{init}}, \Delta)$, where Q is the set of states, Σ is a set of alphabets, $Q_{\text{init}} \subseteq Q$ is the set of initial states, and Δ is a set of rewriting rules of the form

$$Q \times \Sigma \ni (q, \sigma) \mapsto (q_1, \ldots, q_d) \in Q \times \cdots \times Q. \tag{2}$$

We assume that the number of states of the automaton is constant, $|Q| = O(1)$. The automaton is *prefix-closed* if $(q, \sigma) \mapsto (q_1, \ldots, q_d)$ is in Δ then $(q, \sigma) \mapsto (q_1, \ldots, q_{d-1})$ also in Δ.

The *run* of the automaton is defined as follows. Let $T = (V(T), E(T))$ be a rooted tree, and $\sigma : V(T) \to \Sigma$ be labels on the vertices. The automaton first assigns an initial state $q \in Q_{\text{init}}$ to the root of the tree. Then it processes the tree from the top (root) to the bottom (leaves). If vertex $u \in V(T)$ has state $q \in Q$, we choose a rewriting rule $(q, \sigma(u)) \mapsto (q_1, \ldots, q_d)$ and assign the states $q_1, \ldots, q_d$ to the children $v_1, \ldots, v_d \in V(T)$ of u, respectively. Note that, if no rule is applicable to u and q, the run fails. The automaton accepts a labeled tree if there is at least one run from the root to the leaves in which the state of the root is in Q_{init}.

To represent a substructure of a tree using an automaton, we choose the alphabet $\Sigma = \{0, 1\}$ and identify the subgraph $X \subseteq V(T)$ as the labels $\sigma_X : V(T) \to \Sigma$ such that $\sigma_X(u) = 1$ for $u \in X$ and $\sigma_X(u) = 0$ for $u \notin X$. Then, the family of subsets $\mathcal{F}(\mathcal{A}) \subseteq 2^{V(T)}$ represented by this automaton is specified by

$$\mathcal{F}(\mathcal{A}) = \{X \subseteq V(T) : \mathcal{A} \text{ accepts } T \text{ with label } \sigma_X\}. \tag{3}$$

To evaluate the complexity of DP, we introduce the following quantity $\delta(n)$, called the *diversity of transitions*.

$$\delta(n) = \max_{m \le n} \Big| \bigcup_{\text{"}(q,\sigma) \mapsto (q_1, \ldots, q_m)\text{"} \in \Delta} \{(q_1, \ldots, q_m)\} \Big|. \tag{4}$$

By definition, $\delta(n)$ is monotone in n. Intuitively, $\delta(n)$ is the maximum number of subproblems in DP; see Sect. 2.2 below. There is an automaton with exponentially large diversity of transitions, i.e., $\delta(n) = \Theta(|Q|^n)$, and in such case, it looks impossible to obtain $O(\text{poly}(n))$ time algorithm. Therefore, we assume some boundedness of $\delta(n)$. Note that, even $\delta(n) = O(1)$, we can represent some interesting examples, such as independent set constraint (Example 1) and precedence constraint (Example 2).

2.2   Quadratic Pseudo-Polynomial Factor Algorithm

Here, we introduce the standard DP that solves the problem in $O(\text{poly}(n)C^2)$ time if the automaton has a polynomially bounded diversity of transitions [8,14, 16]. We regard this as a baseline algorithm for the problem.

Let $T = (V(T), E(T))$ be a rooted tree. We denote by T_u the subtree of T rooted by $u \in V(T)$. The algorithm computes array $x_{u,q}$ of length $C + 1$ for each $u \in V(T)$ and $q \in Q$, such that

$$x_{u,q}[c] = \max\{p(X) : X \subseteq V(T_u), w(X) = c, \text{ subtree } T_u \text{ with labels } \sigma_X \text{ is}$$
$$\text{accepted by } \mathcal{A}, \text{ where the initial state is } q\}. \tag{5}$$

Once the array for the root vertex $r \in V(T)$ is obtained, the optimal value is computed by $\max_{q \in Q_{\text{init}}, c \in \{0,...,C\}} x_{r,q}[c]$ in $O(|Q_{\text{init}}|C) = O(C)$ time.

We compute these arrays using the bottom-up DP as follows. For each leaf, the array is immediately computed in $O(\delta(0)C) = O(C)$ time. Consider a vertex $u \in V(T)$ with children $v_1, \ldots, v_d \in V(T)$, such that the arrays $x_{v,q}$ are computed for all $v \in \{v_1, \ldots, v_d\}$ and $q \in Q$. Then,

$$x_{u,q}[c] = \max\{x_{v_1,q_1}[c_1] + \cdots + x_{v_d,q_d}[c_d] + w(u)\sigma :$$
$$(q, \sigma) \mapsto (q_1, \ldots, q_d) \in \Delta, c_1 + \cdots + c_d + w(u)\sigma = c\}. \tag{6}$$

Here, we identify symbol $\sigma = $ "0" and "1" as integer 0 and 1, respectively. The maximization with respect to $c_1, \ldots, c_d$ is evaluated by the max-plus convolution; thus, it costs about $O(nC^2)$ time. For the maximization with respect to $(q, \sigma) \rightarrow (q_1, \ldots, q_d) \in \Delta$, we only have to evaluate the formula for distinct $(q_1, \ldots, q_d)$. Therefore, the complexity of evaluating (5) is $O(n\delta(n)C^2)$ time, and the total complexity is $O(n^2\delta(n)C^2) = O(\text{poly}(n)C^2)$.

3   Heavy-Light Recursive Dynamic Programming

In this section, we present the HLRecDP for obtaining an $O(n^{\log(1+\delta(n))}C)$ time algorithm. In Sect. 3.1, we first propose the recursive dynamic programming (RecDP) technique for balanced trees. To handle non-balanced trees, in Sect. 3.2, we combine the heavy-light decomposition to the RecDP.

3.1   Recursive Dynamic Programming for Balanced Trees

Our goal is to compute arrays $\{x_{r,q}\}_{q \in Q_{\text{init}}}$ for the root $r \in V(T)$ of the tree, where $x_{r,q}$ is defined in (5). To avoid quadratic pseudo-polynomial factors, we call the recursive procedure for the children multiple times, instead of merging subtree solutions.

Formally, we design procedure $\text{RecDP}(u, q, a)$, where $u \in V(T)$, $q \in Q$, and a is an array of size $C + 1$. It computes array $y_{u,q,a}$ defined by

$$y_{u,q,a}[c] = \max\{p(X) + a[c'] : X \subseteq V(T_u), w(X) + c' = c, \text{ subtree } T_u \text{ with}$$
$$\text{labels } \sigma_X \text{ is accepted by } \mathcal{A}, \text{ where the initial state is } q\}. \tag{7}$$

The difference between (5) and (7) is that (7) contains the array parameter a, which corresponds to the "initial values" of the DP. More intuitively, it returns an array that is obtained by "adding" items in the subtree T_u optimally to the current solution represented by a. By calling $\text{RECDP}(r, q, [0, -\infty, \ldots, -\infty])$, where $r \in V(T)$ is the root of the tree and $q \in Q_{\text{init}}$, we obtain the desired solution $x_{r,q}$.

Here, $\text{RECDP}(u, q, a)$ is implemented as follows. If $u \in V(T)$ is a leaf, we can compute (7) in $O(C)$ time. Consider a vertex $u \in V(T)$ that has children $v_1, \ldots, v_d \in V(T)$. For each rewriting rule $(q, \sigma) \mapsto (q_1, \ldots, q_d)$, we first call $\text{RECDP}(v_1, q_1, a)$ to obtain array $y_1 = y_{v_1, q_1, a}$. Then, we call $\text{RECDP}(v_2, q_2, y_1)$ to obtain array $y_2 = y_{v_2, q_2, y_1}$, i.e., we use the returned array y_1 as the initial values of the DP to the subtree rooted by v_2. By iterating this process to the last child, we obtain array $y_d = y_{v_d, q_d, y_{d-1}}$. The solution corresponds to this rewriting rule is then obtained by

$$z_{u,q,a}[c] = y_d[c - \sigma w(u)] + \sigma p(u), \quad c \in \{0, \ldots, C\}. \tag{8}$$

By taking the entry-wise maximum of the solutions on all of the rewriting rules, we obtain the solution to $\text{RECDP}(u, q, a)$.

The correctness of the above procedure is easily checked. We evaluate the time complexity. Let $f(n)$ be the complexity of the procedure. Let $n_1, \ldots, n_d$ be the number of vertices on the subtrees rooted by $v_1, \ldots, v_d$. Then we have $n_1 + \cdots + n_d = n - 1$. Because the algorithm calls the procedure recursively to each subtree at most $\delta(n)$ times, the complexity satisfies[2]

$$f(n) \leq \delta(n)(f(n_1) + \cdots + f(n_d)) + O(C). \tag{9}$$

If the tree is balanced, i.e., $n_j \leq n/2$ for all $j = 1, \ldots, d$, this already provides the desired complexity: Without loss of generality, we can assume that $f(n)$ is convex in n. Then, the maximum of the right-hand side is attained at $n_1 = \lceil (n-1)/2 \rceil$, $n_2 = \lfloor (n-1)/2 \rfloor$, and $n_3 = \cdots = n_d = 0$. Therefore,

$$f(n) \leq \delta(n)\left(f(\lceil (n-1)/2 \rceil) + f(\lfloor (n-1)/2 \rfloor)\right) + O(C). \tag{10}$$

By solving this inequality, we have $f(n) = O((2\delta(n))^{\log n} C) = O(n^{1 + \log \delta(n)} C)$.

3.2 Heavy-Light Recursive Dynamic Programming

To obtain an $O(\text{pseudopoly}(n)C)$ time algorithm for general (i.e., non-balanced) trees, we have to make the depth of the recursion to $\tilde{O}(\log n)$. The HLRecDP achieves this by using the *heavy-light decomposition* [17].

First, we introduce the heavy-light decomposition. Let $T = (V(T), E(T))$ be a rooted tree whose edges are directed toward the leaves. An edge $(u, v) \in E(T)$ is a *heavy edge* if v has more descendants than other children of u do

---

[2] The additive term is naturally $O(dC)$; however, it is separated and included in the recursive terms.

(ties are broken arbitrary). An edge is a *light edge* if it is not a heavy edge[3]. $v \in V(T)$ is a *heavy child* of $u \in V(T)$ if (u, v) is a heavy edge. A *light child* is defined similarly. A subtree rooted by a light child is referred to as a *light subtree*. The set of heavy edges forms disjoint paths, called *heavy paths*. The tree is decomposed into the heavy paths, which is referred to as a *heavy-light decomposition*. The heavy-light decomposition is computed in linear time by depth-first search. The most important property of a heavy-light decomposition is that for each $u \in V(T)$, the number of descendants of a light child is at most $|V(T_u)|/2$.

Recall algorithm RECDP(u, q, a) defined in the previous section. We observe that all recursive calls of RECDP(v_1, q_1, a) to the first child has the same initial array a for different q_1. Thus, we can "gather" all recursive calls for the first child into a single recursive call. The HLRecDP sets the heavy child as the first child to avoid an excessive number of recursive calls this child.

Formally, we define procedure HLRECDP(u, a). This returns a set of arrays $\{y_{u,q}\}_{q \in Q}$, where $y_{u,q}$ is defined in (7). For $v_1, \ldots, v_d \in V(T)$ and $q_1, \ldots, q_d \in Q$, we define HLRECDP$(v_1, \ldots, v_d, a)_{q_1, \ldots, q_d}$ as a shorthand notation of the sequential evaluation

$$\text{HLRECDP}(v_d, \text{HLRECDP}(v_{d-1} \cdots \text{HLRECDP}(v_1, a)_{q_1} \cdots)_{q_{d-1}})_{q_d}. \quad (11)$$

Now we describe the procedure. Let $v_1, \ldots, v_d \in V(T)$ be the children of u, where v_1 is the heavy child. First, we call HLRECDP(v_1, a) and store the resulting arrays for all $q \in Q$. Then, for each rewriting rule $(q, \sigma) \mapsto (q_1, \ldots, q_d)$, we call HLRECDP$(v_2, \ldots, v_d, \text{HLRECDP}(v_1, x)_{q_1})_{q_2, \ldots, q_d}$ and add item u if $\sigma = 1$ to obtain the solution to the rewriting rule. By taking the entry-wise maximum over the rewriting rules, we obtain the desired solution; see Algorithm 1.

By construction, HLRECDP gives the same solution as RECDP; thus, it correctly solves the problem. We evaluate the complexity as follows. Let $n_1, \ldots, n_d$ be the number of vertices on the subtrees rooted by $v_1, \ldots, v_d$. As same as the analysis of RECDP, the complexity $f(n)$ of the algorithm satisfies

$$f(n) \leq f(n_1) + \delta(n) \left(f(n_2) + \cdots + f(n_d) \right) + O(C). \quad (12)$$

By the convexity of $f(n)$ and the heavy-light property, i.e., $n_j \leq n/2$ ($j = 2, \ldots, d$), the maximum of the right-hand side is attained at $n_1 = \lceil (n-1)/2 \rceil$, $n_2 = \lfloor (n-1)/2 \rfloor$, and $n_3 = \cdots = n_d = 0$. Thus, we have

$$f(n) \leq f(\lceil (n-1)/2 \rceil) + \delta(n)) f(\lfloor (n-1)/2 \rfloor) + O(C). \quad (13)$$

By solving this inequality, we have $f(n) = O(n^{\log(1+\delta(n))} C)$. $\square$

---

[3] Our definition of the heavy edge is slightly different to the original one: In [17], (u, v) is said to be "heavy" if $2 \times \text{size}(v) > \text{size}(u)$, where $\text{size}(v)$ is the number of descendants of v. Thus, their heavy edge is always our heavy edge, but the converse is not. In particular, in their definition, any internal vertex has *at most* one heavy edge, but in our definition, any internal vertex has *exactly* one heavy edge.

Algorithm 1. Heavy-Light Recursive Dynamic Programming

1: **procedure** HLRECDP(u, a)
2:    $y_{u,q}[c] = -\infty$ for all $c \in \{0, \ldots, C\}$, $q \in Q$
3:    Let $v_1, \ldots, v_d$ be the children of u, where v_1 is the heavy child
4:    Call HLRECDP(v_1, a) and store the arrays for $q \in Q$
5:    **for** "$(q, \sigma) \mapsto (q_1, \ldots, q_d)$" $\in \Delta$ **do**
6:        Let $z = $ HLRECDP($v_2, \ldots, v_d$, HLRECDP($v_1, a)_{q_1})_{q_2, \ldots, q_d}$
7:        **for** $c = 0, \ldots, C$ **do**
8:            $y_{u,q}[c] \leftarrow \max\{y_{u,q}[c], z[c - \sigma w(u)] + \sigma p(u)\}$
9:        **end for**
10:    **end for**
11:    **return** $\{y_{u,q}\}_{q \in Q}$
12: **end procedure**

Remark 1. There is a gap of the tractable classes between the standard DP (Sect. 2.2) and the HLRecDP. The analysis in Sect. 2.2 implies that we can obtain $O(\text{poly}(n)C^2)$ time algorithm if $\delta(n)$ is polynomially bounded. On the other hand, the analysis in this section implies that if $\delta(n)$ is polynomially bounded (rather than bounded by a constant), we can only obtain an algorithm with quasi-polynomial time complexity, i.e., $n^{O(\log n)}C$.

Here, we derive several results for particular cases using our method.

Example 1 (Independent Set Constrained Problem). Let us consider the *independent set constrained* tree knapsack problem whose feasible set contains no adjacent vertices. This constraint is represented by an automaton $\mathcal{A} = (Q, \Sigma, Q_{\text{init}}, \Delta)$, where $Q = Q_{\text{init}} = \{\text{s}, \text{x}\}$ and

$$(\text{s}, 0) \mapsto (\text{s}, \ldots, \text{s}), \quad (\text{s}, 1) \mapsto (\text{x}, \ldots, \text{x}), \quad (\text{x}, 0) \mapsto (\text{s}, \ldots, \text{s}). \tag{14}$$

Here, s means the vertex can be selected and x means the vertex cannot be selected. The diversity of transitions is $\delta(n) = 2$ because the rules for $(\text{s}, 0)$ and $(\text{x}, 0)$ have the same right-hand side; therefore, we can solve the independent set constrained tree knapsack problem in $O(n^{\log(1+\delta(n))}C) = O(n^{\log 3}C) = O(n^{1.585}C)$ time.

Example 2 (Precedence Constrained Problem). Let us consider the *precedence constrained* tree knapsack problem whose feasible set is precedence closed, i.e., if a vertex is contained in a solution, all the precedences are also contained in the solution. This constraint is represented by an automaton $\mathcal{A} = (Q, \Sigma, Q_{\text{init}}, \Delta)$, where $Q = Q_{\text{init}} = \{\text{s}, \text{x}\}$ and

$$(\text{s}, 0) \mapsto (\text{x}, \ldots, \text{x}), \quad (\text{s}, 1) \mapsto (\text{s}, \ldots, \text{s}), \quad (\text{x}, 0) \mapsto (\text{x}, \ldots, \text{x}). \tag{15}$$

Here, state s means the vertex can be selected and state x means the vertex cannot be selected. Since the diversity of transitions is $\delta(n) = 2$, the algorithm runs in $O(n^{\log(1+\delta(n))}C) = O(n^{1.585}C)$ time.

This complexity can be improved further. If a vertex has state x, we cannot select all of the descendants of the vertex; thus we obtain the solution for this case without calling the procedure recursively. Thus, the required number of recursive calls is at most one, which is for $(s, 1)$. Therefore, the algorithm runs in $O(n^{\log(1+1)}C) = O(nC)$ time. Note that this algorithm "coincides" with the left-right DP [11] and the depth-first DP [4] in the sense that these perform the same manipulations.

Example 3 (Connectivity Constrained Problem). Let us consider the *connectivity constrained* tree knapsack problem whose feasible set forms a connected subgraph of a given tree. This constraint is represented by an automaton $\mathcal{A} = (Q, \Sigma, Q_{\text{init}}, \Delta)$, such that $Q = \{s, o, x\}$, $Q_{\text{init}} = \{s\}$ and

$$
\begin{aligned}
(s, 0) &\mapsto (s, x, \dots, x), \quad (s, 0) \mapsto (x, s, \dots, x), \quad \dots, \quad (s, 0) \mapsto (x, x, \dots, s), \\
(s, 1) &\mapsto (o, o, \dots, o), \quad (o, 0) \mapsto (x, x, \dots, x), \quad (o, 1) \mapsto (o, o, \dots, o), \\
(x, 0) &\mapsto (x, x, \dots, x).
\end{aligned}
\tag{16}
$$

Here, state s means the vertex can be selected, state o means the vertex is now selecting, and state x means that the vertex cannot be selected. Note that $\mathcal{A}$ is non-deterministic because there are d rules for $(s, 0)$. Thus, the diversity of transitions is $\delta(n) = n$, which is not bounded by a constant. Thus, the theorem gives only quasi-polynomial time algorithm.

To improve the performance, we make the similar observation to the precedence constraint (Example 2). Then, the number of recursive calls to each subtree is at most twice; one is for $(s, 0)$ and the other is for $(s, 1)$ and $(o, 1)$. Therefore, the algorithm runs in $O(n^{\log(1+2)}C) = O(n^{1.585}C)$ time.

Example 4 (k Connected Component Constrained Problem). Let us consider k *connected component constrained* tree knapsack problem whose feasible solution is k connected components. By using the same technique as the connectivity constrained problem (Example 3), we obtain $n^{O(\log k)}C$ time algorithm for the problem. Note that, if we handle k as a kind of weight, we can derive $O(kn^eC) = O(n^eC)$ time algorithm for some universal constant e.

4   k-Subtree Version Problems

In this section, we consider the k-subtree version problems and prove Theorem 2. We introduce two auxiliary problems: The first one is the *for-all-subtree* problem that requires to solve the problem on each subtree T_u of T rooted by $u \in V(T)$. The second one is the *for-all-subtree-complement* problem that requires to solve the problem on each subtree-complement $T \setminus T_u$ of T for all $u \in V(T)$. These problems can be solved in almost the same time complexity as follows.

Lemma 1. *Suppose that the automaton constrained tree knapsack problem with tree automaton $\mathcal{A}$ can be solved in $f(n) = O(n^eC)$ time by Algorithm 1. Then, the corresponding for-all-subtree version problem can be solved in $O(f(n) \log n)$ time if $e = 1$ and $O(f(n))$ time if $e > 1$.*

Proof. Let us fix a heavy path $u_1, \ldots, u_l$ that starts from the root of the tree, i.e., u_1 is the root and u_l is a leaf. First, we call HLRECDP $(u_1, [0, -\infty, \ldots, -\infty])$ to the root u_1 of the tree. Then, it recursively calls HLRECDP$(u_i, [0, -\infty, \ldots, -\infty])$ to the vertices $u_2, \ldots, u_l$ on the heavy path. This means that this single call gives the solutions to the subtrees rooted by the vertices on the heavy path.

After this computation, we call the procedure recursively to the light subtrees adjacent to the heavy path. The total complexity $g(n)$ satisfies

$$g(n) \leq g(n_1) + \cdots + g(n_s) + f(n), \tag{17}$$

where $n_1, \ldots, n_s$ are the sizes of the subtrees. By definition, $n_1 + \cdots + n_s \leq n - 1$. Also, by the heavy-light property, $n_j \leq n/2$ $(j = 1, \ldots, s)$. Therefore, the maximum of the right-hand side is attained at $n_1 = \lceil (n-1)/2 \rceil$, $n_2 = \lfloor (n-1)/2 \rfloor$, and $n_3 = \cdots = n_s = 0$. Thus,

$$g(n) \leq g(\lceil (n-1)/2 \rceil) + g(\lfloor (n-1)/2 \rfloor) + f(n), \tag{18}$$

By solving this inequality we obtain the desired result. □

Lemma 2. *Suppose that the automaton constrained tree knapsack problem with tree automaton $\mathcal{A}$ can be solved in $f(n) = O(n^e C)$ time by Algorithm 1. Then, the corresponding for-all-subtree-complement version problem can be solved in $O(f(n)(\log n)^2)$ time if $e = 1$ and $O(f(n) \log n)$ time if $e > 1$.*

Proof. For vertex $u \in V(T)$, we define array $x_{u,q}$ of length $C + 1$ that represents the solution on $T \setminus T_u$, where the parent of u has state $q \in Q$. We compute the arrays for all the vertices. We define $x_{r,q} = [0, -\infty, \ldots, -\infty]$ for the root $r \in V(T)$ and all $q \in Q$. Let us fix a heavy path $u_1, \ldots, u_l$ that starts from the root of the tree. We compute the arrays for the vertices on the heavy path, and for the vertices adjacent to the heavy path separately.

Vertices on the heavy path. Suppose that we have $\{x_{u_{i-1},q}\}_{q \in Q}$. Let $v_1, \ldots, v_d$ be children of u_{i-1}, where $v_1 = u_i$. For each rewriting rule $(q, \sigma) \mapsto (q_1, \ldots, q_{d-1}) \in \Delta$, which is a rule of length $d - 1$, which will match to $v_2, \ldots, v_d$, the array corresponds to this rule is obtained by calling HLRECDP$(v_2, \ldots, v_d, x_{u_{i-1},q})_{q_1, \ldots, q_{d-1}}$ and by adding u_i if $\sigma = 1$. By taking the entry-wise maximum of the arrays for different rules, we obtain $\{x_{u_i,q}\}_{q \in Q}$. Since this computation process pays the same computational effort as HLRECDP(u_1, x), the complexity is $f(n)$.

Vertices adjacent to the heavy path. We compute $\{x_{v,q}\}_{q \in Q}$ for all light child v adjacent to the heavy path. It is obtained by calling HLRECDP to all the subtrees except T_v; however, this method involves redundant computations. We reduce the complexity by storing intermediate results by a segment tree-like divide-and-conquer technique.

First, we compute arrays y_{i,j,q_i,q_j} for $i = 0, \ldots, l - 1$, $j = i + 1, \ldots, l$, and $q_i, q_j \in Q$. This stores the vector obtained by calling HLRECDP with initial array $x_{u_1,q}$ for some q to the subtree except the light children of $u_{i+1}, \ldots, u_j$, where the states of u_i and u_i are q_i and q_j, respectively. Initially, we set $y_{0,l,q_0,q_l} = x_{u_1,q_0}$ for all $q_0, q_l \in Q$. If we have $\{y_{i,j,q_i,q_j}\}_{q_i,q_j \in Q}$ for $i+1 < j$, we can compute $\{y_{i,m,q_i,q_m}\}_{q_i,q_m \in Q}$ where $m = \lfloor (i+j)/2 \rfloor$ by calling HLRECDP to the light subtrees of $u_{m+1}, \ldots, u_j$ with initial array z_{i,j,q_i,q_j}. Similarly, we can compute $\{y_{m,j,q_m,q_r}\}_{q_m,q_r \in Q}$. The complexity of computing all the arrays is $O(f(n) \log n)$ since HLRECDP is called to subtree T_{u_i} at most $O(\log n)$ times.

Next, for each u_k $(k = 1, \ldots, l-1)$ on the heavy path, we consider the children $v_1, \ldots, v_d$ of u_k, where v_1 is the heavy child (i.e., $v_1 = u_{k+1}$). For each rewriting rule $(q, \sigma) \mapsto (q_1, \ldots, q_{d-1}) \in \Delta$, we compute arrays z_{i,j,q,q_1} for $i = 1, \ldots, d-1$ and $j = i+1, \ldots, d$. This stores the vector obtained by calling HLRECDP with initial array y_{k-1,k,q,q_1} to the subtrees except $v_{i+1}, \ldots, v_j$, and is computed by the same technique as y. Once the arrays are obtained, we can retrieve $x_{v_i,q}$ by taking the entry-wise maximum of z_{i-1,i,q,q_1} with respect to q_1. Thus, the total complexity of this part is $O(f(n) \log n)$.

After this computation, we call the procedure recursively to the light subtrees adjacent to the heavy path. The total complexity $g(n)$ satisfies

$$g(n) \leq g(n_1) + \cdots + g(n_s) + O(f(n) \log n), \tag{19}$$

where $n_1 + \cdots + n_s \leq n-1$ and $n_j \leq n/2$ $(j = 1, \ldots, s)$. By solving this inequality as similar to Lemma 1, we obtain the desired result. □

Now we provide an outline of the proof of Theorem 2.

Proof (of Theorem 2, outline). We design algorithm CONN(u, k, x) that computes arrays $x_{u,q,b,l}$ where $u \in V(T)$, $q \in Q$, $b \in \{0,1\}$, and $l \in \{0, \ldots, k\}$. The array represents the solution to the subtree T_u such that the root $(= u)$ has state q and is included by a subtree if $b = 1$, and l subtrees are selected. If $k = 0$, the solution is $[0, -\infty, \ldots, -\infty]$. If $k = 1$, we can solve the problem by solving for-all-subtree version problem since the automaton is prefix closed. Thus, in the following, we consider $k \geq 2$. Let $g(n, k)$ be the complexity of the algorithm for n vertices with parameter k. We derive the recursive relation of g. We fix a heavy path, and consider light subtrees adjacent to the heavy path.

Case 1: There is a light subtree T_v that contains at least k/2 components. In this case, the subtree complement $T_u \setminus T_v$ contains at most $k/2$ components. Thus, we guess such subtree T_v and solve the problem on $T_u \setminus T_v$ and T_v separately. We can solve all the subtree complements simultaneously by calling the subtree-complement version of CONN$(u, k/2, *)$. Also, we can solve each subtree by calling CONN$(v, k, *)$. The complexity of this approach is $g(n_1, k) + \cdots + g(n_s, k) + O(g(n, k/2) \log n)$.

Case 2: Otherwise; i.e., all the light subtrees contain at most k/2 components. We call CONN$(v, k/2, *)$ for all subtrees v, sequentially. The complexity of this part is given by $g(n_1, k/2) + \cdots + g(n_s, k/2) \leq g(n, k/2)$.

The total complexity $g(n, k)$ of the algorithm satisfies

$$g(n, k) \leq g(n_1, k) + \cdots + g(n_s, k) + O(g(n, k/2) \log n). \qquad (20)$$

By using $n_1 + \cdots n_s \leq n - 1$ and $n_j \leq n/2$ $(j = 1, \ldots, s)$, we obtain $g(n, k) \leq h(k)f(n)(\log n)^{\log k}$, where $h(k)$ is a polynomial in k. $\qquad \square$

Example 5 (Connected Component Constrained Problem (again)). By using this technique, the connectivity constrained problem can be solved in $O(n \log nC)$ time, and the k connected component constrained problem can be solved in $O(n^{1+e}C)$ time for any $e > 0$, since $(\log n)^k = O(n^e)$ for any $e > 0$.

Acknowledgment. We thank the anonymous reviewers for their helpful comments.

References

1. Backurs, A., Indyk, P., Schmidt, L.: Better approximations for tree Sparsity in nearly-linear time. In: Proceedings of the 28th Annual ACM-SIAM Symposium on Discrete Algorithms (SODA 2017), pp. 2215–2229 (2017)
2. Bremner, D., et al.: Necklaces, convolutions, and $X + Y$. In: Azar, Y., Erlebach, T. (eds.) ESA 2006. LNCS, vol. 4168, pp. 160–171. Springer, Heidelberg (2006). https://doi.org/10.1007/11841036_17
3. Chekuri, C., Pál, M.: A recursive greedy algorithm for walks in directed graphs. In: Proceedings of the 46th Annual Symposium on Foundations of Computer Science (FOCS 2005), vol. 2005, pp. 245–253 (2005)
4. Cho, G., Shaw, D.X.: A depth-first dynamic programming algorithm for the tree knapsack problem. J. Comput. **9**(4), 431–438 (1997)
5. Comon, H., et al.: Tree automata techniques and applications (2007)
6. Cygan, M., Mucha, M., Węgrzycki, K., Włodarczyk, M.: On problems equivalent to (min,+)-convolution. arXiv:1702.07669 (2017)
7. Hirao, T., Yoshida, Y., Nishino, M., Yasuda, N., Nagata, M.: Single-document summarization as a tree knapsack problem. In: Proceedings of the 2013 Conference on Empirical Methods in Natural Language Processing (EMNLP 2013), pp. 1515–1520 (2013)
8. Hochbaum, D.S., Pathria, A.: Node-optimal connected k-subgraphs. University of California, Berkeley, Technical report (1994)
9. Ibarra, O., Kim, C.: Fast approximation algorithms for the knapsack and sum of subset problems. J. ACM **22**(4), 463–468 (1975)
10. Ishihata, M., Maehara, T., Rigaux, T.: Algorithmic meta-theorems for monotone submodular maximization. arXiv:1807.04575 (2018)
11. Johnson, D.S., Niemi, K.: On knapsacks, partitions, and a new dynamic programming technique for trees. Mathe. Oper. Res. **8**(1), 1–14 (1983)
12. Kellerer, H., Pferschy, U., Pisinger, D.: Knapsack Problems. Springer, Heidelberg (2003). https://doi.org/10.1007/978-3-540-24777-7
13. Lawler, E.L.: Fast approximation algorithms for knapsack problems. In: Proceedings of the 18th Annual Symposium on Foundations of Computer Science (FOCS 1977), vol. 4(4), pp. 339–357 (1977)
14. Lukes, J.A.: Efficient algorithm for the partitioning of trees. IBM J. Res. Dev. **18**(3), 217–224 (1974)

15. Van der Merwe, D., Hattingh, J.M.: Tree knapsack approaches for local access network design. Eur. J. Oper. Res. **174**(3), 1968–1978 (2006)
16. Pferschy, U., Schauer, J.: The knapsack problem with conflict graphs. J. Graph Algorithms Appl. **13**(2), 233–249 (2009)
17. Sleator, D.D., Tarjan, R.E.: A data structure for dynamic trees. J. Comput. Syst. Sci. **26**(3), 362–391 (1983)
18. Thatcher, J.W., Wright, J.B.: Generalized finite automata theory with an application to a decision problem of second-order logic. Mathe. Syst. Theor. **2**(1), 57–81 (1968)

Matching Sets of Line Segments

Hyeyun Yang and Antoine Vigneron[⊠][iD]

School of Electrical and Computer Engineering, UNIST, Ulsan, Republic of Korea
{gm1225,antoine}@unist.ac.kr

Abstract. We give approximation algorithms for matching two sets of line segments in constant dimension. We consider several versions of the problem: Hausdorff distance, bottleneck distance and largest common point set. We study these similarity measures under several sets of transformations: translations, rotations about a fixed point and rigid motions. As opposed to previous theoretical work on this problem, we match segments individually, in other words we regard our two input sets as *sets* of segments rather than *unions* of segments.

Keywords: Geometric algorithms · Approximation algorithms Pattern matching

1 Introduction

Line segments pattern matching finds applications in several areas of computer science. For instance, after identifying line features (segments) in two different frames representing the same objects, we may want to match these edges in order to analyze the motion of these objects [7]. It also finds applications to aerial images change detection or updating, or for navigation based on these images [13]. Another application area is bioinformatics. The interface between two proteins can be defined as the set of atoms on different proteins that are within a distance threshold. Hence, an interface can be represented as the set of line segments connecting two interacting atoms. The problem of comparing two protein-protein interfaces [10] thus reduces to matching two sets of line segments.

On the theoretical side, several algorithms have been designed for line segments pattern matching. However, these algorithms consider *unions* of segments $\bigcup S$ and $\bigcup S'$ instead of *sets* of segments S and S': the goal is to find a matching such that each point on any segment of the first set S is close to a segment on the other set S' (See the survey by Alt and Guibas [3]). Figure 1 shows an example where these two notions of similarity differ substantially.

So our goal in this paper is to match sets of line segments, in the sense that two segments are matched if their endpoints are close, and then the sets S_1

This work was supported by Basic Science Research Program through the National Research Foundation of Korea (NRF) funded by the Ministry of Education (2017R1D1A1B04036529).

G. K. Das et al. (Eds.): WALCOM 2019, LNCS 11355, pp. 261–273, 2019.
https://doi.org/10.1007/978-3-030-10564-8_21

and S_1' from Fig. 1 would be considered dissimilar, while S_2 and S_2' would be a good match. Our main contribution is to show that, under a certain model, this problem can be solved using approximation algorithms, which are about as efficient as the currently known algorithms for point-set pattern matching (within a linear factor if we only look at the dependency on the input size).

Fig. 1. On the left, the *union* of the dashed segments $\bigcup S_1$ is similar to the union of the solid segments $\bigcup S_1'$, in the sense any point on a dashed segment is close to a point on a solid segment. However, taken individually, no segment in $s \in S_1$ is similar to any segment $s' \in S_1'$ as they are orthogonal. Our approach ensures that S_1 and S_1' will be considered dissimilar, whereas S_2 and S_2' (right) are similar as the endpoints match closely.

1.1   Problem Statements

We are given two sets of line segments $S = \{s_1, \ldots, s_m\}$ and $S' = \{s_1', \ldots, s_n'\}$ in $\mathbb{R}^d$ such that $m \leqslant n$. Each of these segment s_i or s_j' is directed, and is given by its two endpoints p_i, q_i and p_i', q_i' respectively. We denote by ℓ_i the length of the segment s_i. The restriction to directed segments is only for ease of presentation; we can handle undirected segments without affecting our time bounds, as we will explain in the full version of this paper.

Our goal is to find a matching between S and S' under a set of transformations $\mathcal{F}$. For instance, when $\mathcal{F}$ is the set $\mathcal{T}$ of translations in $\mathbb{R}^d$, we may want to determine whether there exists a translation $\tau \in \mathcal{T}$ such that $\tau(S) \subset S'$. In practice, however, inaccuracies in the data mean that we cannot hope for an exact match (see for instance the applications mentioned above), so we will try to find a translation such that each translated segment of S is close to a segment of S' (See Fig. 2). We therefore need to introduce a similarity measure for line segments.

Matching Two Segments. For any two points $p, q \in \mathbb{R}^d$, we identify the segment from p to q with the pair of points $(p, q) \in \mathbb{R}^{2d}$. The *distance* $d(s, s')$ between two segments $s = (p, q)$ and $s' = (p', q')$ is the Euclidean distance between these two segments regarded as points in $\mathbb{R}^{2d}$, and thus $d(s, s') = \sqrt{\|p' - p\|^2 + \|q' - q\|^2}$. We say that two segments s and s' match if $d(s, s') \leqslant \delta\ell(s)$, where $\ell(s)$ is the length of the segment s, and $\delta > 0$ is a parameter called *tolerance*.

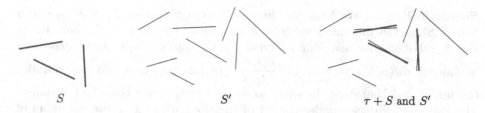

$$S \qquad\qquad S' \qquad\qquad \tau + S \text{ and } S'$$

Fig. 2. Matching two sets of segments under translation. On the right, a translated version of S by a vector τ closely matches a subset of 3 segments of S'.

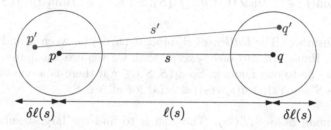

Fig. 3. Two segments at distance $d(s, s') \leqslant \delta\ell(s)$.

We use this matching criterion for two main reasons. First, for small values of δ, two matching segments $s = \overline{pq}$ and $s' = \overline{p'q'}$ are similar in the sense that the distance between their endpoints is small, their angles are close, and their lengths are also close. More precisely, we have $\|p-p'\| \leqslant \delta\ell(s)$, $\|q-q'\| \leqslant \delta\ell(s)$, the angle between s and s' is $O(\delta)$, and $(1 - \delta\sqrt{2})\ell(s) \leqslant \ell(s') \leqslant (1 + \delta\sqrt{2})\ell(s)$ (See Fig. 3). Conversely, if the endpoints are close in the sense that $d(p, p') \leqslant \delta\ell(s)/\sqrt{2}$ and $d(q, q') \leqslant \delta\ell(s)/\sqrt{2}$, then $d(s, s') \leqslant \delta\ell(s)$, and thus the segments match according to our criterion.

The second reason for using this criterion is that it allows us to use approximate near neighbor (ANN) data structures to efficiently compute an approximate nearest segment to a query segment. For instance, using the data structure by Arya et al. [5], the query time is $O((1/\varepsilon^{2d}) \log n)$ for a segment in $\mathbb{R}^d$ identified to a point in $\mathbb{R}^{2d}$. This will help us design efficient approximation algorithms.

Sets of Transformations. We consider different sets $\mathcal{F}$ of transformations. In the *static* case, we do not apply any transformation to our point sets, in other words, we only use the identity transformation. The set $\mathcal{T}$ is the set of *translations* of $\mathbb{R}^d$, so each translation can be represented by a point $\tau \in \mathbb{R}^d$ and it maps any $x \in \mathbb{R}^d$ to $\tau(x) = \tau + x$. We will also consider rotations about a fixed center O in $\mathbb{R}^2$. The rotation about O with angle θ is denoted ρ_θ. Finally, we will consider the set $\mathcal{R}$ of rigid motions in $\mathbb{R}^2$, or more precisely, the set of translations and rotations. We will ignore reflections, as it suffices to run our algorithm on an arbitrary reflected copy of S to cover all possible glide reflections.

Hausdorff Distance. We define the *directed Hausdorff distance* $d_H(S, S')$ between S and S' as the minimum value of δ such that, for all $s \in S$, there exists $s' \in S'$ satisfying our matching criterion $d(s, s') \leqslant \delta\ell(s)$. So it can be expressed as follows: $d_H(S, S') = \max_{s \in S} \min_{s' \in S'} \dfrac{d(s, s')}{\ell(s)}$. In this paper, we do not consider the undirected Hausdorff distance, so we will simply say Hausdorff distance. The Hausdorff distance under the set of transformations $\mathcal{F}$ is the minimum of $d_H(f(S), S')$ over all $f \in \mathcal{F}$. Our approximation algorithms compute a $(1 + \varepsilon)$-approximation of this quantity, for some $0 < \varepsilon < 1$. More precisely, we find a transformation $f^\varepsilon \in \mathcal{F}$ such that $d_H(f^\varepsilon(S), S') \leqslant (1 + \varepsilon) \min_{f \in \mathcal{F}} d_H(f(S), S')$.

Bottleneck Distance. The *bottleneck distance* $d_b(S, S')$ between S and S' is analogous to the Hausdorff distance, except that we require the pairs s, s' to be matched in a one-to-one manner. So $d_b(S, S') \leqslant \delta$ if there is a one-to-one mapping $\sigma : S \to S'$ such that $d(s, \sigma(s)) \leqslant \delta\ell(s)$ for all $s \in S$.

Largest Common Subset (LCS). The goal is to find the largest subset $C \subset S$ such that there exists a transformation $f \in \mathcal{F}$ that matches C to a subset of S'. So there should be a one-to-one matching between C and a subset of S', such that $d(s, s') \leqslant \delta\ell(s)$ for each matching pair (s, s'). We will relax the problem slightly, and return a matching such that $d(s, s') \leqslant (1 + \varepsilon)\delta\ell(s)$ for all matching pair (s, s'), and has cardinality at least the optimal cardinality for the original problem.

1.2 Our Results and Approach

We obtained $(1 + \varepsilon)$-approximation algorithms for all these distance measures under our sets of transformations, when $0 < \varepsilon < 1$. Our algorithms for Hausdorff distance (except for the case of 2D rigid motions) are presented in Sect. 2. The other results are briefly presented in Sect. 3, and more detailed descriptions will be given in the full version of this paper. The time bounds are given in Table 1.

Table 1. Time bounds of our $(1 + \varepsilon)$-approximation algorithms.

	Hausdorff	Bottleneck	LCS
Static	$O((m/\varepsilon^{2d} + n)\log n)$	$O((1/\varepsilon^{2d})n^{1.5}\log n)$	$O((1/\varepsilon^{2d})n^{1.5}\log n)$
Translation	$O((n/\varepsilon^d)(m/\varepsilon^{2d} + n)\log n)$	$O((1/\varepsilon^{3d})n^{2.5}\log n)$	$O((1/\varepsilon^{3d})mn^{2.5}\log n)$
2D rotation	$O((n/\varepsilon)(m/\varepsilon^4 + n)\log n)$	$O((1/\varepsilon^5)n^{2.5}\log n)$	$O((1/\varepsilon^5)mn^{2.5}\log n)$
2D r. motion	$O((n^2/\varepsilon^3)(m/\varepsilon^4 + n)\log n)$	$O((1/\varepsilon^7)n^{3.5}\log n)$	$O((1/\varepsilon^7)m^2n^{3.5}\log n)$

Our algorithms first compute a discretization of the set of transformations, and then solve the problem approximately for each transformation in this set using known algorithms: The ANN data structure by Arya et al. [5] in the case of Hausdorff distance, and a geometric matching algorithm by Efrat et al. [9] for the bottleneck distance and the LCS problem.

To be more precise, we first compute a constant-factor approximation of the solution using a coarse discretization. For translations we use the set of vectors $p'_j - p_1$ where $s_1 = (p_1, q_1)$ is assumed to be the shortest segment in S. For rotations about a fixed center O, we choose the angles that align p_a with each point p'_j, where s_a is the segment with smallest aspect ratio $\alpha_a = \max(\|p_a\|, \|q_a\|)/\ell_a$.

Then we compute a $(1 + \varepsilon)$-approximation by refining these discretizations. In the translation case, we use a uniform grid of $O(1/\varepsilon^d)$ points within a ball of radius proportional to ℓ_1 centered at $p'_j - p_1$. For rotations, we use a set of $(1/\varepsilon)$ equally spaced angles about the angles we chose for obtaining a constant factor approximation, where the spacing is proportional to $1/\alpha_a$ (See Fig. 4). For rigid motions, we discretize the space of rigid motion by combining our discretizations for translations and rotations about a fixed center. The main part of our proof is a careful analysis showing that it yields a $(1+\varepsilon)$-approximation of the optimum.

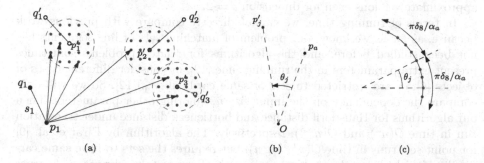

Fig. 4. (a) Discretization of the space of translations. We only drew 6 translation vectors (arrows); for each grid point there is a translation mapping p_1 to it. (b) The angle θ_j used for a constant factor approximation. (c) Discretization of the set of angles around θ_j for obtaining a $(1 + \varepsilon)$-approximation.

1.3   Comparison with Previous Work

As we mentioned earlier, several algorithms are known for matching line segments, but they consider unions of segments instead of sets of segments. The survey by Alt and Guibas [3] mentions several such algorithms that consider unions of objects, instead of sets of objects [2,8]. These algorithms are therefore adapted for matching polygons, seen as unions of segments and their interior, but would not be suitable for the example in Fig. 1. Recent related work presents efficient algorithms for matching polygons or unions of disks, under translations or rigid motions, using the area of overlap as a similarity measure [1,6,11,14].

Point-set pattern matching under translation and rigid motion has also been studied extensively. See again the survey by Alt and Guibas [3]. For instance

Alt et al. [4] gave exact and approximation algorithms for matching point sets under translations and rigid motions. The translation case was improved by Efrat et al. [9]. Heffernan and Schirra [12] considered decision versions of point set matching under translations and rigid motions. Recently, Yon et al. gave approximation schemes for the largest common point set problem (LCP) under the same set of transformations [15].

Our approach is based on discretizing the space of translations using grids, and discretizing the angles of rotations uniformly within appropriate intervals. These techniques have been used to obtain some of the results that we already mentioned [1,6,15]. Our technical contribution is to adapt these methods to the problem of matching sets of line segments, which requires the careful analysis presented in Sect. 2. One difference with the case of point-set pattern matching is that our tolerance δ is weighted by the length $\ell(s)$ of the segment s to be matched, hence the segment s_1 with smallest length, and the segment s_a of smallest aspect ratio play a special role in our algorithms and their analysis. Another issue is that our segments are identified with points in $\mathbb{R}^{2d}$, so exact nearest-neighbor data structures become rather slow, and we need to rely on approximate versions even for dimension $d = 2$.

In terms of running time, we cannot directly compare with previous work because, as far as we know, our problem of matching sets of line segments has not been studied before, and the algorithms for related problems sometimes present extra parameters in the running time [12], or consider different types of objects [6], or are restricted to sets of same cardinality [9,12]. So we will only compare the dependency on the input size n, ignoring other parameters. Then our algorithms for Hausdorff distance and bottleneck distance under translation run in time $O(n^2)$ and $O(n^{2.5})$, respectively. The algorithm by Efrat et al. [9] for point sets runs in time $O(n^{1.5} \log n)$, but requires the sets to have same cardinality, which makes the problem easier as it allows to take advantage of corner points. The algorithm by Heffernan et al. for point sets runs in $O(n^{1.5} \log n)$ and has the same restriction. For LCS of point sets under translation, Yon et al. [15] give an $O(n^{3.5} \log n)$ algorithm, as does ours. Our algorithm for bottleneck distance under rigid motion takes time $O(n^{3.5} \log n)$ while Heffernan and Schirra's algorithm achieves $O(n^{2.5} \log n)$ (still for point sets). So overall, our algorithms have a running time that is similar to previous work on point set pattern matching, at most within an $O(n)$ factor, and our algorithms apply to line segments instead of points.

2 Approximating the Hausdorff Distance

In this section, we give algorithms for approximating the Hausdorff distance between two sets of segments S and S' in $\mathbb{R}^d$. We defined this distance $d_H(S, S')$ as the minimum value of δ such that, for all $s \in S$, there exists $s' \in S'$ satisfying $d(s, s') \leqslant \delta \ell(s)$.

2.1   Static Case

We first give an algorithm for the static case, where S is not subjected to any transformation. Our algorithm starts with recording the segments of S', regarded as points in $\mathbb{R}^{2d}$, in the approximate near neighbor data structure (ANN) by Arya et al. [5]. It is constructed in $O(n \log n)$ time, and allows to report in time $O((1/\varepsilon^{2d}) \log n)$ a $(1 + \varepsilon)$-approximate nearest segment in S'. In other words, it returns a segment $N^\varepsilon(s) \in S'$ that satisfies $d(s, N^\varepsilon(s)) \leqslant (1+\varepsilon) \min_{s' \in S'} d(s, s')$.

Then we compute $N^\varepsilon(s)$ for each segment $s \in S$, and we return $\delta^\varepsilon = \max_{s \in S} d(s, N^\varepsilon(s))/\ell(s)$. By the definition of approximate near neighbors, δ^ε is a $(1 + \varepsilon)$-approximation of $d_H(S, S')$. It follows that:

Theorem 1. *If S and S' are sets of respectively m and n segments in $\mathbb{R}^d$, we can find a tolerance δ^ε such that $d_H(S, S') \leqslant \delta^\varepsilon \leqslant (1 + \varepsilon)d_H(S, S')$ in time $O((m/\varepsilon^{2d} + n) \log n)$.*

2.2   Hausdorff Distance Under Translation

Now we allow translations of S, and we want to minimize its Hausdorff distance to S'. We identify each translation with a point $\tau \in \mathbb{R}^d$. So we denote by $\tau + s$ the copy of s translated by vector τ, and $\tau + S$ denotes the set $\{\tau + s_1, \ldots, \tau + s_m\}$. We denote by τ^* an optimal translation, in the sense that $d_H(\tau^* + S, S') = \min_{\tau \in \mathbb{R}^d} d_H(\tau + S, S')$. The Hausdorff distance under translation $d_H(\tau^* + S, S')$ is denoted by δ^* in this section, and we want to find a translation τ^ε that provides a $(1 + \varepsilon)$-approximation of δ^*.

When we apply a translation τ to a segment, then the corresponding point in $\mathbb{R}^{2d}$ is translated by the vector (τ, τ) that has norm $\sqrt{2}\|\tau\|$. It implies the following.

Proposition 1. *For any two segments s, s' and translation τ, we have $d(\tau + s, s') \leqslant d(s, s') + \sqrt{2}\|\tau\|$.*

Without loss of generality, we assume that s_1 is a shortest segment in S, that is, $\ell_1 = \min_i \ell_i$. The lemma below bounds the variation of the Hausdorff distance after translating a set of segments.

Lemma 1. *For any translation τ, we have $d_H(\tau + S, S') \leqslant d_H(S, S') + \sqrt{2}\|\tau\|/\ell_1$.*

Proof. Proposition 1 implies that $d(\tau + s_i, s'_j) \leqslant d(s_i, s'_j) + \sqrt{2}\|\tau\|$ for all i, j. As s_1 is a shortest segment in S, it follows that $d(\tau + s_i, s'_j)/\ell_i \leqslant d(s_i, s'_j)/\ell_i + \sqrt{2}\|\tau\|/\ell_1$ for all i, j. In particular, for any $s_i \in S$, $\min_j d(\tau + s_i, s'_j)/\ell_i \leqslant \min_j d(s_i, s'_j)/\ell_i + \sqrt{2}\|\tau\|/\ell_1$. By definition of $d_H(S, S')$, there is a segment $s' \in S'$ at distance at most $d_H(S, S')\ell_i$ from any s_i, therefore $\min_j d(s_i, s'_j) \leqslant d_H(S, S')\ell_i$ and thus $\min_j d(\tau + s_i, s'_j)/\ell_i \leqslant d_H(S, S') + \sqrt{2}\|\tau\|/\ell_1$. It means that any segment $\tau + s_i$ has a segment of S' at distance at most $\ell_i(d_H(S, S') + \sqrt{2}\|\tau\|/\ell_1)$, in other words $d_H(\tau + S, S') \leqslant d_H(S, S') + \sqrt{2}\|\tau\|/\ell_1$.

We now present a 3-approximation algorithm. For all j, let τ_j be the translation that maps p_1 to p'_j, in other words $\tau_j = p'_j - p_1$. The lemma below shows that one of these translations yields a $(1 + \sqrt{2})$-approximation of δ^*.

Lemma 2. Let $\hat{\jmath} = argmin_{j=1,\dots,n} d_H(\tau_j + S, S')$. Then $d_H(\tau_{\hat{\jmath}} + S, S') \leqslant (1 + \sqrt{2})\delta^*$.

Proof. The optimal translation τ^* matches s_1 with a segment $s'_k \in S'$ such that $d(\tau^* + s_1, s'_k) \leqslant \delta^*\ell_1$. It implies that $\|p'_k - p_1 - \tau^*\| \leqslant \delta^*\ell_1$, in other words $\|\tau_k - \tau^*\| \leqslant \delta^*\ell_1$. If follows from Lemma 1 that $d_H(\tau_k + S, S') \leqslant d_H(\tau^* + S, S') + \sqrt{2}\|\tau_k - \tau^*\|/\ell_1 \leqslant d_H(\tau^* + S, S') + \sqrt{2}\delta^* = (1 + \sqrt{2})\delta^*$.

We obtain a 6/5-approximation of $d_H(\tau_{\hat{\jmath}} + S, S')$ by running the algorithm of Theorem 1 for each τ_j, with $\varepsilon = 1/5$, and returning the best result. As $(1 + \sqrt{2}) \cdot 6/5 \leqslant 3$, it gives us a 3-approximation.

Lemma 3. *Given S, S', we can compute in time $O(n(m+n)\log n)$ a translation τ and a tolerance δ_3 such that $d_H(\tau + S, S') \leqslant \delta_3 \leqslant 3\delta^*$.*

We now by provide a $(1 + \varepsilon)$-approximate decision algorithm. We will discretize the space of translations based on the following discretization of the unit ball using a uniform grid. Let B^ε denote the set of points in the unit ball whose coordinates are multiples of $\varepsilon/\sqrt{d}$. More generally, for any $\lambda > 0$, we denote $\lambda B^\varepsilon = \{\lambda x \mid x \in B^\varepsilon\}$. This set contains $O(1/\varepsilon^d)$ points and can be constructed in constant time per point:

Proposition 2. *We can construct in time $O(1/\varepsilon^d)$ a set B^ε of $O(1/\varepsilon^d)$ points such that for any point p at distance at most 1 from the origin, there is a point $p' \in B^\varepsilon$ such that $\|p - p'\| \leqslant \varepsilon$.*

Our $(1 + \varepsilon)$-approximation algorithm first computes the 3-approximation δ_3 of δ^* using Lemma 3, in other words we have $\delta^* \leqslant \delta_3 \leqslant 3\delta^*$. For each segment s'_j, we construct the set of translations $T^\varepsilon_j = p'_j - p_1 + \delta_3\ell_1 B^{\varepsilon/9}$, and then $T^\varepsilon = \bigcup_j T^\varepsilon_j$ (See Fig. 4a). We now prove that one translation $\hat{\tau} \in T^\varepsilon$ gives a $(1 + \varepsilon/2)$-approximation of the Hausdorff distance under translation.

Lemma 4. *There is a translation $\hat{\tau} \in T^\varepsilon$ such that $d(\hat{\tau} + S, S') \leqslant (1 + \varepsilon/2)\delta^*$.*

Proof. The optimal translation τ^* maps s_1 to a segment s'_k such that $d(s_1, s'_k) \leqslant \delta^*\ell_1$. So $\tau^* + p_1$ is at distance at most $\delta^*\ell_1$ from p'_k, in other words $\|\tau^* + p_1 - p'_k\| \leqslant \delta^*\ell_1 \leqslant \delta_3\ell_1$. So there is a point $b^\varepsilon \in \delta_3\ell_1 B^{\varepsilon/9}$ such that $\|\tau^* + p_1 - p'_k - b^\varepsilon\| \leqslant \delta_3\ell_1\varepsilon/9$. We let $\hat{\tau} = p'_k - p_1 + b^\varepsilon$, then $\hat{\tau} \in T^\varepsilon_k \subset T^\varepsilon$, and $\|\hat{\tau} - \tau^*\| \leqslant \delta_3\ell_1\varepsilon/9$. By Lemma 1, it implies that

$$d_H(\hat{\tau} + S, S') \leqslant d_H(\tau^* + S, S') + \sqrt{2}\delta_3\ell_1\varepsilon/(9\ell_1)$$
$$= \delta^* + \sqrt{2}\varepsilon\delta_3/9 \leqslant \delta^* + 3\sqrt{2}\varepsilon\delta^*/9 < (1 + \varepsilon/2)\delta^*.$$

We can now describe our approximation scheme for Hausdorff distance under translation. We first compute the sample set of translations T^ε. It consists of $O(n/\varepsilon^d)$ translations and can be constructed in $O(n/\varepsilon^d)$ time. For each translation $\tau \in T^\varepsilon$, we compute in $O((m/\varepsilon^{2d} + n)\log n)$ a $(1 + \varepsilon/3)$-approximation $\delta^\varepsilon(\tau)$ of $d_H(\tau + S, S')$ using Theorem 1. Let τ^ε be the translation that minimizes $\delta^\varepsilon(\tau^\varepsilon)$. Then we have $d_H(\tau^\varepsilon + S, S') \leqslant \delta^\varepsilon(\tau^\varepsilon) \leqslant \delta^\varepsilon(\hat{\tau}) \leqslant (1 + \varepsilon/3)d_H(\hat{\tau} + S, S')$. By Lemma 4, it implies $d(\tau^\varepsilon + S, S') \leqslant (1 + \varepsilon/2)(1 + \varepsilon/3)\delta^* \leqslant (1 + \varepsilon)\delta^*$. It completes the proof of the theorem below.

Theorem 2. *Let $\delta^* = d_H(S, S')$ be the Hausdorff distance between S and S' under translation. We can find a translation τ^ε and a tolerance δ^ε such that $d_H(\tau^\varepsilon + S, S') \leqslant \delta^\varepsilon \leqslant (1 + \varepsilon)\delta^*$ in $O((n/\varepsilon^d)(m/\varepsilon^{2d} + n)\log n)$ time.*

2.3 Hausdorff Distance Under Rotation About a Fixed Center in $\mathbb{R}^2$

In this section, we consider rotations about a fixed center in $\mathbb{R}^2$. Without loss of generality, we assume it to be the origin O. The rotation about O through an angle θ is denoted by ρ_θ. The optimal rotation ρ^* satisfies $d_H(\rho^*(S), S') = \min_{\theta \in [0, 2\pi]} d_H(\rho_\theta(S), S')$, and we denote by θ^* its angle. The Hausdorff distance under rotation $d_H(\rho^*(S), S')$ is denoted by δ^* in this section, and we want to find a rotation ρ^ε that provides a $(1 + \varepsilon)$-approximation of δ^*.

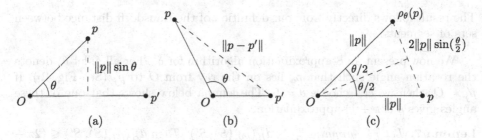

Fig. 5. Proof of Lemma 5 and Proposition 3.

The distance from O to a point $p \in \mathbb{R}^2$ is denoted by $\|p\|$. We will need the following lemma.

Lemma 5. *Let $p, p' \in \mathbb{R}^2$ be two points making an angle $\theta = \angle pOp'$ with the origin. If $\theta \subset [-\pi, \pi]$, then $\|p - p'\| \geq \|p\| \cdot |\theta|/\pi$.*

Proof. If $|\theta| \leqslant \pi/2$, then by concavity of $\sin(\cdot)$ over $[0, \pi/2]$, we have $\sin(|\theta|) \geq 2|\theta|/\pi$. The distance between p and p' is at least the distance between p and its projection onto the line through O and p' (See Fig. 5a). It follows that $\|p - p'\| \geq \|p\| \sin(|\theta|) \geq 2\|p\| \cdot |\theta|/\pi$. On the other hand, if $\pi/2 \leqslant |\theta| \leqslant \pi$, then $\|p - p'\| \geq \|p\|$, and thus $\|p - p'\| \geq \|p\| \cdot |\theta|/\pi$ (See Fig. 5b).

The *aspect ratio* of a segment $s = (p,q)$ is $\alpha(s) = \max(\|p\|, \|q\|)/\ell(s)$. When we apply the rotation ρ_θ to s, then p and q move by a distance $2\sin(|\theta/2|)\|p\|$ and $2\sin(|\theta/2|)\|q\|$, respectively (See Fig. 5c). Therefore, each endpoint p or q move by a distance at most $2\sin(|\theta/2|)\max(\|p\|, \|q\|) = 2\sin(|\theta/2|)\ell(s)\alpha(s) \leqslant |\theta|\ell(s)\alpha(s)$, and therefore:

Proposition 3. *For any two segments s and s', and for any angle θ, we have* $d(\rho_\theta(s), s') \leqslant d(s, s') + 2|\theta|\ell(s)\alpha(s)$.

Let $s_a = (p_a, q_a)$ be the segment in S with largest aspect ratio. So we have $\alpha(s_a) = \max_{s \in S} \alpha(s)$, and we denote $\ell_a = \ell(s_a)$ and $\alpha_a = \alpha(s_a)$. Without loss of generality, we assume that $\|p_a\| \geq \|q_a\|$, and thus $\alpha_a = \|p_a\|/\ell_a$.

Lemma 6. $d_H(\rho_\theta(S), S') \leqslant d_H(\rho_{\theta'}(S), S') + 2\alpha_a|\theta - \theta'|$ *for any two angles θ and θ'.*

Proof. By Proposition 3, we have for any two segments $s \in S$ and $s' \in S'$

$$\frac{d(\rho_{\theta - \theta'}(\rho_{\theta'}(s)), s')}{\ell(s)} \leqslant \frac{d(\rho_{\theta'}(s), s')}{\ell(s)} + 2|\theta - \theta'|\alpha(\rho_{\theta'}(s)).$$

As $\rho_{\theta - \theta'}(\rho_{\theta'}(s)) = \rho_\theta(s)$, $\alpha(\rho_{\theta'}(s)) = \alpha(s)$ and $\alpha(s) \leqslant \alpha_a$, it implies

$$\frac{d(\rho_\theta(s), s')}{\ell(s)} \leqslant \frac{d(\rho_{\theta'}(s), s')}{\ell(s)} + 2|\theta - \theta'|\alpha(s) \leqslant \frac{d(\rho_{\theta'}(s), s')}{\ell(s)} + 2|\theta - \theta'|\alpha_a.$$

The result follows directly from our definition of the Hausdorff distance between sets of segments.

We now present an 8-approximation algorithm for δ^*. For all j, let θ_j denote the rotation angle such that p_a lies on the ray from O to p'_j (See Fig. 4b). If $p'_j = O$, then we can choose $\theta = 0$. The lemma below shows that one of these angles gives a $(2\pi + 1)$-approximation.

Lemma 7. *Let $\hat{\jmath} = \mathrm{argmin}_{j=1,\dots,n} d_H(\rho_{\theta_j}(S), S')$. Then $d_H(\rho_{\theta_{\hat{\jmath}}}(S), S') \leqslant (2\pi + 1)\delta^*$.*

Proof. The optimal rotation ρ^* matches s_a with a segment $s'_j \in S'$ such that $d(\rho^*(s_a), s'_j) \leqslant \delta^*\ell_a$. It implies that $\|\rho^*(p_a) - p'_j\| \leqslant \delta^*\ell_a$. Let $\theta = \theta_j - \theta^*$. In other words, we have $\theta = \angle\rho^*(p_a)Op'_j$. By Lemma 5, we have $\|\rho^*(p_a) - p'_j)\| \geq \|\rho^*(p_a)\| \cdot |\theta|/\pi$, and thus $\delta^*\ell_a \geq \|\rho^*(p_a)\| \cdot |\theta|/\pi$. As $\|\rho^*(p_a)\| = \|p_a\|$, it means that $\pi\delta^* \geq |\theta|\alpha_a$. Then Lemma 6 yields $d_H(\rho_{\theta_j}(S), S') \leqslant d_H(\rho_{\theta^*}(S), S') + 2|\theta|\alpha_a \leqslant d_H(\rho_{\theta^*}(S), S') + 2\pi\delta^* = (2\pi + 1)\delta^*$.

We obtain our 8-approximation of δ^* by applying the algorithm of Theorem 1 with $\varepsilon = (8/(2\pi + 1)) - 1$ to the pair of sets $\rho_{\theta_j}(S)$, S' for $j = 1, \dots, n$, and returning the best result.

Lemma 8. *Given S, S', we can compute in time $O(n(m + n)\log n)$ an angle θ and a tolerance δ_8 such that $d_H(\rho_\theta(S), S') \leqslant \delta_8 \leqslant 8\delta^*$.*

We now present our $(1 + \varepsilon)$-approximation algorithm. We begin with computing the 8-approximation δ_8 from Lemma 8, and then we discretize the set of rotation angles. To this end, we observe that the optimal angle θ^* must be close to an angle θ_j.

Lemma 9. *There exists $j \in \{1, \ldots, n\}$ such that $|\theta^* - \theta_j| \leqslant \pi\delta_8/\alpha_a$.*

Proof. The optimal rotation ρ^* must bring $\rho^*(p_a)$ to a distance at most $\delta^*\ell_a$ of a point p'_j, and thus $\|\rho^*(p_a) - p'_j\| \leqslant \ell_a\delta^*$. As $\angle\rho^*(p_a)Op'_j = |\theta_j - \theta^*|$, Lemma 5 implies that $\|\rho^*(p_a)\| \cdot |\theta^* - \theta_j|/\pi \leqslant \|\rho^*(p_a) - p'_j\|$. As $\|\rho^*(p_a)\| = \|p_a\|$, it yields $\|p_a\| \cdot |\theta^* - \theta_j|/\pi \leqslant \delta^*\ell_a$. The result follows from the facts that $\alpha_a = \|p_a\|/\ell_a$ and $\delta^* \leqslant \delta_8$. $\qquad\blacksquare$

For each $j \in \{1, \ldots, n\}$, let Θ_j^ε denote the set of $O(1/\varepsilon)$ angles sampled uniformly in the interval $[\theta_j - \pi\delta_8/\alpha_a, \theta_j + \pi\delta_8/\alpha_a]$, that is, $\Theta_j^\varepsilon = \{\theta_j + kC_1\varepsilon\delta_8/\alpha_a \mid k \in \mathbb{Z} \text{ and } |kC_1\varepsilon| < \pi\}$, where C_1 is a constant to be determined. Let $\Theta^\varepsilon = \bigcup_j \Theta_j^\varepsilon$ be the union of these n sets. By Lemma 9, there is an angle $\hat{\theta} \in \Theta^\varepsilon$ such that $|\hat{\theta} - \theta^*| \leqslant C_1\varepsilon\delta_8/\alpha_a$. It follows from Lemma 6 that $d_H(\rho_{\hat{\theta}}(S), S') \leqslant d_H(\rho_{\theta^*}(S), S') + 2\alpha_a|\hat{\theta} - \theta^*| \leqslant d_H(\rho_{\theta^*}(S), S') + 2C_1\varepsilon\delta_8 \leqslant d_H(\rho_{\theta^*}(S), S') + 16C_1\varepsilon\delta^* = (1 + 16C_1\varepsilon)\delta^*$. Choosing $C_1 = 1/32$, we obtain the following discretization.

Lemma 10. *Given δ_8, we can compute in time $O(n/\varepsilon)$ a set Θ^ε of $O(n/\varepsilon)$ angles such that one of these angles $\hat{\theta}$ satisfies $d_H(\rho_{\hat{\theta}}(S), S') \leqslant (1 + \varepsilon/2)\delta^*$.*

For each angle in this set, we run the static algorithm of Theorem 1 using an approximation factor $(1 + \varepsilon/3)$, and keep the best tolerance δ^ε. As $\varepsilon < 1$, we have $(1 + \varepsilon/3)(1 + \varepsilon/2) < 1 + \varepsilon$, hence we obtain a $(1 + \varepsilon)$-approximation.

Theorem 3. *Given S, S', we can compute in time $O((n/\varepsilon)(m/\varepsilon^4 + n)\log n)$ an angle $\hat{\theta}$ and a tolerance δ^ε such that $d_H(\rho_{\hat{\theta}}(S), S') \leqslant \delta^\varepsilon \leqslant (1 + \varepsilon)\delta^*$.*

3   Further Results

Due to space limitation, the proof of the result on Hausdorff distance under 2D rigid motions, bottleneck distance, largest common subset (LCS), and undirected line segments, are deferred to the full version of this paper. In this section, we give a brief explanation of these results.

We handle Hausdorff distance under 2D rigid motions by separating the translation parts and the rotation parts of the rigid motions, and combining our discretization schemes for translations and for 2D rotations about a fixed center. Our discretization consists of $O(n/\varepsilon^2)$ translations, and for each one of them, $O(n/\varepsilon)$ rotation angles, so the total size of our discretization is $O(n^2/\varepsilon^3)$, which yields an overall running time $O((n^2/\varepsilon^3)(m/\varepsilon^4 + n)\log n)$.

For bottleneck distance, we use the same discretizations of the space of transformations as we did for the Hausdorff distance. The only difference is that, for the static case, we use the geometric matching algorithm by Efrat et al. [9, Theorem 7.3] which runs in time $O((1/\varepsilon^{2d})n^{1.5}\log n)$ in our case because segments are regarded as points in $\mathbb{R}^{2d}$. Remember that out static algorithm for Hausdorff distance runs in $O((m/\varepsilon^{2d}+n)\log n)$; as it is used as a subroutine for all sets of transformation, this difference in running time applies to the algorithms for translations, rotations and rigid motions.

The algorithms for LCS are similar to the case of bottleneck distance. The main difference is that now the shortest segment s_1 and the segment s_a with smallest aspect ratio are not necessarily part of the largest common subset. So each segment of S is tried instead of just s_1 for candidate translations, i.e. we use translations $p'_j - p_i$ for all pairs i, j instead of just $p'_j - p_1$ for all j. It increases the running time by a factor m for translations and rigid motions. The same happens with the angles defined by s_a, and the running time for rotations and rigid motions increases by a factor m.

Undirected line segments are handled by inserting reversed copies of the input segments in S', and slightly modifying the algorithm for LCS.

References

1. Ahn, H., Cheong, O., Park, C., Shin, C., Vigneron, A.: Maximizing the overlap of two planar convex sets under rigid motions. Comput. Geom. **37**(1), 3–15 (2007)
2. Alt, H., Behrends, B., Blömer, J.: Approximate matching of polygonal shapes. Ann. Math. Artif. Intell. **13**(3), 251–265 (1995)
3. Alt, H., Guibas, L.J.: Discrete geometric shapes: matching, interpolation, and approximation. In: Sack, J.R., Urrutia, J. (eds.) Handbook of Computational Geometry, pp. 121–153. B.V. North-Holland, Amsterdam (2000). Chapter 3
4. Alt, H., Mehlhorn, K., Wagener, H., Welzl, E.: Congruence, similarity, and symmetries of geometric objects. Discrete Comput. Geom. **3**(3), 237–256 (1988)
5. Arya, S., Mount, D.M., Netanyahu, N.S., Silverman, R., Wu, A.Y.: An optimal algorithm for approximate nearest neighbor searching fixed dimensions. J. ACM **45**(6), 891–923 (1998)
6. Cabello, S., de Berg, M., Giannopoulos, P., Knauer, C., van Oostrum, R., Veltkamp, R.C.: Maximizing the area of overlap of two unions of disks under rigid motion. Int. J. Comput. Geom. Appl. **19**(6), 533–556 (2009)
7. Chen, H.H., Huang, T.S.: Matching 3-D line segments with applications to multiple-object motion estimation. IEEE Trans. Pattern Anal. Mach. Intell. **12**(10), 1002–1008 (1990)
8. Chew, L., Goodrich, M.T., Huttenlocher, D.P., Kedem, K., Kleinberg, J.M., Kravets, D.: Geometric pattern matching under Euclidean motion. Comput. Geom. **7**(1), 113–124 (1997)
9. Efrat, A., Itai, A., Katz, M.J.: Geometry helps in bottleneck matching and related problems. Algorithmica **31**(1), 1–28 (2001)
10. Gao, M., Skolnick, J.: iAlign: a method for the structural comparison of protein-protein interfaces. Bioinformatics **26**(18), 2259–2265 (2010)
11. Har-Peled, S., Roy, S.: Approximating the maximum overlap of polygons under translation. Algorithmica **78**(1), 147–165 (2017)

12. Heffernan, P.J., Schirra, S.: Approximate decision algorithms for point set congruence. Comput. Geom. **4**(3), 137–156 (1994)
13. Medioni, G.G., Nevatia, R.: Matching images using linear features. IEEE Trans. Pattern Anal. Mach. Intell. **6**(6), 675–685 (1984)
14. Vigneron, A.: Geometric optimization and sums of algebraic functions. ACM Trans. Algorithms **10**(1), 4:1–4:20 (2014)
15. Yon, J., Cheng, S., Cheong, O., Vigneron, A.: Finding largest common point sets. Int. J. Comput. Geom. Appl. **27**(3), 177–186 (2017)

Miscellaneous

Efficient Algorithm for Box Folding

Koichi Mizunashi[1], Takashi Horiyama[1], and Ryuhei Uehara[2](⊠)

[1] Graduate School of Science and Engineering, Saitama University,
255 Shimo-Okubo, Sakura, Saitama 338-8570, Japan
{mizunashi,horiyama}@al.ics.saitama-u.ac.jp
[2] School of Information Science,
Japan Advanced Institute of Science and Technology (JAIST),
1-1 Asahidai, Nomi, Ishikawa 923-1292, Japan
uehara@jaist.ac.jp

Abstract. For a given polygon P and a polyhedron Q, the folding problem asks if Q can be obtained from P by folding it. This simple problem is quite complicated, and there is no known efficient algorithm that solves this problem in general. In this paper, we focus on the case that Q is a box, and the size of Q is not given. That is, input of the box folding problem is a polygon P, and it asks if P can fold to boxes of certain sizes. We note that there exist an infinite number of polygons P that can fold into three boxes of different sizes. In this paper, we give a pseudo polynomial time algorithm that computes all possible ways of folding of P to boxes.

Keywords: Computational Origami · Computational geometry
Box folding

1 Introduction

In 1525, the German painter Albrecht Dürer published his masterwork on geometry [7], whose title translates as, "On Teaching Measurement with a Compass and Straightedge for lines, planes, and whole bodies." In the book, he presented each polyhedron by drawing a *net* for it: an unfolding of the surface to a planar layout. To this day, it remains an important open problem whether every convex polyhedron has a (non-overlapping) net by cut along edges. When we allow to cut not only along edges, this problem is settled only for tetramonohedron, which is a kind of tetrahedron: any net of a tetramonohedron is characterized by a (non-overlapping) tiling [3].

To understand unfolding, it is interesting to look at the inverse: one folding problem asks what polyhedra can be folded from a given polygonal sheet of paper. For example, the Latin cross, which is one of eleven nets of a cube, can form 23 different convex polyhedra (including doubly covered convex polygons) by 85 distinct ways of folding (and an infinite number of doubly covered concave polygons). Comprehensive surveys of folding and unfolding can be found in [6].

© Springer Nature Switzerland AG 2019
G. K. Das et al. (Eds.): WALCOM 2019, LNCS 11355, pp. 277–288, 2019.
https://doi.org/10.1007/978-3-030-10564-8_22

Recently, Abel et al. investigated the folding problem of bumpy pyramids [1]: For a given petal polygon P (convex n-gon B with n triangular petals), it asks if we can fold to a pyramid (with flat base B) or a convex bumpy pyramid by folding along a certain triangulation of B. In [1], they gave nontrivial linear time algorithms for the problem.

Let us elaborate on this and its related results. Alexandrov's Theorem states that every metric with the global topology, and local geometry required of a convex polyhedron is, in fact, the intrinsic metric of some convex polyhedron. Thus, if P is a net of a convex polyhedron, the shape is uniquely determined. Alexandrov's Theorem was stated in 1942, and a constructive proof was given by Bobenko and Izmestiev in 2008 [4]. A pseudo-polynomial time algorithm for Alexandrov's Theorem was given by Kane, Price, and Demaine in 2009 [9]. However, it runs in $O(n^{456.5} r^{1891}/\epsilon^{121})$ time, where r is the ratio of the largest and smallest distances between vertices, and ϵ is the coordinate relative accuracy. The exponents in the time bound of the result are remarkably huge. As far as the authors know, the results for the bumpy pyramids are the first efficient algorithms for Alexandrov's Theorem for a family of nontrivial convex polyhedra.

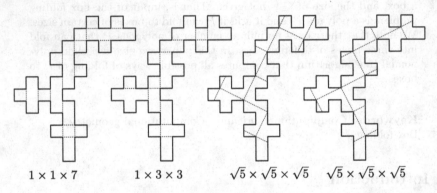

$1 \times 1 \times 7$ $\qquad$ $1 \times 3 \times 3$ $\qquad$ $\sqrt{5} \times \sqrt{5} \times \sqrt{5}$ $\quad$ $\sqrt{5} \times \sqrt{5} \times \sqrt{5}$

Fig. 1. A polygon that can fold into three different boxes of sizes $1 \times 1 \times 7$, $1 \times 3 \times 3$, and $\sqrt{5} \times \sqrt{5} \times \sqrt{5}$ in four different ways [12].

In this paper, for a given polygon P, we consider the box folding problem that asks if P can fold to a box or not. This problem seems to be natural and simple from the viewpoint of our life. We first note that this problem is motivated by counterintuitive polygons. In 1999, Biedl et al. found two polygons that can fold into two different boxes [6, Fig. 25.53]. Later, Mitani and Uehara proved that there exist an infinite number of polygons that can fold into two boxes [10], and Shirakawa and Uehara proved that there exist an infinite number of polygons that can fold into three boxes [11]. So far, a polygon that can fold into three different boxes in four different ways has been found by using a supercomputer (Fig. 1), and it is open that there exists a polygon that can fold into k different boxes for $k > 3$ [12].

In the previous research, they did not solve the box folding problem in general form. The results in [6, Fig. 25.53] and [11] were obtained without a computer.

In [10], they assume that a polygon P is a polyomino, which is a union of unit squares sharing on their edges, and they only search the boxes obtained by folding along the edges of unit squares. In [12], they first obtain the set of all polyominoes of area 30 that can fold into two boxes of sizes $1 \times 1 \times 7$ and $1 \times 3 \times 3$ on the similar model in [10]. Then they solved the box folding problem for the special box of size $\sqrt{5} \times \sqrt{5} \times \sqrt{5}$ on these polyominoes. Namely, it is specialized to this special size (see [12] for the details).

Recently, Horiyama and Mizunashi solved the box folding problem in more general case with supporting parameters [8]; the input polygon P is a polyomino, and the size $a \times b \times c$ of the box Q is also given. Moreover, the matching of edges (that gives us the gluing of the corresponding edges) of P is also given. In this case, the box folding problem can be solved in $O((n + m) \log n)$ time, where m is the maximum number of line segments on an edge of the folded box Q. We here note that this geometric parameter m is independent from the number of vertices in P and Q. Even in a simple example in Fig. 2, m can be arbitrarily large while P and Q have 10 and 8 vertices, respectively.

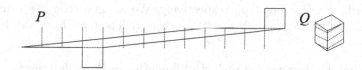

Fig. 2. A geometric parameter m that the number of line segments on an edge of a cube Q. In this example, an edge of Q consists of $m = 4$ line segments, however, it can be arbitrarily large if the slope of P is more slanted.

In this paper, from both viewpoints of theoretical and practical, we give an efficient algorithm for the box folding problem in general. That is, the input is a polygon P with n vertices. Then the output is the set of whole boxes Q folded from P with distinct ways of folding. The algorithm runs in pseudo-polynomial time of n, m and some geometric parameters.

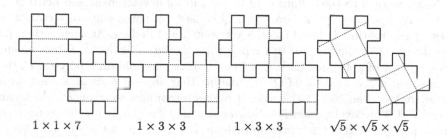

$1 \times 1 \times 7$ $1 \times 3 \times 3$ $1 \times 3 \times 3$ $\sqrt{5} \times \sqrt{5} \times \sqrt{5}$

Fig. 3. Another polygon that can fold into three different boxes of sizes $1 \times 1 \times 7$, $1 \times 3 \times 3$, and $\sqrt{5} \times \sqrt{5} \times \sqrt{5}$ in four different ways not mentioned in [12].

We show a case study of our algorithm for the nine polyominoes shown in [12]. In [12], the authors first compute the set of polyominoes that can fold two boxes of sizes $1 \times 1 \times 7$ and $1 \times 3 \times 3$. There are 1080 polyominoes. Then they solved the box folding problem for the special box of size $\sqrt{5} \times \sqrt{5} \times \sqrt{5}$ on these 1080 polyominoes. Finally, they found nine polyominoes that can fold into three different boxes. We apply our algorithm to this set of nine polyominoes as a case study. Surprisingly, among them, we have another special polyomino that can fold into three different boxes in four different ways of folding (Fig. 3). We will discuss the reason why the authors of [12] missed finding it, while we succeed.

2   Preliminaries

In this section, we first state the box folding problem:

> **Input** : A polygon $P = (p_0, p_1, \ldots, p_{n-1}, p_0)$
> **Output**: A set $S = \{Q_0, Q_1, \ldots, Q_k\}$ of boxes that can be folded from P

Note that S can be an empty set. Let $x(p_i)$ and $y(p_i)$ be the x-coordinate and y-coordinate of a point p_i, respectively. We assume that $x(p_i)$ and $y(p_i)$ are rational numbers for each $i = 0, \ldots, n - 1$. Then we have the following observation:

Observation 1. *Assume that each $x(p_i)$ and $y(p_i)$ in P are described by rational numbers. Let q_{min} be the least common denominator of them. Scaling up by q_{min}, the box folding problem can be solved for $P' = (p'_0, p'_1, \ldots, p'_{n-1}, p'_0)$ such that each $x(p'_i)$ and $y(p'_i)$ in P' are integers in $[0, p_{max} q_{min}]$, where p_{max} is the largest numerator.*

Therefore, hereafter, we assume that each coordinates $x(p_i)$ and y_{p_i} are nonnegative integers without loss of generality. For the polygon P, its *diameter* D is defined by $\max_{i,j} |p_i - p_j| = \max_{i,j} \sqrt{(x(p_i) - x(p_j))^2 + (y(p_i) - y(p_j))^2}$. Here we introduce another geometric parameter m that indicates the number of line segments on an edge of Q. This is independent of the number of vertices in P and Q. In a simple example in Fig. 2, m can be arbitrarily large.

Now, we turn to the definition of a box and its development and net. Let Q be a convex polyhedron. It is a *box* if Q consists of three pairs of rectangular faces. It results that Q has 6 faces, 8 vertices, and 12 edges. At each vertex, its *curvature* is $270°$, and at any other point, its *curvature* is $360°$[1]. When we cut Q along a set of polygonal lines, unfold on a plane, and obtain a polygon P, the set is called a *development* of Q. We assume that any cut ends at a point with curvature less than $360°$. Otherwise, it makes a redundant cut on P, which can be reduced. The development P is called a *net* of Q if and only if P is a connected simple polygon without self-overlap or hole. Let T be the set of cut lines on Q to obtain a net P. Then the following is well known (see [6] for details):

---

[1] We do not give the formal definition of curvature. Intuitively, it indicates the quantity of paper around the point measured by its angle. See [6] for further details.

Theorem 2. *T forms a spanning tree of the vertices of Q.*

In this paper, the following theorem is useful:

Theorem 3. *Let Q be a box and P be a net of Q. Then (1) all vertices of Q appear on the boundary of P, and (2) P has at least two vertices of degree 270°, which correspond to two vertices of Q.*

Proof. By Theorem 2, the set of cut lines on Q forms a spanning tree T. We note that each edge of T appears twice on the boundary of P. If a vertex of Q appears inside of P, P cannot be flat. Therefore, we obtain (1) immediately. When the vertex v of Q has $\deg(v) \geq 2$ on T, the vertex will be cut into $\deg(v)$ pieces and spread on the boundary of P. In this case, v appears $\deg(v)$ times, and P has less than 270° at these points. Let ℓ be a leaf of T. Then ℓ corresponds to a vertex of Q; otherwise, the curvature around ℓ is 360°, and it does not make a boundary of P. Thus around at ℓ, the curvature is 270°. Since T has at least two leaves, we have (2). ◻

As shown in [12, Theorem 2], for any positive integer k, there are a series of boxes Q_i of size $a_i \times b_i \times c_i$ for $i = 1, 2, \ldots, k$ such that all distinct Q_i have the same area. However, once an area is given, we have an upper bound of the number of such boxes:

Observation 4. *Let P be a polygon of area A, and it can fold into some boxes. Then the number of possible edge lengths of boxes is $O(A^{2/\log\log A})$.*

Proof. Let a, b, c be the edge lengths of a box Q of area A with $a \leq b \leq c$. Then we have $2(ab + bc + ca) = A$. Since A is an integer by Observation 1, each of a, b, c can be represented by $i\sqrt{j}$ for some positive integers i and j. By $a \leq b \leq c$ and $2(ab + bc + ca) = A$, we have $6a^2 \leq A$, which implies the number of possible a is $O(A)$ since A is an integer, and a is $i\sqrt{j}$ for some positive integers i and j in general. In fact, the number of divisors of A is $O(A^{1/\log\log A})$. Then the number of possible b is also $O(A^{1/\log\log A})$ since $2b^2 \leq 2(ab + bc + ca) = A$. Once a and b are fixed, c is uniquely determined. Therefore, the number of possible edge lengths (a, b, c) is $O(A^{2/\log\log A})$. ◻

3 Algorithm Description

In this section, we describe our algorithm. The outline of our algorithm is simple:

```
Input  : A polygon P = (p₀, p₁, ..., pₙ₋₁, p₀)
Output : A set S = {Q₀, Q₁, ..., Qₖ} of boxes that can be folded from P
foreach box Q of size a × b × c such that 2(ab + bc + ca) is equal to area
of P do
    for i ← 0 to n − 1 do
        if curvature at pᵢ is 270° then
          | Check if P is a net of Q such that pᵢ is a vertex of Q;
        end
    end
end
```

That is, the algorithm checks all possible points p_i if it makes $270°$. By Theorem 3, if Q can be folded from P, there are at least two points that fold to the vertices of Q. Hereafter, we assume that p_0 is the vertex of Q without loss of generality. This time, for the given P and p_0, the algorithm checks if P can be folded into the box Q. This step is a loop for direction of Q as follows:

1. First, fix the direction of Q on P in an arbitrary way. That is, the algorithm first arbitrarily chooses a direction of Q on P at the vertex p_0.
2. Then the algorithm checks if all vertices of Q are on the boundary of P. This is a necessary condition.
3. If any vertex v of Q is inside of P, the algorithm rotates the direction of Q clockwise at the center p_0 to move v to the boundary of P. Repeat this rotation until no vertex of Q is inside of P.
4. Check if each vertex of Q makes $270°$ in total.
5. Finally, the algorithm checks these vertices can be glued to fold the box Q. If they can be glued to the box Q, output it.
6. The algorithm rotates the direction of Q clockwise to find the next position. If the rotation makes $360°$, the algorithm halts.

Intuitively, the algorithm checks all possible positions of P to fold into Q. There are two major points to be considered. The first point is how to check if P can fold to Q at the position and direction. In order to check this point, we will use a technique named "stamping". The second point is the number of rotations of Q. We will show that the number of rotations can be bounded above by a polynomial of n and some geometric parameters.

Hereafter, we assume that p_0 of P coincides with a vertex of Q, and it makes an angle $270°$. Let $v_0 = p_0$ be a vertex of Q and v_1, v_2, v_3 be three vertices of Q adjacent to v_0 on Q by three edges (or crease lines) v_0v_1, v_0v_2, and v_0v_3 in clockwise. Without loss of generality, we assume that $|v_0v_1| = a$. Then we have two cases that either $|v_0v_2| = b$ and $|v_0v_3| = c$ or $|v_0v_3| = b$ and $|v_0v_2| = c$. The algorithm will check these two cases. Here we suppose that $|v_0v_2| = b$ and $|v_0v_3| = c$ since the other case is symmetry. We describe the details of above two points and show the analysis of the algorithm.

3.1   Stamping

The algorithm first adjusts p_0 of P on v_0 of Q with the assumption the line v_0v_3 is cut. From this position, it rotates the relative position of Q centered at $v_0 = p_0$ to search a proper direction that satisfies the necessary conditions for the vertices of P and Q. We will show that the number of these rotations can be bounded by a polynomial of some geometric parameters. Let denote the angle of rotation by $θ = ∠v_3p_0p_1$ at $p_0 = v_0$. That is, the algorithm starts from $θ = 0°$ and updates $θ$. When $θ ≥ 360°$, it finishes to check.

For a given angle $θ$, the algorithm has to check whether (1) all vertices of Q are on the boundary of P, and (2) for each vertex v_i of Q, the curvatures corresponding to the points on the boundary of P sum up to $270°$. In order to do that, the algorithm has to follow the corresponding points to the vertices of Q on P.

The basic idea is called *stamping* in [2]. In [2], Akiyama rolls a regular tetrahedron on a plane as a *stamper* and obtains a tiling by the stamping. The key property of the stamping in [2] is that a regular tetrahedron has the same direction and position when it returns to the original position, no matter what the route was. Therefore, the cut lines of any net on the surface of a regular tetrahedron tile plane, or the net tiles plane.

We use a box Q as a stamper on P. In this case, when the stamper Q returns to the original position, its face and direction change according to the route. In fact, it is used for designing puzzles, and its complexity is investigated [5]. However, the key properties we use here are that a box Q is orthogonal, and each coordinate is an integer, which are useful to bound the number of possible cases. As used in [2], when we roll Q on an edge e of Q on P, this operation corresponds to develop Q to P. That is, when we draw the cut lines of Q on Q and stamp it on the plane, it corresponds to develop Q into the net P.

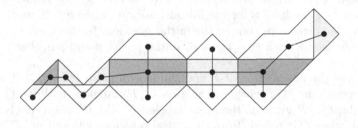

Fig. 4. Tree structure of P: Each face of the box Q is cut into "particles". Then the adjacent relationship of the particles induces a tree on P.

We will use the tree structure of P defined as follows (Fig. 4). Each face of the box Q is cut into "particles" when it is developed to P. In other words, P is partitioned into particles by the edges of Q (or folding lines of P). On P, the particles correspond to the vertices, and two vertices are joined by an edge if and only if the corresponding particles share an edge of positive length on Q. Then since P is a simple polygon and all vertices of Q are on the boundary of P, the resulting graph induces a tree. Essentially, the algorithm performs the breadth first search on this tree by rolling the box Q on P, and it obtains the partition of P into the particles by stamping of Q.

A simple example is given in Fig. 5. The stamper Q has six different labels A, A', B, B', C, and C'. (They are just labels to distinguish with each other, and we do not mind the direction when it is "stamped" on P.) The stamper Q starts at the initial position: It is on the face X that contains $v_0 = p_0$, and the corner (90°) is covered by P without cut. In the case of Fig. 5, the initial position is either on the polygons labeled A or B since the face C of Q is cut into two pieces on P.

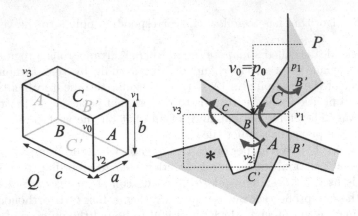

Fig. 5. Rolling a box Q on P. When we roll it left and up, we have B and C. When we roll it up and right, we have C and B'.

We assume that we have the curvature $270°$ at $v_0 = p_0$. Therefore, we always have at least two candidates for the initial position (when $\theta = 0$, we have three candidates). We choose any one as the initial position. In this case, assume that we choose the face A with $v_0 = p_0$ as an initial position and put Q at the initial position.

Let denote the intersection of P and the face of Q on P by $Q \cap P$. Then, at the initial position, $Q \cap P$ gives us the area $(\subset P)$ labeled A. When Q is rolled up and right, $Q \cap P$ gives us the areas labeled C and B', respectively. On the other hand, when Q is rolled (from the initial position) left and up, $P \cap Q$ gives us the areas labeled B and C', respectively.

Here, this stamping (or labeling) should be *continuous*. Precisely, when Q is rolled on an edge e, the resulting polygons in $Q \cap P$ obtain their labels only if they share the edge e with the labeled area before rolling. (In the context of the tree structure shown in Fig. 4, the labeling is done for a polygon in $Q \cap P$ only when it is adjacent to the labeled neighbor.) In the case of Fig. 5, when Q is rolled left from the initial position, the area labeled by B obtains its label since it shares an edge with the area labeled by A. On the other hand, the area with label $*$ does not obtain any label this time.

Intuitively, the stamping sweeps over P from the initial position along P. We repeat rolling from the initial position until all points in P are included in the labeled polygons. When all points in P are in the labeled polygons[2], we say Q *stamps* P. We say that the stamping is *feasible* if no vertex v_i of Q is put inside of P in the stamping. We will consider if P is a net of Q for the current θ. In this situation, P may be a net of Q only if P is feasible by Theorem 3.

---

[2] For sake of simplicity, we do not define the labels of points in P on an edge shared by two rectangles of Q. We also do not define the label of a point p corresponding to the vertex v_i of Q.

Lemma 1. *Assume that Q and θ give us a feasible stamping of P. We also assume that P is a net of Q. Then, each point in P obtains a unique label (except on the edges of Q). That is, the stamping gives us a partition of P along the edges of Q.*

Proof. (Outline) To derive a contradiction, we assume that $p \in P$ obtains two labels. Then there are two different sequences S_1 and S_2 of rolling of Q to stamp on p. Then they produces a hole in P, which contradicts that P is a simple polygon. The details are omitted. □

Lemma 2. *Let D be the diameter of P. Then the total number of stampings of Q on P at the vertex $p_0 = v_0$ by the algorithm is $O((D/a)^6 n)$.*

Proof. (Outline) We first note that the algorithm contains two different steps. First, the algorithm puts Q at the initial position. If the initial position is feasible, it is okay. However, in general, we have to seek the first feasible position. After the first step, we have to seek the next feasible position from the current feasible position.

In the first step at the initial position, when the stamping is not feasible, the algorithm first finds a vertex of Q inside of P. This is done from the initial position by stamping by the breadth first search manner. Let v be the first vertex of Q inside of P. Then the algorithm finds the first boundary of P by rotating δ for the smallest rotation angle. Precisely, after rotation of angle δ, v is moved on the boundary of P. This can be done by computing the locus of v as a part of the circle centered at v_0 with radius $|vp_0|$. After the rotation, the algorithm checks if it has a feasible stamping for the new angle. If it is not feasible by some vertex of Q inside of P, the algorithm repeats the process until there are no such vertices of Q inside of P.

In the second step, now all vertices of Q are on the boundary of P (or outside of P). By the stamping, the algorithm also knows the correspondence between the vertices of Q and the vertices on the boundary of P. Therefore, the algorithm checks whether each vertex of Q has curvature of $270°$ in total. If all vertices are of curvature $270°$, the algorithm performs the check of gluing in the next phase. After checking the gluing, the algorithm has to rotate for finding the next angle of stamping such that no vertex is inside of P after the rotation of, say, δ'. We will show that the total number of this rotation can be bounded above by $O((D/a)^6)$. The crucial point is that the tree structure on P defined by the stamping of Q can be changed after the rotation. Thus we consider all possible vertices that can be produced by the stamping of Q. Essentially, these points form a grid of combinations of three different lengths a, b, c. We estimate the size of grid, and obtain the number of rotations in total. The details are omitted. □

We note that the upper bound $O((D/a)^6)$ of the number of rotations is pessimistic. For example, when $a = b = c$, it is reduced to $O((D/a)^2)$.

We also note that each feasible stamping gives us the whole vertices v_i of Q on the boundary of P. Therefore, we can check if each vertex v_i has a curvature $270°$ in total in linear time of n. Therefore, after the first phase, we know that

P is partitioned into particles of faces of Q with their corresponding labels, and each vertex v_i has a curvature $270°$ in total.

3.2   Check of Gluing

In this phase, we check if we fold to Q by P along the crease lines given in the first phase[3].

Hereafter, we sometimes consider the polygon $P = (p_0, p_1, \ldots, p_{n-1}, p_0)$ consists of vectors $\overrightarrow{p_0 p_1}, \overrightarrow{p_1 p_2}, \ldots, \overrightarrow{p_{n-1} p_0}$ for the sake of simplicity. Then we can deal with "gluing of two edges" by an operation of vectors. For example, in Fig. 5, we assume that we glue two edges $p_0 p_1$ and $p_0 p_{n-1}$. Then, we have three cases after gluing:

(1) $|p_0 p_1| > |p_0 p_{n-1}|$: We obtain $\overrightarrow{p_{n-1} p_1}$ such that $|p_{n-1} p_1| = |p_0 p_1| - |p_0 p_{n-1}|$.
(2) $|p_0 p_1| < |p_0 p_{n-1}|$: We obtain $\overrightarrow{p_{n-1} p_1}$ such that $|p_{n-1} p_1| = |p_0 p_{n-1}| - |p_0 p_1|$.
(3) $|p_0 p_1| = |p_0 p_{n-1}|$: We obtain $p_{n-1} = p_1$.

We here remind that if P is a net of Q, the set of line segments of cut on Q forms a spanning tree T (Theorem 2). Moreover, each edge of T appears twice on the boundary of P. Now we know that $v_0 = p_0$ is the corner of Q of size $a \times b \times c$, and which line is $v_0 v_1$ of length a by θ. Therefore, from this point v_0, we can glue edge by edge and check if Q can be folded from v_0 by P. The details of this part can be found in [8], and it can be done in $O((n + m) \log n)$ time.

3.3   Analysis of Algorithm

The correctness of the algorithm follows from the discussion above. Therefore, we show that the algorithm runs in polynomial time.

Theorem 5. *For a given polygon P of n vertices, the algorithm solves the box folding problem in $O(D^{10}(D^6 n + (n + m) \log n)n)$ time, where D is the diameter of P and m is the maximum number of line segments in an edge of Q for all $Q \in \mathcal{Q}$, where $\mathcal{Q}$ is the set of boxes of the same surface area with P.*

Proof. We first consider the main loop. The number of combinations of the size $a \times b \times c$ of a box is given by $O(A^{2/ \log \log A})$, where A is the area of P. For each trio (a, b, c) with $a \leq b \leq c$, the main loop checks for each point p_i of P of angle $270°$. There are $O(n)$ such angles in the worst case.

For each $p_i = v_0$, the algorithm performs the stamping of Q on P. It is not difficult to see that it can be done in $O(M)$ time, where M is the number of rectangles tiled over P to cover it.

---

[3] Some readers may consider the first phase is enough. However, we have not yet checked if some particles of polygons cause overlap on a face of Q. In other words, we have to check each face is made by particles of polygons by gluing without overlap or hole.

Now we turn to the time complexity of the checking of vertices. First, the algorithm puts a box Q on P so that $v_0 = p_0$ and $\theta = 0$. Then it finds the next feasible stamping if it is not.

The first step is that while there is a vertex of Q inside of P, repeat rotations until all vertices of Q are on P. Finding the first vertex v inside of P takes $O(Mn)$ time; the stamping by the breadth first search takes $O(M)$ rollings and checks if a vertex of the rectangle face is in P takes $O(n)$ time.

Once it finds a position of Q on P such that no vertex is inside of P, the algorithm computes δ', which is the maximum angle that needs to move v_i not inside of P for each v_i. For each vertex v_i, the computation of the corresponding v_i' on the circle centered at v_0 of radius $|v_0 v_i|$ takes $O(n)$ time (along the edges of P). The number of vertices of v_i on the boundary of P is $O(M)$. Therefore, this step takes $O(Mn)$ time.

Once we find a feasible stamping, we have to check if P can be glued to Q. This step takes $O((n+m)\log n)$ time, where m is the maximum number of line segments on an edge of the folded box by using the algorithm in [8].

Thus, the running time of this algorithm is $O(A^{2/\log\log A}(Mn)(Mn + (n+m)\log n))$ time for each phase. By Lemma 2, the total summation of M is $O((D/a)^6)$. Using $A^{2/\log\log A} = O(D^4)$, it is simply that $O(D^4(D/a)^6((D/a)^6 n + (n+m)\log n)n)$ time. Now taking $a = 1$, we have the theorem.   □

4   Case Study

The authors investigated the case that P is an orthogonal polygon. We can assume that P is a polyomino made of unit squares by refining, which simplifies the implementation of the algorithm (the details are omitted). In [12], Xu et al. found nine polyominoes of area 30 that can fold into three boxes of size $1 \times 1 \times 7$, $1 \times 3 \times 3$, and $\sqrt{5} \times \sqrt{5} \times \sqrt{5}$. In our case study, $n < 60$ and each computation takes less than one second. In [12], the authors said that "Interestingly, one of nine such polygons folds into three different boxes $1 \times 1 \times 7$, $1 \times 3 \times 3$, and $\sqrt{5} \times \sqrt{5} \times \sqrt{5}$ in four different ways." The polygon with four different ways of folding is shown in Fig. 1.

However, their claim is not correct. There is another polyomino that has the same property as shown in Fig. 3. That is, the precise claim is as follows. Among polyomino of area 30^4, there are nine polyominoes that can fold into three different boxes of these sizes. Among these nine, two polyominoes have four different ways of folding into three different boxes, and seven polyominoes have three (unique) different ways of folding into three different boxes.

The reason why the authors of [12] missed finding one is hidden in their algorithm. Their first algorithm found all polyominoes that folded into two boxes of sizes $1 \times 1 \times 7$ and $1 \times 3 \times 3$. There are 1080 polyominoes that fold into these two boxes. This time, they did not consider how many ways of folding into two boxes. (Or they never thought that there might have been a polyomino

---

[4] The number of polyomino of area 30 is 2368347037571252.

that could fold into two boxes in three or more different ways.) For these 1080 polyominoes, their second algorithm checked all ways of folding and found nine polyominoes that fold into the third box of size $\sqrt{5} \times \sqrt{5} \times \sqrt{5}$. Since their second algorithm produced all ways of folding, as serendipity, they found that there was a polyomino that folded into the box of size $\sqrt{5} \times \sqrt{5} \times \sqrt{5}$ in two different ways. This is why they concluded that only one polyomino had 4 different ways.

Acknowledgements. A part of this research is supported by JSPS KAKENHI Grant Number JP17H06287 and 18H04091.

References

1. Abel, Z.R., Demaine, E.D., Demaine, M.L., Ito, H., Snoeyink, J., Uehara, R.: Bumpy Pyramid Folding. Computational Geometry: Theory and Applications (2018, accepted). https://doi.org/10.1016/j.comgeo.2018.06.007
2. Akiyama, J.: Tile-Makers and Semi-Tile-Makers. The Mathematical Association of Amerika, Monthly 114, pp. 602–609, August–September 2007
3. Akiyama, J., Nara, C.: Developments of polyhedra using oblique coordinates. J. Indonesia. Math. Soc. **13**(1), 99–114 (2007)
4. Bobenko, A.I., Izmestiev, I.: Alexandrov's theorem, weighted Delaunay triangulations, and mixed volumes. Annales de l'Institut Fourier **58**(2), 447–505 (2008)
5. Buchin, K., et al.: On rolling cube puzzles. In: 19th Canadian Conference on Computational Geometry (CCCG 2007), pp. 141–144 (2007)
6. Demaine, E.D., O'Rourke, J.: Geometric Folding Algorithms: Linkages, Origami, Polyhedra. Cambridge University Press, New York (2007)
7. Dürer, A.: Underweysung der messung, mit den zirckel un richtscheyt, in Linien ebnen unnd gantzen corporen (1525)
8. Horiyama, T., Mizunashi, K.: Folding orthogonal polygons into rectangular boxes. In: Proceedings of the 19th Japan-Korea Joint Workshop on Algorithms and Computation (WAAC 2016) (2016)
9. Kane, D., Price, G.N., Demaine, E.D.: A pseudopolynomial algorithm for Alexandrov's theorem. In: Dehne, F., Gavrilova, M., Sack, J.-R., Tóth, C.D. (eds.) WADS 2009. LNCS, vol. 5664, pp. 435–446. Springer, Heidelberg (2009). https://doi.org/10.1007/978-3-642-03367-4_38
10. Mitani, J., Uehara, R.: Polygons folding to plural incongruent orthogonal boxes. In: Canadian Conference on Computational Geometry (CCCG 2008), pp. 39–42 (2008)
11. Shirakawa, T., Uehara, R.: Common developments of three incongruent orthogonal boxes. Int. J. Comput. Geom. Appl. **23**(1), 65–71 (2013)
12. Xu, D., Horiyama, T., Shirakawa, T., Uehara, R.: Common developments of three incongruent boxes of area 30. Comput. Geom. Theory Appl. **64**, 1–17 (2017)

Analyzing the Quantum Annealing Approach for Solving Linear Least Squares Problems

Ajinkya Borle[✉] and Samuel J. Lomonaco

CSEE Department, University of Maryland Baltimore County,
Baltimore, MD 21250, USA
{aborle1,lomonaco}@umbc.edu

Abstract. With the advent of quantum computers, researchers are exploring if quantum mechanics can be leveraged to solve important problems in ways that may provide advantages not possible with conventional or classical methods. A previous work by O'Malley and Vesselinov in 2016 briefly explored using a quantum annealing machine for solving linear least squares problems for real numbers. They suggested that it is best suited for binary and sparse versions of the problem. In our work, we propose a more compact way to represent variables using two's and one's complement on a quantum annealer. We then do an in-depth theoretical analysis of this approach, showing the conditions for which this method may be able to outperform the traditional classical methods for solving general linear least squares problems. Finally, based on our analysis and observations, we discuss potentially promising areas of further research where quantum annealing can be especially beneficial.

Keywords: Quantum annealing · Simulated annealing
Quantum computing · Combinatorial optimization
Linear least squares · Numerical methods

1 Introduction

Quantum computing opens up a new paradigm of approaching computational problems that may be able to provide advantages that classical (i.e. conventional) computation cannot match. A specific subset of quantum computing is the quantum annealing meta-heuristic, which is aimed at optimization problems.

Quantum annealing is a hardware implementation of exploiting the effects of quantum mechanics in hopes to get as close to a global minimum of the objective function [4]. One popular model of an optimization problem that quantum annealers are based upon is the Ising Model [12]. It can be written as:

$$F(h, J) = \sum_a h_a \sigma_a + \sum_{a<b} J_{ab} \sigma_a \sigma_b \tag{1}$$

© Springer Nature Switzerland AG 2019
G. K. Das et al. (Eds.): WALCOM 2019, LNCS 11355, pp. 289–301, 2019.
https://doi.org/10.1007/978-3-030-10564-8_23

where $\sigma_a \in \{-1, 1\}$ represents the qubit (quantum bit) spin and h_a and J_{ab} are the coefficients for the qubit spins and couplers respectively [7]. The quantum annealer's job is to return the set of values for σ_as that would correspond to the smallest value of $F(h, J)$.

There have been various efforts by different organizations to make non Von Neumann architecture computers based on the Ising model such as D-wave Systems [13] and IARPA's QEO effort [1] that are attempting to make quantum annealers. Other similar efforts are focussed towards making 'quantum like' optimizers for the Ising model, such as Fujitsu's Digital Annealer chip [3] and NTT's photonic quantum neural network [11]. The former is a quantum inspired classical annealer and the latter uses photonic qubits for doing its optimization. At the time of writing this document, D-wave computers are the most prominent Ising model based quantum annealers.

In order to solve a problem on a quantum annealer, the programmers first have to convert their problem for the Ising model. A lot of work has been done showing various types of problems running on D-wave machines [2,14,16].

In 2016, O'Malley and Vesselinov [15] briefly explored using the D-wave Quantum Annealer for the linear least squares problem for binary and real numbers. In this paper, we shall study their approach in more detail. Section 2 is devoted to the necessary background and related work for our results. Section 3 is a review of the quantum annealing approach where we introduce one's and two's complement encoding for a quantum annealer. Section 4 deals with the runtime cost analysis and comparison where we define necessary conditions for expecting a speedup. Section 5 is dedicated to theoretical accuracy analysis. Based on our results, Sect. 6 is a discussion which lays out potentially promising future work. We finally conclude our paper with Sect. 7. The D-wave 2000Q and the experiments performed on them are elaborated upon in Appendices A and B respectively in arXiv:1809.07649.

2   Background and Related Work

2.1   Background

Before we get started, we shall first lay out the terms and concepts we will use in the rest of the paper.

Quantum Annealing: The Quantum Annealing approach aims to employ quantum mechanical phenomena to traverse through the energy landscape of the Ising model to find the ground state configuration of σ_a variables from Eq. (1). The σ_a variables are called as qubits spins in quantum annealing, essentially being quantum bits.

The process begins with all qubits in equal quantum superposition: where all qubits are equally weighted to be either -1 or $+1$. After going through a quantum-mechanical evolution of the system, given enough time, the resultant state would be the ground state or the global minimum of the Ising objective function. During this process, the entanglement between qubits (in the form of couplings) along with quantum tunneling effects (to escape being stuck in configurations of local minima) plays a part in the search for the global minimum. A more detailed description can be found in the book by Tanaka et al. [18]. For our purposes however, we will focus on the computational aspects related to quantum annealing: cost of preparing the problem for the annealer, cost of annealing and the accuracy of the answers. From an accuracy perspective, a quantum annealer is essentially trying to take samples of a Boltzmann distribution whose energy is the Ising objective function [2]

$$P(\sigma) = \frac{1}{Z} e^{-F(h,J)} \tag{2}$$

$$\text{where } Z = exp\left(\sum_{\{\sigma_a\}} \left[\sum_a h_a \sigma_a + \sum_{a<b} J_{ab} \sigma_a \sigma_b \right] \right) \tag{3}$$

Equation (2) tells us that the qubit configuration of the global minimum would have the highest probability to be sampled. The D-wave 2000Q is one such quantum annealer made by D-wave Systems. Its description is in the Appendix A of arXiv:1809.07649.

Quadratic Unconstrained Binary Optimization (QUBO): These are minimization problems of the type

$$F'(v, w) = \sum_a v_a q_a + \sum_{a<b} w_{ab} q_a q_b \tag{4}$$

where $q_a \in \{0, 1\}$ are the qubit variables returned by the machine after the minimization and v_a and w_{ab} are the coefficients for the qubits and the couplers respectively [7]. The QUBO model is equivalent to the Ising model by the following relationship between σ_a and q_a

$$\sigma_a = 2q_a - 1 \tag{5}$$

$$\text{and } F(h, J) = F'(v, w) + \text{offset} \tag{6}$$

Since the offset value is a constant, the actual minimization is only done upon F or F'. We use this model for the rest of the paper.

Linear Least Squares: Given a matrix $A \in \mathbb{R}^{m \times n}$, a column vector of variables $x \in \mathbb{R}^n$ and a column vector $b \in \mathbb{R}^m$ (Where $m > n$). The linear least squares problem is to find the x that would minimize $\|Ax - b\|$ the most. In other words, it can be described as:

$$\arg\min_x \|Ax - b\| \tag{7}$$

Various classical algorithms have been developed over the time in order to solve this problem. Some of the most prominent ones are (1) Normal Equations by Cholesky Factorization, (2) QR Factorization and the (3) SVD Method [6].

2.2 Related Work

The technique of using quantum annealing for solving linear least squares problems for real numbers was created by O'Malley and Vesselinov [15]. In that work, they discovered that because the time complexity is in the order of $O(mn^2)$ (with certain assumptions mentioned in Sect. 4.2), which is the same time complexity class as the methods mentioned above, the approach might be best suited for binary and sparse least squares problem. In a later work, O'Malley et al. applied binary linear least squares using quantum annealing for the purposes of nonnegative/binary matrix factorization [16].

The problem of solving linear least squares has been well studied classically. Like mentioned above, the most prominent methods of solving the problems are Normal Equations (using Cholesky Factorization), QR Factorization and by Singular Value Decomposition [6]. But other works in recent years have tried to get a better time complexity for certain types of matrices, such as the work by Drineas et al. that presents a randomized algorithm in $O(mn \log n)$ for $(m >> n)$ [9]. The iterative approximation techniques such as Picani and Wainwright's work [17] of using sketch Hessian matrices to solve constrained and unconstrained least squares problems. But since the approach by O'Malley and Vesselinov is a direct approach to solve least squares, we shall focus on comparisons with the big 3 direct methods mentioned above.

Finally, it is important to note that algorithms exists in the gate-based quantum computation model like the one by Wang [21] that runs in $poly(\log(N), d, \kappa, 1/\epsilon)$ where N is the size of data, d is the number of adjustable parameters, κ represents the condition number of A and ϵ is the desired precision. However, the results of this algorithm are in a quantum superposition state and not directly available for us to observe.

3   Quantum Annealing for Linear Least Squares

In order to solve Eq. (7), let us begin by writing out $Ax - b$

$$Ax - b = \begin{pmatrix} A_{11} & A_{12} & ... & A_{1n} \\ A_{21} & A_{22} & ... & A_{2n} \\ \vdots & \vdots & \vdots & \vdots \\ A_{m1} & A_{m2} & ... & A_{mn} \end{pmatrix} \begin{pmatrix} x_1 \\ x_2 \\ \vdots \\ x_n \end{pmatrix} - \begin{pmatrix} b_1 \\ b_2 \\ \vdots \\ b_m \end{pmatrix} \tag{8}$$

$$Ax - b = \begin{pmatrix} A_{11}x_1 + A_{12}x_2 + ... + A_{1n}x_n - b_1 \\ A_{21}x_1 + A_{22}x_2 + ... + A_{2n}x_n - b_2 \\ \vdots \\ A_{m1}x_1 + A_{m2}x_2 + ... + A_{mn}x_n - b_m \end{pmatrix} \tag{9}$$

Taking the 2 norm square of the resultant vector of Eq. (9), we get

$$\|Ax - b\|_2^2 = \sum_{i=1}^{m} (|A_{i1}x_1 + A_{i2}x_2 + ... + A_{in}x_n - b_i|)^2 \tag{10}$$

Because we are dealing with real numbers here, $(|.|)^2 = (.)^2$

$$\|Ax - b\|_2^2 = \sum_{i=1}^{m}(A_{i1}x_1 + A_{i2}x_2 + ... + A_{in}x_n - b_i)^2 \tag{11}$$

Now if we were solving binary least squares [15,16] then each x_j would be represented by the qubit q_j. The coefficients in Eq. (4) are found by expanding Eq. (11) to be

$$v_j = \sum_i A_{ij}(A_{ij} - 2b_i) \tag{12}$$

$$w_{jk} = 2\sum_i A_{ij}A_{ik} \tag{13}$$

But for solving the general version of the least squares problem we need to represent x_j, which is a real number, in its equivalent radix 2 approximation by using multiple qubits. Let Θ be the set of powers of 2 we use to represent every x_j, defined as

$$\Theta = \{2^l : l \in [o, p] \wedge l, o, p \in \mathbb{Z}\} \tag{14}$$

Here, it is assumed that l represents contiguous values from the interval of $[o, p]$. The values of o and p are the user defined lower and upper limits of the interval. In the work by O'Malley and Vesselinov [15], the radix 2 representation of x_j is given by

$$x_j \approx \sum_{\theta \in \Theta} \theta q_{j\theta} \tag{15}$$

But this would mean that only approximations of positive real numbers can be done, so we need to introduce another set of qubits q_j^*, to represent negative real numbers

$$x_j \approx \sum_{\theta \in \Theta} \theta q_{j\theta} + \sum_{\theta \in \Theta} -(\theta q_{j\theta}^*) \tag{16}$$

Which means that representing a (fixed point approximation of) real number that can be either positive or negative would require $2|\Theta|$ number of qubits.

However, we can greatly reduce the amount of qubits to be used in Eq. (16) by introducing a sign bit q_{j0}

$$x_j \approx \vartheta q_{j0} + \sum_{\theta \in \Theta} \theta q_{j\theta} \tag{17}$$

$$\text{where } \vartheta = \begin{cases} -2^{p+1}, & \text{for two's complement} \\ -2^{p+1} + 2^o, & \text{for one's complement} \end{cases} \tag{18}$$

Where p and o are the upper and lower limits of the exponents used for powers of 2 present in Θ. In other words, Eq. (17) represents an approximation of a real number in one's or two's complement binary. Combining Eqs. (11) and (17), we get

$$\|Ax - b\|_2^2 = \sum_{i=1}^{m}(A_{i1}(\vartheta q_{10} + \sum_{\theta \in \Theta} \theta q_{10})$$

$$+ A_{i2}(\vartheta q_{20} + \sum_{\theta \in \Theta} \theta q_{20}) + ... + A_{in}(\vartheta q_{n0} + \sum_{\theta \in \Theta} \theta q_{n0}) - b_i)^2 \quad (19)$$

Which means that the v and w coefficients of the qubits for the general version of the least squares problem would be

$$v_{js} = \sum_i sA_{ij}(sA_{ij} - 2b_i) \quad (20)$$

$$w_{jskt} = 2st \sum_i A_{ij}A_{ik} \quad (21)$$

where $s, t \in \vartheta \cup \Theta$

4   Cost Analysis and Comparison

In order to analyze the cost incurred using a quantum annealer to solve a problem, one good way is by combining together the (1) time required to prepare the problem (so that it is in the QUBO/Ising model that the machine would understand) and (2) the runtime of the problem on the machine.

We can calculate the first part of the cost concretely. But the second part of the cost depends heavily upon user parameters and heuristics to gauge how long should the machine run and/or how many runs of the problem should be done. Nonetheless, we can set some conditions that must hold true if any speedup is to be observed using a quantum annealer for this problem.

4.1   Cost of Preparing the Problem

As mentioned in O'Malley and Vesselinov [15] the complexity class of preparing a QUBO problem from A,x and b is $O(mn^2)$, which is the same as all the other prominent classical methods to solve linear least squares [6]. However, for numerical methods, it is also important to analyze the floating point operation cost as they grow with the data. This is because methods in the same time complexity class may be comparatively faster or slower.

So starting with Eq. (20), we assume that the values of the set $\vartheta \cup \Theta$ are preprocessed. Matrix A has m rows and n columns, the variable vector x is of the length n. Let $c = |\Theta| + 1$. To calculate the expression $sA_{ij}(sA_{ij} - 2b_i)$, we can compute $2b_i$ for m rows first and use the results to help in all the future computations. After that, we see that it takes 3 flops to process the expression for 1 qubit of the radix 2 representation of a variable, per row. This expression has to be calculated for n variables each requiring c qubits for the radix 2 representation, over m rows. That is: it would take $3cmn$ operations. On top of that, we need to sum up the resulting elements over all the m rows, which requires an additional cmn operations. Hence we have

$$\text{Total cost of computing } v_{js} = 4cmn + m \tag{22}$$

Now for the operation costs associated with terms processed in Eq. (21). Let us consider a particular subset of those terms:

$$w_{j1k1} = 2 \sum_i A_{ij} A_{ik} \tag{23}$$

By computing Eq. (23) first, we create a template for all the other w_{jskt} variables and also reduce the computation cost. Each $A_{ij}A_{ik}$ operation is 1 flop. There are $\binom{n}{2}$ pairs of (j,k) for $j < k$, But we also need to consider pair interactions for qubits when $j = k$. Hence we have $\binom{n}{2} + n$ operations, which comes out to $0.5(n^2 + n)$. Furthermore, these w coefficients are computed for m rows with m summations, which brings the total up to: $m(n^2 + n)$. After this, we need to multiply 2 to all the resultant $0.5(n^2 + n)$ variables. Making the total cost:

$$\text{Cost of all } w_{j1k1} = m(n^2 + n) + 0.5(n^2 + n) \tag{24}$$

Now we can use w_{j1k1} for the next step. Without loss in generality, we assume that $\forall s,t \in \vartheta \cup \Theta,\ s \times t$ is preprocessed. This would mean that we would have $\binom{c}{2} + c$ qubit to qubit interactions for each pair of variables in x. From the previous step, we know that we'll have to do this for $\binom{n}{2} + n$ variables. Which means that the final part of the cost for w_{jskt} is $0.25(c^2 + c)(n^2 + n)$. Summing up all the costs, we get the total cost to prepare the entire QUBO problem:

$$\text{Cost of tot. prep} = mn^2 + mn(4c + 1) + 0.25(n^2 + n)(c^2 + c + 2) + m \tag{25}$$

4.2   Cost of Executing the Problem

Let τ be the cost of executing the QUBO form of a given problem. It can be expressed as

$$\tau = a_t r \tag{26}$$

Where a_t is the anneal time per run and r is the number of runs or samples. However, for ease of analysis, we need to interprot τ in a way where we can study the runtime in terms of the data itself. The nature of the Ising/Qubo problem is such that we need $O(\exp(\gamma N^\alpha))$ time classically to get the ground state with a probability of 1. As stated in Boxio et al. [4], we don't yet know what's going to be the actual time complexity class under quantum annealing for the Ising problem, but a safe bet is that it won't reduce the complexity of the Ising to a polynomial one, only the values of α and γ would be reduced. But because quantum annealing is a metaheuristic for finding the ground state, we needn't necessarily run it for $O(\exp(\gamma N^\alpha))$. We make the following assumption:

$$\tau^* = poly(cn) \tag{27}$$

$$\text{and } \deg(poly(cn)) = \beta \tag{28}$$

Here, τ^* represents the combined operations required for annealing as well as post-processing on the returned samples. The assumption is that τ^* is a polynomial in cn with β as its degree. From Eq. (19), we know that m (number of rows of the matrix A) doesn't play a role in deciding the size of the problem embedded inside the quantum annealer.

The reason for this assumption is the fact that linear least squares is a convex optimization problem. Thus, even if we don't get the global minimum solution, the hope is to get samples on the convex energy landscape that are close to the global minimum (Based on the observations in [8]). Using those samples, and exploiting the convex nature of the energy landscape, the conjecture is that there exists a polynomial time post-processing technique with $\beta < 3$ by which we can converge to the global minimum very quickly. We shall see in Sect. 4.3 why it is important for $\beta < 3$ for any speed improvement over the standard classical methods. Just as in Dorband's MQC technique [8] for generalized Ising landscapes, we hope that such a technique would be able to intelligently use the resultant samples. The difference being that we have the added advantage of convexity in our specific problem.

4.3   Cost Comparison with Classical Methods

In the following table, we compare the costs of the most popular classical methods for finding linear least squares [6] and the quantum annealing approach.

Table 1. Comparison of the classical methods and QA

Method for least squares	Operational cost
Normal equations	$mn^2 + n^3/3$
QR factorization	$2mn^2 - 2n^3/3$
SVD	$2mn^2 + 11n^3$
Quantum annealing	$mn^2 + mn(4c+1) + poly(cn) + 0.25(n^2+n)(c^2+c+2) + m$

When it comes to theoretical runtime analysis, because c doesn't necessarily grow in direct proportion to the number of rows m or columns n, we shall consider c to be a constant for our analysis.

From Table 1, Let us define $Cost_{NE}$, $Cost_{QR}$, $Cost_{SVD}$ and $Cost_{QA}$ as the costs for the methods of finding least squares solution by Normal Equations,

QR Factorization, Singular Value Decomposition and Quantum Annealing (QA) respectively.

$$Cost_{NE} = mn^2 + n^3/3 \tag{29}$$

$$Cost_{QR} = 2mn^2 - 2n^3/3 \tag{30}$$

$$Cost_{SVD} = 2mn^2 + 11n^3 \tag{31}$$

$$Cost_{QA} = mn^2 + mn(4c+1) + poly(cn) + 0.25(n^2+n)(c^2+c+2) + m \tag{32}$$

The degree of Eqs. (29)–(32) is 3. Since $m > n$, we can assess that $mn^2 > n^3$

The next thing we need to do is to define a range for β (degree of $pol(cn)$) in such a way that Eq. (32) will be competitive with the other methods. This is another assumption upon which a speedup is conditional.

$$\boxed{0 < \beta < 3} \tag{33}$$

This is done so that we do not have another term of degree 3. Now, we turn our attention to the terms of the type kmn^2 where k is the coefficient, we can see that the cost of the QA method is lesser than QR Factorization and SVD by a factor of 2. However, we still have to deal with the cost of the Normal Equations Method. Let us define $\Delta Cost_{NE}$ and $\Delta Cost_{QA}$ as

$$\Delta Cost_{NE} = Cost_{NE} - mn^2 = n^3/3 \tag{34}$$

$$\Delta Cost_{QA} = Cost_{QA} - mn^2 = mn(4c+1) + poly(cn)$$
$$+ 0.25(n^2+n)(c^2+c+2) + m \tag{35}$$

The degree of $\Delta Cost_{NE}$ is 3 while that of $\Delta Cost_{QA}$ is < 3. For simplicity (and without loss of generality) we consider $mn(4c+1)$ and can ignore all the other similar and lower degree terms. The reason we can do this is because those terms grow comparatively slower than $n^3/3$, but the relationship between $n^3/3$ and $mn(4c+1)$ has to be clearly defined. Thus our simplified cost difference for the QA method is $\Delta Cost_{QA}^*$

$$\Delta Cost_{QA}^* = (4c+1)mn \tag{36}$$

We need to analyze the case where the quantum annealing method is more cost effective, i.e

$$\Delta Cost_{QA}^* < \Delta Cost_{NE} \tag{37}$$

$$\text{or, } mn(4c+1) < n^3/3 \tag{38}$$

For this comparison, we need to define m in terms of n

$$m = \lambda n \tag{39}$$

Using Eq. (39) in Eq. (38), we get

$$\lambda n^2 (4c+1) < \frac{n^3}{3} \tag{40}$$

$$\text{or, } \lambda(4c+1) < \frac{n}{3} \tag{41}$$

$$\text{or, } \lambda < \frac{n}{3(4c+1)} \tag{42}$$

Combining Eq. (42) with the fact that $m > n$, we get :

$$\boxed{1 < \lambda < \frac{n}{3(4c+1)}} \tag{43}$$

$$\text{given } \frac{n}{3(4c+1)} > 1 \tag{44}$$

Thus, the Quantum Annealing method being faster than the Normal Equations method is for when Eqs. (33, 43, 44) holds true and is conditional on the conjectures described in Eqs. (27, 28, 33).

The above condition makes our speed advantage very limited to a small number of cases. However, it is important to note that the Normal Equations method is known to be numerically unstable due to the $A^T A$ operation involved [6]. The quantum annealing approach does not seem to have such types of calculations that would make it numerically unstable to the extent of Normal Equations method (because of the condition number of A), assuming the precision of qubit and coupler coefficients is not an issue. Thus for most practical cases, it competes with the QR Factorization method rather than the Normal Equations method.

5 Accuracy Analysis

Because quantum annealing is a physical metaheuristic, it is important to analyze the quality of the results obtained from it. The results from our experiments are in Appendix B of arXiv:1809.07649. We can define the probability of getting the global minimum configuration of qubits in the QUBO form by using Eqs. (2 and 3)

$$P(q) = \frac{1}{Z'} e^{-F'(v,w)} \tag{45}$$

$$\text{where } Z' = exp\left(\sum_{\{q_a\}} \left[\sum_a v_a q_a + \sum_{a<b} w_{ab} q_a q_b \right] \right) \tag{46}$$

Which means that the set of solutions corresponding to the global minimum $\hat{Q}$ have the highest probability of all the possible solution states. A problem arises when we need to use more qubits for better precision. This would mean that the set of approximate solutions Q', would also increase as a result. The net result would be that $\sum_{\hat{q} \in \hat{Q}} P(\hat{q}) < \sum_{q' \in Q'} P(q')$, which means that as the number of qubits used for precision increases, it would be harder to get the best solution directly from the machine. But like discussed in Sect. 4.2, if the conjecture for a polynomial time post-processing technique with degree< 3 holds true, we should be able to get the best possible answer, in a competitive amount of time, by using the results of a quantum annealer.

Another area of potential problems is the fact that quantum annealing happens on the actual physical qubits and its connectivity graph (see Appendix A of arXiv:1809.07649). This means that the energy landscape for the physical qubit graph is bigger than the one for the logical qubit graph. This problem should be alleviated to a degree when and if the next generation of quantum annealers have a more dense connectivity between their physical qubits.

6 Discussion and Future Work

Based on our theoretical and experimental results (see Appendix B of arXiv:1809.07649), we can see that there are potential advantages as well as drawbacks to this approach. Our work outlines the need of a polynomial time post-processing technique for convex problems (with degree< 3),only then will this approach have any runtime advantage. Whether such a post-processing technique exists is an interesting open research problem. But based on our work, we can comment on few areas where quantum annealing may have a potential advantage. We have affirmed the conjecture that machines like the D-wave find good solutions (that may not be optimal) in a small amount of time [16] (Appendix B, arXiv:1809.07649).

It may be useful to use quantum annealing for least squares-like problems, i.e. problems that require us to minimize $\|Ax - b\|$, but are time constrained in nature. Two such problems are: (i) The Sparse Approximate Inverse [10] (SPAI) type preconditioners used in solving linear equations and (ii) the Anderson acceleration method for iterative fixed point methods [20]. Both of these methods require approximate least squares solutions, but under time constraints. It would be interesting to see if quantum annealing can be potentially useful there.

Another area of work could be to use quantum annealing within the latest iterative techniques for least squares approximation itself. Sketch based techniques like the Hessian sketch by Pilanci and Wainwright [17] may be able to use quantum annealing as a subroutine.

Finally, just like O'Malley and Vesselinov mentioned in their papers [15, 16], quantum annealing also has potential in specific areas like the Binary [19] and box-constrained integer least squares [5] where classical methods struggle.

7 Concluding Remarks

In this paper, we did an in-depth theoretical analysis of the quantum annealing approach to solve linear least squares problems. We proposed a one's complement and two's complement representation of the variables in qubits. We then showed that the actual annealing time does not depend on the number of rows of the matrix, just the number of columns/variables and number of qubits required to represent them. We outlined conditions for which quantum annealing will have a speed advantage over the prominent classical methods to find least squares. An accuracy analysis shows how as precision bits are added, it is harder to get the 'best' least square answer, unless any post-processing is applied. Finally, we outline possible areas of interesting research work that may hold promise.

Acknowledgement. The authors would like to thank Daniel O'Malley from LANL for his feedback. A special thanks to John Dorband, whose suggestions inspired the development the one's/two's complement encoding. Finally, the authors would like to thank Milton Halem of UMBC and D-wave Systems for providing access to their machines.

References

1. Quantum enhanced optimization (qeo). https://www.iarpa.gov/index.php/research-programs/qeo
2. Adachi, S.H., Henderson, M.P.: Application of quantum annealing to training of deep neural networks. arXiv preprint arXiv:1510.06356 (2015)
3. Aramon, M., Rosenberg, G., Miyazawa, T., Tamura, H., Katzgraber, H.G.: Physics-inspired optimization for constraint-satisfaction problems using a digital annealer. arXiv preprint arXiv:1806.08815 (2018)
4. Boixo, S., et al.: Evidence for quantum annealing with more than one hundred qubits. Nat. Phys. **10**(3), 218 (2014)
5. Chang, X.W., Han, Q.: Solving box-constrained integer least squares problems. IEEE Trans. Wirel. Commun. **7**(1), 277–287 (2008)
6. Do, Q.L.: Numerically efficient methods for solving least squares problems (2012)
7. Dorband, J.E.: Stochastic characteristics of qubits and qubit chains on the D-wave 2X. arXiv preprint arXiv:1606.05550 (2016)
8. Dorband, J.E.: A method of finding a lower energy solution to a QUBO/Ising objective function. arXiv preprint arXiv:1801.04849 (2018)
9. Drineas, P., Mahoney, M.W., Muthukrishnan, S., Sarlós, T.: Faster least squares approximation. Numer. math. **117**(2), 219–249 (2011)
10. Grote, M.J., Huckle, T.: Parallel preconditioning with sparse approximate inverses. SIAM J. Sci. Comput. **18**(3), 838–853 (1997)
11. Honjo, T., Inagaki, T., Inaba, K., Ikuta, T., Takesue, H.: Long-term stable operation of coherent Ising machine for cloud service. In: CLEO: Science and Innovations, pp. JTu2A-87. Optical Society of America (2018)
12. Kadowaki, T., Nishimori, H.: Quantum annealing in the transverse Ising model. Phys. Rev. E **58**(5), 5355 (1998)
13. Karimi, K., et al.: Investigating the performance of an adiabatic quantum optimization processor. Quantum Inf. Process. **11**(1), 77–88 (2012)
14. OGorman, B., Babbush, R., Perdomo-Ortiz, A., Aspuru-Guzik, A., Smelyanskiy, V.: Bayesian network structure learning using quantum annealing. Eur. Phys. J. Spec. Top. **224**(1), 163–188 (2015)
15. O'Malley, D., Vesselinov, V.V.: ToQ. jl: A high-level programming language for D-wave machines based on Julia. In: 2016 IEEE High Performance Extreme Computing Conference (HPEC), pp. 1–7. IEEE (2016)

16. O'Malley, D., Vesselinov, V.V., Alexandrov, B.S., Alexandrov, L.B.: Nonnegative/binary matrix factorization with a D-wave quantum annealer. arXiv preprint arXiv:1704.01605 (2017)

17. Pilanci, M., Wainwright, M.J.: Iterative Hessian sketch: fast and accurate solution approximation for constrained least-squares. J. Mach. Learn. Res. **17**(1), 1842–1879 (2016)

18. Tanaka, S., Tamura, R., Chakrabarti, B.K.: Quantum Spin Glasses, Annealing and Computation. Cambridge University Press, Cambridge (2017)

19. Tsakonas, E., Jaldén, J., Ottersten, B.: Robust binary least squares: relaxations and algorithms. In: 2011 IEEE International Conference on Acoustics, Speech and Signal Processing (ICASSP), pp. 3780–3783. IEEE (2011)

20. Walker, H.F., Ni, P.: Anderson acceleration for fixed-point iterations. SIAM J. Numer. Anal. **49**(4), 1715–1735 (2011)

21. Wang, G.: Quantum algorithm for linear regression. Phys. Rev. A **96**(1), 012335 (2017)

Data Structures

Greedy Consensus Tree and Maximum Greedy Consensus Tree Problems

Wing-Kin Sung[1,2]([envelope])

[1] School of Computing, National University of Singapore,
13 Computing Drive, Singapore 117417, Singapore
ksung@comp.nus.edu.sg
[2] Genome Institute of Singapore, 60 Biopolis Street, Singapore 138672, Singapore

Abstract. Consensus tree is a phylogenetic tree that summarizes the branching information of a set of conflicting phylogenetic trees. Computing consensus tree is a major step in phylogenetic tree reconstruction. It also finds application in predicting a species tree from a set of gene trees. Here, we focus our study on one of the most frequently used consensus tree problem, called greedy consensus tree problem. Given k phylogenetic trees leaf-labeled by n taxa, previous best known algorithm for constructing a greedy consensus tree of these k trees runs in $O(kn^{1.5} \log n)$ time. Here, we describe an $O(k^2 n)$-time solution. Our method is the fastest when $k = O(\sqrt{n} \log n)$.

Existing greedy consensus tree methods may not report the most resolved greedy consensus tree. Here, we propose a new computational problem called the maximum greedy consensus tree problem that aims to find the most resolved greedy consensus tree. We showed that this problem is NP-hard for $k \geq 3$. We also give a polynomial time solution when $k = 2$ and an approximation algorithm for $k = 3$.

1 Introduction

A phylogenetic tree is a rooted, unordered, leaf-labeled tree in which every internal node has at least two children and all leaves have different labels, representing the set of taxa. It is used to describe the evolutionary relationship among the taxa.

Phylogenetic tree of a set of taxa is reconstructed by analyzing their available data (like their DNA sequences). Using different data sources, conflicting phylogenetic trees are reconstructed. To summarize the set of conflicting trees, consensus tree problem is proposed. Precisely, the input is a set of k conflicting phylogenetic trees leaf-labeled by the same set of taxa L, where $n = |L|$. The consensus tree is a phylogenetic tree T leaf-labeled by L that summarizes the branching information of all k conflicting trees. Consensus tree has been widely used in two applications: reconstructing phylogenetic tree and constructing species tree from a set of gene trees.

The consensus tree problem was first proposed by Adam [1]. After that, many different consensus tree definitions have been proposed. They include Adam's

© Springer Nature Switzerland AG 2019
G. K. Das et al. (Eds.): WALCOM 2019, LNCS 11355, pp. 305–316, 2019.
https://doi.org/10.1007/978-3-030-10564-8_24

consensus tree [1], strict consensus tree [17], loose consensus tree [2], majority-rule consensus tree [12], asymmetric median tree [13], greedy consensus tree [3,6], R* consensus tree [3], etc. Please refer to the surveys in [3], Chap. 30 in [5], and Chap. 8.4 in [18] for more details about different consensus trees and their advantages and disadvantages.

As n and k can be large, efficient algorithms for constructing consensus trees are required. Recently, a number of break-through results are proposed (e.g. [8,10,11,15]). Here, we focused our discussion on greedy consensus tree.

The greedy consensus tree problem is proposed by Felsenstein in 1989. It is formally defined in Sect. 2. Greedy consensus tree is an extension of the majority-rule consensus tree, and is sometimes called the majority-rule extended consensus tree. A naive implementation implied by the definition immediately yields an algorithm of time complexity $O(kn^3)$. Jansson et al. [10] gives an $O(kn^2)$-time algorithm. (This algorithm is implemented in FACT package [9].) Gawrychowski et al. [7] recently further improves the running time to $O(kn^{1.5} \log n)$. Algorithm for constructing greedy consensus tree has implemented in popular phylogenetics software packages like PHYLIP [6], PAUP*[19] and MrBayes [16].

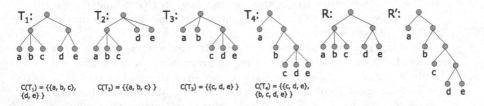

Fig. 1. Consider a set of 4 trees $T = \{T_1, T_2, T_3, T_4\}$, each is leaf-labeled by $\{a, b, c, d, e\}$. $C(T_i)$ gives the set of non-trival clusters of T_i for $i = 1, 2, 3, 4$. Two non-trival clusters $\{a, b, c\}$ and $\{c, d, e\}$ have two occurrences in T. The greedy consensus tree for T is not unique. If we include the non-trival cluster $\{a, b, c\}$ first, we obtain the greedy consensus tree R; otherwise, if we include the non-trival cluster $\{c, d, e\}$ first, we obtain the greedy consensus tree R'.

Greedy consensus tree is not unique. For the example in Fig. 1, there are two possible greedy consensus trees for $\{T_1, T_2, T_3, T_4\}$. We sometimes want to obtain the most refined greedy consensus tree (i.e. a greedy consensus tree that has the maximum number of nodes). For the example in Fig. 1, we may prefer R' since it has more nodes.

Hence, we propose another computational problem called the maximum greedy consensus tree problem. For this problem, given a set of phylogenetics trees T, we aim to find a greedy consensus tree R that maximizes the number of internal nodes. A naive solution is to enumerate all possible greedy consensus trees and to identify a tree that maximizes the number of nodes. There is no better solution in the literature.

This manuscript has 4 contributions. First, we give an $O(k^2 n)$-time algorithm to build a greedy consensus tree. Our algorithm is faster than all existing methods when $k = O(\sqrt{n} \log n)$. Second, we show that the maximum greedy consensus tree problem is NP-hard for $k \geq 3$. Third, we give a polynomial time algorithm that constructs a maximum greedy consensus tree for two trees $\{T_1, T_2\}$. Fourth, we give a polynomial time approximation algorithm for constructing maximum greedy consensus tree of three trees.

This paper is organized as follows. Section 2 gives preliminary. Section 3 describes an $O(k^2 n)$-time algorithm for constructing a greedy consensus tree of $\mathcal{T}$. Section 4 shows that the maximum greedy consensus tree problem is NP-hard for $k \geq 3$. Section 5 gives a polynomial time algorithm for finding a maximum greedy consensus tree of two trees. Section 6 gives a polynomial time approximation algorithms for finding a maximum greedy consensus tree of three trees.

2   Preliminary

A phylogenetic tree is a rooted, unordered, leaf-labeled tree in which every internal node has at least two children and all leaves have different labels. To simplify the presentation, phylogenetic trees are referred to as "tree" from here on, and every leaf in a tree is identified with its (unique) label.

Let T be a tree leaf-labeled by L. The set of all nodes in T is denoted by $V(T)$. For any $u \in V(T)$, the subtree of T rooted at u is denoted at $T[u]$. Let $\Lambda(T[u])$ denotes the set of all leaves in $T[u]$. The set $\Lambda(T[u])$ is called a cluster in T. If u is not a leaf or not a root of T, the set $\Lambda(T[u])$ is called a non-trival cluster in T. Otherwise, $\Lambda(T[u])$ is a trival cluster. A trival cluster either equals L (if u is the root of T) or is a singleton set containing one taxa in L (if u is a leaf). We denote $\mathcal{C}(T)$ be the set of all non-trival clusters of T. For example, in Fig. 1, $\mathcal{C}(T_1) = \{\{a, b, c\}, \{d, e\}\}$.

Consider any two clusters C_1 and C_2, C_1 and C_2 are compatible if either $C_1 \subseteq C_2$, $C_1 \supseteq C_2$ or $C_1 \cap C_2 = \emptyset$. For example, in Fig. 1, the cluster $\{d, e\}$ in T_1 is compatible with another cluster $\{c, d, e\}$ in T_4. However, the cluster $\{a, b, c\}$ in T_1 is not compatible with the cluster $\{c, d, e\}$ in T_4. All clusters in a tree are pairwise compatible. In fact, a tree T uniquely defines a set of pairwise compatible clusters $\mathcal{C}(T)$, and vice versa.

Consider a cluster C and a tree T, C is said to be compatible with T if C is compatible with all clusters in $\mathcal{C}(T)$. Consider two trees T_1 and T_2, T_1 and T_2 are said to be compatible if for all clusters C_1 in T_1, for all clusters C_2 in T_2, C_1 and C_2 are compatible. For example, in Fig 1, $\{d, e\}$ is compatible with T_2 but $\{c, d, e\}$ is not compatible with T_2. T_1 is compatible with T_2 but T_4 is not compatible with T_2.

Given a set of k trees $\mathcal{T} = \{T_1, T_2, \ldots, T_k\}$, For any cluster C, denote $count_{\mathcal{T}}(C)$ be the number of times C is compatible with T_i for $i = 1, 2, \ldots, k$. Precisely, $count_{\mathcal{T}}(C) = |\{T_j \mid C \in \mathcal{C}(T_j), j = 1, \ldots, k\}|$. For example, in Fig. 1, $\mathcal{T} = \{T_1, T_2, T_3, T_4\}$. We have: $count_{\mathcal{T}}(\{a, b, c\}) = 2$, $count_{\mathcal{T}}(\{d, e\}) = 1$ and $count_{\mathcal{T}}(\{c, d\}) = 0$.

Now, we formally define the greedy consensus tree problem. Consider a set $T = \{T_1, T_2, \ldots, T_k\}$ of k phylogenetics trees leaf-labeled by L where $|L| = n$. Let $\mathcal{X}$ be a list of clusters that occur in at least one tree in T (i.e. $\mathcal{X} = \bigcup_{i=1}^{k} \mathcal{C}(T_i)$), sorted in decreasing order of $count_T(C)$ for all $C \in \mathcal{X}$ (clusters with the same counts are ordered arbitrarily). Construct a set $\mathcal{Y}$ of clusters as follows: Initialize $\mathcal{Y} = \emptyset$. Then, traverse the list $\mathcal{X}$ and for each cluster C encountered in this order, check if C and C' are pairwise compatible for all $C' \in \mathcal{Y}$; if yes then let $\mathcal{Y} = \mathcal{Y} \cup \{C\}$. A *greedy consensus tree* of T is a tree T such that $\Lambda(T) = L$ and $\mathcal{C}(T) = \mathcal{Y}$.

Note that greedy consensus tree is not unique. The reason is that when two clusters C and C' have the same count, their ordering in the sorted list will affect the final greedy consensus tree we obtained. Figure 1 gives an example.

Below, we state a few technical lemmas that are useful in our paper.

We first define strict consensus tree. Let T_1 and T_2 be two trees with $\Lambda(T_1) = \Lambda(T_2) = L$. The strict consensus tree, denoted as $strict(T_1, T_2)$, of T_1 and T_2 is a tree T such that $\mathcal{C}(T) = \mathcal{C}(T_1) \cap \mathcal{C}(T_2)$. Note that strict consensus tree always exists and is unique.

Lemma 1 ([4]). *Let T_1 and T_2 be two trees with $\Lambda(T_1) = \Lambda(T_2) = L$ and $|L| = n$. $strict(T_1, T_2)$ can be computed in $O(n)$ time.*

Next, we define asymmetric median tree. Let $T = \{T_1, T_2, \ldots, T_k\}$, where $\Lambda(T_1) = \ldots = \Lambda(T_k) = L$ and $|L| = n$. The asymmetric median tree of T, denoted as $median(T)$, is a tree T that maximizes $\sum_{i=1}^{k} |strict(T, T_i)|$. Note that asymmetric median tree always exists, but it is not unique. Below lemma give a polynomial time algorithm to find $median(T_1, T_2)$.

Lemma 2 ([15]). *Let T_1 and T_2 be two trees with $\Lambda(T_1) = \Lambda(T_2) = L$ and $|L| = n$. $median(T_1, T_2)$ can be computed in $O(n^{1.5} \log^3 n)$ time.*

Below lemma gives an approximation solution of $median(T)$ when $k \geq 3$.

Lemma 3 ([14]). *Let $T = \{T_1, T_2, \ldots, T_k\}$, where $\Lambda(T_1) = \ldots = \Lambda(T_k) = L$ and $|L| = n$. Using $O(k^2 n^{2.5})$ time, we can find a tree T such that $\mathcal{C}(T) \subseteq \bigcup_{i=1}^{k} \mathcal{C}(T_i)$ and the size of T is at least $\frac{2}{k}$ of the size of the asymmetric median tree of T.*

Next, we define majority-rule consensus tree. Let $T = \{T_1, T_2, \ldots, T_k\}$. The majority-rule consensus tree of T, denoted as $majority(T)$ is a tree T such that $\mathcal{C}(T) = \{C \in \mathcal{C}(T_i) \mid count_T(C) \geq \lfloor k/2 \rfloor + 1\}$. Note that majority-rule consensus tree always exists and is unique.

Lemma 4 ([10]). *Let $T = \{T_1, T_2, \ldots, T_k\}$ be a set of k trees, each leaf-labeled by L, $n = |L|$. $majority(T)$ can be computed in $O(kn)$ time.*

Let T_1 and T_2 be two trees with $\Lambda(T_1) = \Lambda(T_2) = L$. One-Way_Compatible($T_1, T_2$) reports the tree formed by all clusters in T_1 that are compatible to T_2. Precisely, One-Way_Compatible(T_1, T_2) reports a tree

T such that $\mathcal{C}(T) = \{C \in \mathcal{C}(T_1) | C$ is compatible with $T_2\}$. Note that One-Way_Compatible(T_1, T_2) may not equal One-Way_Compatible(T_2, T_1). Figure 2 gives an example to illustrate one-way compatible. We have the following lemma.

Lemma 5 ([10]). *Let T_1 and T_2 be two trees with $\Lambda(T_1) = \Lambda(T_2) = L$. One-Way_Compatible$(T_1, T_2)$ can be computed in $O(n)$ time, where $n = |L|$.*

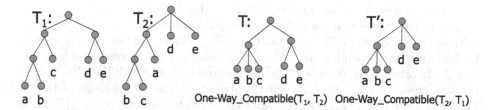

Fig. 2. T_1 and T_2 are two incompatible trees (i.e. some clusters in T_1 are not compatible to T_2, and vice versa). The rightmost two trees are One-Way_Compatible(T_1, T_2) and One-Way_Compatible(T_2, T_1), respectively. Observe that these two trees are different.

Let T_1 and T_2 be two trees with $\Lambda(T_1) = \Lambda(T_2) = L$. Suppose T_1 and T_2 are compatible. $merge(T_1, T_2)$ reports the tree T that contains all clusters in $\mathcal{C}(T_1) \cup \mathcal{C}(T_2)$. Figure 3 gives an example to illustrate the merge operation. We have the following lemma.

Lemma 6 ([10]). *Let T_1 and T_2 be two trees with $\Lambda(T_1) = \Lambda(T_2) = L$. Suppose T_1 and T_2 are compatible. $merge(T_1, T_2)$ can be computed in $O(n)$ time, where $n = |L|$.*

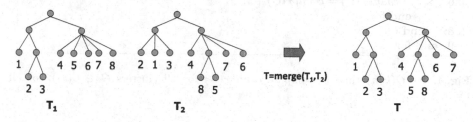

Fig. 3. T_1 and T_2 are two compatible trees (i.e. all clusters in T_1 are compatible to T_2, and vice versa). $T = merge(T_1, T_2)$.

3 $O(k^2 n)$ Time Algorithm

This section presents an $O(k^2 n)$-time algorithm to compute a greedy consensus tree for $\mathcal{T} = \{T_1, \ldots, T_k\}$ where $\Lambda(T_i) = L$ and $|L| = n$.

Our solution has two phases. Phase 1 computes $count_{\mathcal{T}}(C)$, the number of trees containing the cluster C, for every cluster $C \in \bigcup_{i=1}^{k} \mathcal{C}(T_i)$. Then, Phase 2 builds a greedy consensus tree of $\mathcal{T}$.

Now, we discuss Phase 1. Phase 1 applies Day's algorithm (see Lemma 1) to compute $count_{\mathcal{T}}(C)$ for all clusters C in $\mathcal{C}(T_i)$, for $i = 1, 2, \ldots, k$.

For a fixed i, we first show how to compute $count_{\mathcal{T}}(C)$ for all $C \in \mathcal{C}(T_i)$. Initially, we set $count_{\mathcal{T}}(C) = 0$ for all $C \in \mathcal{C}(T_i)$. The algorithm iterates k steps. In the t-th step ($t \le k$), for every $C \in \mathcal{C}(T_i)$, $count_{\mathcal{T}}(C)$ is incremented by 1 if C is in T_t. In other word, $count_{\mathcal{T}}(C)$ is incremented by 1 if $C \in \mathcal{C}(T_i) \cap \mathcal{C}(T_t)$. After the t-th step, we have the invariant that $count_{\mathcal{T}}(C)$ equals the size of the set $\{T_j \mid 1 \le j \le t, C \in \mathcal{C}(T_j)\}$. Hence, after the k-th step, $count_{\mathcal{T}}(C) = |\{T_j \mid C \in \mathcal{C}(T_j), j = 1, \ldots, k\}|$. The t-th step needs to compute $\mathcal{C}(T_i) \cap \mathcal{C}(T_t)$ by Lemma 1, which takes $O(n)$ time. Hence, all k steps take $O(kn)$ time.

We perform the above steps for all $i = 1, \ldots, k$. Hence, we can count the occurrences of every cluster in $\bigcup_{i=1}^{k} \mathcal{C}(T_i)$ using $O(k^2 n)$ time. Figure 4 gives the detail of this algorithm.

Algorithm CountCluster($\mathcal{T}$)
Require: A set of k trees $\mathcal{T} = \{T_1, T_2, \ldots, T_k\}$ where $\Lambda(T_i) = L$
Ensure: Compute $Count(T_i[u])$ for every $u \in V(T_i)$, for every $i = 1, \ldots, k$.
1: **for** $i = 1$ to k **do**
2:     Set $count_{\mathcal{T}}(T_i[u]) = 0$ for every $u \in V(T_i)$;
3:     **for** $j = 1$ to k **do**
4:         By Day's algorithm (Lemma 1), we find a subset $Z = \mathcal{C}(T_j) \cap \mathcal{C}(T_i)$;
5:         **for** every $C \in Z$ **do**
6:             $count_{\mathcal{T}}(C) = count_{\mathcal{T}}(C) + 1$;
7:         **end for**
8:     **end for**
9: **end for**

Fig. 4. An $O(k^2 n)$ time to compute $count_{\mathcal{T}}(C)$ for all clusters $C \in \mathcal{C}(T_i)$, for all $i = 1, \ldots, k$.

The second phase builds a greedy consensus tree of $\mathcal{T}$. We need a few lemmas. For any $i = 1, \ldots, k$ and any $s = 1, \ldots, k$, let $T_i^{(s)}$ be the tree formed by the set of non-trival clusters $\{C \in \mathcal{C}(T_i) \mid count_{\mathcal{T}}(C) = s\}$. Also, let $T_i^{(s..k)}$ be the tree formed by the set of non-trival clusters $\{C \in \mathcal{C}(T_i) \mid k \ge count_{\mathcal{T}}(C) \ge s\}$.

Let $T^{(s..k)}$ be a greedy consensus tree of $T_1^{(s..k)}, \ldots, T_k^{(s..k)}$. Our aim is to build $T^{(1..k)}$, which is a greedy consensus tree of $T_1, \ldots, T_k$. We have the following lemma that let us build $T^{(s..k)}$ from $T^{(s+1..k)}$.

Lemma 7. $T^{(s..k)}$ *is a tree formed by (1) all clusters in* $T^{(s+1..k)}$ *and (2) a maximal set of clusters in* $\bigcup_{i=1}^{k} \mathcal{C}(T_i^{(s)})$ *that are compatible with* $T^{(s+1..k)}$.

Proof. Recall that $T^{(s..k)}$ is a greedy consensus tree of $T_1^{(s..k)}, \ldots, T_k^{(s..k)}$. To build $T^{(s..k)}$, we include clusters C one by one (if compatible) in decreasing order of $count_T(C)$. After we process all clusters C with $count_T(C) \geq s + 1$, we obtain a greedy consensus tree for $T_1^{(s+1..k)}, \ldots, T_k^{(s+1..k)}$, which is $T^{(s+1..k)}$. Then, we need to include maximal number of clusters C with $count_T(C) = s$ that are compatible with $T^{(s+1..k)}$. This means that we need to include a maximal set of clusters in $\bigcup_{i=1}^{k} \mathcal{C}(T_i^{(s)})$ that are compatible with $T^{(s+1..k)}$. The lemma follows. $\qquad\square$

Lemma 7 gives an idea to build $T^{(1..k)}$. The algorithm first initalizes the tree $T^{(k+1..k)}$ as a star tree, which has no non-trival cluster. Then, the algorithm iteratively builds $T^{(s..k)}$ for s from k down to 1. In the s-th iteration, the algorithm builds $T^{(s..k)}$ from $T^{(s+1..k)}$ by including a maximal subset of clusters in $\bigcup_{i=1}^{k} \mathcal{C}(T_i^{(s)})$ that are compatible with $T^{(s+1..k)}$. Below lemma shows a way to find such a maximal subset.

Lemma 8. *Given* $T^{(s+1..k)}$, $T_1^{(s)}, \ldots, T_k^{(s)}$, *below algorithm computes* $T^{(s..k)}$ *in* $O(kn)$ *time.*

1: $T^{(s..k)} = T^{(s+1..k)}$;
2: **for** $i = 1$ to k **do**
3:    $T^{(s..k)} = merge(T^{(s..k)}, One\text{-}Way\_Compatible(T_i^{(s)}, T^{(s..k)}))$;
4: **end for**

Proof. Recall that $T_i^{(s)}$ is the tree formed by all non-trivial clusters C in T_i with $count_T(C) \neq s$. To build $T^{(s..k)}$, we first initialize $T^{(s..k)} = T^{(s+1..k)}$. Then, for $i = 1, \ldots, k$, we try to include maximal subset of clusters in $\mathcal{C}(T_i^{(s)})$ which are compatible with $T^{(s..k)}$ into $T^{(s..k)}$. The maximal subset of clusters can be computed by One-Way_Compatible$(T_i^{(s)}, T^{(s..k)})$. To include this maximal subset into $T^{(s..k)}$, we set $T^{(s..k)} = merge(T^{(s..k)}, One\text{-}Way\_Compatible(T_i^{(s)}, T^{(s..k)}))$. Hence, the algorithm correctly constructs $T^{(s..k)}$.

For running time, by Lemmas 6 and 5, merge$(T^{(s..k)}, One\text{-}Way\_Compatible$ $(T_i^{(s)}, T^{(s..k)}))$ can be computed in $O(n)$ time. Since we need to run this step k times, the total running time is $O(kn)$. $\qquad\square$

The algorithm FastGreedy is presented in Fig. 5. Step 1 runs Count Cluster(T) to obtain $count_T(C)$ for all clusters C in T_i, for all $i = 1, 2, \ldots, k$. This step takes $O(k^2n)$ time. Then, Steps 2 to 6 builds $T_i^{(s)}$ for $i = 1, \ldots, k$ and $s = 1, \ldots, k$. This step also takes $O(k^2n)$ time. Steps 7 to 14 iteratively build $T^{(s..k)}$ for s from k down to 1 by Lemma 8. Each iteration takes $O(kn)$ time. We can build a greedy consensus tree $T^{(1..k)}$ after k iterations, which takes $O(k^2n)$ time in total.

Algorithm FastGreedy($\mathcal{T}$)
Require: A set of k trees $\mathcal{T} = \{T_1, T_2, \ldots, T_k\}$ where $\Lambda(T_i) = L$
Ensure: A greedy consensus tree of $T_1, \ldots, T_k$.
1: CountCluster($\mathcal{T}$);
2: **for** $i = 1$ to k **do**
3:     **for** $s = 1$ to k **do**
4:         Set $T_i^{(s)}$ be the tree formed by the set of non-trival clusters $\{C \in \mathcal{C}(T_i) \mid count_T(C) = s\}$;
5:     **end for**
6: **end for**
7: Let $T^{(k+1..k)}$ be a star tree leaf-labeled by L;
8: **for** $s = k$ downto 1 **do**
9:     $\{/*$ Build $T^{(s..k)}$ from $T^{(s+1..k)}$, $T_1^{(s)}, \ldots, T_k^{(s)}$ $*/\}$
10:     $T^{(s..k)} = T^{(s+1..k)}$;
11:     **for** $i = 1$ to k **do**
12:         $T^{(s..k)} = \text{merge}(T^{(s..k)}, \text{One-Way\_Compatible}(T_i^{(s)}, T^{(s..k)}))$;
13:     **end for**
14: **end for**
15: Return $T^{(1..k)}$;

Fig. 5. An $O(k^2 n)$-time algorithm for computing a greedy consensus tree.

4 Maximum Greedy Consensus Tree Problem Is NP-hard and Difficult to Approximate

This section gives a reduction from the maximum independent set problem to the maximum greedy consensus tree problem. Based on the reduction, we give a NP-hardness result and an inapproximability result.

We first define the maximum independent set problem. Given an undirected graph $G = (V, E)$, an independet set is a subset $V' \subseteq V$ such that there is no edge connecting any two nodes in V'. A maximal independent set V' of G is an independent set which is not a proper subset of any other independent set of G. A maximum independent set of G is an independent set of G of maximum size.

Next, we define k-partite graph. An undirected graph $G = (V, E)$ is called a k-partite graph if V can be partitioned into k subsets, say, $\{V_1, \ldots, V_k\}$, such that no edge in E connecting two vertices from the same subset. The maximum k-partite graph independent set problem is the maximum independent set problem for k-partite graph. The maximum k-partite graph independent set problem is known to be NP-hard for $k \geq 3$ [14].

For every node $u \in V$, we define E_u be the set of edges adjacent to u and C_u be $\{u\} \cup E_u$. For $i = 1, 2, \ldots, k$, for all $u \in V_i$, C_u satisfies the following property.

Lemma 9. *For $i = 1, 2, \ldots, k$, we have $C_u \cap C_v = \emptyset$ for any two nodes $u, v \in V_i$.*

Proof. For any two nodes $u, v \in V_i$, $C_u = \{u\} \cup E_u$ and $C_v = \{v\} \cup E_v$. Since $u \neq v$, we have $E_u \cap E_v = \emptyset$. Hence, the lemma follows. $\qquad\square$

Here, we show that, for $k \geq 3$, the maximum greedy consensus tree problem for k trees is NP-hard by giving a reduction from the maximum k-partite graph problem.

Given a k-partite graph $G = (V, E)$ where V is partitioned into k subsets $\{V_1, \ldots, V_k\}$, we transform G into a set of k trees $\mathcal{T}_G = \{T_1, \ldots, T_k\}$ where each tree T_i is leaf-labeled by $V \cup E$ as follows.

For $i = 1, 2, \ldots, k$, T_i is the tree formed by a set of non-trival clusters $C_u = \{u\} \cup E_u$ for all $u \in V_i$.

By Lemma 9, all T_i are valid trees since C_u for all $u \in V_i$ are pairwise compatible. Hence the above transformation is valid. Furthermore, the above transformation takes $O(|E|)$ time.

Below lemma gives the properties of $\mathcal{T}_G$.

Lemma 10. *Given an undirected graph $G = (V, E)$, the trees in $\mathcal{T}_G$ satisfy the following properties.*

(a) *The number of leaf labels is $|L| = |V| + |E|$.*
(b) *The total number of non-trival clusters in all trees is $|V|$*
(c) *For any $u \in V$, $u \in C_u$ and $v \notin C_u$ for $v \in V - \{u\}$*
(d) *For any distinct $u, v \in V$, $(u, v) \in E$ if and only if $C_u \cap C_v \neq \emptyset$.*
(e) *For any $(u, v) \in E$, C_u and C_v are incompatible.*

Proof. For (a), since the set of leaf labels is $L = V \cup E$. Hence, the number of leaf labels is $|V| + |E|$. For (b), we have a cluster C_u for every $u \in V$. Hence, the number of clusters is $|V|$.

For (c), by definition, for any $u \in V$, C_u only contains one node from V, which is u. Hence, property (c) follows.

For if-statement of (d), if $(u, v) \in E$, then $e \in C_u$ and $c \in C_v$, hence, $C_u \cap C_v \neq \emptyset$. For only-if-statement of (d), we shows the contra-positive, if $(u, v) \notin E$, the edges attached to u and the edges attach to v are different. This implies that $C_u \cap C_v = \emptyset$.

For (e), since $e = (u, v) \in E$, we have (1) $e \in C_u \cap C_v$, (2) $u \in C_u - C_v$ and (3) $v \in C_v - C_u$. This implies C_u and C_v are incompatible.   □

Below lemma relates a maximal independent set of G and a greedy consensus tree of $\mathcal{T}_G$.

Lemma 11. *For any subset $V' \subseteq V$, let $T_{V'}$ be a tree whose set of non-trival clusters is $\{C_u \mid u \in V'\}$. We have: V' is a maximal independent set of G if and only if $T_{V'}$ is a greedy consensus tree of $\mathcal{T}_G$.*

Proof. ($\rightarrow$) By the definition of $\mathcal{T}_G$ and Lemma 10(c), we have every cluster appears in exactly one tree in $\mathcal{T}_G$. Hence, $T_{V'}$ is a greedy consensus tree of $\mathcal{T}_G$ if (1) C_u and C_v are compatible for all $u, v \in V'$ and (2) C_w is not compatible with $T_{V'}$ for $w \in V - V'$. Below, we show that both (1) and (2) are true.

For (1), since V' is an independent set of G, there is no edge connect u and v for all $u, v \in V'$. By Lemma 10(d), $C_u \cap C_v = \emptyset$ for all $u, v \in V'$. This implies that C_u and C_v are pairwise compatible for all $u, v \in V'$.

For (2), since V' is a maximal independent set of G, for every $w \in V - V'$, there exists some $u \in V'$ such that $(w, u) \in E$; then, by Lemma 10(e), C_w and C_u are not compatible.

($\leftarrow$) Since $T_{V'}$ is a greedy consensus tree, V' is a maximal set such that C_u and C_v are pairwise compatible for every $u, v \in V'$. By Lemma 10(c, d), we have $(u, v) \notin E$ for every $u, v \in V'$. Hence, V' is an independent set in V'.

We claim that, V' is a maximal independent set of G. By contrary, there exists some $w \in V - V'$ such that $(w, u) \notin E$ for all $u \in V'$. Lemma 10(c) implies that C_w and C_u are compatible for all $u \in V'$. Then, C_u and C_v are pairwise compatible for every $u, v \in V' \cup \{w\}$. This contradict with the fact that $T_{V'}$ is a greedy consensus tree. We arrived at contradiction. $\square$

By the above lemma, we can show that the maximum greedy consensus tree problem is NP-hard.

Lemma 12. *Consider a set of tress $T = \{T_1, \ldots, T_k\}$. When $k \geq 3$, the maximum greedy consensus tree problem for T is NP-hard.*

Proof. Note that the maximum k-partite graph independent set problem is NP-hard. Since $G = (V, E)$ is an undirected k-partite graph, V can be partitioned into k subsets $\{V_1, \ldots, V_k\}$ such that every edge in E connecting two vertices from the two different subsets. We can transform G into a set of k trees T_G where each tree in T_G is leaf-labeled by $V \cup E$. The transformation takes $O(|E|)$ time.

By Lemma 11, we know that V' is a maximum independent set of G if and only if $T_{V'}$ is a maximum greedy consensus tree of T_G.

Since maximum k-partite independent set problem is NP-hard, we conclude that the maximum greedy consensus tree problem for k trees is also NP-hard. $\square$

Below lemma gives the inapproximability result.

Lemma 13. *If $P \neq NP$, there is no polynomial time algorithm that guarantee to construct a greedy consensus tree whose size is at least $25/26$ of the maximum greedy consensus tree.*

Proof. Will be given in the full paper. $\square$

5   Finding Maximum Greedy Consensus Tree of Two Trees

Consider two trees T_1 and T_2, both leaf-labeled by L, $n = |L|$. Below, we show that the maximum greedy consensus tree of T_1 and T_2 is the same as the asymmetric median tree of T_1 and T_2.

Lemma 14. *The maximum greedy consensus tree of T_1 and T_2 is the same as the asymmetric median tree of T_1 and T_2.*

Proof. Let T be an asymmetric median tree of T_1 and T_2.

To show that T is also a maximum greedy consensus tree, T needs to satisfy the following conditions:

(1) T contains all clusters in strict(T_1, T_2), i.e., $Q \subseteq \mathcal{C}(T)$ where $Q = \mathcal{C}(T_1) \cap \mathcal{C}(T_2)$.

(2) Let S be the set of all clusters that appear in either T_1 or T_2, i.e., $S = \mathcal{C}(T_1) \cup \mathcal{C}(T_2) - Q$. $\mathcal{C}(T) - Q$ is a maximum subset of S such that all clusters in $\mathcal{C}(T) - Q$ are pairwise compatible.

For (1), we show that $Q \subseteq \mathcal{C}(T)$. By contrary, suppose there exists $C \in Q$ that is not a cluster in $\mathcal{C}(T)$. Since C is a cluster in both T_1 and T_2, we have C is compatible with all clusters in $\mathcal{C}(T)$. Then, let T' be a tree such that $\mathcal{C}(T') = \{C\} \cup \mathcal{C}(T)$. We have $|strict(T', T_1)| + |strict(T', T_2)| = (1 + |strict(T, T_1)|) + (1 + |strict(T, T_2)|) > |strict(T, T_1)| + |strict(T, T_2)|$. This means that T is not an asymmetric median tree of T_1 and T_2. We arrived at contradiction.

For (2), by contrary, suppose there exists another subset $S' \subseteq S$ such that $|S'| > |\mathcal{C}(T) - Q|$ and all clusters in S' are pairwise compatible. Let T' be a tree such that $\mathcal{C}(T') = S' \cup Q$. Then, $|strict(T', T_1)| + |strict(T', T_2)| = |S'| + 2|Q| > |\mathcal{C}(T) - Q| + 2|Q| = |strict(T, T_1)| + |strict(T, T_2)|$. This means that T is not an asymmetric median tree of T_1 and T_2. We arrived at contradiction.

Hence, the lemma follows. $\square$

By the above lemma, we have:

Lemma 15. *The maximum greedy consensus tree of T_1 and T_2 can be computed using $O(n^{1.5} \log^3 n)$ time.*

Proof. By Lemma 14, computing a maximum greedy consensus tree of T_1 and T_2 is the same as computing an asymmetric median tree of T_1 and T_2.

By Lemma 2, we can compute an asymmetric median tree of T_1 and T_2 using $O(n^{1.5} \log^3 n)$ time. The lemma follows. $\square$

6   Approximation Algorithms for Finding Maximum Greedy Consensus Tree of Three Trees

As it is NP-hard to reconstruct maximum greedy consensus tree for k trees where $k \geq 3$, this section gives an approximation algorithm for reconstructing maximumm greedy consensus tree when $k = 3$. Due to space limit, we just summarize the result. The detail will be given in full paper.

Lemma 16. *Consider three trees T_1, T_2 and T_3 leaf-labeled by L, where $n = |L|$. Using $O(n^{1.5} \log^3 n)$ time, we can compute a greedy consensus tree for T_1, T_2, T_3 whose size is at least $\frac{2}{3}$ of that of a maximum greedy consensus tree for T_1, T_2, T_3.*

References

1. Adams III, E.N.: Consensus techniques and the comparison of taxonomic trees. Syst. Biol. **21**(4), 390–397 (1972)
2. Bremer, K.: Combinable component consensus. Cladistics **6**(4), 369–372 (1990)
3. Bryantm, D.: A classification of consensus methods for phylogenetics. In: Janowitz, M.F., Lapointe, F.-J., McMorris, F.R., Mirkin, B., Roberts, F.S. (eds.) Bioconsensus. DIMACS Series in Discrete Mathematics and Theoretical Computer Science, vol. 61, pp. 163–184. American Mathematical Society (2003)
4. Day, W.H.E.: Optimal algorithms for comparing trees with labeled leaves. J. Classif. **2**, 7–28 (1985)
5. Felsenstein, J.: Inferring Phylogenies. Sinauer Associates, Inc., Sunderland (2004)
6. Felsenstein, J.: PHYLIP, version 3.6. Software package, Department of Genome Sciences, University of Washington, Seattle, USA (2005)
7. Gawrychowski, P., Landau, G.M., Sung, W.-K., Weimann, O.: A faster construction of greedy consensus trees. In: 45th International Colloquium on Automata, Languages, and Programming, ICALP 2018, Prague, Czech Republic, pp. 63:1–63:14, 9–13 July 2018
8. Jansson, J., Rajaby, R., Shen, C., Sung, W.-K.: Algorithms for the majority rule (+) consensus tree and the frequency difference consensus tree. IEEE/ACM Trans. Comput. Biol. Bioinform. September 2016
9. Jansson, J., Shen, C., Sung, W.-K.: Fast algorithms for consensus trees (FACT) (2013). http://compbio.ddns.comp.nus.edu.sg/~consensus.tree
10. Jansson, J., Shen, C., Sung, W.-K.: Improved algorithms for constructing consensus trees. J. ACM **63**(3), 1–24 (2016)
11. Jansson, J., Sung, W.-K., Vu, H., Yiu, S.-M.: Faster algorithms for computing the R* consensus tree. In: Ahn, H.-K., Shin, C.-S. (eds.) ISAAC 2014. LNCS, vol. 8889, pp. 414–425. Springer, Cham (2014). https://doi.org/10.1007/978-3-319-13075-0_33
12. Margush, T., McMorris, F.R.: Consensus n-trees. Bull. Math. Biol. **43**(2), 239–244 (1981)
13. Phillips, C., Warnow, T.J.: The asymmetric median tree — a new model for building consensus trees. Discrete Appl. Math. **71**(1–3), 311–335 (1996)
14. Phillips, C., Warnow, T.J.: The asymmetric median tree — a new model for building consensus trees. In: Hirschberg, D., Myers, G. (eds.) CPM 1996. LNCS, vol. 1075, pp. 234–252. Springer, Heidelberg (1996). https://doi.org/10.1007/3-540-61258-0_18
15. Rajaby, R., Sung, W.-K.: Computing asymmetric median tree of two trees via better bipartite matching algorithm. In: Brankovic, L., Ryan, J., Smyth, W.F. (eds.) IWOCA 2017. LNCS, vol. 10765, pp. 356–367. Springer, Cham (2018). https://doi.org/10.1007/978-3-319-78825-8_29
16. Ronquist, F., Huelsenbeck, J.P.: MrBayes 3: Bayesian phylogenetic inference under mixed models. Bioinformatics **19**(12), 1572–1574 (2003)
17. Sokal, R.R., Rohlf, F.J.: Taxonomic congruence in the Leptopodomorpha reexamined. Syst. Zool. **30**(3), 309–325 (1981)
18. Sung, W.-K.: Algorithms in Bioinformatics: A Practical Introduction. Chapman & Hall/CRC, Boca Raton (2010)
19. Swofford, D.L.: PAUP*, Version 4.0. Software Package. Sinauer Associates Inc., Sunderland (2003)

A Two Query Adaptive Bitprobe Scheme Storing Five Elements

Mirza Galib Anwarul Husain Baig$^{(\boxtimes)}$, Deepanjan Kesh, and Chirag Sodani

Indian Institute of Technology Guwahati, Guwahati, India
{mirza.baig,deepkesh,chirag.sodani}@iitg.ac.in

Abstract. We are studying the adaptive *bitprobe* model to store an arbitrary subset S of size at most five from a universe of size m and answer the membership queries of the form "Is x in S?" in two *bitprobes*. In this paper, we present a data structure for the aforementioned problem. Our data structure takes $\mathcal{O}(m^{10/11})$ space. This result improves the non-explicit result by Garg and Radhakrishnan [6] which takes $\mathcal{O}(m^{20/21})$ space, and the explicit result by Garg [5] which takes $\mathcal{O}(m^{18/19})$ space for the aforementioned set and query sizes.

Keywords: Set membership problem · Bitprobe model
Data structures

1 Introduction

In the static membership problem, we are g iven a universe of size m and our task is to design a data structure that can store arbitrary subset S of size n such that the membership queries of the form "Is x in S?" can be answered correctly. We study this problem in the *bitprobe* model of computation. The complexity of the static membership problem in this model is measured in terms of the size of the data structure denoted by s, and the number of bits of the data structure accessed denoted by t. It is the later of the two properties which lend its name *bitprobe* model. In this model all other operations are free. Solutions to the above mentioned problems in this model are termed as schemes. Each scheme consists of two parts, one is the storage scheme, and the other is query scheme. Storage scheme maps an arbitrary subset of cardinality n from a universe of size m given to be stored to the s bits of the data structure. Query scheme maps every element belonging to the universe m to the t locations of the data structure and it decides the membership of the query element by reading those t locations. The storage and query scheme together gives a (n, m, s, t)-scheme which stores a set of size at most n from a universe of size m and uses s bits in such a way that membership query can be answered in t probes. This is a well studied problem over several decades and it has been discussed in [1,4,6,7,11–14].

A (n, m, s, t)-scheme is said to be adaptive if the location of the probes depends upon the bit returned by the prior probes. Whereas in a non-adaptive scheme location of the probes are fixed and it does not depend upon the bit returned by prior probes.

© Springer Nature Switzerland AG 2019
G. K. Das et al. (Eds.): WALCOM 2019, LNCS 11355, pp. 317–328, 2019.
https://doi.org/10.1007/978-3-030-10564-8_25

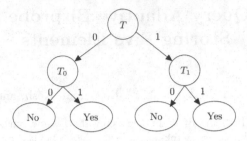

Fig. 1. A decision tree for the two adaptive *bitprobe* model

1.1 Two Adaptive *Bitprobe* Model

In this section we will discuss a two adaptive *bitprobe* model in the context of two adaptive *bitprobe* scheme. A two adaptive *bitprobe* scheme in this model consist of three tables namely T, T_0 and T_1 as shown in Fig. 1. Furthermore, as discussed earlier the data structures in this model consist of two schemes a storage scheme and a query scheme. Storage scheme maps an arbitrary subset given to be stored to the three tables mentioned earlier. Query scheme decides the membership of a query element by probing two location of the data structure. Given a query element, the first probe is made into the table T. The next query depends upon whether the bit returned by the table T is zero or one. If the bit returned by the table T is zero it makes next query to the table T_0 otherwise to the table T_1. We say that a query element is part of the set if and only if the last query returns one.

1.2 The Problem Statement

In this paper, we are dealing with the design of explicit adaptive scheme in the *bitprobe* model to store an arbitrary subset of size at most five from a universe of size m and answer the membership query in two adaptive bit probes. In other words our objective is to design an adaptive $(5, m, s, 2)$-scheme in the *bitprobe* model.

1.3 Previous Results

As we are going to study a two adaptive *bitprobe* scheme, let us discuss some existing results in the context of this problem. For the set of size one ($n = 1$), there exist a trivial scheme which takes $\mathcal{O}(m^{1/2})$ space. The space requirement for this scheme matches the lower bound of $\Omega(m^{1/2})$ [4]. For the set of size two ($n = 2$), Radhakrishnan et al. [13] came up with a scheme which takes $\mathcal{O}(m^{2/3})$ space. Radhakrishnan et al. [14] conjectured that this scheme is asymptotically tight but it has not been resolved yet. For the set of size three ($n = 3$), Baig and Kesh [2] came up with a scheme which takes $\mathcal{O}(m^{2/3})$ space. This scheme has been proved asymptotically tight by Kesh [9].

Moreover, for the set of size four ($n = 4$), Baig et al. [10] have given a scheme which takes $\mathcal{O}(m^{5/6})$ space. This scheme improves upon the non-explicit $(n, m, c \cdot m^{1 - \frac{1}{4n+1}}, 2)$-scheme by Garg and Radhakrishnan [6] for the set of size four ($n = 4$). For the given set size their scheme takes $\mathcal{O}(m^{16/17})$ space. Our scheme also improves upon the explicit $(n, m, c \cdot m^{1 - \frac{1}{4n-1}}, 2)$-scheme given by Garg [5] for the set of size four ($n = 4$). His scheme takes $\mathcal{O}(m^{14/15})$ space for the given set size.

In this paper, we have come up with a scheme for the set of size five ($n = 5$). Our scheme takes $\mathcal{O}(m^{10/11})$ space. This scheme improves upon the non-explicit scheme by Garg and Radhakrishnan [6] for the set of size five ($n = 5$). For the set of size five ($n = 5$) their scheme takes $\mathcal{O}(m^{20/21})$ space. Our scheme also improves upon the explicit scheme given by Garg [5] for the set of size five. ($n = 5$). His scheme takes $\mathcal{O}(m^{18/19})$ space for the set of size ($n = 5$).

2   The Approach to the Problem

Our scheme has borrowed the idea of the geometric arrangement of elements on a three-dimensional cube from Kesh [8]. Kesh in his paper used the idea of geometric arrangements of elements on high dimensional cubes to come up with $(2, m, c \cdot m^{1/(t-2^{-1})}, t)$-scheme for $t \geq 2$. We have also used the idea of dividing the universe into blocks and superblocks from Radhakrishnan et al. [13]. The combination of the aforesaid ideas was used by Baig and Kesh [2] to come up with a tight explicit adaptive scheme for $n = 3$. Baig et al. [10] used the similar idea to map elements on a square grid to come

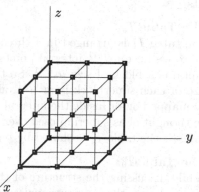

Fig. 2. Blocks of superblocks placed on the integral points of a cube

up with an improved scheme for $n = 4$. We in this paper use this geometrical technique to map the blocks of elements from superblocks to the integral point of a three-dimensional cube of suitable size as shown in Fig. 2.

We divide the universe of size m into blocks and superblocks similar to the Radhakrishnan et al. [13]. For the $(5, m, \mathcal{O}(m^{10/11}), 2)$-scheme we divide the universe into blocks of size $m^{1/11}$, so we will have $m^{10/11}$ blocks. We then merge the $m^{9/11}$ consecutive blocks to form a superblock of size $m^{10/11}$. So we will have $m^{1/11}$ superblocks of size $m^{10/11}$.

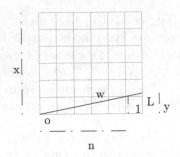

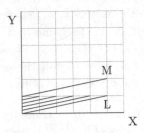

Fig. 3. A line with slope $1/n$ in the bottom most layer of the cube

Fig. 4. Figure showing number of lines drawn between two same slope lines

The Table T
This table consists of one bit of space for each block. Therefore its size is $m^{10/11}$ bits.

The Table T_1
The table T_1 is arranged in a three-dimensional cube of side $m^{3/11}$. So in this cube, we have $m^{9/11}$ integral points. Each integral point on or inside the cube contains a block of size $m^{1/11}$. So the size of the table T_1 is $m^{10/11}$. Since the size of each superblock is $m^{10/11}$, all the blocks belonging to a superblock can be mapped on or inside the integral point of the cube. All other superblocks can be thought of as superimposed over each other in the cube. So each point in the cube or the table T_1 is shared by blocks of $m^{1/11}$ superblocks.

The Table T_0
While discussing the structure of the table T_1, we saw that each superblock is mapped on a three-dimensional cube in such a way that all of them are superimposed. Now, for the nth superblock, we first draw a family of lines in the bottom-most layer of the cube in the XY-plane with slope $1/n$ in such a way that all the integral points are covered by the lines.

Lemma 1. *The number of lines passing through all the integral points of a square grid with slope $1/n$ is $2x + (n-1)(x-1) - 1$, where x is the length of the square grid.*

Proof. As shown in Fig. 4, if the slope of the line M and L is $1/n$ then between them there can be only $n-1$ lines of slope $1/n$ passing through integral points of the grid. So the total number of lines that we can draw with slope $1/n$ is x lines from integral points on X-axis, $x-1$ lines from the integral points on Y axis and $(n-1)(x-1)$ lines between lines from the integral points on Y-axis. So we have $2m^{1/2} + (n-1)(m^{1/2} - 1) - 1$ lines with slope $1/n$.

Using Lemma 1, we can say that the total number of lines with slope $1/n$ in the bottom most layer can be $c \cdot nx$, where c is a constant and x is the side of the cube. Now we cut slices of the cube along these lines and perpendicular to the XY plane. So the total number of slices for the nth superblock will be equal to the number of lines drawn in the bottom most layer of the cube for the nth superblock i.e. $c \cdot nx$. All the slices have height equal to length of the cube i.e. x. Let us now calculate the maximum width of a slice of slope $1/n$. Width of the slice formed by line segment OL as shown in Fig. 3 can be calculated to be $x/n\sqrt{1+n^2}$. We can see from Fig. 3 that all other slices with this slope

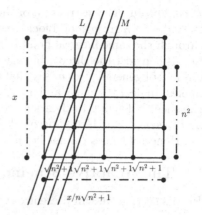

Fig. 5. A slice belonging to nth superblock

will have width less than or equal to $x/n\sqrt{1+n^2}$. So a slice belonging to nth superblock will have length x and width less than equal to $x/n\sqrt{1+n^2}$.

We now draw a family of lines in all the slices of all the superblocks. For the slices belonging to the nth superblock, we draw lines with slope $n^2/\sqrt{1+n^2}$ as shown in Fig. 5. The lines are drawn in such a way that all the integral points are covered. Let us now calculate the maximum number of lines drawn on a slice belonging to the nth superblock. We can see from Fig. 5 that the total number of lines drawn from Z-axis is equal to x. Also, number of integral points on the width of the slice is equal to $x/n\sqrt{1+n^2}/\sqrt{1+n^2} = x/n$. So we can draw $\frac{x}{n}$ lines through those integral points on the width of the slice. Now from Fig. 5, we can see that the number of lines that can pass between two consecutive integral points on the width of the slice is $n^2 - 1$. So the total number of lines drawn on this slice with slope $n^2/\sqrt{1+n^2}$ is equal to $x + x/n + n^2 \cdot x/n$ i.e. $c \cdot nx$. We say that a slice is having slope $1/n$ if it's projection on the XY-plane has slope $1/n$. Now let us bound the total number of lines drawn on slices whose projections on the XY-plane makes slope $1/n$. The total number of lines should be less than the product of the number of lines drawn on a slice of a maximum width of slope $1/n$ and the total number of slices of slope $1/n$. So the total number of lines drawn for the nth superblock is less than $c_1 \cdot n^2 x^2$. We need to sum this for all the superblocks to get the total number of lines drawn. So the total number of lines drawn is

$$\sum_{i=1}^{m^{1/11}} c_1 \cdot i^2 (m^{3/11})^2 \leq c \; m^{9/11}. \tag{1}$$

For each line drawn in a slice, we have a block of space in the table T_0. So the total size of the table T_0 is $c \cdot m^{10/11}$ bits. Now we will prove the following lemma.

Lemma 2. *No three lines passing through an integral point of the cube lies in the same plane.*

Proof. From the construction of the table, we can see that all the lines drawn in the cube for a given superblock are parallel to each other. So the lines which pass through the same integral point of the cube belongs to the different superblocks. Let us consider the three arbitrary superblocks to which our lines belong. Without loss of generality let us say that the projection of these slices on the XY-plane makes angle $1/n_1, 1/n_2$ and $1/n_3$ with the X-axis. Our lines lie completely in the slices belonging to their superblock. While drawing lines in the slices for the first, second and third superblock we are going up in Z direction by n_1^2, n_2^2 and n_3^2. Hence our lines cannot lie in the same plane.

3   The Adaptive Scheme for Five Elements

In this section we will present our $(5, m, \mathcal{O}(m^{10/11}), 2)$-scheme.

3.1   Our Data Structure

As mentioned earlier our data structure has three tables T, T_0 and T_1. In the first table, we have one bit of space for all the blocks. Now, since the block size as mentioned earlier is $m^{1/11}$ therefore size of the table T is $m/m^{1/11} = m^{10/11}$. In the table T_1, we have space for a superblock which can be seen as a three-dimensional cube. As the size of superblock is $m^{10/11}$, so the space of the table T_1 is $m^{10/11}$. We have also mentioned earlier that in the table T_1 all the superblocks are superimposed. The table T_0 has one block of space for all the lines drawn in all the superblocks. From the Eq. 1, we have the total number of lines drawn is $c \cdot m^{9/11}$. So the size of the table T_0 is equal to $cm^{1/11} \cdot m^{9/11} = c \cdot m^{10/11}$. So the size of our data structure is $\mathcal{O}(m^{10/11})$.

From the structure of the tables, we may draw the following conclusion. In general two blocks having elements should not map at the same location in the table T_1 or T_0. Otherwise, we may make mistake on the query belonging to these blocks. Further, if the block having element is mapped in the table T_1 or T_0 then no other block should be sent to that table whose position is matched with the block having an element. So if a block having an element on a line is mapped to the table T_0 then all other blocks lying on that line should be sent to the table T_1. Since for each line we have only one block of space in the table T_0. On the contrary, if the block having an element from a line is sent to the table T_1 then other blocks lying on the line which contains this block can be sent to table T_0 or T_1. The blocks which are not having any elements given to be stored can be mapped at the same location in table T_1 or T_0. Also, in the rest of the paper whenever we say the line passing through a block or the line having block, we always mean the line drawn in the superblock to which the block belongs.

3.2   The Query Scheme

Given a query element, we find the block and superblock to which it belongs. Then as mentioned earlier, we query the first table, if the first table returns zero we query to the table T_0 else we query to the table T_1. We say that element is part of the set to be stored if and only if the last query returns one.

3.3   The Storage Scheme

In this section we talk about the way bits of the tables are set to store the subset of size at most five from a universe of size m. We divide the storage scheme into various cases depending upon the way blocks having elements are distributed on the line belonging to their superblock. To generate the cases, first of all, we partition the number five, then we put those many elements into different superblocks. Further, the positions of the blocks having elements on the line belonging to their superblocks are considered. While handling cases we see the intersections of the lines which contains blocks having element given to be stored. We then decide which block to send to the Table T_0 and which to the T_1. As in our data structure, we always send a block to either table T_0 or T_1 and we store its bit vector there, so we will always assume that elements which are given to be stored lies in the different block. Proving the results for elements belonging to different blocks proves the result when many elements belong to the same block. Keeping in mind the page limit, here we would be discussing a few simple cases and few tricky cases. Also, for the sake of completeness, we have added rest of the cases in the paper [3] in arXiv.

Case 1. If all the elements of S lie in one superblock then we send the blocks having elements to the table T_1 and all the empty blocks to the table T_0.

Case 2. If four elements $S_1 = \{n_1, n_2, n_3, n_4\}$ lie in one superblock and one element $S_2 = \{n_5\}$ in other superblock then we can have two cases, either the block containing the element n_5 coincides with one of the block containing element from S_1 in the table T_1 or it does not coincides. So if the block containing the element n_5 coincides with one of the blocks containing an element from S_1 then we send the block having the element n_5 to the table T_1 and send the block from which it was coinciding to the table T_0. All other blocks of superblock which contain elements from S_1 are sent to the table T_1. All other blocks of superblock which contain the element n_5 are sent to the table T_0. Rest all the empty blocks are sent to the table T_0.

On other hand if the block containing element n_5 do not coincide with any of the block having element from S_1 in the table T_1 then we send all the blocks having elements from S_1 and S_2 to the table T_1 and rest all the empty blocks to the table T_0.

Case 3. If three elements $S_1 = \{n_1, n_2, n_3\}$ lie in one superblock and two elements $S_2 = \{n_4, n_5\}$ in other superblock then we store according to following scheme.

Case 3.1. All the blocks to which elements from S_1 belong lies on the same line of their superblock. From here onwards whenever we say line passing through a block or blocks lying on a line, we mean the line drawn in the superblock to which these blocks belongs.

Case 3.1.1. Two blocks to which elements from S_2 belong coincides with the blocks corresponding to the elements from S_1 in the table T_1. In this case, we send the blocks having elements from S_2 to the table T_0. Also, we send empty blocks lying on the lines to which elements from S_2 belongs to the table T_1. We send all the blocks which contain elements from S_1 in the table T_1. We send the rest of the empty blocks to table T_0.

Case 3.1.2. Only one block which contains an element from S_2 coincides with the block corresponding to the elements from S_1 in the table T_1. In this case, we send all the blocks which contain elements from S_1 to the table T_1. We send the coinciding block of the element from S_2 to the table T_0 and the rest of the blocks which lies on the line containing this block to the table T_1. If after this other nonempty block having an element from S_2 is still unassigned then we send it to the table T_1, and all the empty blocks lying on the line containing this block to the table T_0. Rest all the empty blocks are sent to the table T_0.

Case 3.1.3. None of the blocks which contain an element from S_2 coincides with the block which contains an element from S_1 in the table T_1. In this case, we send all the nonempty blocks to the table T_1 and all the empty blocks to the table T_0.

Case 3.2. Two blocks which contain elements from S_1 lies on the same line and other lies on a different line.

Case 3.2.1. All the blocks which contain an element from S_1 lies in the same slice. From here onward whenever we say blocks belonging to a slice, we mean the slice drawn in a superblock to which these blocks belongs.

Case 3.2.1.1. All the blocks which contain elements from S_2 coincides with blocks which contain elements from S_1 in the table T_1. In this case, we send the blocks which contain elements from S_2 to the table T_0, and the rest of the blocks lying on the lines containing these blocks to the table T_1. All the blocks which contain elements from S_1 are sent to the table T_1. Rest all the empty blocks are sent to the table T_0.

Case 3.2.1.2. Only one block which contains an element from S_2 coincides with the block which contains an element from S_1 in the table T_1. In this case, we send the coinciding block which contains an element from S_2 to the table T_0 and the rest of the block which lies on the line containing this block to table T_1. If after this assignment other block having the element from S_2 is still unassigned then we send the other block which contains the element from S_2 to the table T_1 and the rest of the block which lies on the line containing this block to table T_0. Also, we send all the blocks which contain elements from S_1 to the table T_1. We send the rest of the blocks which do not contain any elements to the table T_0.

Case 3.2.1.3. None of the blocks which contain an element from S_2 coincides with the block which contains an element from S_1 in the table T_1. This case is the same as Case 3.1.3.

Case 3.2.2. Two blocks which contain elements from S_1 lies in a slice and another block which contains an element from S_1 in another slice.

Case 3.2.2.1. All the blocks which contain elements from S_2 coincides with blocks which contain elements from S_1 in the table T_1. If the coinciding blocks lie in the same slices as that of two blocks which contain elements from S_1 then we send both the coinciding blocks which contain the elements from S_2 to the table T_0. Also, we send all the empty blocks lying on the lines which contain these blocks to table T_1. Two blocks which contain the elements from S_1 and lying in the same slice are sent to the table T_1. The remaining block which contains the element is sent to the table T_0 and all the empty blocks lying on the line containing this block to the table T_1. Rest all the empty blocks are sent to the table T_0.

Now consider the case in which the coinciding blocks which contains the elements from S_2 lies in the different slices of the superblock which contains the elements from S_1. In this case, we send the two blocks which contain the elements from S_1 lying in a slice to the table T_1. Rest of the block which contains the element from S_1 lying in the other slice is sent to the table T_0 and the empty blocks which lie on the line containing this block is sent to the table T_1. One of the blocks which contains an element from S_2 and is lying in the slice containing two elements from S_1 is sent to the table T_0 and the rest of the blocks on this line is sent to the table T_1. If after this assignment other block having the element from S_2 is still unassigned then we send it to the table T_1 and the empty blocks on the line containing this block to the table T_0. Rest all the empty blocks are sent to the table T_0.

Case 3.2.2.2. Only one block which contains an element from S_2 coincides with a block which contains an element from S_1 in the table T_1. Let us first consider the case where coinciding block having element form S_2 lies in the slice which contains two blocks having elements from S_1. Without loss of generality let us say that the blocks having elements n_1 and n_2 lies in the same slice and the block having the element n_4 coincide with the block having the element n_1. In this case, we send the block having the element n_4 to the table T_0 and all the blocks on the line containing this block is sent to the table T_1.

If the block which contains an element n_3 lies on the line which contains the block having the element n_4 then we send the block having n_3 to the table T_0 and empty blocks of the line which contains block having n_3 to the table T_1. Now if the block having the element n_5 is still unassigned then we send the block having the element n_5 to the table T_0 and empty blocks on the line containing this block to the table T_1. We send the block having the element n_2 to the table T_0 and all the blocks which lie on the line containing this block the table T_1. Rest all the empty blocks are sent to the table T_0.

Further, let us consider the case where the block which contains the element n_3 does not lie on the line which contains the block having the element n_4. In this case, we send the block having n_3 to the table T_1, and the empty blocks on the line containing this block are sent to the table T_0. Now again if the block having the element n_5 is still unassigned then we send it to the table T_1 and the empty blocks lying on the line containing this block to the table T_0. We send the blocks having element from S_1 to the table T_1 and all the empty blocks lying on the line containing these blocks to the table T_0. Rest all the empty blocks are sent to the table T_0.

Now we are left with the case where coinciding block of S_2 having an element n_4 coincides with the block having the element n_3. In this case, we send the block having the element n_3 to the table T_0 and empty block which lies on the line containing this block to the table T_1. We send the block having the element n_4 to the table T_1 and empty blocks which lies on the line containing this block to the table T_0.

Now we see the position of the block having the element n_5. If the block having the element n_5 lies on the line which contains block having the element n_3 then we send the block having the element n_5 to the table T_0. Also, we send the empty blocks of the line which contains block having the element n_5 to the table T_1. Now, consider the case where the line containing the block having the element n_5 passes through one of the blocks having the element n_1 or n_2. Without loss of generality let us say that the line containing the block having the element n_5 passes through the block having the element n_1. In this case, we send the block having the element n_1 to the table T_0 and rest all the blocks lying on this line to the table T_1. Rest all the empty blocks are sent to the table T_0. If the line which contains the block having element n_5 does not pass through the block having element n_1 or n_2, in this case we can send both the blocks having elements n_1 and n_2 to the table T_1. Rest all the empty blocks are sent to the table T_0.

Now consider the case where block having the element n_5 does not lie on the line which contains block having the element n_3. In this case, we send the block having the element n_5 to the table T_1 and all the empty blocks lying on the line containing this block to the table T_0. Blocks having elements n_1 and n_2 are sent to the table T_1 and rest all the empty blocks are sent to the table T_0.

Case 3.2.2.3. None of the blocks which contain elements from S_2 coincide with blocks having elements from S_1 in the table T_1. This case is the same as Case 3.1.3.

Case 3.3. All the blocks which contain elements from S_1 lies on the different line.

Case 3.3.1. Both the blocks having elements n_4 and n_5 coincides with the blocks having elements from S_1 in the table T_1. Without loss of generality let us say that block having element n_1 coincides with the block having the element n_4 and the block having the element n_2 coincide with the block having the element n_5. In this case, we send the blocks having elements n_1, n_2 and n_3 to the table T_0 and all the empty blocks lying on the lines which contain these blocks to the table T_1. Also, we send the blocks having the element n_4 and n_5 to the table T_1. Rest all the empty blocks are sent to the table T_0.

Case 3.3.2. Only one of the block having element say n_4 from S_2 coincides with blocks having element from S_1 in the table T_1. Without loss of generality let us say that block having the element n_1 coincides with the block having the element n_4. Similar to the last case in this case also we send the blocks having elements n_1, n_2 and n_3 to the table T_0 and all the empty blocks lying on the lines which contain these blocks to the table T_1. We send the block having element

n_4 to the table T_1 and the block having element n_5 to the table T_0. Rest all the empty blocks are sent to the table T_1.

Case 3.3.3. None of the blocks which contain an element from S_2 coincides with the block which contains an element from S_1. This case is the same as Case 3.1.3.

Correctness. The correctness of the scheme relies on the fact that blocks having the elements do not coincide in the table T_1 or in the table T_0. Also, the blocks which are not having the elements are not sent to the place where block having elements are placed.

We summaries the conclusion of this section as follows.

Theorem 1. *There is a two probe explicit adaptive scheme which stores an arbitrary subset of size at most five from a universe of size m and uses $\mathcal{O}(m^{10/11})$ bits of space.*

4 Conclusion

In this paper we have come up with an explicit adaptive $(5, m, \mathcal{O}(m^{10/11}), 2)$-scheme, which improves upon the non-explicit scheme by Garg and Radhakrishnan [6] and the explicit scheme by Garg [5] for the given set and query sizes. We have borrowed the idea of the geometrical arrangement of elements on the three-dimensional cube from Kesh [8] and the idea of dividing the universe into blocks and superblocks from Radhakrishnan et al. [13]. Using these ideas there are improved schemes for the set of size three, four and five. We believe that this idea can be further generalized to improve the existing results for arbitrary subsets of size n.

References

1. Alon, N., Feige, U.: On the power of two, three and four probes. In: Proceedings of the twentieth Annual ACM-SIAM Symposium on Discrete Algorithms, pp. 346–354. SIAM (2009)
2. Baig, M.G.A.H., Kesh, D.: Two new schemes in the bitprobe model. In: Rahman, M.S., Sung, W.-K., Uehara, R. (eds.) WALCOM 2018. LNCS, vol. 10755, pp. 68–79. Springer, Cham (2018). https://doi.org/10.1007/978-3-319-75172-6_7
3. Baig, M.G.A.H., Kesh, D., Sodani, C.: A two query adaptive bitprobe scheme storing five elements. arXiv preprint arXiv:1810.13331v1 (2018)
4. Buhrman, H., Miltersen, P.B., Radhakrishnan, J., Venkatesh, S.: Are bitvectors optimal? SIAM J. Comput. **31**(6), 1723–1744 (2002)
5. Garg, M.: The bit-probe complexity of set membership. Ph.D. thesis, School of Technology and Computer Science, Tata Institute of Fundamental Research, Homi Bhabha Road, Navy Nagar, Colaba, Mumbai 400005, India (2016)
6. Garg, M., Radhakrishnan, J.: Set membership with a few bit probes. In: Proceedings of the Twenty-Sixth Annual ACM-SIAM Symposium on Discrete Algorithms, pp. 776–784. SIAM (2014)

7. Garg, M., Radhakrishnan, J.: Set membership with non-adaptive bit probes. arXiv preprint arXiv:1612.09388 (2016)
8. Kesh, D.: On adaptive bitprobe schemes for storing two elements. In: Gao, X., Du, H., Han, M. (eds.) COCOA 2017. LNCS, vol. 10627, pp. 471–479. Springer, Cham (2017). https://doi.org/10.1007/978-3-319-71150-8_39
9. Kesh, D.: Space Complexity of Two Adaptive Bitprobe Schemes Storing Three Elements (2018). http://www.iitg.ac.in/deepkesh/paper32.pdf
10. Kesh, D.: Space Complexity of Two Adaptive Bitprobe Schemes Storing Three Elements (2018). http://www.iitg.ac.in/deepkesh/A42.pdf
11. Lewenstein, M., Munro, J.I., Nicholson, P.K., Raman, V.: Improved explicit data structures in the bitprobe model. In: Schulz, A.S., Wagner, D. (eds.) ESA 2014. LNCS, vol. 8737, pp. 630–641. Springer, Heidelberg (2014). https://doi.org/10.1007/978-3-662-44777-2_52
12. Nicholson, P.K., Raman, V., Rao, S.S.: A survey of data structures in the bitprobe model. In: Brodnik, A., López-Ortiz, A., Raman, V., Viola, A. (eds.) Space-Efficient Data Structures, Streams, and Algorithms. LNCS, vol. 8066, pp. 303–318. Springer, Heidelberg (2013). https://doi.org/10.1007/978-3-642-40273-9_19
13. Radhakrishnan, J., Raman, V., Srinivasa Rao, S.: Explicit deterministic constructions for membership in the bitprobe model. In: auf der Heide, F.M. (ed.) ESA 2001. LNCS, vol. 2161, pp. 290–299. Springer, Heidelberg (2001). https://doi.org/10.1007/3-540-44676-1_24
14. Radhakrishnan, J., Shah, S., Shannigrahi, S.: Data structures for storing small sets in the bitprobe model. In: de Berg, M., Meyer, U. (eds.) ESA 2010. LNCS, vol. 6347, pp. 159–170. Springer, Heidelberg (2010). https://doi.org/10.1007/978-3-642-15781-3_14

Applications of V-Order: Suffix Arrays, the Burrows-Wheeler Transform & the FM-index

Ali Alatabbi[1], Jacqueline W. Daykin[1,2,3,4], Neerja Mhaskar[6(✉)],
M. Sohel Rahman[5], and W. F. Smyth[1,6,7]

[1] Department of Informatics, King's College London, London, UK
{ali.alatabbi,jackie.daykin}@kcl.ac.uk
[2] Department of Computer Science, Aberystwyth University, Aberystwyth, Wales
[3] Department of Computer Science, Aberystwyth University, Flic en Flac, Mauritius
[4] Department of Information Science, Stellenbosch University,
Stellenbosch, South Africa
[5] Department of CSE, BUET, ECE Building, West Palasi, Dhaka 1205, Bangladesh
msrahman@cse.buet.ac.bd
[6] Algorithms Research Group, Department of Computing and Software,
McMaster University, Hamilton, Canada
{pophlin,smyth}@mcmaster.ca
[7] School of Engineering and Information Technology, Murdoch University, Perth,
Western Australia

Abstract. V-order is a total order on strings that determines an instance of Unique Maximal Factorization Families (UMFFs), a generalization of Lyndon words. The fundamental V-comparison of strings can be done in linear time and constant space. V-order has been proposed as an alternative to lexicographic order (lexorder) in the computation of suffix arrays and in the suffix-sorting induced by the Burrows-Wheeler transform (BWT). In line with the recent interest in the connection between suffix arrays and the Lyndon factorization, we in this paper make a first attempt to obtain similar results for the V-order factorization. Indeed, we show that the results describing the connection between suffix arrays and the Lyndon factorization are matched by analogous V-order processing. We then apply the V-BWT to implement pattern matching in V-order after suitably modifying the FM-index.

Keywords: Combinatorics · Lexorder · String comparison · V-order
Suffix sorting · V-BWT · Pattern matching · FM-index

1 Introduction

This paper extends current knowledge on applications of a non-lexicographic global order known as V-order [DD96]. It is an intriguing question, now investigated for more than 20 years, whether such a counter-intuitive ordering might

© Springer Nature Switzerland AG 2019
G. K. Das et al. (Eds.): WALCOM 2019, LNCS 11355, pp. 329–338, 2019.
https://doi.org/10.1007/978-3-030-10564-8_26

nevertheless yield algorithmic efficiencies, or other theoretical/computational benefits, compared to lexorder. Here are some applications that have been considered.

The central problem of efficient V-ordering of strings was first studied in [DDS11, DDS13], leading to a simple, linear-time, constant-space algorithm [ADK+16], further improved in [ADM+18]. Other V-order applications include a variant (V-BWT) of the standard lexicographic Burrows-Wheeler transform, introduced in [DS14] — instances of enhanced data clustering were demonstrated. In this context, a linear-time algorithm based on [DDS11, DDS13] for V-sorting of all the rotations of a string was described, along with linear-time suffix-sorting based on Ko-Aluru [KA03]. In work closely related to this paper, [DDS11, DDS13] also describe efficient Lyndon-like factorization of a string into V-words.

In this paper, we modify ideas given in [MRRS14] that relate the Lyndon factorization to suffix arrays and the Burrows-Wheeler transform in order to obtain similar results for the V-order factorization (Sect. 3). We go on to introduce FM-index type functions in V-order and apply them to pattern matching in V-order (Sect. 4). The differences between lexorder and V-order are intriguing: while the latter generally appears trickier to work with, on the other hand computing suffix arrays in V-order is trivial.

2   Preliminaries

We are given a finite totally ordered set of cardinality $\sigma = |\Sigma|$, called the *alphabet*, whose elements are *characters* (equivalently *letters*). A *string* is a sequence of zero or more characters over Σ. A string $x = x_1 x_2 \cdots x_n$ of *length* $|x| = n$ is represented by $x[1..n]$, where $x[i] \in \Sigma$ for $1 \le i \le n$. The set of all non-empty strings over the alphabet Σ is denoted by Σ^+. The *empty string* of zero length is denoted by ε, with $\Sigma^* = \Sigma^+ \cup \varepsilon$. If $x = uwv$ for strings $u, w, v \in \Sigma^*$, then u is a *prefix*, w a *substring* or *factor*, and v a *suffix* of x. If $x = u^k$ (a concatenation of k copies of u) for some nonempty string u and some integer $k > 1$, then x is said to be a *repetition*; otherwise, x is *primitive*. For further stringological definitions, theory and algorithmics see [CHL07, Smy03].

Some of our applications are derived from Lyndon words, which we now introduce. A string $y = y[1 \ldots n]$ is a *conjugate* (or *cyclic rotation*) of $x = x[1 \ldots n]$ if $y[1 \ldots n] = x[i \ldots n]x[1 \ldots i-1]$ for some $1 \le i \le n$ (for $i = 1$, $y = x$). A *Lyndon word* is a primitive word which is minimum in lexicographic order (lexorder $<$) over its conjugacy class.

Theorem 1. *[CFL58] Any word x can be written uniquely as a non-increasing product $x = u_1 u_2 \cdots u_k$ of Lyndon words.*

We now define a non-lexicographic order V-order and some of its notable properties.

Let $x = x_1 x_2 \cdots x_n$ be a string over Σ. Define $h \in \{1, \ldots, n\}$ by $h = 1$ if $x_1 \le x_2 \le \cdots \le x_n$; otherwise, by the unique value such that $x_{h-1} > x_h \le$

$x_{h+1} \le x_{h+2} \le \cdots \le x_n$. Let $\boldsymbol{x}^* = x_1 x_2 \cdots x_{h-1} x_{h+1} \cdots x_n$, where the star $*$ indicates deletion of the letter x_h. Write $\boldsymbol{x}^{s*}$ for $(\ldots(\boldsymbol{x}^*)^* \ldots)^*$ with $s \ge 0$ stars. Let $g = \max\{x_1, x_2, \ldots, x_n\}$, and let k be the number of occurrences of g in $\boldsymbol{x}$. Then the sequence $\boldsymbol{x}, \boldsymbol{x}^*, \boldsymbol{x}^{2*}, \ldots$ ends $g^k, \ldots, g^1, g^0 = \varepsilon$. From all strings $\boldsymbol{x}$ over Σ, we form the *star tree*, where each string $\boldsymbol{x}$ labels a vertex, and there is a directed edge upward from $\boldsymbol{x}$ to $\boldsymbol{x}^*$, with the empty string ε as the root.

Definition 1 (V-Order [DD96]).   *We define V-order $\prec$ between distinct strings $\boldsymbol{x}$, $\boldsymbol{y}$. First $\boldsymbol{x} \prec \boldsymbol{y}$ if in the star tree $\boldsymbol{x}$ is in the path $\boldsymbol{y}, \boldsymbol{y}^*, \boldsymbol{y}^{2*}, \ldots, \varepsilon$. If $\boldsymbol{x}, \boldsymbol{y}$ are not in a path, there exist smallest s, t such that $\boldsymbol{x}^{(s+1)*} = \boldsymbol{y}^{(t+1)*}$. Let $\boldsymbol{s} = \boldsymbol{x}^{s*}$ and $\boldsymbol{t} = \boldsymbol{y}^{t*}$; then $\boldsymbol{s} \ne \boldsymbol{t}$ but $|\boldsymbol{s}| = |\boldsymbol{t}| = m$ say. Let $j \in [1..m]$ be the greatest integer such that $\boldsymbol{s}[j] \ne \boldsymbol{t}[j]$. If $\boldsymbol{s}[j] < \boldsymbol{t}[j]$ in Σ then $\boldsymbol{x} \prec \boldsymbol{y}$; otherwise, $\boldsymbol{y} \prec \boldsymbol{x}$. Clearly $\prec$ is a total order on all strings in Σ^*.*

Definition 2 (V-form [DD96, DD03, DDS11, DDS13]).
   The V-form of a string $\boldsymbol{x}$ is defined as

$$V_k(\boldsymbol{x}) = \boldsymbol{x} = x_0 g x_1 g \cdots x_{k-1} g x_k$$

for strings x_i, $i = 0, 1, \ldots, k$, where g is the largest letter in $\boldsymbol{x}$ — thus we suppose that g occurs exactly k times. For clarity, when more than one string is involved, we use the notation $\mathcal{L}_x = g$, $\mathcal{C}_x = k$.

Lemma 1. *[DD96, DD03, DDS11, DDS13] Suppose we are given distinct strings $\boldsymbol{x}$ and $\boldsymbol{y}$ with corresponding V-forms as follows:*

$$\boldsymbol{x} = x_0 \mathcal{L}_x x_1 \mathcal{L}_x x_2 \cdots x_{j-1} \mathcal{L}_x x_j,$$
$$\boldsymbol{y} = y_0 \mathcal{L}_y y_1 \mathcal{L}_y y_2 \cdots y_{k-1} \mathcal{L}_y y_k,$$

where $j = \mathcal{C}_x$, $k = \mathcal{C}_y$.
   Let $h \in 0..\max(j, k)$ be the least integer such that $x_h \ne y_h$. Then $\boldsymbol{x} \prec \boldsymbol{y}$ if, and only if, one of the following conditions holds:

(C1) $\mathcal{L}_x < \mathcal{L}_y$
(C2) $\mathcal{L}_x = \mathcal{L}_y$ and $\mathcal{C}_x < \mathcal{C}_y$
(C3) $\mathcal{L}_x = \mathcal{L}_y$, $\mathcal{C}_x = \mathcal{C}_y$ and $x_h \prec y_h$.

Lemma 2. *[DDS11, DDS13] For given strings $\boldsymbol{x}$ and $\boldsymbol{v}$, if $\boldsymbol{v}$ is a proper subsequence of $\boldsymbol{x}$, then $\boldsymbol{v} \prec \boldsymbol{x}$.*

This remarkable observation has many consequences, not least the trivial sorting of suffixes in V-order discussed in Sect. 3.

Theorem 2. *[ADK+16] For any strings $\boldsymbol{u}$, $\boldsymbol{v}$, $\boldsymbol{x}$, $\boldsymbol{y}$: $\boldsymbol{x} \prec \boldsymbol{y} \Leftrightarrow \boldsymbol{u}\boldsymbol{x}\boldsymbol{v} \prec \boldsymbol{u}\boldsymbol{y}\boldsymbol{v}$.*

According to this result, a comparison of two strings can ignore equal prefixes (or suffixes) — see Algorithm COMPARE in [ADK+16, ADM+18].

Lemma 3. *[DDS11, DDS13, ADK+16, ADM+18] V-comparison requires linear time and constant space.*

Definition 3 (V-Word [DD03]**).** *A string w over Σ is said to be a V-word if it is the unique minimum in V-order $\prec$ in the conjugacy class of w.*

Thus, like a Lyndon word, a V-word is necessarily primitive.

Example 1. We can apply Definition 1, equivalently the structure given by Lemma 1, to conclude that

$$u = 7173 \prec 7371 \prec 1737 \prec 3717,$$

so that 7173 is a V-word, while 1737 is a Lyndon word. Similarly, $v = 71727174$ and $w = 818382$ are V-words, while conjugates 17271747 and 183828 are Lyndon words.

Every word $w \in \Sigma^+$ can be uniquely factorized into V-words, $w = w_1 w_2 ... w_m$, such that $w_1 \geq_V w_2 \geq_V \cdots \geq_V w_m$. This factorization is called the *V-order factorization*. Note that in this definition for any two consecutive factors, $w_i \geq_V w_{i+1}$ implies that $w_i w_{i+1}$ is not a V-word; that is, concatenation of these factors is not possible and hence implies factoring (See [DDS09]). Also note that $w_i \prec w_{i+1}$ does not always hold (see Example 2).

Lemma 4. *[DDS11, DDS13] Using only linear time and space, a string x can be factored uniquely into V-words $x = x_1 x_2 \cdots x_m$, with $x_1 \geq_V x_2 \geq_V \cdots \geq_V x_m$.*

Example 2. For $x = 33132421$, the Lyndon decomposition is $3 \geq 3 \geq 13242 \geq 1$, while the V-order factorization is $33132 \geq_V 421$. Similarly, from Example 1, the string

$$x = uvw = (7173)(71727174)(818382)$$

has the unique V-order factorization $u \geq_V v \geq_V w$.

The on-line algorithm VF, that computes the V-order factorization of a string x in linear time and space, is described in detail in [DDS11, DDS13].

Definition 4. *[DD03, DDS13, DS14] Suppose that according to some factorization F, two strings $u, v \in \Sigma^+$ are expressed in terms of nonempty factors:*

$$u = u_1 u_2 \ldots u_m, \quad v = v_1 v_2 \ldots v_n.$$

Then $u \prec_{LEX(F)} v$ if and only if one of the following holds:

(1) u is a proper prefix of v (that is, $u_i = v_i$ for $1 \leq i \leq m < n$); or
(2) for some $i \in 1..min(m, n)$, $u_j = v_j$ for $j = 1, 2, \ldots, i - 1$, and $u_i \prec v_i$.

In other words, $u_1 u_2 \ldots u_m \prec_{LEX(F)} v_1 v_2 \ldots v_n$ in lexicographic extension (lex-extension) order, using not < but $\prec$.

Since the factorization F that we are interested in here is into V-words (V-order factorization), therefore instead of $\prec_{LEX(F)}$ we write $\prec_{LEX(V)}$. Also, analogous to the traditional definition of BWT, V-BWT is *BWT* with respect to V-order.

3   Suffix-Sorting and the Burrows-Wheeler Transform

In this section we show how to translate the results of Mantaci et al. [MRRS14] from lexorder into V-order. Mantaci et al. describe a strategy for obtaining the suffix array SA_x of a string x from its Lyndon factorization LF_x. To do the equivalent calculation in V-order — SA_x from the V-order factorization of x (defined below) — is not however of interest, because sorting the suffixes of x in V-order is trivial! By virtue of Lemma 2, the V-order of the suffixes is

$$x[n] \prec x[n-1..n] \prec \cdots \prec x[1..n],$$

so that $\mathrm{SA}_x[i] = n - i + 1$.

Thus, in the V-order version of this problem, we replace the suffixes with the "extended suffixes" (conjugates of x) employed to determine the V-BWT; that is,

$$x[n-i+1..n] \longrightarrow x[n-i+1..n]x[1..n-i]. \tag{1}$$

The following example, from [DS14], shows that the ordering of the extended suffixes is no longer straightforward:

Example 3. For $x = 9191919293$, consider suffixes 919293 and 91919293, with $919293 \prec 91919293$ by Lemma 2. However, extending these suffixes and applying Lemma 1 (C3) yields the opposite order:

$$9191929391 \prec 9192939191.$$

Thus we show here how to use the V-order factorization of x to generate the sorted conjugates of x.

Now we turn to the idea of "compatibility" of sorted suffixes as introduced in [MRRS14]. Let $x = x[1..n]$ be a string and $u = x[i..j]$, $1 \le i \le j \le n$, a substring of x. Then the sorting of suffixes $s_p = x[p..j]$, $s_q = x[q..j]$ of u is *compatible* with the sorting of the corresponding suffixes $x[p..n]$, $x[q..n]$ of x if these two (p, q) pairs have the same order. In lexorder compatibility of arbitrarily chosen u and v does not alway hold [MRRS14], but does hold when they are substrings of Lyndon factors in the Lyndon factorization of x. However, in V-order, compatibility holds for every choice of p, q. Moreover, the shorter suffix is always lesser, thus allowing comparison in terms of indexes:

Lemma 5. *Let $x \in \Sigma^+$ and u be a substring of x with s_1 a suffix of u. If s_2 is a proper suffix of s_1 then $s_2 \prec s_1$ with respect to both u and x.*

Proof. Immediate from Lemma 2.   □

As we have seen, Lemma 5 is not sufficient for extended suffixes, but since each rotation has the same number of maximum g's, condition (C3) implicitly applies. However, as a corollary we obtain the following V-order analogy of Theorem 3.2 given in [MRRS14] for Lyndon decomposition.

Theorem 3. *Let $x \in \Sigma^+$ with V-order factorization $x = v_1 \cdots v_k$, and let $u = v_i \cdots v_j$, for $1 \leq i \leq j \leq k$. Then the sorting of the suffixes of u is compatible with the sorting of the corresponding suffixes of x.*

Equipped with this theorem, we can modify the clever incremental suffix-sorting/BWT strategy of [MRRS14] to construct the extended V-order suffix array of $x = x[1..n]$. We first outline the steps:

Step 1: Identify the first factor v_1 in the V-order factorization $v_1 \geq_\mathcal{V} \cdots \geq_\mathcal{V} v_k$ of x in linear time [DDS11, DDS13].

Step 2: Compute the lex-extension order suffix array $\text{SA}(v_1)$ of v_1 in linear time [DS14].

Step 3: Extract $\text{BWT}(v_1)$ from $\text{SA}(v_1)$.

Step 4: For factor $v_i[1..t]$, $1 < i \leq k$, insert each suffix $v_i[t], v_i[t - 1..t], \ldots, v_i[1..t]$ into the current SA in their V-order – so for $i = 2$, one by one the suffixes of v_2 are inserted into $\text{SA}(v_1)$ giving $\text{SA}(v_1 v_2)$. Further, as each v_i is processed we can extract $\text{BWT}(v_1 \cdots v_i)$: for a suffix $x[i..j]$ the BWT letter is $x[i - 1]$; if an end-marker \$ is appended to v_i then its BWT letter is \$.

Note: After the above steps all the V-word factors will have been incrementally processed and $\text{BWT}(x)$ computed. The method avoids merging the suffix arrays of both the SA computed to date and that of the current factor being processed due to the complexity incurred.

We proceed to describe the suffix insertion (Step 4) in more detail. For this we apply Lemma 3.16 in [DDS13] which states that, unlike Lyndon words, the order of the set $\mathcal{V}$ of V-words is in some cases the same as V-order while in other cases it is reversed. This leads to two cases for Step 4: firstly, the next factor to be processed has a larger maximum letter g than the maximum letter in all the factors processed so far, and secondly the maximum letters are the same. However, the following example shows that the method doesn't work in the first case of distinct maximum letters.

Example 4. Let $x = v_1 \geq_\mathcal{V} v_2$ with $v_1 = 321$ and $v_2 = 5152$. The ordered conjugates of v_1 are: $321 \prec 132 \prec 213$, but note that the order of these conjugates is changed after processing v_2. We list the ordered conjugates of $x = 3215152$ below with complete factors shown in square brackets and the conjugates of 321 underlined.

$$5152[321]$$
$$52[321]51$$
$$\underline{1}[5152]\underline{32}$$
$$152[321]5$$
$$\underline{21}[5152]\underline{3}$$
$$\underline{321}[5152]$$
$$2[321]515$$

On the other hand, as we have seen, we are constrained to extended suffixes relating to a conjugacy class where each rotation has the same maximum letter g and occurring with the same frequency. In particular, condition (C3) ensures that the order of an existing suffix array is preserved during this iterative processing of factors. So we go on to deal with the case of the same maximum letters.

Suppose a prefix p of p factors of x has been processed resulting in SA(p) with an associated set of ordered conjugates, where all conjugates have the same maximum letter. The task is now to insert the suffixes of factor v_{p+1} into SA(p). Applying properties established in [DS14], if $g = \mathcal{L}_w$ and $k = \mathcal{C}_w$ for a string w, then in V-order, the first k conjugates of w start with g.

- The technique for calculating the index for inserting each proper suffix of v_{p+1} (increasing one letter at a time from right to left) into the extending suffix array is given with the FM-index in Sect. 4.1.
- Inserting the last improper suffix, v_{p+1}, is straightforward. We establish that analogously to the lexorder case in [MRRS14] this suffix is least in V-order and can be entered directly into the suffix array. We have $v_1 \geq_V v_2$, and assume that SA(v_1) has been computed; since v_1 is a V-word it is least in SA(v_1). Then when commencing the processing of v_2, v_1 must be concatenated with v_2 forming the suffix $v_1 v_2$. Applying Lemma 2, we have $v_2 \prec v_1 v_2$ in the suffix array – this holds iteratively as we continue processing the factors of x.

As expressed in [MRRS14] for the Lyndon case, this technique is suitable for integration with the on-line V-order factoring algorithm: suffix-sorting can proceed in tandem as soon as the first V-factor is identified.

4   BWT-type Pattern Matching with V-order

We outline here how to modify the well-known BWT-related backward search technique so as to implement pattern matching with V-order – we assume that the given string is a V-word and the V-BWT matrix (sorted conjugates in V-order) of an input string/text has been computed using the technique in Sect. 3. The process successively refines the search with respect to a current interval - all occurrences of the pattern will be located or it will be determined that the pattern does not occur in the string. We proceed to calculate the indexes of occurrences of letters in the matrix using modifications to the lexorder FM-index [FM00].

4.1   Applying FM-index

Following the approach in [MRRS14], we will use the well-known FM-index functions $RANK$ and C introduced in [FM00], which are defined as follows: For any character $\lambda \in \Sigma$, let $C(v, \lambda)$ denote the number of letters in the given string v that are smaller than λ, and let $RANK(v, t, \lambda)$ denote the number of occurrences of λ in the prefix of length t of v. These functions will enable backward search where a pattern is processed repeatedly from right to left: the first column of the V-BWT matrix is queried for existence of the current pattern letter in the given text string.

In order to adapt the FM functions $RANK$ and C, from lexorder applications to those for V-order, we make the following observations:

Obs 1 Analogous to the lexorder BWT, the i-th occurrence of a letter λ in the last column L of the V-BWT matrix is the same letter as the i-th occurrence of λ in the first column F. To see this, consider two rows in the matrix, $u\lambda_1$ and $v\lambda_2$, with $\lambda_1 = \lambda_2$. Since the matrix is based on a V-word, which is a primitive string, these rows must be distinct, so w.l.o.g. assume $u\lambda_1 \prec v\lambda_2$; the primitive condition, along with all rows being the same length, also implies $u, v \neq \varepsilon$. From the sufficiency in Theorem 2 we have $u \prec v$. Next consider the conjugates $\lambda_1 u$ and $\lambda_2 v$ in the conjugacy class, that is rows in the matrix. From the necessity in Theorem 2, we have $\lambda_1 u \prec \lambda_2 v$. Therefore, we cannot have λ_1 before λ_2 in L and λ_2 before λ_1 in F.

Obs 2 The k maximum letters g in the column L occur as the first k letters in the column F [DS14].

Obs 3 By Obs 1 and the structure of V-order, Definition 1 and Lemma 1, the first k letters in L which are the suffix letters of the first k rows beginning g, will occur in F in their alphabet order (but not necessarily adjacent and therefore a subsequence; also consider g to be least in Σ here) – with one rotation each of these letters will occur as a prefix of rows in the form λg.

Obs 4 By Obs 2 and lex-extension order Definition 4 together with condition (C1), using the x_0 substring of each row's V-form, Definition 2, we see that the V-BWT matrix rows $1..n$ are partitioned into groups. Traversing the rows from row $k+1$ to n, the maximum letter(s) in the x_0 prefixes (or first distinct x_i substrings), in each part, that is group, in the partition Π is less than the maximum letter(s) in those prefixes (substrings) in the neighbouring group. The set S of k suffix letters in Obs 3 yields up to k groups. Let q be the largest letter in S, then the $r \geq 0$ distinct letters in L which are greater than q but less than g yield r more groups. However, any other non-maximum letters in L that are not in S do not yield parts – by Lemma 2, on rotation, these letters belong to the x_0 prefixes of existing groups in Π.

Given a letter λ in L, with $L[i] = \lambda$, the goal is to calculate the index of λ in F. For this we first find the size of each group $\mathcal{G}$ using Obs 3 & 4, and then find the position of λ in the relevant group using Obs 1.

The first group, after those containing g, where we suppose its maximum letter is w, starts at $F[k+1]$. Consider the largest h such that the corresponding substring $L[k+1..h]$ does not contain a letter larger than w (except g), then the size of this first group is $h - k$ if $h > k$, otherwise its size is 1. Subsequently, with the next group starting at $F[h+1]$, the sizes of all groups are determined in this way. Note that these sizes of groups will be used in a similar way to the original FM function C. So for each maximum letter μ (not equal to g) in a group we define the function $VC(\mu)$ to be the total number of rows in all groups with maximum letter less than μ. We also define a corresponding VC array indexed by Σ.

Next suppose that some λ has an index i in L which corresponds to a group $\mathcal{G}_v$ with maximum letter v. From the previous step we have determined the interval I of indices in F and thus in L of $\mathcal{G}_v$. Let ℓ be the number of occurrences of λ in the interval I of L with index less than i, then using Obs 1 we know that there are ℓ occurrences of λ in the interval I of F. Additionally, if $\lambda \neq v$ there are V occurrences of v in $L[1..i-1]$ added to the rank plus the number of remaining letters in I with index less than i. This is similar to the original FM function $RANK$. Clearly, Obs 1 shows that we can use the FM function $RANK(v, t, \lambda)$ almost directly: we define $VRANK(v, t, \lambda)$ to be the rank of λ in the substring t of v, where v is the V-BWT, that is L, and the substring is given by I.

For a letter $\lambda = g$ in L, to calculate its index in F then Obs 1 & 2 apply. Otherwise, for $\lambda \neq g$ with L index i, its F index is given by: the number k of maximum letters g plus the array entry $VC[\mu]$, where μ is the maximum letter in the associated group of λ, plus $VRANK(L, I, \lambda)$.

Procedure BACKWARDSEARCH in Fig. 1 shows that the usual BWT backward search technique for lexorder also works for V-order following some modifications to the classic FM functions C and $RANK$. The inputs to the procedure are the pattern of length m, the V-BWT of length n and the arrays VC and $VRANK$. The backward search proceeds similarly to the Last First function and repeatedly calculates the interval in F of letters in the pattern which is processed from right to left.

procedure BACKWARDSEARCH(p, m, V-BWT, n, VC, $VRANK$)
$\quad (i, j, h) \leftarrow (1, n, m - 1)$ $\qquad\qquad$ ▷ h indexes the pattern p of length m
$\quad$ **while** $i \leq j$ and $h \geq 0$ **do**
$\qquad \lambda \leftarrow p[h]$ $\qquad\qquad\qquad\qquad$ ▷ λ is the current letter of the pattern

$\qquad (i, j) \leftarrow (k + VC[\lambda] + VRANK(V - BWT, I_i, \lambda),$
$\qquad\qquad\qquad k + VC[\lambda] + VRANK(V - BWT, I_j, \lambda))$
$\qquad h - -$

$\quad$ **if** $i \leq j$ **then** $\qquad\qquad\qquad\qquad$ ▷ all the pattern has been processed
$\qquad$ return (i, j)
$\quad$ **else**
$\qquad$ return $\emptyset$

Fig. 1. Searching for all occurrences of a pattern using backward search.

The output of procedure BACKWARDSEARCH is an interval $I = [i, j]$ containing all the occurrences of each pattern, possibly as a subsequence. Specifically, I indexes rows in the V-BWT matrix corresponding to values in the extended suffix array, which index the starting positions of the pattern in the text. On termination, if the end j value is less than the start value i, then the pattern does not occur in the text.

Note that we have considered the case of V-words; the computation of the V-transform and inverse for an arbitrary input is given in [DS14].

Acknowledgements. The third and fifth authors were funded by NSERC Grant Number: 10536797. The fourth author was partially supported by a grant from Pubali Bank Ltd., Bangladesh. The second author was part-funded by the European Regional Development Fund through the Welsh Government, Grant Number 80761-AU-137 (West):

References

ADK+16. Alatabbi, A., Daykin, J.W., Kärkkäinen, J., Rahman, M.S., Smyth, W.F.: V-Order: new combinatorial properties & a simple comparison algorithm. Discrete Appl. Math. **215**, 41–46 (2016)

ADM+18. Alatabbi, A., Daykin, J.W., Mhaskar, N., Rahman, M.S., Smyth, W.F.: A faster V-Order string comparison algorithm. In: Proceedings of Prague Stringology Conference, pp. 38–48 (2018)

CFL58. Chen, K.T., Fox, R.H., Lyndon, R.C.: Free differential calculus, iv - the quotient groups of the lower central series. Ann. Math. **68**, 81–95 (1958)

CHL07. Crochemore, M., Hancart, C., Lecroq, T.: Algorithms on Strings. Cambridge University Press, New York (2007)

DD96. Danh, T.-N., Daykin, D.E.: The structure of V-Order for integer vectors. In: Hilton, A.J.W. (ed.) Congressus Numerantium, Utilitas Mat. Pub. Inc., Winnipeg, Canada, vol. 113, pp. 43–53 (1996)

DD03. Daykin, D.E., Daykin, J.W.: Lyndon-like and V-Order factorizations of strings. J. Discrete Algorithms **1**(3–4), 357–365 (2003)

DDS09. Daykin, D.E., Daykin, J.W., Smyth, W.F.: Combinatorics of unique maximal factorization families (UMFFs). Fund. Inform. **97**(3). Special Issue on Stringology Janicki, R., Puglisi, S.J., Rahman, M.S. (eds.) pp. 295–309 (2009)

DDS11. Daykin, D.E., Daykin, J.W., Smyth, W.F.: String comparison and Lyndon-like factorization using V-order in linear time. In: Giancarlo, R., Manzini, G. (eds.) CPM 2011. LNCS, vol. 6661, pp. 65–76. Springer, Heidelberg (2011). https://doi.org/10.1007/978-3-642-21458-5_8

DDS13. Daykin, D.E., Daykin, J.W., Smyth, W.F.: A linear partitioning algorithm for hybrid Lyndons using V-Order. Theoret. Comput. Sci. **483**, 149–161 (2013)

DS14. Daykin, J.W., Smyth, W.F.: A bijective variant of the Burrows- Wheeler transform using V-Order. Theoret. Comput. Sci. **531**, 77–89 (2014)

FM00. Ferragina, P., Manzini, G.: Opportunistic data structures with applications. In: Proceedings of 41st Annual Symposium on Foundations of Computer Science, (FOCS 2000), pp. 390–398 (2000)

KA03. Ko, P., Aluru, S.: Space efficient linear time construction of suffix arrays. In: Baeza-Yates, R., Chávez, E., Crochemore, M. (eds.) CPM 2003. LNCS, vol. 2676, pp. 200–210. Springer, Heidelberg (2003). https://doi.org/10.1007/3-540-44888-8_15

MRRS14. Mantaci, S., Restivo, A., Rosone, G., Sciortino, M.: Suffix array and Lyndon factorization of a text. J. Discrete Algorithms **28**, 2–8 (2014)

Smy03. Smyth, B.: Computing Patterns in Strings. Pearson/Addison-Wesley, Harlow (2003)

Parallel and Distributed Algorithms

Towards Work-Efficient Parallel Parameterized Algorithms

Max Bannach[1], Malte Skambath[2](✉), and Till Tantau[1]

[1] Institute of Theoretical Computer Science, Universität zu Lübeck,
Lübeck, Germany
{bannach,tantau}@tcs.uni-luebeck.de
[2] Department of Computer Science, Kiel University, Kiel, Germany
malte.skambath@email.uni-kiel.de

Abstract. Parallel parameterized complexity theory studies how fixed-parameter tractable (fpt) problems can be solved in parallel. Previous theoretical work focused on parallel algorithms that are very fast in principle, but did not take into account that when we only have a small number of processors (between 2 and, say, 1024), it is more important that the parallel algorithms are *work-efficient*. In the present paper we investigate how work-efficient fpt algorithms can be designed. We review standard methods from fpt theory, like kernelization, search trees, and interleaving, and prove trade-offs for them between work efficiency and runtime improvements. This results in a toolbox for developing work-efficient parallel fpt algorithms.

Keywords: Parallel computation · Fixed-parameter tractability
Work efficiency

1 Introduction

Since its introduction by Downey and Fellows [9] about thirty years ago, parameterized complexity theory has been successful at identifying which problems are fixed-parameter tractable (fpt), but has also had high practical impact. Efforts to formalize and devise *parallel* fpt algorithms date back twenty years [6,7], but a lot of the theoretical research is quite recent [1,2,10]. The findings can be summarized, very briefly, as follows: It is possible to classify the problems in FPT according to how well they can be solved in parallel, and we find natural parameterized problems on all levels — from problems in FPT that are inherently sequential to problems that can be solved in constant (!) parallel time.

One aspect that the existing research lacks – and which may also explain the small number of actual implementations – is a fine-grained analysis of the *work* done by parallel fpt algorithms, which is defined as the total number of computational steps done by an algorithm summed over all processing units (in particular, for a sequential algorithm, its work equals its runtime). Unfortunately, "the work must be done": on a machine with p processors, a parallel algorithm

© Springer Nature Switzerland AG 2019
G. K. Das et al. (Eds.): WALCOM 2019, LNCS 11355, pp. 341–353, 2019.
https://doi.org/10.1007/978-3-030-10564-8_27

with $W(n)$ work cannot finish faster than in time $W(n)/p$ on length-n inputs. Since real-life values of p are small (between 2 and perhaps 1024), a large $W(n)$ can lead to actual runtimes ("wall clock runtimes") that are larger than those of sequential algorithms.

A common pattern in the design and analysis of parallel algorithms is that as we try to decrease the work $W(n)$ in order to get down the quotient $W(n)/p$, the "theoretical" parallel runtime $T(n)$ rises. This pattern is repeated in the fpt setting: Table 1 shows the work and time needed by different parallel algorithms for p-VERTEX-COVER. Note that we will never be able to reduce the work of a parallel algorithm below the work of the fastest sequential algorithm and we call an algorithm *work-optimal* if it matches this lower bound.

Table 1. Faster parallel algorithms for p-VERTEX-COVER entail more work. We can achieve a runtime of $O(1)$ at the cost of the expensive use of color coding [1]. If we allow $O(\log n)$ time, a parallel Buss kernelization in conjunction with a simple brute force algorithm reduces the work. The next two lines are based on shallow search trees, discussed in Sect. 3.2, and the work starts to become competitive with sequential algorithms. The last lines show that being work-competitive to the best known sequential algorithms implies larger and larger runtimes.

Work	Parallel time
$O(kn + 2^{2^k + k})$	$O(1)$
$O(kn + 2^{k^2} \cdot k^2)$	$O(\log n)$
$O(kn + 3^k k^2)$	$O(\log n + \log^2(k))$
$O(kn + 2^k)$	$O(\log n + \log^4(k))$
$O(kn + 1.6181^k)$	$O(\log n + k\log(k))$
$O(kn + 1.4656^k)$	$O(\log n + k\log(k))$
$O(kn + 1.2738^k)$	$O(\log n + k^4\sqrt{k}))$

Our Contributions. Many fpt algorithms are based on the *search tree technique*, which recursively traverses a search tree whose depth and degree are bounded by the parameter, resulting in a sequential runtime of the form c^k or perhaps $(ck)^k$ for some constant c. Intuitively, search tree algorithms should be easy to parallelize since the different branches of the search tree can be processed independently. We show that this intuition is correct and we provide precise conditions for search tree algorithms under which they can be turned into work-efficient parallel algorithms. A parallel search tree algorithm still has to process, and thus construct, all branches of the tree, leading to a parallel runtime that is proportional to the depth of the search tree, which is normally $\Theta(k)$. This theoretical runtime is typically much smaller than the actual wall-clock time $W(n)/p = (c^k + O(n))/p \gg k$. However, we show that in some cases there is room for improvement and the runtime of $\Omega(k)$ can be replaced by $O(\log k)$ without increasing the work. The idea is to modify the search tree such that it "branches aggressively," thereby reducing the depth to $O(\log k)$.

A second tool of parameterized complexity theory are kernelizations: mappings from input instances to membership-equivalent instances whose size is bounded by the parameter. Some problems admit more than one kernelization and we have to determine an execution order of them. In the sequential case this is quite unambiguous: First apply the fastest kernel, which may however result in a still rather large instance. Then apply a slower kernel with a smaller output – the high runtime matters less since it is applied to a smaller input. Such *kernel cascades* are also possible in the parallel setting, but here kernelizations may have incomparable work, runtime, and output size. We provide a general procedure to combine a set of parallel kernelizations into a work-efficient and fast kernelization that minimizes the output size.

A third tool is *interleaving:* Instead of using a kernelization just as a preprocessing procedure, during a search tree traversal call the kernel algorithm at each tree node to ensure that the intermediate instances are small. In the sequential setting this has the desirable effect of turning a runtime of the form $O(k^c \cdot \xi^k + n^c)$ into one of the form $O(\xi^k + n^c)$ [14]. We show that interleaving is also possible in the parallel setting in a work-efficient manner, including the mentioned depth-$O(\log k)$ search trees that do not arise in the sequential setting.

Related Work. First efforts to formalize *parallel* fpt algorithms are due to Cesati and Di Ianni [6], though the definitions were rather ad hoc. Around the same time, Cai et al. [5] investigated *space bounded* fpt algorithms – and since logarithmic space is related to parallel computations, these algorithms can be seen as parallel fpt results. A first experimental analysis of a parallel fpt algorithm for vertex cover is due to Cheetham et al. [7]. Recent work on a theoretical framework for parallel fpt has mainly been done by Bannach et al. [1,2] and Elberfeld et al. [10]. These papers establish hierarchies of parallel parameterized complexity classes and place well-known problems in them, but do not consider work-efficiency. Many algorithms in the cited papers are based on the expensive *color-coding* technique, which needs work $O(n \log^2 n \log c \cdot c^{k^2} \cdot k^4)$ and results in unpractical algorithms.

Organization of This Paper. Following the preliminaries, we investigate, in order, parallel search trees, parallel kernels, and parallel interleaving. Due to lack of space, some proofs are only included in the technical report version.

2 Preliminaries

A *parameterized problem* Q is a set $Q \subseteq \Sigma^* \times \mathbb{N}$, where in an instance $(x, k) \in \Sigma^* \times \mathbb{N}$ the number k is called the *parameter*. A parameterized problem is *fixed-parameter tractable* (in FPT) if there is an algorithm that decides the problem for all $(x, k) \in \Sigma^* \times \mathbb{N}$ in time $f(k) \cdot |x|^c$. Here, and in the following, f is always a computable function and c a constant. As model of parallel computation we use standard PRAMs (rather than circuits), see for instance [12]. For a PRAM program, let $T_p(n)$ denote the maximum time the program needs on inputs of length n when p processors are available. Let $T(n) = \inf_{p \to \infty} T_p(n)$ and let $W(n)$ denote the maximum number of computational steps (summed over all non-idle processors) performed by the algorithm on inputs of length n. It is well-known that $T_p(n) \leq W(n)/p + T(n)$ holds when the set of non-idle processors is easily computable for each step (so a compiler can schedule the to-be-done work for each step when less processors are available than there is work to be done) [12]. We have $T_p(n) \geq W(n)/p$ and $T_p(n) \geq T(n)$. Since for fast parallel algorithms we have $W(n)/p \gg T(n)$, the work of a parallel algorithm is the dominating factor. We say an algorithm is *work-optimal* if its work is the best possible among all algorithms. This definition hinges, to a certain degree, on the fact that there are clear notions of "minimal work" and "minimal runtime". In the parameterized world, however, this is no longer the case: it is not clear which of the terms $3^k n$, $2^k n^2$, $n^3 + 2^k$, and n^k is "minimal." Depending on the values of n and k, any of the terms may be more desirable than the others. For this reason, we strive for optimality only with respect to the following notion (throughout the paper, we assume that functions like $W(n, k)$ or $T(n, k)$ are monotone with respect to both parameters): An algorithm A is *work-competitive to a function* f if $W_A \in O(f)$, that is, if $W_A(n, k) \leq c \cdot f(n, k)$ for all $n \geq n_0$ and $k \geq k_0$ for some constants c, n_0, and k_0. An algorithm A is *work-competitive to an algorithm* B if it is work-competitive to the function W_B.

3   Work-Efficient Parallel Search Tree Algorithms

For a parameterized problem Q and an instance (x, k), a *search tree* algorithm invokes a *branching rule* (or *branching algorithm*) to determine a sequence (x_1, k_1), ..., (x_m, k_m) of new instances such that $(x, k) \in Q$ if, and only if, we have $(x_i, k_i) \in Q$ for at least one i. Crucially, each k_i must be smaller than k, that is, $d_i = k - k_i > 0$. (Let us also require $|x_i| \leq |x|$ to simplify the presentation, but this is less crucial.) The search tree algorithm recursively calls itself on these new instances (unless it can directly decide the instance for "trivial" k or for "trivial" x_i). An example of a search tree algorithm is the branching algorithm for the vertex cover problem where we "branch on an arbitrary edge": Map (G, k) to $(G - \{u\}, k - 1)$ and $(G - \{v\}, k - 1)$ for an arbitrary edge $\{u, v\}$ (we have $d_1 = d_2 = 1$ and $m = 2$). Another example is the branching rule "branch on the maximum-degree vertex and either take it into the vertex cover or all of its neighbors," meaning that we map (G, k) to $(G - \{u\}, k - 1)$ and $(G - N(u), k - |N(u)|)$ where $N(u)$ is the neighborhood of u. This leads to $d_1 = 1$ and $d_2 = |N(u)|$; and since we can solve the vertex cover problem directly in graphs of maximum degree 2, we have $d_1 = 1$ and $d_2 \geq 3$.

3.1   Simple Parallel Search Trees

As mentioned in the introduction, parallelizing a search tree is more or less trivial, since we can process all resulting branches in parallel. Of course, it may now become important how well the *branching rule* can be parallelized, since we have to invoke it on each level of the tree. In detail, for a set D of vectors $d = (d_1, \ldots, d_m)$, a *D-branching algorithm* B *for* Q is an algorithm that on input (x, k) either correctly outputs "$(x, k) \in Q$", "$(x, k) \notin Q$", or instances $(x_1, k - d_1), \ldots, (x_m, k - d_m)$ for some $d \in D$ such that $(x, k) \in Q$ if, and only if, $(x_i, k - d_i) \in Q$ for some $i \in \{1, \ldots, m\}$. Let SeqSearchTree-B and ParSearchTree-B denote the sequential and parallel search tree algorithms based on B, respectively. Note that both algorithms traverse the same tree on an input (x, k). Let $\mathsf{size}_B(n, k)$ and $\mathsf{depth}_B(n, k)$ denote the maximum number of nodes and the maximum depths of the search trees traversed by the algorithms on inputs of length n, respectively.

From a sequential perspective, the objective in the design of search tree algorithms is to reduce the size of the search tree since this will be the dominating factor in the runtime. From the parallel perspective, however, we will also be interested in the *depth* of the search tree since, intuitively, this depth corresponds to the parallel time needed by the algorithm.

Theorem 3.1. *Let* B *be a branching algorithm. Then*

$$T_{\mathsf{SeqSearchTree}\text{-}B}(n, k) = W_{\mathsf{SeqSearchTree}\text{-}B}(n, k) = O(\mathsf{size}_B(n, k) \cdot W_B(n, k)),$$
$$T_{\mathsf{ParSearchTree}\text{-}B}(n, k) = O(\mathsf{depth}_B(n, k) \cdot T_B(n, k)),$$
$$W_{\mathsf{ParSearchTree}\text{-}B}(n, k) = O(\mathsf{size}_B(n, k) \cdot W_B(n, k)).$$

Of course, a lot is known concerning the size of search trees resulting from D-branching algorithms: If $s(k) = \mathsf{size}_B(n, k)$ is independent of n, we always have $s(k) \leq \max_{(d_1, \ldots, d_m) \in D}(s(k - d_1) + \cdots + s(k - d_m) + 1)$ and it is known [14] how to compute a number ξ_D such that $s(k) = \Theta(\xi_D^k)$ is a minimal solution of the inequality. For instance, for the simple branching algorithm for the vertex cover problem with $D = \{(1, 1)\}$ we have $\xi_D = 2$ and the search tree has size 2^k, while for $D = \{(1, 3)\}$ from the branch-on-a-degree-3-vertex algorithm[1] we have $\xi_D \leq 1.4656$. Regarding the depth of the search tree, it is clearly upper-bounded by $k/\min d$ for the "worst $d \in D$" since in each recursive call we decrease k by at least the minimal d_i in d. In summary, we see that ParSearchTree-B is always work-competitive to SeqSearchTree-B and $T_{\mathsf{ParSearchTree}\text{-}B}(n, k) = \frac{k}{\max_{d \in D} \min_i d_i} \cdot T_B(n, k)$ and $W_{\mathsf{ParSearchTree}\text{-}B}(n, k) = \xi_D^k \cdot W_B(n, k)$.

3.2 Shallow Parallel Search Trees

If we wish to find faster work-optimal parallel search tree algorithms, a closer look at Theorem 3.1 shows that there are two lines of attack: First, we can try to decrease $T_B(n, k)$ while keeping $W_B(n, k)$ optimal. Second, we can try to decrease the depth of the search trees without increasing their size.

Regarding the first line of attack, there is often "little that we can do" since $T_B(n, k)$ will often already be optimal. For instance, the branching algorithm

---

[1] Of course, we actually branch on a vertex of degree *at least* 3, meaning that $D = \{(1, 3); (1, 4); (1, 5); \ldots\}$ holds, but $d = (1, 3)$ clearly leads to the largest and deepest search trees and it suffices to only consider this "worst d.".

"pick an arbitrary edge" can be implemented optimally in parallel time $O(1)$ assuming an appropriate memory access model; and for the branch-on-a-degree-3-vertex algorithm, both finding a degree-3 vertex and solving the instance if no such vertex exists can be done work-optimally in polylogarithmic time.

Regarding the second line, however, new algorithmic ideas *are* possible and lead to work-efficient algorithms whose runtime is *logarithmic* in the parameter instead of linear. A word of caution, however, before we proceed: We improve runtimes from $O(k + \log^{O(1)} n)$ to $O(\log k + \log^{O(1)} n)$, where the $O(\log n)$ is needed already for many pre- and postprocessing operations on the input. Clearly, the improvement in the runtime is rather modest since we generally think of k being something very small. Nevertheless, achieving even this modest speedup optimally is highly nontrivial for many problems.

To get some intuition for the idea, consider once more the vertex cover problem, but let us now try to find *ten* arbitrary edges that form a matching. Then every vertex cover of the input graph must contain at least one endpoint from each of these ten edges and we get the following new branching rule: Branch to all 1024 possible ways of choosing one vertex from each of the ten edges, each time reducing the size of sought vertex cover by 10. This corresponds to a branching vector $d' = (10, 10, \ldots, 10)$ of length 1024; compared to the vector $d = (1, 1)$ if we branch over a single edge. In the sequential setting this idea only complicates things since $\xi_{d'} = \xi_d = 2$ and this new algorithm produces a search tree of the same size as before. In contrast, in the parallel setting we make progress as the *depth* of the search tree is decreased by a factor of 10, without an increase in the work being done. Naturally, a factor-10 speedup is just a constant speedup, but we can extend the idea to move from a runtime of k to $\log k$:

Theorem 3.2. *There is an algorithm that solves* p-VERTEX-COVER *in time* $T(n, k) = O(\log k \cdot \log^3 n)$ *and work* $W(n, k) = O(2^k n)$ *on a* CRCW-PRAM.

Proof (Sketch). On input (G, k) determine a maximal matching M of G. This can be done in time $O(\log^3 n)$ [11]. In case $|M| < k/2$ or $|M| > k$ we are done, otherwise branch on all $2^{|M|}$ ways of choosing endpoints in this matching.   □

Let us try to generalize the ideas underlying the above theorem and its proof.

- *Branch structures:* The original branching algorithm first found "a substructure on which to branch." For example, the vertex cover branching algorithm normally finds "an arbitrary edge;" the branch-on-degree-at-least-3 algorithm finds "a high-degree vertex."
- *Conflict-free branch structures:* If the original branching algorithm has the choice among several possible substructures on which it could branch and if the substructures are disjoint, we can also branch on these structures "in parallel." In Theorem 3.2, "disjoint substructures that are edges" are matchings and we can branch on them in parallel; for the branch-on-degree-at-least-3 algorithm we can branch in parallel on any star forest.
- *A large number of conflict-free branch structures:* Lastly, we need to be able to find a large enough collection of such disjoint substructures quickly and

work-efficiently. Its size needs to be at least a fraction of the parameter to ensure that we get a depth that is logarithmic in the parameter.

Since formalizing the above notions can easily lead to rather technical definitions, we suggest a formalization that is not as general as it could be, but that nicely captures the essential ideas. We only consider *vertex search problems* Q on simple graphs where the objective is to find a parameter-sized subset of the vertices that has a certain property. Concerning branching rules, we only consider rules that identify a subset of the vertices and then branch over different ways in which some of these vertices can be added to the partial solution:

Definition 3.3 (Local branching rule). *Let Q be a vertex search problem. A local branching rule is a partial mapping that gets a tuple as input consisting of a graph $G = (V, E)$, a parameter k, an already computed partial solution $P \subseteq V$, and a set $S \subseteq V \setminus P$ on which we would like to branch. If defined, it outputs a family F of nonempty subsets of S such that for every solution $Y \supseteq P$ for (G, k) the intersection $Y \cap S$ is a superset of an element of F.*

The local branching rule for the vertex cover algorithm maps the tuple $(G, k, P, \{u, v\})$ with $\{u, v\} \in E$ and $u, v \notin P$ to $\{\{u\}, \{v\}\}$ and is undefined otherwise. For the branch-on-degree-at-least-3 rule, if S is the closed neighborhood in $G - P$ of some vertex v of degree 3 in $G - P$, we map (G, k, P, S) to $\{\{v\}, S \setminus \{v\}\}$. Returning to the three ingredients of the proof of Theorem 3.2, the sets S in the definition of a local branching rule are exactly the sought "branching structures." A collection M of such sets is "conflict-free" if all members of M are pairwise disjoint. In the proof of Theorem 3.2 such an M was simply a matching in the graph; but given any collection N of sets S, any maximal set packing $M \subseteq N$ will be conflict-free. Maximal set packings can be obtained efficiently and quickly in parallel by building a conflict graph over the sets and computing a maximal independent set [13]. Therefore, in a general setting it suffices to compute a polynomial-size set N of sets S that has a set packing $M \subseteq N$ whose size at least a fraction of k. Algorithm 1.1 makes these ideas precise.

Definition 3.4. *An implementation of a local branching rule consists of three algorithms* decide, choices, *and* branches *with the following properties:*

1. *On inputs (G, k, P) for which there is no S such that the local branching rule is defined for (G, k, P, S), algorithm* decide *must correctly output "yes" or "no" depending on whether P is a partial solution.*
2. *For all other inputs (G, k, P), the algorithm* choices *must output a nonempty set N such the local branching rule is defined on all (G, k, P, S) for $S \in N$.*
3. *For all (G, k, P, S) for which the local branching rule is defined,* branches *must output the corresponding family F of branches.*

Theorem 3.5. *Given an implementation* (decide, choices, branches) *for a local branching rule for some Q, algorithms* B_1 *and* B_* *from Algorithm 1.1 satisfy:*

1. ParSearchTree-B_* *is work-competitive to* SeqSearchTree-B_1 *if $W_{\mathsf{decide}}(n, k) + W_{\mathsf{choices}}(n, k) + W_{\mathsf{branches}}(n, k) \in \Omega(n^3)$.*

2. *If the size of the maximal set packings M computed by B_* is always at least $\varepsilon(k - |P|)$, then $T_{\mathsf{ParSearchTree\text{-}B}_*}(n, k) = O(\log k \cdot (T_{\mathsf{decide}}(n, k) + T_{\mathsf{choices}}(n, k) + T_{\mathsf{branches}}(n, k) + \log^4 n))$.*

Algorithm 1.1. For an implementation (decide, choices, branches), B_1 is the resulting standard branching rule. The new parallel branch algorithm B_* first computes a set packing M of the set N of possible branch structures and then branches on all of them simultaneously. Let s be the maximum size of any X produced in B_1 on any input.

```
1   algorithm B₁(G, k, P)
2       if decide(G, k, P) ∈ {yes, no} then return decide(G, k, P)
3       N ← choices(G, k, P) // for vertex cover, N is the edge set of G − P
4       S ← an arbitrary element of N // for vertex cover, S = {u, v} for some edge in N
5       for each X ∈ branches(G, k, P, S) par do
6           output in parallel (G, k, P ∪ X)
7
8   algorithm B∗(G, k, P)
9       if decide(G, k, P) ∈ {yes, no} then return decide(G, k, P) // Recursion break
10      N ← choices(G, k, P) // for vertex cover, N is the edge set of G − P
11      M ← a maximal set packing of N among those of size at most (k − |P|)/(s + 1)
12      {S₁, …, Sₘ} ← M // name the elements of M
13      for each X₁ ∈ branches(G, k, P, S₁), …, Xₘ ∈ branches(G, k, P, Sₘ) par do
14          output in parallel (G, k, P ∪ X₁ ∪ · · · ∪ Xₘ)
```

4   Work-Efficient Parallel Kernels

Kernels are self-reductions that map instances to new instances whose size is bounded in terms of the parameter. Like search trees, they are basic concepts of fpt theory. Unlike search trees, kernels are often hard to parallelize: They are typically described in terms of *reduction rules*, which locally change an input instance in such a way that it gets a bit smaller without changing problem membership and such that at least one rule is still applicable as long as the instance size is not bounded in terms of the parameter. Unfortunately, it is known that some sets of reduction rules are "inherently sequential," meaning that computing the result of applying them exhaustively is complete for sequential polynomial time [4]. On the other hand, some reduction rules can easily be applied in parallel just as well as sequentially, leading to kernelization algorithms running in polylogarithmic time or even in constant time [3].

While it seems hard to characterize which sets of reduction rules yield parallel kernels, the situation is more favorable when we consider a sequence of kernels (a *kernel cascade*). In the sequential setting, the situation is simple: Given several kernelizations for the same problem, the asymptotically fastest way to compute a minimum-size kernel is simply to apply them in sequence starting with the fastest and ending with the slowest. In the parallel setting, the situation is also simple when we can parallelize all kernels of a cascade optimally. However, even when this is not the case, we may still get a fast parallel algorithm and there is an intriguing dependence on the parallel runtime and the kernel size: Theorem 4.2 states that it suffices to parallelize the kernels in a cascade until the *kernel size* equals the desired *parallel runtime* – while later kernels need not be parallelized.

4.1 Sequential Kernel Cascades

A *kernelization* for a parameterized problem $Q \subseteq \Sigma^* \times \mathbb{N}$ is a polynomial-time computable function $K \colon \Sigma^* \times \mathbb{N} \to \Sigma^* \times \mathbb{N}$ such that (a) $(x, k) \in Q$ if, and only if, $K(x, k) \in Q$ for the *kernel* $K(x, k)$ and such that (b) for some computable function s_K we have $|K(x, k)| \le s_K(k)$ for all x and k. We call the kernelization *polynomial* if s_K is a polynomial. A *kernel algorithm* is an algorithm K that computes a kernelization K.

As indicated earlier, there can be several kernelizations (and, hence, kernel algorithms) for the same problem and they may differ regarding their runtime and their kernel sizes. For instance, on input (G, k) the *Buss kernelization* of the vertex cover problem removes all vertices of degree larger than k (which must be in a vertex cover) and then removes all isolated vertices (which are not needed for a vertex cover). It yields kernels of size $s_{\mathsf{Buss}}(k) = k^2$ and can be computed very quickly. In contrast, the *linear program kernelization* [8] for the vertex cover problem solves a linear program in order to compute a kernel of size $2k$, but solving the linear program takes more time. It now makes sense to *first* compute a Buss kernel *followed* by an application of the linear program kernelization since we then apply a "slow" algorithm only to an already reduced input size (from originally n to only k^2).

In general, let a *kernel cascade* be a sequence $C = (K_1, \ldots, K_t)$ of kernel algorithms for the same parameterized problem Q sorted in strictly increasing order of runtime (that is, we require $T_{K_i} \in o(T_{K_{i+1}})$ and thereby implicitly rule out situations where runtimes are incomparable) and strictly decreasing order of kernel sizes (that is, we require $s_{K_i}(k) > s_{K_{i+1}}(k)$ for all but finitely many k). The *cascaded kernel algorithm* K_C of a cascade C will, on input (x, k), first apply K_1 to (x, k), then applies K_2 to the result, then K_3 and so on, and output the result of the last K_t. Clearly, the following holds:

Observation 4.1 *Let $C = (K_1, \ldots, K_t)$ be an kernel cascade. Then $s_{K_C} = s_{K_t}$ and the runtime of K_C is $T_{K_C}(n, k) = T_{K_1}(n) + T_{K_2}(s_{K_1}(k)) + T_{K_3}(s_{K_2}(k)) + \cdots + T_{K_t}(s_{K_{t-1}}(k))$. Furthermore, no subsequence C' of C with $s_{K_{C'}} = s_{K_t}$ achieves an asymptotically faster runtime.*

4.2 Parallel Kernel Cascades

Faced with the problem that kernels based on reduction rules are often difficult to parallelize, parallelizing a whole kernel cascade in a work-optimal way seems even more challenging: Observation 4.1 states that for a given cascade the asymptotically fastest runtime is achieved by applying all kernels in the cascade in sequence. Since "work optimal" means, by definition, "parallel work equal to the fastest sequential runtime," we also must apply work-optimal parallel versions of all kernels in the cascade in sequence in the parallel setting.

It turns out that it may *not* be necessary to parallelize all kernels in a cascade: Suppose we only parallelize the first kernel in a cascade, that is, suppose we find a work-optimal algorithm for K_1 with runtime $O(\log^{O(1)} n)$ and then apply this parallel algorithm *followed by the unchanged sequential kernels* K_2 *to* K_t. The work of the resulting cascade will be identical to the runtime of the original sequential cascade (since K_1 is work-optimal and nothing else is changed). The runtime, however, will now be $O(\log^{O(1)} n)$ plus some function *that depends only on* k (since all later kernels are applied to inputs whose size depends only on k). Assuming that we consider a runtime of the form $O(\log^{O(1)} n + f(k))$ "acceptable," we see that we can turn any sequential kernel cascade into a parallel one by parallelizing only the *first* kernel. Of course, there are functions f that we might not consider "acceptable"; for instance, f might be exponential. Intuitively, we then need to "parallelize more kernels of the sequence."

Theorem 4.2. *Let* $C = (K_1, \ldots, K_t)$ *be a kernel cascade and for some* $r \leq t$ *let* $K'_1, \ldots, K'_r$ *be parallel implementations of* $K_1, \ldots, K_r$, *that is, for* $i \in \{1, \ldots, r\}$ *let* K'_i *be a work-competitive parallel implementation of* K_i *with runtime* $T_{K'_i} \in O(\log^{O(1)} n)$. *Let* $C' = (K'_1, \ldots, K'_r, K_{r+1}, \ldots, K_t)$. *Then*

1. $K_{C'}$ *is work-competitive to* K_C *and*
2. $T_{K_{C'}}(n, k) = \log^{O(1)} n + s_{K'_r}(k)^{O(1)}$.

As a concluding example, consider once more p-VERTEX-COVER. We mentioned already that there is a size-k^2 kernel algorithm Buss for this problem, which is easy to implement in linear sequential time, but also in logarithmic parallel time and linear work (and, thus, optimally). There is also a size-$2k$ kernel algorithm LP based on [8] that needs sequential time $O(|E|\sqrt{|V|})$. For this kernel, no work-optimal (deterministic) polylogarithmic time implementation is known (indeed, *any* parallel implementation is difficult to achieve [4]). By Observation 4.1, there is a *sequential* kernel algorithm for the vertex cover problem that runs in time $O(n + k^2\sqrt{k^2}) = O(n + k^3)$. By Theorem 4.2, there is a *parallel* kernel algorithm that is work-competitive and needs time $O(\log n + k^3)$.

5   Work-Efficient Parallel Interleaving

Interleaving is a method to combine a branching algorithm B and kernel algorithm K to "automatically" reduce the runtime of SeqSearchTree-B: During the recursion, the algorithm SeqInterleave-B-K applies K at the beginning of each recursive call (thus, calls to the kernel algorithm are "interleaved" with the recursive calls, hence the name of the method). Intuitively, at the start of the recursion, calling a kernel algorithm is superfluous (the input is typically already kernelized) and only adds to the runtime, but deeper in the recursion it will ensure that the inputs are kept small. Since the bulk of all calls are "deep inside the recursion" we can hope that "keeping things small there" has more of a positive effect than the negative effect caused by the superfluous calls at the beginning. Niedermeier and Rossmanith have show that this intuition is correct:

Fact 5.1 ([14]) *Let* K *be an arbitrary kernel algorithm that produces kernels of polynomial size. Let* B *be a D-branching algorithm running in polynomial time. Then* $T_{\text{SeqInterleave-B-K}}(n, k) = \text{size}_B(n, k) + n^{O(1)} \leq \xi_D^k + n^{O(1)}$.

5.1 Simple Parallel Interleaving

Interleaving also helps to reduce the work of parallel search tree algorithms: Consider the algorithm ParInterleave-B-K, the version of ParSearchTree-B that applies K at the beginning of each recursive call. First applying K and then computing branch instances using B is itself a branching algorithm and, thus, Theorem 3.1 tells us that $T_{\mathsf{SeqInterleave\text{-}B\text{-}K}}(n, k) = W_{\mathsf{ParInterleave\text{-}B\text{-}K}}(n, k)$ holds. This observation suggests that in order to minimise the work, we have to choose the most work-efficient kernel algorithm K available to us. However, it turns out that we have more options in the parallel setting: The work of K is only relevant at the very beginning, when the input size still depends on n. Later on, all remaining computations get inputs whose size depends only on the parameter. For these calls, the work of K is no longer relevant – it is "drowned out" by ξ_D^k. This suggests the following strategy: We use *two* kernels, namely an *initial kernel* whose job is to quickly and, more importantly, work-efficiently reduce the input size once (how such kernels can be constructed was exactly what we investigated in Sect. 4); and then use an *interleaving kernel* during the actual interleaving, whose job is just to kernelize the intermediate instances as quickly as possible – but we need no longer care about the work! Let us write A|B for the sequential concatenation of algorithms A and B.

Theorem 5.2. *Let* B *be a D-branching algorithm, and let* $K_{\mathsf{init}}, K_{\mathsf{interleave}}$ *be polynomial-sized kernels. Then* $W_{K_{\mathsf{init}}|\mathsf{ParInterleave\text{-}B\text{-}K_{inteleave}}}(n, k) \in O(W_{K_{\mathsf{init}}}(n, k) + \xi_D^k)$.

5.2 Shallow Parallel Interleaving

At the end of Sect. 3 we introduced the idea of *shallow* search trees as a method to speedup parallel search tree algorithms. However, shallow search trees are not necessarily compatible with the interleaving technique: From the parallel point of view, a "perfect" branching algorithm would branch on input (G, k) in constant time to $m = \xi_D^k$ simple instances $(G_1, 1), \ldots, (G_m, 1)$, all of which can then be processed in parallel. Applying a kernel at this point is "too late": The work will be something like $m = \xi_D^k$ times work of the kernel, which is decidedly not of the form ξ_D^k *plus* the work of the kernel.

What goes wrong here is, of course, that we parallelize "too much": we must ensure that the kernel algorithm gets a chance to kick in while the inputs still have a *large* enough size. On inputs of (still) large parameter k, all branches have to have a parameter of size *at least* εk (normally, we want a parameter *at most* εk). We remark that it does not follow from [14] that interleaving is possible here since [14] considers only the case where the number of branch instances is bounded by a constant. For the following theorem, let us write $d(x, k)$ for the branching vector d used by B on input (x, k) and $|d(x, k)|$ for its length.

Theorem 5.3. *Let* B *be a D-branching algorithm such that for all inputs* (x, k), *(a) the work done by* B *is at most* $|d(x, k)| \cdot |x|^{O(1)}$ *and (b) the maximum value in* $d(x, k)$ *is at most* $(1-\varepsilon)k + O(1)$. *Let* K *be a polynomially-sized kernel algorithm. Then* $W_{\mathsf{ParInterleave\text{-}B\text{-}K}}(n, k) = O(W_K(n, k) + \xi_D^k)$.

Note that the search trees arising from the branching rule B_* always have property (b), that is, they never "parallelize too well" since we capped to size m of M to $(k - |P|)/(s + 1)$ and, thus, $k - |P| - |X_1| - \cdots - |X_m| \geq k - |P| - s(k - |P|)/(s + 1) = (k - |P|)/(s + 1)$, meaning that we can set $\varepsilon = 1/(s + 1)$.

6 Conclusion and Outlook

We have extended the field of parallel parameterized algorithms with respect to work-optimality. This is a first step towards the aim of closing the gap between theoretical parallel algorithms (which are fast but produce massive work) and algorithms that work well in practice. To that end we provided a framework that allows to transform sequential search tree algorithms as well as kernelizations into parallel algorithms that are work-efficient. Furthermore, we have shown that combining both techniques via interleaving is still possible in the parallel setting. There are multiple paths to extend this line of research: It would be interesting to know if the presented algorithms do, in fact, lead to competitive parallel implementations. From the theory point of view, a natural next step is to study which other fpt techniques allow work-optimal implementations.

References

1. Bannach, M., Stockhusen, C., Tantau, T.: Fast parallel fixed-parameter algorithms via color coding. In: IPEC 2015, pp. 224–235 (2015)
2. Bannach, M., Tantau, T.: Parallel multivariate meta-theorems. In: IPEC 2016, pp. 4:1–4:17 (2016)
3. Bannach, M., Tantau, T.: Computing hitting set kernels by AC^0-circuits. In: STACS 2018, pp. 9:1–9:14 (2018)
4. Bannach, M., Tantau, T.: Computing kernels in parallel: Lower and upper bounds. In: IPEC 2018, pp. 13:1–13:14 (2018)
5. Cai, L., Chen, J., Downey, R.G., Fellows, M.R.: Advice classes of parameterized tractability. Ann. Pure Appl. Logic 84(1), 119–138 (1997)
6. Cesati, M., Di Ianni, M.: Parameterized parallel complexity. In: Pritchard, D., Reeve, J. (eds.) Euro-Par 1998. LNCS, vol. 1470, pp. 892–896. Springer, Heidelberg (1998). https://doi.org/10.1007/BFb0057945
7. Cheetham, J., Dehne, F., Rau-Chaplin, A., Stege, U., Taillon, P.J.: Solving large fpt problems on coarse-grained parallel machines. JCSS 67(4), 691–706 (2003)
8. Chen, J., Kanj, I.A., Jia, W.: Vertex cover: further observations and further improvements. J. Algorithms 41(2), 280–301 (2001)
9. Downey, R., Fellows, M.R.: Parameterized Complexity. Springer, New York (1999). https://doi.org/10.1007/978-1-4612-0515-9
10. Elberfeld, M., Stockhusen, C., Tantau, T.: On the space and circuit complexity of parameterized problems: classes and completeness. Algorithmica 71(3), 661–701 (2015)
11. Han, Y.: An improvement on parallel computation of a maximal matching. IPL 56(6), 343–348 (1995)
12. JáJá, J.: An Introduction to Parallel Algorithms. Addison-Wesley, Reading (1992)

13. Karp, R.M., Wigderson, A.: A fast parallel algorithm for the maximal independent set problem. J. ACM **32**(4), 762–773 (1985)
14. Niedermeier, R., Rossmanith, P.: A general method to speed up fixed-parameter-tractable algorithms. IPL **73**(3), 125–129 (2000)

Arbitrary Pattern Formation on Infinite Grid by Asynchronous Oblivious Robots

Kaustav Bose[(✉)][(iD)], Ranendu Adhikary[(iD)], Manash Kumar Kundu[(iD)],
and Buddhadeb Sau

Department of Mathematics, Jadavpur University, Kolkata, India
{kaustavbose.rs,ranenduadhikary.rs,
manashkrkundu.rs}@jadavpuruniversity.in,
bsau@math.jdvu.ac.in

Abstract. The ARBITRARY PATTERN FORMATION problem asks to design a distributed algorithm that allows a set of autonomous mobile robots to form any specific but arbitrary geometric pattern given as input. The problem has been extensively studied in literature in continuous domains. This paper investigates a discrete version of the problem where the robots are operating on a two dimensional infinite grid. The robots are assumed to be autonomous, identical, anonymous and oblivious. They operate in Look-Compute-Move cycles under a fully asynchronous scheduler. The robots do not agree on any common global coordinate system or chirality. We have shown that a set of robots can form any arbitrary pattern, if their starting configuration is asymmetric.

Keywords: Distributed algorithm · Autonomous robots
Arbitrary Pattern Formation · Grid · Asynchronous
Look-Compute-Move cycle

1 Introduction

1.1 Motivation

Distributed coordination of autonomous mobile robot systems has attracted considerable research interest in recent years owing to its potential applications in a wide range of real-world problems. The problem of forming an arbitrary geometric pattern is a fundamental coordination task for autonomous robot swarms. The pattern formation problem has been extensively investigated in continuous domains under different assumptions. In the continuous setting, the robots are assumed to be able to execute accurate movements in arbitrary directions and by arbitrarily small amounts. Hence, even in densely crowded situations, the robots can maneuver avoiding collisions. Certain models also permit the robots to move along curved trajectories, in particular, circumference of a circle. For robots with weak mechanical capabilities, it may not be possible to execute such intricate movements with precision. This motivates us to consider the problem

© Springer Nature Switzerland AG 2019
G. K. Das et al. (Eds.): WALCOM 2019, LNCS 11355, pp. 354–366, 2019.
https://doi.org/10.1007/978-3-030-10564-8_28

in a grid based terrain where the movements of the robots are restricted only along grid lines and only by a unit distance in each step. Grid type floor layouts can be easily implemented in real life robot navigation systems using magnets or optical guidances.

1.2   Earlier Works

The ARBITRARY PATTERN FORMATION problem was first studied by Suzuki and Yamashita [13,14] on the Euclidean plane. In these papers, a complete characterization of the class of formable patterns has been provided for autonomous and anonymous robots with an unbounded amount of memory. They characterized the class of formable patterns by using the notion of *symmetricity* which is essentially the order of the cyclic group that acts on the initial configuration. In [9], Flocchini et al. investigated the solvability of the problem for fully asynchronous and oblivious robots. Initially, the robots are in arbitrary positions, with the only requirement that no two robots are in the same position. They showed that if the robots have no common agreement on coordinate system, they cannot form an arbitrary pattern. If the robots have one-axis agreement, then any odd number of robots can form an arbitrary pattern, but an even number of robots cannot, in the worst case. If the robots agree on both X and Y axes, then any set of robots can form any pattern. They also proved that it is possible to elect a leader for $n \geq 3$ robots if it is possible to form any pattern. In [7,8], the authors studied the relationship between ARBITRARY PATTERN FORMATION and LEADER ELECTION among robots in asynchronous scheduler. They provided algorithms to form an arbitrary pattern starting from any geometric configuration wherein the leader election is possible. More precisely, their solutions work for four or more robots with chirality and for at least five robots without chirality. Combined with the result in [9], they deduced that ARBITRARY PATTERN FORMATION and LEADER ELECTION are equivalent, i.e., it is possible to solve ARBITRARY PATTERN FORMATION for $n \geq 4$ with chirality (resp. $n \geq 5$ without chirality) if and only if LEADER ELECTION is solvable. While all the previous works considered robots with unlimited visibility, Yamauchi et al. [15] first studied the problem with limited visibility. Randomized pattern formation algorithms were studied in [3,16]. In [6], Das et al. investigated the problem of forming a sequence of patterns in a given order. In [4,10], the problem was studied allowing the pattern to have multiplicities. In [5,10] the so-called EMBEDDED PATTERN FORMATION problem was studied where the pattern to be formed is provided as a set of visible points in the plane. Recently in [11], the pattern formation problem was studied on an infinite grid for robots with limited visibility. The problem was studied in synchronous setting for robots with constant size memory, and having a common coordinate system. Furthermore, robots were given a fixed point on the grid so that they can form a connected configuration containing it. Other specific types of formation problems that have been studied in the infinite grid set up, are the *Gathering* problem [12], i.e., the point formation problem and the *Mutual Visibility* problem [1], where a set of opaque robots have to form a pattern in which no three robots are collinear.

2 Model and Definition

2.1 The Model

Robots: The robots are *autonomous* (there is no central control), *homogeneous* (they execute the same distributed algorithm), *anonymous* (they have no unique identifiers), *identical* (they are indistinguishable by their appearance) and *oblivious* (they have no memory of past configurations and previous actions). The robots cannot explicitly communicate with each other. The robots have *global visibility* which means that they can observe the entire grid and the positions of all the robots. The robots do not have access to any common global coordinate system. In particular, they do not have a common notion of direction or chirality. Each robot has its own local view of the world with respect to its local Cartesian coordinate system. All the robots are initially positioned on distinct grid points.

Movement: The movement of the robots are restricted only along grid lines from one grid point to one of its four neighboring grid points. Traditionally in discrete domains, robot movements are assumed to be instantaneous. For simplicity of analysis, we also assume the movements to be instantaneous. This implies that the robots are always seen on grid points, not on edges. However, our strategy will also work without this assumption (by asking the robots to wait i.e, do nothing, if they see a robot on a grid edge).

Look-Compute-Move Cycles: The robots, when active, operate according to the so-called LOOK-COMPUTE-MOVE cycle. In each cycle, a previously idle or inactive robot wakes up and executes the following steps. In the LOOK phase, the robot takes the snapshot of the positions of all the robots, represented in its own local coordinate system. Based on the perceived configuration, the robot performs computations according to a deterministic algorithm to decide whether to stay put or to move to an adjacent grid point. Based on the outcome of the algorithm, the robot either remains stationary or makes an instantaneous move to an adjacent grid point.

Scheduler: We assume that the robots are controlled by a fully asynchronous adversarial scheduler ($ASYNC$). This implies that the amount of time spent in LOOK, COMPUTE, MOVE and inactive states by different robots is finite but unbounded and unpredictable. As a result, the robots do not have a common notion of time and the configuration perceived by a robot during the LOOK phase may significantly change before it actually makes a move.

2.2 Basic Geometric Definitions

Consider a team of a finite number of robots placed on the vertices of a simple undirected connected graph $G = (V, E)$. Define a function $f : V \longrightarrow \mathbb{N} \cup \{0\}$, where $f(v)$ is the number of robots on the vertex v^1. The pair (G, f) is called a

---

[1] Since we have assumed that the robots are initially positioned on distinct grid points and our algorithm guarantees collisionless movements, $f(v)$ is always either 0 or 1.

configuration of robots on G, or simply a *configuration*. Given a configuration of robots $\mathscr{C}$, let $\mathscr{R}$ denote the smallest grid-aligned rectangle that contains all the robots.

An *automorphism* of a graph $G = (V, E)$ is a bijection $\varphi : V \longrightarrow V$ such that for all $u, v \in V$, u, v are adjacent if and only if $\varphi(u), \varphi(v)$ are adjacent. The set of all automorphisms of G forms a group, called the *automorphism group* of G and is denoted by $Aut(G)$. The definition of automorphism of graphs can be extended to robot configurations on graphs. An *automorphism of a configuration* (G, f) is an automorphism φ of G such that $f(v) = f(\varphi(v))$ for all $v \in V$. The set of all automorphisms of (G, f) also forms a group that will be denoted by $Aut(G, f)$. We shall refer to an automorphism of a configuration as a *symmetry*. We shall call a symmetry *trivial* if $\varphi(v) = v$, for all $v \in V$ with $f(v) \neq 0$. If a configuration admits no non-trivial symmetries, then it is called an *asymmetric configuration*, and otherwise, a *symmetric configuration*.

An infinite path is the graph $P = (\mathbb{Z}, E)$, where $E = \{(i, i + 1) \mid i \in \mathbb{Z}\}$. An infinite grid can be defined as the Cartesian product $G = P \times P$. Assume that the infinite grid G is embedded in the Cartesian plane $\mathbb{R}^2$. It is not difficult to see that $Aut(G)$ is generated by three types of automorphisms: translations, reflections and rotations. A translation shifts all the vertices of G by the same amount. Since a configuration (G, f) has only finite number of robots, it is not difficult to see that $Aut(G, f)$ has no translations. Reflections are defined by an axis of reflection. The axis can be horizontal or vertical or diagonal. The angle of rotation can be of 90 or 180°, and the center of a rotation can be a vertex, or the center of an edge, or the center of the unit square.

The solvability of the arbitrary pattern formation problem depends on the symmetries of the initial configuration of the robots. This paper exclusively considers only asymmetric initial configurations. Some impossibility results regarding symmetric configurations are briefly discussed in Sect. 5.

2.3 The Arbitrary Pattern Formation Problem

A swarm of k robots is arbitrarily deployed on the vertices of the infinite grid. We assume that the initial configuration $\mathscr{C}_{init}$ is asymmetric, and no two robots are in the same position. The goal of the ARBITRARY PATTERN FORMATION problem is to design a distributed algorithm that guides the robots to form an arbitrary geometric pattern $\mathscr{C}_{target}$. The pattern $\mathscr{C}_{target}$ is a set of k (distinct) vertices in the grid given in an arbitrary Cartesian coordinate system. The pattern $\mathscr{C}_{target}$ is given to all robots in the system as input. Due to absence of a common global coordinate system, the robots decide that the pattern is formed when their present configuration becomes 'similar' to $\mathscr{C}_{target}$ with respect to translations, rotations, reflections. We say that a pattern formation algorithm is *collision-free*, if, at any time t, there are no two robots that occupy the same grid point. Avoiding collisions is a necessary requirement of the problem under this model. This is because, if two robots at any point in time, occupy the same grid point, they can not be deterministically separated thereafter, as they both execute the same deterministic algorithm.

3   Pattern Formation on a Finite Grid

In this section, we will discuss a related problem that will be used in the main algorithm. Consider a set of k robots deployed on an $m \times n$ finite grid. Starting from any arbitrary (symmetric or asymmetric) configuration, they are required to form a given arbitary pattern. Unlike our original problem, we assume that the robots agree on a common global coordinate system. The input $\mathscr{C}_{target}$ is also given in this coordinate system. Hence, the given input corresponds to a fixed set $\mathscr{T}$ of k grid points on the grid and our problem is to place a robot on each of these grid points. All the other assumptions from our original problem, stated in Sect. 2.1, are retained.

We shall first consider the case where $m = 1$, i.e., the grid is just a discretized line segment. Since the robots have a common global coordinate system, they agree on left and right. Hence, in the starting configuration, the robots can be labeled as $r_1, \ldots, r_k$ from left to right. If we can devise a swap-free (the act of two adjacent robots exchanging their positions is called a swap) and collision-free movement strategy, then the labels will remain unchanged throughout the algorithm. Note that, in the asynchronous setting, a collision-free algorithm is necessarily swap-free. We can also label the grid points in $\mathscr{T}$ as $t_1, \ldots, t_k$ from left to right. Our strategy is to simply ask each r_i to go to t_i. In order to avoid collisions, a robot will move to an adjacent grid point only if it is empty. A pseudocode description of the strategy is given in Algorithm 1.

Algorithm 1: Pattern formation on a $1 \times n$ grid

---

1 **Procedure** PFONPATH()
2     $s \in \{left, right\}$
3     $r_i \leftarrow$ me
4   **if** *I am not at* t_i **then**
5   **if** t_i *is on my s* **then**
6             $u \leftarrow$ the adjacent grid point on my s
7   **if** u *is empty* **then**
8   Move to u

Now we consider the general $m \times n$ finite grid. An $m \times n$ finite grid can be seen as a coiled up path as shown in Fig. 1. To be precise, an $m \times n$ grid has a spanning subgraph isomorphic to the finite path P_{mn}. But there are many such spanning subgraphs. The common global coordinate system allows the robots to agree on a particular subgraph as shown in Fig. 1. Hence, pattern formation on a finite grid reduces to pattern formation on a path, which can be solved by algorithm PFONPATH().

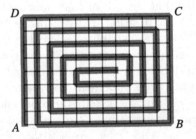

Fig. 1. A coiled up path in a finite grid.

4   The Main Algorithm

Consider a configuration $\mathscr{C}$, where $\mathscr{R} = ABCD$ is an $m \times n$ rectangle with $|AB| = n \geq m = |AD|$ (See Fig. 2). Here, the size of a side is defined as the number of grid points on it. If all robots in $\mathscr{C}$ lie on one grid line, then $\mathscr{R}$ is just a line segment. In this case, when we say $\mathscr{R} = ABCD$, it is to be regarded as a $1 \times n$ 'rectangle' with $A = D$, $B = C$ and $|AD| = |BC| = 1$. Let us first assume that $ABCD$ is a non-square rectangle with $|AB| = n > m = |AD| > 1$. We associate a binary string of length mn to each corner of $\mathscr{R}$. The binary string associated to a corner A is defined as follows. Scan the grid from A along the shorter side AD to D and sequentially all grid lines parallel to AD in the same direction. For each grid point, put a 0 or 1 according to whether it is empty or occupied. We denote this string by λ_{AD}. The three other strings λ_{BC}, λ_{CB} and λ_{DA} are defined similarly. If $ABCD$ is a square, i.e., $m = n$, then we have to associate two strings to each corner. In that case, the two sequences associated with A will be denoted by λ_{AD} and λ_{AB}. If any two of these strings are equal, then it implies that the configuration has a (reflectional or rotational) symmetry. Hence, if the configuration is asymmetric, then all the strings are distinct and we can find a unique lexicographically largest string. Assume that λ_{AD} is the lexicographically largest string. Then A will be called the *leading corner*. Once we have the unique lexicographically largest string λ_{AD}, the robots can agree on a common coordinate system as follows. The leading corner A is taken as origin and X-axis $= \overrightarrow{AB}$, Y-axis $= \overrightarrow{AD}$. Unless mentioned otherwise, any asymmetric configuration $\mathscr{C}$ will be expressed in this coordinate system. In the case where $m = 1$, λ_{AD} and λ_{DA} essentially refers to the same string. Hence, in this case, we have only two binary strings to compare. Again, if they are equal then the configuration is symmetric. Hence, if the configuration is asymmetric[2], then we shall have a leading corner, say $A = D$. Then A will be taken as origin and X-axis $= \overrightarrow{AB}$. But there will be no agreement on the Y-axis. However, as all the robots lie on the X-axis, the points in $\mathscr{C}$ can still be unambiguously expressed in coordinates.

---

[2] In the $m = 1$ case, we already have a reflectional symmetry with respect to $\overleftrightarrow{AB}$. But this is a trivial symmetry, and is to be ignored by our definition of asymmetric configurations.

Therefore, we see that in an asymmetric configuration, all the robots can agree on a common global coordinate system. By 'up' (resp. 'right') and 'down' (resp. 'left'), we shall refer to the positive and negative direction of Y (resp. X) axis of this coordinate system respectively. Also, given any asymmetric configuration $\mathscr{C}$, the robots corresponding to the first and the last 1 in the lexicographically largest string, will be called the *head* and the *tail* respectively. The remaining robots will be called *interior* robots. $\mathscr{C}'$ and $\mathscr{C}''$ will denote the sets $\mathscr{C} \setminus \{tail\}$ and $\mathscr{C} \setminus \{head,\ tail\}$ respectively.

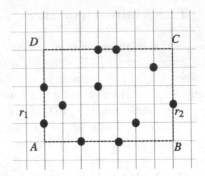

Fig. 2. In this configuration, the lexicographically largest string is $\lambda_{AD} =$ 010100001000100000001011000010100 00000010001000. The head and the tail are respectively r_1 and r_2.

The configuration $\mathscr{C}_{target}$, given to the robots as an input, is expressed in some arbitrary coordinate system. We can take the smallest enclosing rectangle of $\mathscr{C}_{target}$, call it $\mathscr{R}_{target}$. Assume that $\mathscr{R}_{target}$ is an $M \times N$ rectangle, with $N \geq M$. Now associate binary strings to its corners in the same manner as we did for $\mathscr{R}$. We shall assume that $\mathscr{C}_{target}$ is expressed in a coordinate system where the origin is the leading corner and the positive Y axis is along the side corresponding to the lexicographically largest string. No generality is lost, as the robots can always perform such a coordinate transformation on the input. However, unlike the previous case, we may not have a unique lexicographically largest string. This is because, the configuration $\mathscr{C}_{target}$ can have symmetries. In that case, for the coordinate transformation, we have to choose one among the largest strings to define the coordinate system. Notice that any choice leads to the same set of values. Therefore, in general, we shall assume that the origin of the coordinate system of $\mathscr{C}_{target}$ is *one of the* leading corners, and the positive Y axis is along the side corresponding to *one of the* lexicographically largest strings. We shall call this coordinate system the *canonical coordinate system*. Given $\mathscr{C}_{target}$ in the canonical coordinate system, we define $h_{target}, t_{target} \in \mathscr{C}_{target}$ as the points, corresponding to the first and the last 1 of the binary string that starts from the origin and goes along the Y axis respectively. Also, define $\mathscr{C}'_{target} = \mathscr{C}_{target} \setminus \{t_{target}\}$ and $\mathscr{C}''_{target} = \mathscr{C} \setminus \{h_{target}, t_{target}\}$.

We can logically divide the algorithm into seven phases. The starting configuration of the robots can fall into any one of the phases. These phases will be described in detail in the following subsections. Since the robots are oblivious, in each LOOK-COMPUTE-MOVE cycle, it has to infer from the perceived configuration, which phase it is currently in. It does so by checking if certain conditions are fulfilled or not. These conditions can be expressed in terms of Boolean variables listed in Fig. 3.

The main algorithmic difficulty of the problem arises from the restrictions imposed on the movements of the robots. In the continuous setting the robots can

C_0	$\mathscr{C} = \mathscr{C}_{target}$
C_1	$\mathscr{C}' = \mathscr{C}'_{target}$
C_2	Y-coordinate of the tail in $\mathscr{C} = Y$-coordinate of t_{target} in $\mathscr{C}_{target}$
C_3	$n \geq \max\{M, m\} + 2$
C_4	$n \geq 2 \cdot \max\{N, H\}$, where H is the length of the horizontal side of the smallest enclosing rectangle of $\mathscr{C}'$
C_5	The head in $\mathscr{C}$ is at the origin
C_6	$m \geq \max\{M, V\} + 1$, where V is the length of the vertical side of the smallest enclosing rectangle of $\mathscr{C}'$
C_7	$\mathscr{C}'' = \mathscr{C}''_{target}$
C_8	$\mathscr{C}'$ has a non-trivial reflectional symmetry with respect to a horizontal line

Fig. 3. The Boolean variable on the left is true if and only if the condition on the right is satisfied.

freely move in any direction by arbitrarily small amounts and in some models, along any curve. Therefore, techniques used in the previous works on continuous space (e.g., [7–9]) are not immediately portable in the discrete setting. Collision less movement is a major challenge in the grid model due to movement restriction. To resolve this, the tail expands the initial smallest enclosing rectangle (in Phase 1 and 3) making enough room for the interior robots to reconfigure themselves inside the rectangle without colliding. Our main idea is to utilize the asymmetry of the configuration to reach an agreement on a coordinate system, and try to keep the coordinate system invariant during the movements. To achieve this, in the first three phases, the head is put at the origin and the smallest enclosing rectangle is large enough so that the interior robots are confined in an appropriately small finite subgrid. Any movement by the interior robots restricted inside the finite subgrid keeps the coordinate system unaltered. So in Phase 4, the interior robots will rearrange themselves inside the finite subgrid to partially form the given pattern. In the final three phases, the head and the tail will move to their prescribed positions. Despite the apparent simplicity of the final three phases, designing movements is somewhat complicated as the coordinate system may change or the agreement in the coordinate system may be lost in some cases in the final phases.

We briefly describe the algorithm in the following subsection. The readers are referred to the full version [2] of the paper for a more detailed description of the algorithm and the proofs of the claims.

4.1 Description of the Algorithm

Phase 1. A robot infers that it is in phase 1, if $\neg(C_1 \wedge C_2) \wedge \neg(C_3 \wedge C_4)$ is true. In this case, the tail will move to the right and all other robots will remain static. Our aim is to make both C_3 and C_4 true. It can be proved that after finite number of moves by the tail, phase 1 completes with $\neg(C_1 \wedge C_2) \wedge (C_3 \wedge C_4) =$ true.

Phase 2. The algorithm is in phase 2, when either $C_3 \wedge C_4 \wedge \neg C_5 \wedge \neg C_7$ or $\neg C_2 \wedge C_3 \wedge C_4 \wedge \neg C_5 \wedge C_7$ is true. Our aim is to take the head to the origin. Hence, in this phase, the head will move down towards the origin. After finite number of moves by the head, phase 2 completes with $C_3 \wedge C_4 \wedge C_5 \wedge \neg C_7$ or $\neg C_2 \wedge C_3 \wedge C_4 \wedge C_5 \wedge C_7$ true.

Phase 3. The algorithm is in phase 3, if $C_3 \wedge C_4 \wedge C_5 \wedge \neg C_6 \wedge \neg C_7$ is true. In this phase, there are two cases to consider. The robots will check if C_8 is true or false. If C_8 is false, the tail will move upwards and the rest will remain static. Now assume that C_8 is true, i.e., $\mathscr{C}'$ has a non-trivial reflectional symmetry with respect to a horizontal line L. Again, let the smallest enclosing rectangle be $\mathscr{R} = ABCD$, with $|AD| = m$ and $|AB| = n$, $n > m$. Let λ_{AD} be the lexicographically largest string. Let E be the point of BC where it intersects with L. Let the smallest enclosing rectangle of $\mathscr{C}'$ be $\mathscr{R}' = AB'C'D'$. There are two cases to consider: $D \neq D'$ (case 1) and $D = D'$ (case 2). In case 1, the tail will move upwards, and in case 2, it will move downwards. It can be proved that after finite number of moves by the tail, we shall have $C_3 \wedge C_4 \wedge C_5 \wedge C_6 \wedge \neg C_7 =$ true.

Phase 4. If the configuration satisfies $C_3 \wedge C_4 \wedge C_5 \wedge C_6 \wedge \neg C_7 =$ true, then the algorithm is in phase 4. In this phase, the head and the tail will remain static. Let $\mathscr{F}$ be the subgrid of $\mathscr{R}$ of size $(m-1) \times \lfloor \frac{n}{2} \rfloor$ with coinciding bottom-left corners (See Fig. 4). $\mathscr{F}$ can be considered as a finite line segment $\mathscr{L}$ as shown in Fig. 4. The interior robots will execute the protocol PFONPATH() on $\mathscr{L}$ to achieve $\mathscr{C}'' = \mathscr{C}''_{target}$, i.e., $C_7 =$ true. After finite number of moves by the interior robots, Phase 4 completes with $C_3 \wedge C_4 \wedge C_5 \wedge C_6 \wedge C_7 =$ true.

Phase 5. The algorithm is in phase 5, if $\neg C_2 \wedge C_3 \wedge C_4 \wedge C_5 \wedge C_7$ is true. In this phase, the tail will move along the vertical grid line in order to make C_2 true.

If we have an asymmetric configuration $\mathscr{C}$ which is in phase 5, then depending on whether C_8 is true or false, there are two cases to consider.

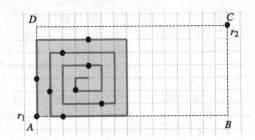

Fig. 4. A configuration in phase 4.

Let the smallest enclosing rectangle of $\mathscr{C}$ be $\mathscr{R} = ABCD$, with $|AD| = m$ and $|AB| = n$, $n > m$. Let A be the leading corner, and hence X-axis $= \overrightarrow{AB}$ and Y-axis $= \overrightarrow{AD}$. Now, let us plot the points of $\mathscr{C}_{target}$ in this coordinate system. Except h_{target} and t_{target}, all other points of $\mathscr{C}_{target}$ are occupied by the robots.

Let the smallest enclosing rectangle of $\mathscr{C}'$ be $\mathscr{R}' = AB'C'D'$ (See Fig. 5). Hence, the tail is currently on BC, and all the remaining robots are inside the region $AB'C'D'$.

Case-1. First, assume that C_8 is true. Since the head is at A, D' is also occupied due to the symmetry. Let C'' be the grid point where the grid lines $\overleftrightarrow{D'C'}$ and $\overleftrightarrow{BC}$ intersect. Let C''' be the middle point of BC''. C''' is a grid point if $|BC''|$ is odd. If the tail is on BC'', then $C = C''$ and $D = D'$. Note that the tail can not be on $[C''', C'']$, because then we shall have $\lambda_{DA} \geq \lambda_{AD}$. Hence, the tail is on $[B, C''')$ or (C'', ∞).

In this phase, we want to make C_2 true. This means that the tail needs to go to the grid point on $\overleftrightarrow{BC''}$ that is on the same horizontal line with t_{target}. Let us call this point $\tilde{t}_{target}$. Consider the case where $\tilde{t}_{target} \in [B, C'']$. In this case, the upper left corner of the smallest enclosing rectangle of $\mathscr{C}_{target}$ is D', which is occupied by a robot. Since the input is given in canonical coordinates, the bottom left corner (origin) of the smallest enclosing rectangle of $\mathscr{C}_{target}$, i.e., A, must be the leading corner. Therefore, A must be occupied in the final configuration. Since A is already occupied, it implies that C_1 is currently true. Also note that $\tilde{t}_{target} \notin (C''', C'']$, as A is the leading corner in $\mathscr{C}_{target}$. Hence, $\tilde{t}_{target}$ is on $[B, C''']$ or (C'', ∞).

Case 1A: tail $\in [B, C''')$ and $\tilde{t}_{target} \in [B, C''']$

The tail will move towards $\tilde{t}_{target}$. During the movements, the coordinate system remains invariant. However, if $\tilde{t}_{target}$ is at C''', a horizontal symmetry will be created when it reaches $\tilde{t}_{target}$.

Case 1B: tail $\in (C'', \infty)$ and $\tilde{t}_{target} \in (C'', \infty)$

The tail will move towards $\tilde{t}_{target}$. Again it is easy to see that the coordinate system remains invariant during the movements.

Case 1C: tail $\in (C'', \infty)$ and $\tilde{t}_{target} \in [B, C''']$

In this case, the tail will move downwards. When r reaches C'', the coordinate system flips. The new coordinate system has origin at D', X-axis $= \overrightarrow{D'C''}$ and Y-axis $= \overrightarrow{D'A}$. In the new coordinate system, r requires to place itself in $[C'', C''']$. Hence, the case is reduced to the situation similar to case 1A. Thus r achieves $C_2 = $ true without going beneath C'''.

Case 1D: tail $\in [B, C''')$ and $\tilde{t}_{target} \in (C'', \infty)$

In this case, the tail will move downwards. When r goes beneath B, the coordinate system flips. The new coordinate system has origin at D', X-axis $= \overrightarrow{D'C''}$ and Y-axis $= \overrightarrow{D'A}$. Clearly, the case is reduced to the situation similar to case 1B.

Case-2. Now assume that C_8 is false. It is easy to see that where ever $\tilde{t}_{target}$ is on $\overleftrightarrow{BC}$, the binary string attached to A is lexicographically strictly largest as the tail moves towards it. Hence, the movement of the tail in phase 5 does not change the coordinate system. Clearly, after finite number of moves, C_2 becomes true. Then phase 5 is completed with $C_2 \wedge C_3 \wedge C_4 \wedge C_5 \wedge C_7$ true.

Therefore, we can conclude that after finite number of steps, phase 5 completes with $C_2 \wedge C_3 \wedge C_4 \wedge C_5 \wedge C_7 = $ true.

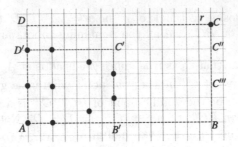

Fig. 5. Illustration of case 1 of phase 5.

Phase 6. If we have $\neg C_1 \wedge C_2 \wedge C_3 \wedge C_4 \wedge C_7 =$ true, then the algorithm is in phase 6. Consider an asymmetric configuration $\mathscr{C}$ which is in phase 6. Let $ABCD$ be the smallest enclosing rectangle, with λ_{AD} being the (strictly) largest string. Let H and T be the position of the head and the tail respectively. H and T are clearly on AD and BC respectively. Plot the points of $\mathscr{C}_{target}$ on the grid with respect to the current coordinate system (X-axis $= \overrightarrow{AB}$ and Y-axis $= \overrightarrow{AD}$). The smallest enclosing rectangle of these points is $AB'C'D$ (See Fig. 6). Let H' and T' be the points h_{target} and t_{target}. Therefore, if the head moves from H to H' and the tail moves from T to T', then the given pattern is formed. H' and T' are clearly on AD and $B'C'$ respectively, with T and T' being on the same horizontal line. The aim of this phase is to move the head from H to H'. It can be shown that the head can reach H' in finite number of moves. Therefore, phase 6 completes with $C_1 \wedge C_2 \wedge C_3 \wedge C_4 =$ true, and hence, $\neg C_0 \wedge C_1 \wedge C_2 =$ true.

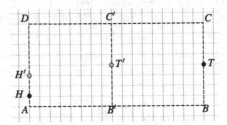

Fig. 6. Illustration of phase 6.

Phase 7. Finally, the algorithm is in phase 7, if we have $\neg C_0 \wedge C_1 \wedge C_2 =$ true. In this phase, the tail will move horizontally towards $T = t_{target}$. After finite number of moves by the tail, C_0 will become true.

It is not difficult to verify (See the full version [2]) that any configuration with $C_0 =$ false, belongs to one of the seven phases that we have discussed. Starting from any asymmetric configuration, our algorithm can form any given pattern in finite time (See the full version [2] for a phase transition diagram of our proposed algorithm). Hence, we can conclude the following theorem.

Theorem 1. ARBITRARY PATTERN FORMATION *is solvable in ASYNC from any asymmetric initial configuration.*

5   Concluding Remarks

We have proved that any arbitrary pattern is formable by a set of asynchronous robots if the initial configuration is asymmetric. The immediate course of future research would be to characterize the patterns formable from symmetric configurations. It can be proved that if a configuration $\mathscr{C}$ admits symmetry φ such that no robot lies on the axis of reflection or the center of rotation, then any configuration formable from $\mathscr{C}$ necessarily has the same symmetry φ. This is however not true, if the axis of reflection or the center of rotation contains a robot r. The symmetry may be broken by asking the robot r to move. However, this is not straightforward especially in a crowded situation. It would be also interesting to consider randomized algorithms. Another direction of future research would be to extend our work for patterns allowing multiplicities.

Acknowledgements. The first three authors are supported by NBHM, DAE, Govt. of India, CSIR, Govt. of India and UGC, Govt. of India respectively. We would like to thank the anonymous reviewers for their valuable comments which helped us improve the quality and presentation of this paper.

References

1. Adhikary, R., Bose, K., Kundu, M.K., Sau, B.: Mutual visibility by asynchronous robots on infinite grid. In: 14th International Symposium on Algorithms and Experiments for Wireless Networks (ALGOSENSORS 2018), Helsinki, Finland, 23–24 August 2018 (Forthcoming)
2. Bose, K., Adhikary, R., Kundu, M.K., Sau, B.: Arbitrary pattern formation on infinite grid by asynchronous oblivious robots. CoRR abs/1811.00834 (2018). http://arxiv.org/abs/1811.00834
3. Bramas, Q., Tixeuil, S.: Brief announcement: probabilistic asynchronous arbitrary pattern formation. In: Proceedings of the 2016 ACM Symposium on Principles of Distributed Computing, PODC 2016, Chicago, IL, USA, pp. 443–445, 25–28 July 2016. https://doi.org/10.1145/2933057.2933074
4. Cicerone, S., Di Stefano, G., Navarra, A.: Asynchronous arbitrary pattern formation. the effects of a rigorous approach. Distrib. Comput., 1–42 (2018) https://doi.org/10.1007/s00446-018-0325-7
5. Cicerone, S., Di Stefano, G., Navarra, A.: Embedded pattern formation by asynchronous robots without chirality. Distrib. Comput., 1–25 (2018). https://doi.org/10.1007/s00446-018-0333-7
6. Das, S., Flocchini, P., Santoro, N., Yamashita, M.: Forming sequences of geometric patterns with oblivious mobile robots. Distrib. Comput. **28**(2), 131–145 (2015). https://doi.org/10.1007/s00446-014-0220-9

7. Dieudonné, Y., Petit, F., Villain, V.: Leader election problem versus pattern formation problem. CoRR abs/0902.2851 (2009). http://arxiv.org/abs/0902.2851

8. Dieudonné, Y., Petit, F., Villain, V.: Leader election problem versus pattern formation problem. In: Lynch, N.A., Shvartsman, A.A. (eds.) DISC 2010. LNCS, vol. 6343, pp. 267–281. Springer, Heidelberg (2010). https://doi.org/10.1007/978-3-642-15763-9_26

9. Flocchini, P., Prencipe, G., Santoro, N., Widmayer, P.: Arbitrary pattern formation by asynchronous, anonymous, oblivious robots. Theor. Comput. Sci. **407**(1–3), 412–447 (2008). https://doi.org/10.1016/j.tcs.2008.07.026

10. Fujinaga, N., Yamauchi, Y., Ono, H., Kijima, S., Yamashita, M.: Pattern formation by oblivious asynchronous mobile robots. SIAM J. Comput. **44**(3), 740–785 (2015). https://doi.org/10.1137/140958682

11. Lukovszki, T., Meyer auf der Heide, F.: Fast collisionless pattern formation by anonymous, position-aware robots. In: Aguilera, M.K., Querzoni, L., Shapiro, M. (eds.) OPODIS 2014. LNCS, vol. 8878, pp. 248–262. Springer, Cham (2014). https://doi.org/10.1007/978-3-319-14472-6_17

12. Stefano, G.D., Navarra, A.: Gathering of oblivious robots on infinite grids with minimum traveled distance. Inf. Comput. **254**, 377–391 (2017). https://doi.org/10.1016/j.ic.2016.09.004

13. Suzuki, I., Yamashita, M.: Distributed anonymous mobile robots. In: SIROCCO 1996, The 3rd International Colloquium on Structural Information & Communication Complexity, Siena, Italy, pp. 313–330, 6–8 June 1996

14. Suzuki, I., Yamashita, M.: Distributed anonymous mobile robots: formation of geometric patterns. SIAM J. Comput. **28**(4), 1347–1363 (1999). https://doi.org/10.1137/S009753979628292X

15. Yamauchi, Y., Yamashita, M.: Pattern formation by mobile robots with limited visibility. In: Moscibroda, T., Rescigno, A.A. (eds.) SIROCCO 2013. LNCS, vol. 8179, pp. 201–212. Springer, Cham (2013). https://doi.org/10.1007/978-3-319-03578-9_17

16. Yamauchi, Y., Yamashita, M.: Randomized pattern formation algorithm for asynchronous oblivious mobile robots. In: Kuhn, F. (ed.) DISC 2014. LNCS, vol. 8784, pp. 137–151. Springer, Heidelberg (2014). https://doi.org/10.1007/978-3-662-45174-8_10

Packing and Covering

Packing 2D Disks into a 3D Container

Helmut Alt[1]([✉]), Otfried Cheong[2], Ji-won Park[2], and Nadja Scharf[1]

[1] Institut für Informatik, Freie Universität Berlin, Berlin, Germany
alt@mi.fu-berlin.de, nadja.scharf@fu-berlin.de
[2] School of Computing, KAIST, Daejeon, South Korea
otfried@kaist.edu, wldnjs1727@kaist.ac.kr

Abstract. In this article, we consider the problem of finding in three dimensions a minimum volume axis-parallel box into which a given set of unit size disks can be packed under translations. The problem is neither known to be NP-hard nor to be in NP. We give a constant factor approximation algorithm based on reduction to finding a shortest Hamiltonian path in a weighted graph. As a byproduct, we can show that there is no finite size container into which all unit disks can be packed simultaneously.

1 Introduction

Packing a set of geometric objects in a nonoverlapping way into a minimum size container is an intriguing problem and because of its practical significance it has been widely investigated. For a survey see [1,9] and the references therein. Even simple variants like packing a set of rectangles into a rectangular container turn out to be NP-hard [5]. Whereas that simple problem is in NP, in many cases not much is known about the true complexity of the problem.

Constant factor approximation algorithms of polynomial running time have been found for many variants of packing, in particular for finding minimum size rectangular or convex containers for a set of convex polygons under translations [2], i.e., the objects may be translated but rotations are not allowed. Also, approximation algorithms for rigid motions (translations and rotations) are known in this case, see e.g. [8].

In three dimensions, most previous results are concerned with "regular" packing problems where objects to be packed are axis-parallel boxes, see e.g. [4,7].

In addition, approximation algorithms for packing rectangular cuboids or convex polyhedra into minimum volume rectangular cuboid or convex containers are known if rigid motions are allowed [3].

Whether this is possible for translations only seems to be a much more difficult problem which remains open. In this paper, we give a positive answer for

N. Scharf was partially supported by a fellowship within the FITweltweit program and H. Alt by the Johann-Gottfried-Herder program, both of the German Academic Exchange Service (DAAD). J. Park was supported in part by starlab project (IITP-2015-0-00199) and NRF-2017M3C4A7066317.

G. K. Das et al. (Eds.): WALCOM 2019, LNCS 11355, pp. 369–380, 2019.
https://doi.org/10.1007/978-3-030-10564-8_29

a restricted set of possible objects, namely disks of unit radius and axis-parallel box containers (see Fig. 1). So far, our approximation factor is forbiddingly high but it should be of theoretical interest that the problem, which is neither known to be NP-hard nor to be in NP, can be approximated in polynomial time at all.

Packing disks in 3D is meant in the following sense: We say that two disks *touch* if their intersection contains only one point and that two disks *intersect* if their intersection consists of more than one point. By *nonoverlapping*, we mean that no two disks intersect whereas it is allowed that two disks touch. The main problem we study in this work is then defined as follows:

Definition 1 (disk-packing). *Given a set of unit disks in $\mathbb{R}^3$ by their unit normal vectors in $\mathbb{S}^2$. The goal is to find*

- *an axis-parallel box of minimum volume such that all disks can be packed without overlapping under translation inside the box*
- *and the actual packing of the disks inside the box.*

We assume that no two disks are the same, i.e., no two unit normal vectors are identical.

For an example see Fig. 1.

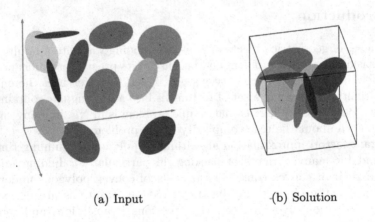

(a) Input   (b) Solution

Fig. 1. Example for disk-packing

We will reduce approximating this problem with a constant factor to approximating the following problem with a constant factor.

Definition 2 (disk-stabbing). *Given a set of nonidentical unit disks by their normal vectors in three dimensional space and an additional vector defining the direction of a line. The goal is to find an ordering of the disks with the following property: If the disks are placed nonoverlappingly with their centers in this order on the line, the distance from the center of the first disk to the last is minimum, in which case we call the distance length of the ordering.*

See Figs. 2(a) and (b) for 3D-disk-stabbing and Fig. 3 for 2D-disk-stabbing.

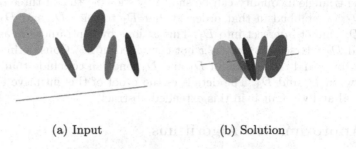

(a) Input   (b) Solution

Fig. 2. Example for 3D-disk-stabbing

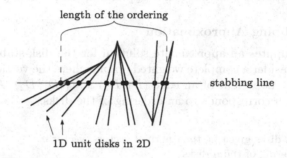

Fig. 3. Example for 2D-disk-stabbing. Here, the unit disks are unit line segments. Note the length of the ordering.

This problem then again will be reduced to finding the shortest Hamiltonian path in a complete weighted graph. To define the weights in the complete graph, we will use the function defined in the following.

Let $a \in S^2$ be a unit length vector in $\mathbb{R}^3$. Define $h_a(D_1, D_2)$ to be the distance of the centers of the disks D_1 and D_2 when placed with their centers on a line parallel to the vector a such that D_1 and D_2 touch. For a special case, if D_1 and D_2 have the same normal vectors, we define $h_a(D_1, D_2) = 0$. $h_a(D_1, D_2)$ can be computed easily from the normal vectors of D_1 and D_2 and it can be shown:

Lemma 1. *For any $a \in S^2$, h_a is a metric on the set of unit disks (modulo translation).*

If D_1 and D_2 have different normal vectors, it is clear that $h_a(D_1, D_2) > 0$; Otherwise, $h_a(D_1, D_2) = 0$ by definition.

Symmetry also can easily be observed: Assume that D_1 and D_2 are stabbed (without loss of generality by the x-axis as stabbing line) in that order so that they touch. Then a point reflection about the origin will preserve the orientation of the disks (invert their normal vectors) and the distance of their centers, whereas their order on the stabbing line is reversed.

The triangle inequality can be shown by showing that if three disks D_1, D_2, and D_3 are stabbed in that order so that D_1 touches D_2 and D_2 touches D_3 then D_3 cannot intersect into D_1. This is done by contradiction, assuming that D_1 and D_3 intersect in a point x not contained in D_2. Considering the triangle formed by x and the centers of D_1 and D_3, one can conclude that D_2 does not fit between D_1 and D_3. The details of the proof of this intuitive fact are quite technical and we omit it in this extended abstract.

2   Approximation Algorithms

Next, we will show how to reduce the disk-stabbing problem to finding the shortest Hamiltonian path in a complete weighted graph and obtain a constant factor approximation in this way. Afterwards we will use this approximation algorithm to compute a constant factor approximation for disk-packing.

2.1   Disk-Stabbing Approximation

Algorithm 1 computes an approximate solution for the disk-stabbing problem. The idea is to consider a complete weighted graph, where the vertices correspond to the disks and the weight of an edge (D_1, D_2) is $h_a(D_1, D_2)$. A Hamiltonian path in this graph corresponds to an ordering of the disks.

Input: n unit disks given by their normal vectors, vector a
Output: Ordering of the n disks
1 Generate complete weighted graph G with n vertices:
2 Set the weight of the edge (i, j) to $h_a(D_i, D_j)$ for all $1 \leq i, j \leq n, i \neq j$;
3 For all $1 \leq i, j \leq n$ with $i \neq j$, approximate shortest Hamiltonian path on the graph with endpoints i and j with Hoogeveen's algorithm [6] and determine the overall shortest path;
4 **return** the ordering of the overall shortest path;

Algorithm 1: Approximation algorithm for disk-stabbing

Theorem 1. *Algorithm 1 computes a $\frac{5}{3}$-approximation for disk-stabbing in polynomial time.*

Proof. By Lemma 1 the triangle inequality holds in G. Let $O = D_{i_1}, D_{i_2}, \ldots D_{i_n}$ be the optimal ordering for the input instance and OPT the length of O when stabbed by a line in direction a in order O. Then, there is a path in G from vertex i_1 to i_n visiting each vertex exactly once, i.e., a Hamiltonian path, of length OPT. Therefore, the algorithm of Hoogeveen finds a Hamiltonian path of length at most $\frac{5}{3}$OPT and this yields an ordering of length at most $\frac{5}{3}$OPT. Since the algorithm of Hoogeveen runs in polynomial time, Algorithm 1 runs in polynomial time.

In the next section, we will use Algorithm 1 to approximate disk-packing.

2.2 Disk-Packing Approximation

The idea for the approximation algorithm for disk-packing is as follows. We divide the disks into three subsets corresponding to the three axes such that the disks are almost orthogonal to the assigned axis. Then, we use Algorithm 1 to compute disk-stabbings of the three sets on the corresponding axes. The result can be interpreted as three containers, of which one is possibly very wide, one very high and the third very deep. The other two dimensions are relatively small. The last step is to divide these three boxes into pieces and arrange those pieces such that they form one single axis-parallel box. To describe the details of the algorithm, we use the following definitions.

We define $w_{\max}, d_{\max}, h_{\max}$ to be the maximum extent of any disk in x-,y-, and z-direction respectively and, thus, the minimum width, depth, and height any container for the disks must have. Let $w_{box} = s \cdot w_{\max}$ and $d_{box} = s \cdot d_{\max}$ for a constant $s > 1$ to be defined later. Algorithm 2 gives the details of the idea described above.

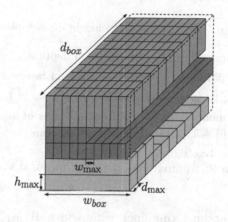

Fig. 4. Example container for, e.g., $s = 10.5$. The green boxes are the enlarged pieces obtained by dividing the container-box computed by Algorithm 1 for the disks in $\mathcal{X}$. Here, they form two layers. The blue boxes contain disks from $\mathcal{Y}$ and the orange boxes contain disks from $\mathcal{Z}$.

To analyze Algorithm 2 we first give a bound on W, D, and H as defined in Algorithm 2. Observe that the angle between the normal vector of a disk and the axis it gets stabbed by in Algorithm 2 can be at most $\varphi = \arccos(\frac{1}{\sqrt{3}})$.

Lemma 2. *It holds that*

$$W \leq 109 \cdot \frac{\mathrm{OPT}}{d_{\max} h_{\max}}, D \leq 109 \cdot \frac{\mathrm{OPT}}{w_{\max} h_{\max}}, H \leq 109 \cdot \frac{\mathrm{OPT}}{w_{\max} d_{\max}},$$

where OPT *is the volume of an optimal container.*

Input: n unit disks given by their normal vectors

Output: Nonoverlapping packing of the disks into an axis-parallel box

1 Partition the n disks into three sets $\mathcal{X}, \mathcal{Y}, \mathcal{Z}$ according to the axis their normal vectors form the smallest angle with;

2 Call Algorithm 1 for the disks in $\mathcal{X}$ and vector $(1,0,0)$. If L_x is the length of the returned ordering, this can be interpreted as a packing of the disks in $\mathcal{X}$ into an axis-parallel box of width $W = L_x + w_{\max}$, depth $d_{\max}$, and height $h_{\max}$;

3 Analogously to Step 2 apply Algorithm 1 for the disks in $\mathcal{Y}$ and $\mathcal{Z}$ giving lengths L_y and L_z, respectively. This can be seen as packing $\mathcal{Y}$ and $\mathcal{Z}$ into boxes of dimensions $w_{\max} \times D \times h_{\max}$ and $w_{\max} \times d_{\max} \times H$, respectively, where $D = L_y + d_{\max}$ and $H = L_z + h_{\max}$;

4 Divide the box obtained for $\mathcal{X}$ into pieces of width $w_{box} - w_{\max}$;

5 Assign each disk to the piece its point with smallest x-coordinate lies in;

6 Enlarge each piece from width $w_{box} - w_{\max}$ to width w_{box} such that all disks that are assigned to a piece are completely contained in that piece;

7 Divide the box obtained for $\mathcal{Y}$ into pieces of depth d_{box} analogously to Steps 4 to 6;

8 Divide the box obtained for $\mathcal{Z}$ into $\left\lfloor \frac{w_{box}}{w_{\max}} \right\rfloor \left\lfloor \frac{d_{box}}{d_{\max}} \right\rfloor$ pieces of width $w_{\max}$ and depth $d_{\max}$;

9 Analogously to Steps 5 and 6, enlarge the height of each piece from step 8 by $h_{\max}$;

10 Arrange all pieces into a box of width w_{box} and depth d_{box}, so that the pieces containing disks of $\mathcal{X}$ form $\left\lceil \left\lceil \frac{W}{w_{box} - w_{\max}} \right\rceil \Big/ \left\lfloor \frac{d_{box}}{d_{\max}} \right\rfloor \right\rceil$ layers of height $h_{\max}$, the pieces containing disks of $\mathcal{Y}$ form $\left\lceil \left\lceil \frac{D}{d_{box} - d_{\max}} \right\rceil \Big/ \left\lfloor \frac{w_{box}}{w_{\max}} \right\rfloor \right\rceil$ layers of height $h_{\max}$, and the pieces containing disks from $\mathcal{Z}$ form one layer of height $H \Big/ \left(\left\lfloor \frac{w_{box}}{w_{\max}} \right\rfloor \left\lfloor \frac{d_{box}}{d_{\max}} \right\rfloor \right) + h_{\max}$ (See Figure 4 for an example);

11 **return** the resulting box with the packed disks;

Algorithm 2: Approximation algorithm for disk-packing

Proof. Consider an optimal container with width W_{OPT}, depth D_{OPT}, and height H_{OPT} and let $\mathcal{X}, \mathcal{Y}, \mathcal{Z}$ be the partition of disks into subsets as in Algorithm 2. Furthermore consider a square grid of grid cells with side length g on the x-z-plane and lines parallel to the y-axis through the grid cell centers (see Fig. 5(a) for illustration). Then, each point has distance at most $\frac{g}{\sqrt{2}}$ to the closest line. So, in the optimal packing, every disk in $\mathcal{Y}$ is stabbed by a line in a point of distance at most $\frac{g}{\sqrt{2}\sin\left(\frac{\pi}{2} - \varphi\right)}$ from the disk center if g is small enough, i.e., $cg < 1$, where $c = \frac{1}{\sqrt{2}\sin\left(\frac{\pi}{2} - \varphi\right)} = \sqrt{\frac{3}{2}}$. See Fig. 5(b) for illustration.

Therefore, each disk in $\mathcal{Y}$ contains a disk of radius $1 - cg$ stabbed by a line through its center. So, by placing the $\left\lceil \frac{H_{\text{OPT}}}{g} \right\rceil \left\lceil \frac{W_{\text{OPT}}}{g} \right\rceil$ line segments of length D_{OPT} that are the intersection of the container and the lines behind each other so that they touch, we get a solution to the disk-stabbing problem for the disks in $\mathcal{Y}$ but with radius $1 - cg$. By stretching this solution by $1/(1 - cg)$, we get a solution for disks of radius 1. Let L_{OPT_y} be the length of an optimal solution for

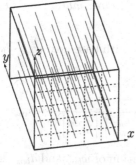

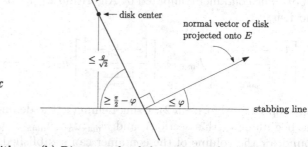

(a) Optimal container with grid and line segments stabbing the disks.

(b) Distance of a disk center to the stabbing line. The figure shows the cross section with the plane E that contains the stabbing line and the center of the disk.

Fig. 5. Aids for disk-packing.

the disk-stabbing problem for the disks in $\mathcal{Y}$. Then, this length can be at most the length of our solution, i.e.,

$$L_{\text{OPT}_\mathcal{Y}} \le \left\lceil \frac{H_{\text{OPT}}}{g} \right\rceil \left\lceil \frac{W_{\text{OPT}}}{g} \right\rceil D_{\text{OPT}} \cdot \frac{1}{1 - cg}.$$

By using $w_{\max}, h_{\max} \le 2$ and $W_{\text{OPT}} \ge w_{\max}, H_{\text{OPT}} \ge h_{\max}$, it follows that

$$L_{\text{OPT}_\mathcal{Y}} \le \frac{(g+2)^2}{g^2(1-cg)} \cdot \frac{\text{OPT}}{w_{\max} h_{\max}}. \tag{1}$$

Since we use Algorithm 1 to compute a disk-stabbing solution for $\mathcal{Y}$, we get by Theorem 1

$$D \le \frac{5}{3} \cdot L_{\text{OPT}_\mathcal{Y}} + d_{\max},$$

where the extra term $d_{\max}$ comes from the fact that the length of a disk-stabbing is defined as the distance of the center of the first disk to the center of the last disk and we are interested in the total depth of the packing. By inequality (1),

$$D w_{\max} h_{\max} \le \left(\frac{5(y+2)^2}{3g^2(1-cg)} + 1 \right) \text{OPT}.$$

Optimizing for g yields $g = \sqrt{\frac{1}{3}\left(27 + 4\sqrt{6}\right)} - 3 \ (\approx 0.5022)$ and a factor of approximately 108.49. The calculations for W and H are analogous. This implies the lemma.

Now, we are ready to state the main theorem of this article.

Theorem 2. *Algorithm 2 computes a constant factor approximation for disk-packing in polynomial time.*

Proof. The container computed by Algorithm 2 is a box with base area $w_{box} \cdot d_{box}$ and height

$$\left\lceil \frac{\left\lceil \frac{W}{w_{box}-w_{max}} \right\rceil}{\left\lfloor \frac{d_{box}}{d_{max}} \right\rfloor} \right\rceil h_{max} + \left\lceil \frac{\left\lceil \frac{D}{d_{box}-d_{max}} \right\rceil}{\left\lfloor \frac{w_{box}}{w_{max}} \right\rfloor} \right\rceil h_{max} + \frac{H}{\left(\left\lfloor \frac{w_{box}}{w_{max}} \right\rfloor \left\lfloor \frac{d_{box}}{d_{max}} \right\rfloor \right)} + h_{max}$$

(see step 10 in Algorithm 2). Using Lemma 2, the definition of w_{box} and d_{box} (see the beginning of this section), and $w_{max}d_{max}h_{max} \leq$ OPT the following upper bound for the volume of the container can established by some rearrangements and simple estimations:

$$s^2 \left(\frac{2 \cdot \frac{109}{s-1} + 1}{s-1} + \frac{109}{(s-1)^2} + 3 \right) \text{OPT}.$$

Optimizing for s gives an approximation factor that is slightly smaller than 593 for $s \approx 5.7334$, the real root of $6s^3 - 17s^2 + 15s - 658$.

3   Unbounded Containers Are Necessary

At first glance, the question may seem strange whether all, uncountably many, unit disks in three-space can be packed into a finite volume container. However, in dimension two, obviously all unit length line segments can be packed nonoverlapping into a rectangle of area 2, as Fig. 6 shows (There are even smaller containers). Observe that no two distinct segments intersect in interior points.

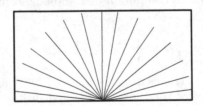

Fig. 6. How to pack all unit length segments into a container of area 2.

However, as a corollary of our previous results we will conclude that there is no bounded size container into which all unit disks can be packed. More precisely, we will show that even for a subset of all disks there is no bounded size container into which all unit disks from that subset can be packed. To do so, we need the following lemma.

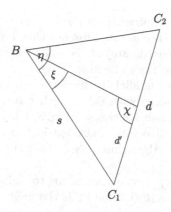

Fig. 7. Triangle formed by C_1, C_2, and B.

Lemma 3. *Let D_1 and D_2 be two disks whose normal vectors form an angle ξ smaller than $\frac{\pi}{2}$. Then their distance $h_a(D_1, D_2)$ when stabbed by a line in direction a is at least $\sin \xi$ for any $a \in S^2$.*

Proof. Let C_1 and C_2 be the positions of the two disk centers on the line. Consider the shortest path P from C_1 to C_2 on the planes σ_1 containing D_1 and σ_2 containing D_2. It is easy to see that P contains only one bend on the intersection line l of σ_1 and σ_2. We refer to this point by B. Observe that C_1 or C_2 must have distance at least 1 to B since otherwise D_1 and D_2 would intersect in B. Furthermore, P forms an angle η of at least ξ at B. To see that, suppose C_1 is fixed and C_2 can move parallel to l inside σ_2. Observe that the smallest angle P can form in this way is ξ.

Now, we consider the triangle formed by C_1, C_2, and B. Let without loss of generality the distance s of C_1 to B be at least 1. Then within the plane through C_1, C_2, and B we have the situation shown in Fig. 7. Consider the ray in this plane emanating from B that has an angle of ξ with the line segment $\overline{C_1 B}$ which hits the line segment $\overline{C_1 C_2}$ since $\eta \geq \xi$. By the law of sines:

$$\frac{s}{\sin \chi} = \frac{d'}{\sin \xi}$$

hence

$$d \geq d' = \frac{s \cdot \sin \xi}{\sin \chi} \geq \sin \xi$$

where the last inequality holds since $s \geq 1$ and $\sin \chi \leq 1$.

Theorem 3. *Packing a set of n unit disks requires a container of size $\Omega(\sqrt{n})$ in the worst case.*

Proof. In the following, we will show that $\Omega(\sqrt{n})$ is a lower bound for the container constructed by Algorithm 2 which is within a constant factor of the optimal container. From that the theorem follows immediately.

We identify every unit disk with its normal vector in the upper half of a unit sphere S^2 centered at the origin. Observe that for every normal vector in the lower half (negative z-coordinate) of the unit sphere there is a vector in the upper half corresponding to the same disk.

Consider the projection parallel to the z-axis of a $c \times c$-square partitioned into a square grid of grid cells with side length ε in the xy-plane centered at the origin onto the upper half sphere, see Fig. 8. Choose the constant c to be sufficiently small so that all points contained in the projection correspond to disks contained in set $\mathcal{Z}$ in Algorithm 2. Note that the grid contains $n = \Omega(\frac{1}{\varepsilon^2})$ points.

For any two grid-points p_1, p_2 corresponding to disks D_1 and D_2 it holds that $h_a(D_1, D_2) \geq \sin \xi$ with $a = (0,0,1)$ and ξ is the angle between the two normal vectors by Lemma 3.

By construction, see Fig. 9, the projected points $s_1, s_2 \in S^2$ have Euclidean distance $2 \cdot \sin(\xi/2)$ which is at least ε. So we have

$$\varepsilon \leq 2\sin(\xi/2) \leq 2\sin\xi$$

So by Lemma 3, for any two disks D_1, D_2 corresponding to grid-points in the $c \times c$-square we have $h_a(D_1, D_2) \geq \varepsilon/2$. Therefore, no matter in which order these disks are stabbed they will occupy a segment of length $\Omega((1/\varepsilon^2) \cdot \varepsilon)$, i.e., $\Omega(1/\varepsilon)$ of the stabbing line which is $\Omega(\sqrt{n})$. From Theorem 2 it follows that this is also a lower bound for the volume of a container computed by Algorithm 2, and, since that is within a constant factor of the optimum, of any container for the set of disks.

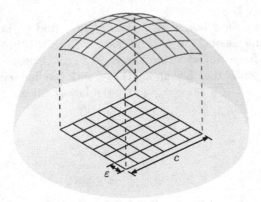

Fig. 8. Projecting a grid onto the unit sphere.

From Theorem 3, we obtain immediately

Corollary 1. *There is no bounded size container into which all unit disks can be packed.*

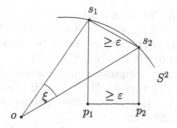

Fig. 9. Two grid points p_1, p_2 and their projection onto S^2 with center o.

4   Other Objects, Open Problems

Observe that our approximation algorithm can be extended to any arbitrary fixed planar shape S, provided that S can be enclosed by some disk D (i.e., is bounded) and contains some disk d (i.e., it has nonempty interior). More precisely, if we are given finite set of congruent copies of S in three dimensions we can approximate the smallest axis-parallel box into which it can be packed by translations.

This can be done by just applying our algorithm to the corresponding set of copies of D. Since it gives a constant factor approximation of the optimal packing of the D's it also gives an approximation of the optimal packing of the d's. Observe however, that the approximation factor is multiplied by r^3 where r is the ratio between the radii of D and d. Since the optimal packing of the S's provides some packing of the d's its container must be at least as large from which we obtain an approximation for the S's.

Notice however, that the approximation factor obtained this way depends on the shape of S. For standard shapes such as squares ($r = \sqrt{2}$), equilateral triangles ($r = 2$) etc., we can directly compute it from our approximation factor.

It remains an open problem whether an optimal packing of disks of different radii can be efficiently approximated.

In particular, approximating the packing of arbitrarily oriented boxes or convex polyhedra seems to be much more difficult.

References

1. Alt, H.: Computational aspects of packing problems. Bull. EATCS **118**, 28–42 (2016). http://eatcs.org/images/bulletin/beatcs118.pdf
2. Alt, H., de Berg, M., Knauer, C.: Approximating minimum-area rectangular and convex containers for packing convex polygons. JoCG **8**(1), 1–10 (2017)
3. Alt, H., Scharf, N.: Approximating Smallest Containers for Packing Three-Dimensional Convex Objects. Int. J. Comput. Geom. Appl. **28**(2), 111–128 (2018). https://doi.org/10.1142/S0218195918600026
4. Diedrich, F., Harren, R., Jansen, K., Thöle, R., Thomas, H.: Approximation algorithms for 3D orthogonal knapsack. J. Comput. Sci. Technol. **23**(5), 749–762 (2008). https://doi.org/10.1007/s11390-008-9170-7

5. Fowler, R.J., Paterson, M., Tanimoto, S.L.: Optimal packing and covering in the plane are NP-complete. Inf. Process. Lett. **12**(3), 133–137 (1981)
6. Hoogeveen, J.A.: Analysis of christofides' heuristic: some paths are more difficult than cycles. Oper. Res. Lett. **10**(5), 291–295 (1991)
7. Jansen, K., Prädel, L.: A new asymptotic approximation algorithm for 3-dimensional strip packing. In: Geffert, V., Preneel, B., Rovan, B., Štuller, J., Tjoa, A.M. (eds.) SOFSEM 2014. LNCS, vol. 8327, pp. 327–338. Springer, Cham (2014). https://doi.org/10.1007/978-3-319-04298-5_29
8. von Niederhäusern, L.: Packing Polygons. Master's thesis, EPFL Lausanne, FU Berlin (2014)
9. Scheithauer, G.: Zuschnitt- und Packungsoptimierung: Problemstellungen. Lösungsmethoden. Vieweg+Teubner Verlag, Modellierungstechniken (2008)

Covering and Packing of Rectilinear Subdivision

Satyabrata Jana[1](✉) and Supantha Pandit[2](✉)

[1] Indian Statistical Institute, Kolkata, India
satyamtma@gmail.com
[2] Stony Brook University, Stony Brook, NY, USA
pantha.pandit@gmail.com

Abstract. We study a class of geometric covering and packing problems for bounded closed regions on the plane. We are given a set of axis-parallel line segments that induce a planar subdivision with bounded (rectilinear) faces. We are interested in the following problems.

(P1) STABBING-SUBDIVISION: Stab all closed bounded faces by selecting a minimum number of points in the plane.

(P2) INDEPENDENT-SUBDIVISION: Select a maximum size collection of pairwise non-intersecting closed bounded faces.

(P3) DOMINATING-SUBDIVISION: Select a minimum size collection of bounded faces such that every other face has a non-empty intersection (i.e., sharing an edge or a vertex) with some selected face.

We show that these problems are NP-hard. We even prove that these problems are NP-hard when we concentrate only on the rectangular faces of the subdivision. Further, we provide constant factor approximation algorithms for the STABBING-SUBDIVISION problem.

Keywords: Planar subdivision · Set cover · Independent set Dominating set · NP-hard · PTAS

1 Introduction

The Set Cover and Independent Set problems are two well-studied problems in many fields. In the Set Cover problem, we are given a set of points and a set of geometric objects such that their union contains the set of points, and the goal is to find a minimum cardinality collection of objects that covers all of the given set of points. In the Independent Set problem, we are given a set of objects and seek a maximum cardinality subset of objects that are pairwise non-intersecting. The Dominating Set problem is a variation of the Set Cover problem in which we are given a set of objects and seek a minimum cardinality subset of objects such that every object has a non-empty intersection with one of the chosen objects.

S. Pandit—Partially supported by the Indo-US Science & Technology Forum (IUSSTF) under the SERB Indo-US Postdoctoral Fellowship scheme with grant number 2017/94, Department of Science and Technology, Government of India.

G. K. Das et al. (Eds.): WALCOM 2019, LNCS 11355, pp. 381–393, 2019.
https://doi.org/10.1007/978-3-030-10564-8_30

In this paper, we study variations of the Set Cover, Independent Set, and Dominating Set problems. We are given m axis-parallel line segments that induce a planar subdivision $\mathcal{P}$ with a set F of n bounded rectilinear faces. Further, we consider each bounded face to be a closed region, i.e. including the boundary. We formally define these problems as follows.

(P1) STABBING-SUBDIVISION: Given a planar subdivision having n bounded faces F, find a minimum cardinality set of points in the plane such that each face in F is stabbed (intersected) by one of the selected points.

(P2) INDEPENDENT-SUBDIVISION: Given a planar subdivision having n bounded faces F, find a maximum cardinality subset $F' \subseteq F$ of faces such that any pair of faces in F' is non-intersecting.

(P3) DOMINATING-SUBDIVISION: Given a planar subdivision having n bounded faces F, find a minimum cardinality subset $F' \subseteq F$ of faces such that any face in $F \setminus F'$ has a non-empty intersection with a face in F'.

A special case of the STABBING-SUBDIVISION problem has an application to the art gallery problem [4]. Suppose a rectangular art gallery is given. The gallery is subdivided into rectangular rooms. The art gallery problem seeks to find the fewest guards (points) so that every room (face) is protected (stabbed) by a guard point. This problem is precisely the STABBING-SUBDIVISION problem in which the input faces are all rectangular. More generally, we consider the case of rectilinear rooms (the original input of the STABBING-SUBDIVISION problem), not just rectangular rooms, and ask the same question, to find the fewest guards to protect all of the rectilinear rooms.

In this paper, we sometimes use the term "rectangle" and "rectangular face" (of a subdivision) interchangeably.

1.1   Previous Work

The Set Cover, Independent Set, and Dominating Set problems are NP-hard for simple geometric objects such as disks [5], squares [5], rectangles [5], etc. There is a long line of research of these problems and its various variants and special cases [1–3,7,10,12–15].

Recently, Korman et al. [9] studied an interesting variation of the Set Cover problem, the Line-Segment Covering problem. In this problem, they cover all the cells of an arrangement formed by a set of line segments in the plane using a minimum number of line segments. They showed that the problem is NP-hard, even when all segments are axis-aligned. In fact, they also proved that it is NP-hard to cover all rectangular cells of the arrangement by a minimum number of axis-parallel line segments.

In [6], Gaur et al. studied the rectangle stabbing problem. Here given a set of rectangles, the objective is to stab all rectangles with a minimum number of axis-parallel lines. They provided a 2-approximation for this problem.

Czyzowicz et al. [4] considered the guarding problem in rectangular art galleries. They showed that if a rectangular art gallery divided into n rectangular rooms, then $\lceil n/2 \rceil$ guards are always sufficient to protect all rooms in that rectangular art gallery. They also extend their result in non-rectangular galleries and 3-dimensional art galleries [4].

1.2 Our Results

In this paper, we present the following results.

➡ We first prove that the STABBING-SUBDIVISION problem is NP-hard when we stab all the rectangular faces of the subdivision. Next, we show that the STABBING-SUBDIVISION problem is NP-hard. Further, we provide a 2.083-approximation and a PTAS for this problem. (Sect. 2)

➡ We prove that the INDEPENDENT-SUBDIVISION problem is NP-hard when we consider only the rectangular faces. Then we prove that the INDEPENDENT-SUBDIVISION problem is NP-hard. (Sect. 3)

➡ We prove that the DOMINATING-SUBDIVISION problem is NP-hard by considering only the rectangular faces. Next, we prove that the DOMINATING-SUBDIVISION problem is NP-hard. (Sect. 4)

2 STABBING-SUBDIVISION

2.1 NP-hardness

We first prove that the STABBING-SUBDIVISION problem is NP-hard when we are restricted to stab only rectangular faces of the subdivision. Next, we modify the construction to show that the STABBING-SUBDIVISION problem is NP-hard. We give a reduction from the *Rectilinear Planar 3SAT (RP3SAT) Problem*. Lichtenstein [11] proved that the Planar 3SAT problem is NP-complete. Later, Knuth and Raghunathan [8] showed that every Planar 3SAT problem can be expressed as an *RP3SAT* problem. We define the *RP3SAT* problem as follows. We are given a *3-SAT* formula ϕ with n variables $x_1, x_2, \ldots, x_n$ and m clauses $C_1, C_2, \ldots, C_m$ where each clause contains exactly 3 literals. For each variable or clause take a rectangle. The variable rectangles are placed on a horizontal line such that no two of them intersect. The clause rectangles are placed above and below this horizontal line such that they form a nested structure. The clause rectangles connect to the variable rectangles by vertical lines such that no two lines intersect. The objective is to decide whether there is a truth assignment to the variables that satisfies ϕ. See Fig. 1(a) for an instance of the *RP3SAT* problem.

Variable gadget: The gadget of x_i consists of $8m + 4$ vertical and 4 horizontal line segments. See Fig. 1(b) for the construction of the gadget. The 4 segments

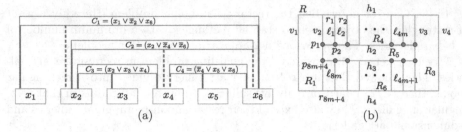

Fig. 1. (a) An instance of the *RP3SAT* problem. We only show the clauses which connect to the variables from above. The solid (resp. dotted) lines represent that the variable is positively (resp. negatively) present in the corresponding clauses. (b) Structure of a variable gadget.

v_1, v_4, h_1, and h_4 together form a rectangular region R. Next, the 2 vertical segments v_2 and v_3 partition R vertically into 3 rectangles R_1, R_2, and R_3. Further, two horizontal segments h_2 and h_3 partition R_2 horizontally into three rectangles R_4, R_5, and R_6. Finally, the $4m$ vertical segments $l_1, l_2, \ldots, l_{4m}$ partition R_4 vertically into $4m+1$ small rectangles $r_1, r_2, \ldots, r_{4m+1}$. Similarly, the $4m$ vertical segments $l_{4m+1}, l_{4m+2}, \ldots, l_{8m}$ partition R_6 vertically into $4m+1$ small rectangles $r_{4m+2}, r_{4m+3}, \ldots, r_{8m+2}$. Finally we have the total of $8m+5$ rectangles $R_1, R_3, R_5, r_1, r_2, \ldots, r_{8m+2}$ inside R. Clearly, these rectangles except R_5 form a cycle of size $8m+4$. Observe that any point along the cycle can stab at most two consecutive regions. Therefore there are two optimal solutions $P_1^i = \{p_1, p_3, \ldots, p_{8m+3}\}$ and $P_2^i = \{p_2, p_4, \ldots, p_{8m+4}\}$ each of size $4m+2$ (Note that these points are not as a part of the input, they are one set of canonical points.). These two solutions are corresponding to the truth value of x_i.

Clause gadget: The gadget for clause C_α consists of a single rectangle r_α that is formed by four line segments. The rectangle r_α can be interpreted as the same rectangle as C_α in the *RP3SAT*-problem instance.

Interaction: Now we describe how the clause gadgets interact with the variable gadgets. Observe that the description for the clauses which connect to the variables from above are independent with the clauses which connect to the variables from below. Therefore, we only describe the construction for the clauses which connect to the variables from above. Let $C_1^i, C_2^i, \ldots, C_\tau^i$ be the left to right order of the clauses which connect to x_i. Then we say that C_k^i is the kth clause for x_i. For example, C_3, C_2, and C_4 are the 1st, 2nd, and 3rd clause for the variable x_4 in Fig. 1(a). Let C_α be a clause containing the variable x_i, x_j, x_t. We say that clause C_α is the k_1, k_2, and $k_3{}^{\text{th}}$ clause for variable x_i, x_j, and x_t respectively based on the above ordering. For example, C_3 is the 3rd, 1st, and 1st clause for variable x_2, x_3, and x_4 respectively in Fig. 1(a). Let r_α be the rectangle corresponding to C_α. Now we have the following cases.

- If x_i appears as a positive literal in C_α, then extend the 3 segments l_{4k_1-3}, l_{4k_1-2}, and l_{4k_1-1} vertically upward such that it touches the bottom boundary of r_α. Move p_{4k_1-1} vertically upward to the bottom boundary of r_α.

- If x_i appears as a negative literal in C_α, then extend the 3 segments l_{4k_1-2}, l_{4k_1-1}, and l_{4k_1} vertically upward such that it touches the bottom boundary of r_α. Move p_{4k_1} vertically upward to the bottom boundary of r_α.

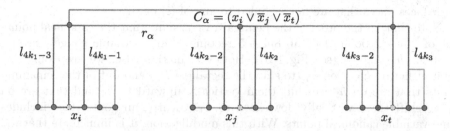

Fig. 2. Variable clause interaction.

The similar construction can be done for x_j and x_t by replacing k_1 with k_2 and k_3 respectively. The whole construction is shown in Fig. 2. Note that, we break the horizontal segment h_1 in the variable gadgets into smaller intervals and shifted the intervals vertically along with the extension of the vertical lines. This completes the construction and clearly, it can be done in polynomial time.

Lemma 1. ϕ *is satisfiable if and only if there is a solution to the* STABBING-SUBDIVISION *problem to stab only rectangular faces with* $n(4m+2)$ *points.*

Proof. Assume that ϕ is satisfiable i.e., we have a truth assignment of the variables in ϕ. Now consider a variable x_i. If x_i is true, we select the set P_1^i, otherwise we select the set P_2^i. Clearly, the $n(4m+2)$ selected points corresponding to all variable gadgets stab all the rectangular faces of the construction.

On the other hand, assume that STABBING-SUBDIVISION problem has a solution with $n(4m+2)$ points. Observe that at least $(4m+2)$ points are needed to stab all the faces of a variable gadget. Since the rectangular faces of variable gadgets are disjoint from each other, exactly $(4m+2)$ points must be selected from each variable gadget. Now there are exactly two solutions of size $(4m+2)$, either P_1^i or P_2^i. Therefore, we set variable x_i to be true if P_1^i is selected from the gadget of x_i, otherwise we set x_i to be false. Note that for each clause C_α the six faces corresponding to three literals it contains, touches the rectangle r_α. Since r_α is stabbed, at least one of the selected points must be chosen in the solution. Such a point is either in one of the sets P_1^i or P_2^i of the corresponding variable gadget based on whether the variable is positively or negatively present in that clause. Hence, the above assignment is a satisfying assignment. □

Theorem 1. *The* STABBING-SUBDIVISION *problem is* NP-*hard for stabbing only rectangular faces of a subdivision.*

The STABBING-SUBDIVISION problem for stabbing all rectilinear faces: We now prove that it is also NP-hard to stab all (rectilinear) faces of a subdivision. To get this result, we modify the NP-hardness result for the rectangular faces that is described above. Note that, after embedding the gadgets on the plane, the subdivision creates three types of faces, (i) variable faces: the faces interior to a variable gadget (note that all variable faces are rectangular), (ii) clause faces: rectangular faces associated with clause gadgets, and (iii) outer faces: faces that are not included to any of (i) or (ii).

Note that, in the proof of the Lemma 1, we assume that the canonical points (set of $4m + 2$ points to stab $8m + 5$ rectangles in a variable gadget) are on the lines h_2 and h_3 (see Fig. 1(b)). However, in this case, we keep only one point either on h_2 or on h_3 (to stab the rectangle R_5) and out of the remaining points that are on h_2 we shift them vertically upward to h_1 and that are on h_3 we shift them vertically downward to h_4. Clearly, any outer face includes some variable canonical points. With this modification, it is immediate that the Lemma 1 is true even when we are intended to stab all the faces (rectilinear) of the subdivision.

2.2   Approximation Algorithms

Factor 2.083 Approximation. We are given m axis-parallel line segments that induce a planar subdivision $\mathcal{P}$ with a set F of n bounded rectilinear faces. To provide the approximation algorithm, we transform any instance of the STABBING-SUBDIVISION problem into an instance of the Set Cover problem where the size of each set is at most 4. Observe that, there exists an optimal solution to the STABBING-SUBDIVISION problem that only contains vertices of $\mathcal{P}$ (we can call them as corner points of F). Also, any corner point of F can stab at most 4 rectilinear faces in $\mathcal{P}$.

We now create an instance of the Set Cover problem as follows. The set of elements is the set of all faces and the collection is all sets of faces corresponding to the corner points of F. Note that each set in the collection is of size at most 4, since any corner point can stab at most 4 faces. This Set Cover instance admits a 2.083 (H_4 i.e., the harmonic series sum of the first 4 terms) factor approximation [16]. Hence we have the following theorem.

Theorem 2. *There exists a 2.083 factor approximation algorithm for* STABBING-SUBDIVISION *problem in a planar subdivision by rectilinear line segments.*

2.3   PTAS via Local Search Algorithm

In this section, we show that a local search framework [14] leads to a PTAS for the STABBING-SUBDIVISION problem. We are given a planar subdivision with a set F of n bounded faces. Note that, we can choose points only from the vertex set V of the subdivision. Therefore, $\mathcal{R} = (V, F)$ be the given range space. Clearly V is a feasible solution to the STABBING-SUBDIVISION problem. We apply the k-level local search (k is a given parameter) as follows.

1. Let X be some feasible solution to the STABBING-SUBDIVISION problem (initially take X as V).
2. Do the following:
   (a) Search for $X' \subseteq X$ and Y such that $|Y| \subseteq V$, $|Y| < |X'| \leqslant k$ and $(X \setminus X') \cup Y$ is a feasible solution.
   (b) If such X' and Y exist, update X with $(X \setminus X') \cup Y$ and repeat the above step. Otherwise, return X and stop.

It is easy to see that the running time of the algorithm is polynomial in n and k. Further, the local search algorithm always returns a local optimum solution. A feasible solution X is said to be a local optimum if no X' exists in Step 2(a) in the above algorithm. We show that given any $\epsilon > 0$, a $O(1/\epsilon^2)$-level local search returns a hitting set of size at most $(1 + \epsilon)$ times an optimal hitting set for $\mathcal{R}$.

Locality condition ([14]): A range space $\mathcal{R} = (V, F)$ satisfies the locality condition if for any two disjoint subsets $R, B \subseteq V$, it is possible to construct a planar bipartite graph $G = (R \cup B, E)$ with all edges going between R and B such that for any $f \in F$, there exist two vertices $u \in f \cap R$ and $v \in f \cap B$ such that edge $(u, v) \in E$.

Theorem 3 [14]. *Let $\mathcal{R} = (V, F)$ be a range space satisfying the locality condition. Let $R \subseteq V$ be an optimal hitting set for F, and $B \subseteq V$ be the hitting set returned by a k-level local search. Furthermore, assume $R \cap B = \emptyset$. Then there exists a planar bipartite graph $G = (R \cup B, E)$ such that for every subset $B' \subseteq B$ of size at most k, $|N_G(B')| \geq |B'|$ where $N_G(W)$ denotes the set of all neighbours of the vertices of W in G.*

The following lemma implies that given any $\epsilon > 0$, a k-level local search with $\epsilon = \dfrac{c}{\sqrt{k}}$ gives a $(1 + \epsilon)$-approximation for the STABBING-SUBDIVISION problem.

Lemma 2 [14]. *Let $G = (R \cup B, E)$ be a bipartite planar graph on red and blue vertex sets R and B, $|R| \geq 2$, such that for every subset $B' \subseteq B$ of size at most k, where k is a large enough number, $|N_G(B')| \geq |B'|$. Then $|B| \leq (1 + \dfrac{c}{\sqrt{k}})|R|$, where c is a constant.*

PTAS for the STABBING-SUBDIVISION problem: Let R (red) and B (blue) be disjoint subsets of the vertices in planar subdivision $\mathcal{P}$ where R and B be an optimum solution and the solution returned by the k-level local search respectively. For simplicity, we assume that $R \cap B = \emptyset$. Otherwise, we can remove the common elements from each of R and B, and then do a similar analysis. As we remove the same number of elements from both R and B, the approximation ratio of the original instance is at most the approximation ratio of the restricted one. We construct the required graph G on the vertices $R \cup B$ in the following way. Since R and B are feasible solutions of the STABBING-SUBDIVISION problem, every face $f \in F$ must contain at least one red and one blue point. We simply join exactly one pair of red and blue points by an edge for each face $f \in F$. Clearly, the edge for a face $f \in F$ lies completely inside f. Therefore G becomes a planar bipartite graph and hence $\mathcal{R}$ satisfies the locality condition. Therefore, from Theorem 3 and Lemma 2, we say that the STABBING-SUBDIVISION problem admits a PTAS.

3   INDEPENDENT-SUBDIVISION

In this section, we prove that the INDEPENDENT-SUBDIVISION problem is NP-hard by giving a reduction from the *RP3SAT* problem. The reduction follows the same line of the reduction presented in Sect. 2. We construct an instance I of the INDEPENDENT-SUBDIVISION problem from an instance ϕ of the *RP3SAT* problem and prove that the construction is correct.

Variable gadget: The variable gadget is similar to the variable gadget that is described in the Sect. 2. See Fig. 3 for the construction of a variable gadget. The difference of this variable gadget from the gadget in the Sect. 2 is that we partition R_4 into $4m - 2$ smaller rectangles $r_1, r_2, \ldots, r_{4m-2}$ and R_6 into $4m - 2$ smaller rectangles $r_{4m-1}, r_{4m}, \ldots, r_{8m-4}$. Finally, we have the total of $8m - 1$ rectangles $R_1, R_3, R_5, r_1, r_2, \ldots, r_{8m-4}$ inside R. Notice that, these rectangles except R_5 form a cycle of size $8m - 2$. Therefore there are exactly two optimal solutions $S_1^i = \{R_3, r_1, r_3, \ldots, r_{4m-3}, r_{4m}, r_{4m+2}, \ldots, r_{8m-4}\}$ and $S_2^i = \{R_1, r_2, r_4, \ldots, r_{4m-2}, r_{4m-1}, r_{4m+1}, \ldots, r_{8m-5}\}$, each with size $4m - 1$. These two solutions are corresponding to the truth values of the variable x_i.

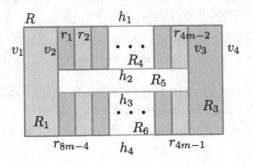

Fig. 3. Structure of a variable gadget.

Clause gadget: The gadget of the clause C_α includes 9 rectangles $r_\alpha^1, r_\alpha^2, \ldots, r_\alpha^9$ (see green rectangles in Fig. 4. The six rectangles $r_\alpha^4, r_\alpha^5, \ldots, r_\alpha^9$ are placed inside the rectangle of C_α in the *RP3SAT*-problem instance and the other three rectangles $r_\alpha^1, r_\alpha^2, r_\alpha^3$ are corresponding to the three vertical legs between C_α and the three variables it contains. Note that there is another rectangle present in the clause gadget bounded by the above 9 rectangles. However, this rectangle has no effect in the reduction, since picking this rectangle makes other 9 rectangles invalid (cannot be selected).

Interaction: Here also we describe the construction for the clauses that connect to the variables from above, since the construction is similar and independent from the clauses that connect to the variables from below. Let C_α be a clause containing the variables x_i, x_j, x_t. Also assume that this is the left to right order of these variables in which they appear in ϕ. Using the similar way as before (Sect. 2), we say that the clause C_α is the k_1, k_2, and k_3^{th} clause for the variables x_i, x_j, and x_t respectively.

- If x_i appears as a positive literal in the clause C_α, then attach the rectangle r_α^1 to the rectangle r_{4k_1-3}.
- If x_i appears as a negative literal in clause C_α, then attach the rectangle r_α^1 to the rectangle r_{4k_1-2}.

The similar construction can be done for x_j by replacing r_α^1 and k_1 with r_α^2 and k_2 respectively and for x_t by replacing r_α^1 and k_1 with r_α^3 and k_3 respectively. The whole construction is depicted in Fig. 4. Clearly, the construction can be done in polynomial time. We now prove the correctness of the construction.

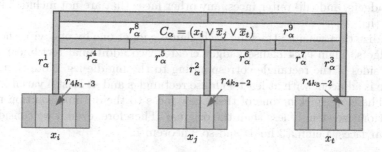

Fig. 4. Variable clause interaction.

Lemma 3. *ϕ is satisfiable if and only if there is a solution of size $n(4m-1)+4m$ to* INDEPENDENT-SUBDIVISION *problem while considering only rectangular faces.*

Proof. Assume that ϕ has a satisfying assignment. For the variable x_i, if x_i is true, select the set S_2^i, otherwise select the set S_1^i. Since each set is of cardinality $(4m-1)$, clearly we select $n(4m-1)$ independent rectangles across all variable gadgets. Now let C_α be a clause containing variables x_i, x_j, x_t. Since C_α is satisfiable at least one of the three rectangles $r_\alpha^1, r_\alpha^2, r_\alpha^3$ is free to choose in a solution. This implies we can select exactly 4 rectangles from the gadget of C_α. We can picked 4 rectangles independently from each clause gadget. Hence, in total we can select $n(4m-1)+4m$ rectangles.

On the other hand, assume that the INDEPENDENT-SUBDIVISION problem has a solution S with $n(4m-1)+4m$ rectangles. Note that for each variable gadget the size of an optimal independent set is $(4m-1)$, either the set S_1^i or S_2^i. We set the variable x_i to be true if S_2^i is selected from the gadget of x_i, otherwise we set x_i to be false. Now we have to show that this assignment is a satisfying assignment for ϕ i.e., each clause of ϕ is satisfied. Since the variable gadgets are independent, there are at most $n(4m-1)$ rectangles from the variable gadgets belongs to S. Also since the size of the solution is $n(4m-1)+4m$, from each clause gadget exactly 4 rectangles are is in S. Let C_α be a clause containing variables x_i, x_j, x_t. As there are 4 independent rectangles from the set $\{r_\alpha^1, r_\alpha^2, \ldots, r_\alpha^9\}$, so one must be from the set $\{r_\alpha^1, r_\alpha^2, r_\alpha^3\}$ that is in the given solution. W.l.o.g. let r_α^1 be present, then surely x_i is a true variable as our assignment. Hence the above assignment is a satisfying assignment. □

Theorem 4. *The* INDEPENDENT-SUBDIVISION *problem is* NP-*hard by considering only rectangular faces of a subdivision.*

The INDEPENDENT-SUBDIVISION problem for all rectilinear faces: We now prove that it is also NP-hard to find a maximum independent set of rectilinear faces in a subdivision. After embedding the construction on the plane, the subdivision creates three types of faces, (i) variable faces: The faces that are interior to a variable gadget, (ii) clause faces: the faces associated with the clause gadgets, and (iii) outer faces: any other faces that are not included in any of (i) or (ii).

Visualize that we are attaching each clause gadget one by one with the variable gadgets. Then each clause gadget creates two additional rectilinear faces, on both sides of the rectangle corresponding to the middle leg. Note that, each such face is adjacent with at least 4 clause rectangles and at least 4 variable rectangles. Therefore, picking one of these new faces to the optimal solution makes the solution size strictly less than the original. Therefore, even if we consider all rectilinear faces, Lemma 3 holds and so Theorem 4.

4   DOMINATING-SUBDIVISION

In this section, we prove that the DOMINATING-SUBDIVISION problem is NP-hard. We give a reduction from the *RP3SAT* problem similar to Sect. 3.

We construct an instance I of the DOMINATING-SUBDIVISION problem from an instance ϕ of the *RP3SAT* problem and prove that the construction is correct.

Variable gadget: Variable gadgets are similar to the variable gadgets that are described in Sect. 2. The difference between this variable gadget and that of in Sect. 2 is as follows. We partition R_4 into $3m+1$ small rectangles $r_1, r_2, \ldots, r_{3m+1}$ and R_6 into $3m+1$ small rectangles $r_{3m+4}, r_{3m+5}, \ldots, r_{6m+4}$. We partition R_1 into two rectangles r_{6m+6}, r_{6m+5} and R_3 into r_{3m+2}, r_{3m+3}. Next we take $2m+2$ mutually independent rectangles $s_1, s_2, \ldots, s_{2m+2}$ inside R_5 such that rectangle s_i touches the two regions r_{3i-2} and r_{3i-1}, for $1 \leq i \leq 2m+2$. Finally we have a total of $8m+8$ rectangles $r_1, r_2, \ldots, r_{6m+6}, s_1, s_2, \ldots, s_{2m+2}$ inside R. Figure 5 illustrate the construction of a variable gadget just described.

Lemma 4. *There exists exactly two optimal dominating sets of rectangles, $D_1^i = \{r_1, r_4, \ldots, r_{6m+4}\}$ and $D_2^i = \{r_2, r_5, \ldots, r_{6m+5}\}$, for the gadget of x_i.*

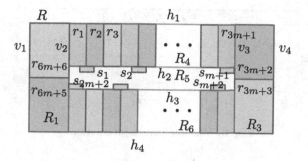

Fig. 5. Structure of a variable gadget.

Proof. There is no rectangle that can dominate more than 4 rectangles. Since there are in total $(8m + 8)$ rectangles, any dominating set cannot have size less than $(2m + 2)$. Further, both D_1^i and D_2^i, each of size $(2m + 2)$, dominate all the faces of the subdivision and hence they are optimal solutions. Now we show that there is no other optimal solution.

Clearly, no rectangle of the form r_{3k} or s_k where $1 \leq k \leq (2m + 2)$ can be a part of an optimal solution, since each of them dominates exactly 3 rectangles. As a result, any optimal solution contains only rectangles of the form r_{3k-1} or r_{3k-2}, for $1 \leq k \leq (2m + 2)$. Also, two rectangles, one of the form r_{3k-1} and other of the form r_{3k-2}, together cannot be a part of an optimal solution. $\square$

Clause gadget: The gadget for the clause C_α is a rectangle r_α (Fig. 6).

Interaction: Here we describe the construction for the clauses that connect to the variables from above. A similar construction can be done for the clauses that connect to the variables from below. As before, we interpret C_α that contains variables x_i, x_j, and x_t as the k_1, k_2, and k_3^{th} clause for the variables x_i, x_j, and x_t respectively.

- If x_i appears as a positive literal in the clause C_α, then we extend the rectangle r_{3k_1-1} vertically upward such that it touches the rectangle r_α.
- If x_i appears as a negative literal in the clause C_α, then we extend the rectangle r_{3k_1-2} vertically upward such that it touches the rectangle r_α.

We make the similar construction for x_j and x_t by replacing k_1 with k_2 and k_3 respectively. The whole construction is depicted in Fig. 6. Clearly, the construction can be done in polynomial time. We now prove the correctness.

Lemma 5. *ϕ is satisfiable if and only if there is a solution of size $n(2m + 2)$ to the* DOMINATING-SUBDIVISION *problem while considering only rectangular faces.*

Proof. Assume that ϕ is satisfiable i.e., we have a truth assignment to the variables of ϕ. For the variable x_i, if x_i is true we select the set D_2^i, otherwise we select the set D_1^i. Clearly, the $n(2m + 2)$ selected rectangles corresponding to all the variable gadgets dominate all the rectangular faces of the subdivision.

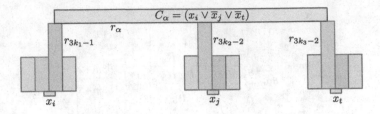

Fig. 6. Variable clause interaction.

On the other hand, assume that the DOMINATING-SUBDIVISION problem has a solution with $n(2m + 2)$ rectangles. Observe that at least $(2m + 2)$ rectangles are needed to dominate all the rectangular faces of a variable gadget. Since the rectangular faces of variable gadgets are disjoint from each other and the size of the solution is $n(2m+2)$, from each variable gadget exactly $(2m+2)$ rectangles must be selected. Therefore, we set variable x_i to be true if D_2^i is selected from the gadget of x_i, otherwise we set x_i to be false. Note that for each clause C_α the three rectangles corresponding to the three literals it contains attach to the rectangle r_α. Since r_α is dominated, at least one of these three rectangles is chosen in the solution. Such a rectangle is either in D_2^i or D_1^i of the corresponding variable gadget based on whether the variable is positively or negatively present in that clause. Hence, the above assignment is a satisfying assignment. □

Theorem 5. *The* DOMINATING-SUBDIVISION *problem is* NP-*hard when we are constrained to dominate all the rectangular faces of a subdivision.*

The DOMINATING-SUBDIVISION problem for all rectilinear faces: We only modify the variable gadgets such that it has exactly two distinct optimal solutions and the rest of the construction and the proofs remain the same. We take $2m + 2$ rectangles $b_1, b_2, \ldots, b_{2m+2}$. We place the rectangle b_i in between the rectangles r_{3i-2} and r_{3i-1}, for $1 \leq i \leq 2m + 2$ of the variable gadget shown in Fig. 5 (see Fig. 7). These additional rectangles enforce not to choose R_5 in an optimal solution. Now it is easy to verify that the Lemma 4 remains true for this modified gadget even when we consider all the bounded faces of the subdivision.

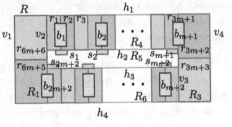

Fig. 7. Modified variable gadget.

References

1. Adamaszek, A., Wiese, A.: Approximation schemes for maximum weight independent set of rectangles. In: FOCS, pp. 400–409 (2013)
2. Chan, T.M., Har-Peled, S.: Approximation algorithms for maximum independent set of pseudo-disks. Discret. Comput. Geom. **48**(2), 373–392 (2012)
3. Chuzhoy, J., Ene, A.: On approximating maximum independent set of rectangles. In: FOCS, pp. 820–829 (2016)
4. Czyzowicz, J., Rivera-Campo, E., Santoro, N., Urrutia, J., Zaks, J.: Guarding rectangular art galleries. Discret. Appl. Math. **50**(2), 149–157 (1994)
5. Fowler, R.J., Paterson, M.S., Tanimoto, S.L.: Optimal packing and covering in the plane are NP-complete. Inf. Process. Lett. **12**, 133–137 (1981)
6. Gaur, D.R., Ibaraki, T., Krishnamurti, R.: Constant ratio approximation algorithms for the rectangle stabbing problem and the rectilinear partitioning problem. J. Algorithms **43**(1), 138–152 (2002)
7. Hochbaum, D.S., Maass, W.: Approximation schemes for covering and packing problems in image processing and VLSI. J. ACM **32**(1), 130–136 (1985)
8. Knuth, D.E., Raghunathan, A.: The problem of compatible representatives. SIAM J. Discret. Math. **5**(3), 422–427 (1992)
9. Korman, M., Poon, S.H., Roeloffzen, M.: Line segment covering of cells in arrangements. Inf. Process. Lett. **129**, 25–30 (2018)
10. van Leeuwen, E.J.: Optimization and approximation on systems of geometric objects. Ph.D. thesis, University of Amsterdam (2009)
11. Lichtenstein, D.: Planar formulae and their uses. SIAM J. Comput. **11**(2), 329–343 (1982)
12. Mudgal, A., Pandit, S.: Covering, hitting, piercing and packing rectangles intersecting an inclined line. In: Lu, Z., Kim, D., Wu, W., Li, W., Du, D.-Z. (eds.) COCOA 2015. LNCS, vol. 9486, pp. 126–137. Springer, Cham (2015). https://doi.org/10.1007/978-3-319-26626-8_10
13. Mustafa, N.H., Raman, R., Ray, S.: Settling the APX-hardness status for geometric set cover. In: FOCS, pp. 541–550 (2014)
14. Mustafa, N.H., Ray, S.: Improved results on geometric hitting set problems. Discret. Comput. Geom. **44**(4), 883–895 (2010)
15. Pandit, S.: Dominating set of rectangles intersecting a straight line. In: Canadian Conference on Computational Geometry, CCCG, pp. 144–149 (2017)
16. Vazirani, V.V.: Approximation Algorithms. Springer, Hecidelberg (2001)

Minimum Membership Covering and Hitting

Joseph S. B. Mitchell and Supantha Pandit$^{(\boxtimes)}$

Stony Brook University, Stony Brook, NY, USA
joseph.mitchell@stonybrook.edu, pantha.pandit@gmail.com

Abstract. Set cover is a well-studied problem with application in many fields. A well-known variation of this problem is the Minimum Membership Set Cover problem. In this problem, given a set of points and a set of objects, the objective is to cover all points while minimizing the maximum number of objects that contain any one point. A dual of this problem is the Minimum Membership Hitting Set problem. In this problem, given a set of points and a set of objects, the objective is to stab all of the objects while minimizing the maximum number of points that an object contains. We study both of these variations in a geometric setting with various types of geometric objects in the plane, including axis-parallel line segments, axis-parallel strips, rectangles that are anchored on a horizontal line from one side, rectangles that are stabbed by a horizontal line, and rectangles that are anchored on one of two horizontal lines (i.e., each rectangle shares at least one boundary edge (top or bottom) with one of the input horizontal lines). For each of these problems either we prove NP-hardness or design a polynomial-time algorithm. More precisely, we show that it is NP-complete to decide whether there exists a solution with depth exactly 1 for either the Minimum Membership Set Cover or the Minimum Membership Hitting Set problem. We also provide approximation algorithms for some of the problems. In addition, we study a generalized version of the Minimum Membership Hitting Set problem.

Keywords: Minimum Membership Set Cover
Minimum Membership Hitting Set · Rectangles · NP-hard · Segments
Strips · Depth of a point

1 Introduction

The set cover problem is one of the fundamental problems in computer science and combinatorial optimization. This problem and its many variations play

J.S.B. Mitchell—Partially supported by the National Science Foundation (CCF-1526406) and the US-Israel Binational Science Foundation (project 2016116).
S. Pandit—Partially supported by the Indo-US Science & Technology Forum (IUSSTF) under the SERB Indo-US Postdoctoral Fellowship scheme with grant number 2017/94, Department of Science and Technology, Government of India.

G. K. Das et al. (Eds.): WALCOM 2019, LNCS 11355, pp. 394–406, 2019.
https://doi.org/10.1007/978-3-030-10564-8_31

an important role in modelling various problems arising in practical scenarios. One of its variations is the *Minimum Membership Set Cover (MMSC)* problem, which is defined in a geometric setting as follows.

Minimum Membership Set Cover ($MMSC$): Given a point set P and a set O of objects (regions), cover all the points in P with a subset $O' \subseteq O$ of objects such that the maximum depth of a point is minimized, where the *depth of a point* $p \in P$ is the number of objects in O' that contain it. We say that O' is a *cover* of P, and we let $d(O')$ denote the maximum depth of any point $p \in P$ with respect to O'.

A related problem that is "dual" to the *MMSC* problem is the *Minimum Membership Hitting Set (MMHS)* problem, defined as follows.

Minimum Membership Hitting Set ($MMHS$): Given a point set P and a set O of objects (regions) determine a subset $P' \subseteq P$ of points stabbing (intersecting) all objects O such that the maximum depth of an object is minimized, where the *depth of an object* $o \in O$ is the number of points in P' that stab it. We say that P' is a *hitting set* of O, and we let $d(P')$ denote the maximum depth of any object $o \in O$ with respect to P'.

In addition to the above two problems, we consider a *generalized* version of the *MMHS* problem, the *Generalized Minimum Membership Hitting Set (GMMHS)* problem, where, instead of a point set and a object set, we are given two sets R ("red") and B ("blue") of objects. The objective is to stab (intersect) all of the objects in B using a subset $R' \subseteq R$ such that the maximum number of red objects in R' hitting any single object in B is minimized.

1.1   Previous Work

The standard set cover problem is NP-hard. A simple greedy heuristic gives a $O(\log n)$-factor approximation, and it is NP-hard to compute an approximation better than logarithmic [11]. The Minimum Membership Set Cover variation was first introduced by Kuhn et al. [6]. They showed that the problem cannot be approximated better than $O(\log n)$ and gave an approximation factor that matches this lower bound. Erlebach and van Leeuwen [3] considered the geometric variation of the problem, proving that for unit squares and unit disks the problem is NP-hard and there does not exist a polynomial-time factor 2 approximation algorithm, unless P = NP. Further, for unit squares, they provided a factor 5 approximation for the case in which the optimum objective value is bounded by a constant. Recently, Nandy et al. [9] reconsidered the same problem and gave polynomial-time algorithms for both unweighted and weighted intervals on the real line. Recently, Narayanaswamy et al. [10], considered the problem of hitting a set of horizontal segments with vertical segments while minimizing the number of times a vertical segment is hit by the chosen horizontal segments. They showed that this problem is NP-hard and cannot be approximated better than factor 2. Further, if the segments are of unbounded length (i.e., they are lines), then it can be solved in polynomial time (see also [2] for this algorithm and some generalizations of this problem).

1.2 Our Contributions: Overview

Minimum Membership Set Cover (*MMSC*) problem

We give a polynomial-time algorithm for deciding if there exists a cover with depth one for the *MMSC* problem with objects that are rectangles anchored on a horizontal line. In contrast, we show that if the objects are rectangles that intersect a horizontal line (versus that are anchored, sharing a side with a horizontal line), the *MMSC* problem is NP-hard. We also prove NP-hardness for the cases of objects that are axis-parallel strips or rectangles anchored on two horizontal lines.

Minimum Membership Hitting Set (*MMHS*) problem

We give a polynomial-time algorithm for deciding if there exists a hitting set with depth one for the *MMHS* problem with objects that are rectangles anchored on a horizontal line. In contrast, we show that if the objects are rectangles that intersect a horizontal line, the *MMHS* problem is NP-hard. We also prove NP-hardness for the cases of objects that are axis-parallel strips or rectangles anchored on two horizontal lines.

Generalized Minimum Membership Hitting Set (*GMMHS*) problem

We show that *GMMHS*, with objects R, B given as horizontal/vertical line segments, is NP-hard; even deciding if a solution exists with depth one is NP-complete. We also give a 5-approximation algorithm if the optimal objective function is bounded by a constant.

Equivalence of MMSC and MMHS with Unit Disks/Squares. There is a connection (equivalence) between the *MMSC* and *MMHS* problems where the input objects are either unit disks or unit squares. Consider the case of unit squares. Given an instance $C = (P, T)$ of the *MMSC* problem, with a set P of points and a set T of unit squares, we consider a "dual" instance, H, of a *MMHS* problem whose regions are specified by the set of unit squares centered on the points $p \in P$, and whose points are specified as the centerpoints of the squares $t \in T$. We then note that determining a solution to the *MMSC* problem C is equivalent to determining a solution to the *MMHS* problem H. Thus, we conclude, by applying the results in [3,9]: The *MMHS* problem is NP-complete with unit squares and unit disks and there exists a 5-approximation for the *MMHS* problem with unit squares where the optimal objective value is bounded by a constant.

1.3 Definitions and Notations

In a *3SAT* problem we are give a *CNF* formula ϕ with n variables $\mathcal{X} = x_1, x_2, \ldots, x_n$ and m clauses $\mathcal{C} = \{C_1, C_2, \ldots, C_m\}$ where each clause is a disjunction of exactly 3 literals, and the objective is to decide whether there is a truth assignment to variables such that ϕ is satisfiable. This problem is known to be NP-complete [4]. In a planar version of this problem, each variable or clause represents a vertex and there is an edge between a variable vertex and a clause vertex if and only if the corresponding clause contains the corresponding literal. Finally, the resulting bipartite graph is planar. This problem is called the *Planar-3SAT* problem and Lichtenstein [7] proved that this problem is also NP-complete. Later on, Knuth and Raghunathan [5] showed that every *Planar-3SAT* problem can be represented using the following rectilinear representation. The variables are placed on a horizontal line and the clauses containing 3 legs each connecting those variables either from above or below the horizontal line such that no two clause legs intersect. This problem is called the *Rectilinear-Planar-3SAT* problem and is also NP-complete [5]. A *Positive-1-in-3SAT* problem is a *3SAT* problem, however the objective is different: Here, the objective is to decide whether there is a truth assignment to the variables such that exactly one literal per clause is true. Schaefer [12] proved that this problem is NP-complete. This problem can be represented using the rectilinear representation as defined above; we refer to it as the *Rectilinear-Positive-Planar-1-in-3SAT* problem (see Fig. 1). Surprisingly, Mulzer and Rote [8] proved that it is also NP-complete.

We now define some terminology. Let $\mathcal{C}_{above} \subseteq \mathcal{C}$ be the set of clauses in a *PP1in3SAT* formula ϕ that connect to the variables from above. Similarly, let $\mathcal{C}_{below} \subseteq \mathcal{C}$ be the set of clauses that connect to the variables from below. For each variable x_i, $1 \leq i \leq n$,

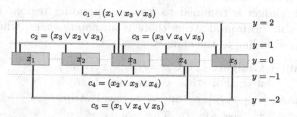

Fig. 1. Representation of a *Rectilinear-Positive-Planar-1-in-3SAT* problem.

we order the clauses in $\mathcal{C}_{above}$ left to right that connect x_i. Let $C_\ell \in \mathcal{C}_{above}$ be a clause containing the three variables x_i, x_j, and x_k. Then, according to the ordering defined above, we assume that C_ℓ is the ℓ_1-, ℓ_2-, and ℓ_3-th clause for the variables x_i, x_j, and x_k, respectively. For example, the clause C_3 is a 3-rd, 1-st, and 1-st clause for the variables x_3, x_4, and x_5, respectively, in the *PP1in3SAT* instance in Fig. 1. We also say that the clause C_ℓ connects to x_i by left, to x_j by middle, and to x_k by right legs.

2   Minimum Membership Set Cover Problem

2.1   Rectangles Anchored on a Horizontal Line

In polynomial time, one can decide if there exists a cover of depth one for the *MMSC* problem with rectangles anchored on a horizontal line from one side (*MMSCRAHL*), as follows. Let the weight of a rectangle be the number of points it contains. Now, apply the algorithm of [1] to compute a maximum weight independent set of rectangles (no two of them share an input point). Then, to see if there is a cover of points having depth exactly 1, check if the total weight of the independent set is equal to the number of input points.

2.2   Axis-Parallel Strips

In this section we prove that the *MMSC* problem with axis-parallel strips (*MMSCS*) is NP-hard. We give a reduction from the Positive-1-in-3SAT (*P1in3SAT*) problem (see Sect. 1.3 for the definition). Let ϕ be a *P1in3SAT* formula. We generate an instance $Z(S, P)$ of the *MMSCS* problem from ϕ in the following way, where S is a set of strips and P is a set of points.

Variable Gadget: For variable x_i, the gadget consists of one vertical strip v_i, one horizontal strip h_i, and a point p_i. The point is covered by both v_i and h_i (see Fig. 2). Clearly, either v_i or h_i will cover p_i with depth one. We assume that choosing h_i makes x_i true, while choosing v_i makes x_i false.

Overall Structure: We place the variable gadgets (points) along a diagonal line. For each clause we take a vertical bounded *region*. The clause gadgets are placed sequentially one by one to the right of the variable gadgets, and each gadget is confined to its corresponding region. Between two consecutive variable horizontal strips there is an *empty space*, where we place some points corresponding to the clauses.

Clause Gadget: Let $C_\ell = (x_i \vee x_j \vee x_k)$ be a clause. For this clause, we take 5 points $p_i^\ell, p_j^\ell, p_k^\ell, p_1^\ell, p_2^\ell$ and 4 vertical strips $q^\ell, r^\ell, s^\ell, t^\ell$ (see Fig. 2). The points $p_i^\ell,\ p_j^\ell$, and p_k^ℓ are corresponding to the variables x_i, x_j and x_k respectively and are placed inside the strips h_i, h_j, and h_k respectively. The other two points p_1^ℓ and p_2^ℓ are placed in any empty space between the variable horizontal strips of x_i, x_j (i.e., between h_i and h_j) and x_j, x_k (i.e., between h_j and h_k) respectively. Points $\{p_i^\ell, p_1^\ell\}$ are contained in q^ℓ. Similarly,

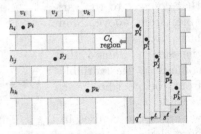

Fig. 2. Gadgets of variables x_i, x_j, x_k, and clause C_ℓ and their interaction.

$\{p_1^\ell, p_j^\ell\}$, $\{p_j^\ell, p_2^\ell\}$, and $\{p_2^\ell, p_k^\ell\}$ are contained in r^ℓ, s^ℓ, and t^ℓ, respectively. These 5 points and 4 rectangles are strictly contained inside the vertical region of C_ℓ (Fig. 2).

Theorem 1. *The MMSCS problem is NP-hard.*

Proof. We prove that, ϕ is satisfiable (i.e., at least one literal is true per clause) if and only if $Z(P, S)$ has a solution of depth one. Assume that ϕ has a satisfying assignment. If x_i is true, take h_i; otherwise, take v_i. Now, for each clause, exactly one of $p_i^\ell, p_j^\ell, p_k^\ell$ is covered by the solution. Hence, the remaining 4 points are covered by exactly two strips with depth one.

On the other hand, assume that there is a cover of the points with depth one. Now, for each variable gadget, to cover p_i we need one of the two strips h_i

or v_i. We set variable x_i to be true if h_i is in the solution; otherwise, we set x_i to be false. Now consider any clause C_ℓ. Since the depth of the solution (indeed a cover of all points) is one, exactly one of $p_i^\ell, p_j^\ell, p_k^\ell$ corresponding to C_ℓ is covered by a variable horizontal strip. We set this variable to be true. Hence, exactly one literal per clause is true in ϕ.   □

Corollary 1. *The MMSC problem with rectangles anchored on two orthogonal lines (MMSCRATOL) is NP-hard. (Take a vertical and a horizontal line both at* $-\infty$ *to restrict the axis-parallel strips.)*

2.3   Rectangles Intersecting a Horizontal Line

In this section we prove that the *MMSC* problem with rectangles intersecting a horizontal line (*MMSCRIHL*) is NP-hard. The reduction is from the *PP1in3SAT* problem [8]. From an instance ϕ of the *PP1in3SAT* problem, we generate an instance Z, where the rectangles in Z intersect a horizontal line L.

Variable Gadget: The gadget for the variable x_i consists of $12m$ rectangles $\{1, 2, \ldots, 12m\}$ and $12m - 1$ points $\{p_1, p_2, \ldots, p_{12m-1}\}$ (see Fig. 3(a)). The points are along the top edge of the rectangles. The 1-st and the $12m$-th rectangles contain the points p_1 and p_{12m-1}, respectively, and the j-th rectangle contains the p_{j-1}-th and p_j-th points, for $2 \leq j \leq 12m - 1$. We note that the first $6m$ rectangles $\{1, 2, \ldots, 6m\}$ are responsible for the clauses in C_{above}, whereas the next $6m$ rectangles $\{6m + 1, 6m2, \ldots, 12m\}$ are responsible for the clauses in C_{below}. All of the rectangles are intersecting a horizontal line L. Now, in order to cover all of the points while minimizing the depth, we have only two distinct optimal solutions: Either all even-numbered or all odd-numbered rectangles with depth exactly one. This gives the truth value of the variable x_i.

Clause Gadget: We first modify the *PP1in3SAT* problem in the following way. Note that the variables of ϕ are placed on a horizontal line ($y = 0$). We move the variables vertically up such that they are placed on a horizontal line $y = m + 1$ (above the y-values of all the clauses in C_{above}) (see Fig. 4). The clauses in C_{above}

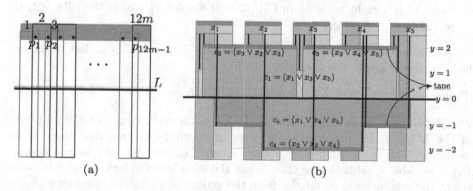

Fig. 3. (a) A variable gadget. (b) Position of the clause gadgets.

are placed above L and below the line $y = m + 1$ while connecting the same set of variables as before. Note that these clauses now connect the variables from below. On the contrary, the clauses in $\mathcal{C}_{below}$ are placed below L and still connect to the same set of variables from below.

Let us now consider the set $\mathcal{C}_{above}$ of clauses. Notice that, in the definition of the $PP1in3SAT$ problem these clauses can be ordered in increasing y-direction (see Fig. 1). Here we reverse the order of the clauses (see Fig. 3(b)). Now for each clause $C \in \mathcal{C}_{above}$ we take a *rectangular box* whose top boundary is the segment of C in the modified construction. The bottom boundary of the box touches the line L. Each box has a thin strip along the top edge of that box, called the *tape* of that clause. Similarly, we reverse the order of the clauses in $\mathcal{C}_{below}$ and for each clause C we take a box whose bottom boundary is the segment of C in the modified construction. The top boundary of the box touches the line L. Now here the tape is along the bottom boundary of each box.

Let $C_\ell = (x_i \vee x_j \vee x_k)$ be a clause in $\mathcal{C}_{above}$. We say that x_i is a *left*, x_j is a *middle*, and x_k is a *right* variable for C_ℓ. We take 5 points; point p_i^ℓ corresponding to x_i, points $p_j^\ell, q_j^\ell, r_j^\ell$ corresponding to x_j, and point p_k^ℓ corresponding to x_k; and 4 rectangles $s_1^\ell, s_2^\ell, s_3^\ell, s_4^\ell$. The rectangle s_1^ℓ covers the points $\{p_i^\ell, p_j^\ell\}$, s_2^ℓ covers the points $\{p_i^\ell, q_j^\ell\}$, s_3^ℓ covers the points $\{p_j^\ell, p_k^\ell\}$, and s_4^ℓ covers the points $\{r_j^\ell, p_k^\ell\}$ (see Fig. 4). The rectangles are placed inside the box and the points are placed inside the tape of C_ℓ.

Variable and Clause Interaction: We now describe the placement of the clause rectangles and points with respect to the variable rectangles. Let $1, 2, \ldots$ be the left to right order the clauses in $\mathcal{C}_{above}$ which connects to the variable x_i. In this order, assume that C_ℓ be the ℓ_1-, ℓ_2-, and ℓ_3-th clause for the variables x_i, x_j, and x_k respectively. Then we do the following.

⤳ Since x_i is a left variable in C_ℓ, place the point p_i^ℓ inside the $(6\ell_1 - 2)$-th rectangle of the gadget of x_i.

⤳ Since x_j is a middle variable in C_ℓ, place the point p_j^ℓ inside the $(6\ell_2 - 2)$-th rectangle of the gadget of x_j. Also place the point q_j^ℓ and r_j^ℓ inside the $(6k - 3)$-th and $(6k - 1)$-th rectangles of the gadget of x_j.

⤳ Since x_k is a right variable in C_ℓ, place the point p_k^ℓ inside the $(6\ell_3 - 2)$-th rectangle of the gadget of x_k.

A similar construction can be made for the clauses in $\mathcal{C}_{below}$, but using the last $6m$ rectangles in the variables. See Fig. 4.

Theorem 2. *The MMSCRIHL problem is NP-hard.*

Proof. We prove that exactly one literal is true in every clause of ϕ if and only of the *MMSCRIHL* problem has a cover of depth 1. Assume that there is an assignment to the variables of ϕ that satisfies exactly one literal per clause. For a variable x_i, if it is true then select the even indexed rectangles otherwise select the odd indexed rectangles from the gadget of x_i. Let us consider a clause $C_\ell = (x_i \vee x_j \vee x_k)$. Since exactly one literal per clause is true, exactly one of p_i^ℓ

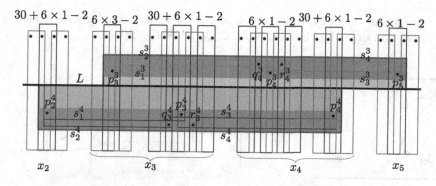

Fig. 4. Interaction with the variable and clause gadgets. We demonstrate the interaction of C_3 and C_4 with the variables in the *P1in3SAT* instance in Fig. 1.

or p_j^ℓ, or p_k^ℓ is covered by a variable rectangle. Clearly, the remaining points in the clause gadget are covered by the clause rectangles with depth one.

In the reverse direction, assume that the *MMSCRIHL* problem has a cover of depth 1. To cover the points in a variable gadget and in order to make their depth 1, there are only two possibilities to select the rectangles. We set the variable x_i to be true if all even indexed rectangles are selected from the gadget of x_i, otherwise set x_i to be false. Now consider a clause $C_\ell = (x_i \vee x_j \vee x_k)$. Now in C_ℓ, if more than one literal is true then the depth of a point in the gadget of C_ℓ will be more than 1. If the clause is not satisfiable then also either at least one point is not covered of there will be a point whose depth will be more than one. The only possibility is exactly one literal per clause is true. Hence, the theorem. □

2.4   Rectangles Anchored on Two Horizontal Lines

We prove that the *MMSC* problem with rectangles anchored on two horizontal lines (*MMSCRATHL*) is NP-hard by a reduction from *PP1in3SAT* problem [8].

Variable Gadget: For the variable gadget of x_i, we consider $12m$ points in two horizontal lines l_1 and l_2 each contains $6m$ points. We also consider $12m$ rectangles such that each rectangles i covers exactly two points p_i and p_{i+1}, for $1 \le i \le 12m-1$ and the rectangles $12m$ covers points p_{12m} and p_1 (see Fig. 5(a)). Rectangles $1, 2, \ldots, 6m$ are anchored on line l_1 and the remaining Rectangles are anchored on line l_2. Now in order to cover all the points while minimizing the depth, we have only two different optimal solutions. Either all even numbered or all odd numbered rectangles with depth exactly 1. This gives the truth value of the variable x_i.

Clause Gadget: We first consider the set C_{below} of clauses in ϕ. These clauses can be ordered in decreasing y-direction (see Fig. 1). Now for each clause $C \in C_{below}$ we take a *rectangular box* whose top boundary is the segment of C. The bottom boundary of the box touches the line l_1. Each box has a thin strip along the top edge of that box, called the *tape* of that clause. Similarly, we construct the boxes and tapes for the clauses for $\mathcal{C}_{above}$. See Fig. 5(b).

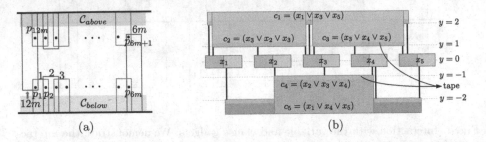

Fig. 5. (a) A variable gadget. (b) Position of the clause gadgets.

The placement of the clause points and rectangles is similar to the placement of the clause points and rectangles described in Sect. 2.3. The clause structure is exactly the same as in Sect. 2.3. For a clause $C_\ell = (x_i \vee x_j \vee x_k)$ in $\mathcal{C}_{below}$ with x_i, x_j, and x_k as *left*, *middle*, and *right* variable, we take 5 points; point p_i^ℓ corresponding to x_i, points $p_j^\ell, q_j^\ell, r_j^\ell$ corresponding to x_j, and point p_k^ℓ corresponding to x_k; and 4 rectangles $s_1^\ell, s_2^\ell, s_3^\ell, s_4^\ell$. The rectangle s_1^ℓ cover the points $\{p_i^\ell, p_j^\ell\}$, s_2^ℓ cover the points $\{p_i^\ell, q_j^\ell\}$, s_3^ℓ cover the points $\{p_j^\ell, p_k^\ell\}$, and s_4^ℓ cover the points $\{r_j^\ell, p_k^\ell\}$. The rectangles are placed inside the box and the points are placed inside the tape of C_ℓ.

Variable and Clause Interaction: The interaction of the variables and the clauses is similar to that in Sect. 2.3, but now here we consider a clause $C \in \mathcal{C}_{below}$. As in the proof of Theorem 2, we conclude:

Theorem 3. *The MMSCRATHL problem is NP-hard.*

3   Minimum Membership Hitting Set Problem

3.1   Rectangles Anchored on a Horizontal Line

Similar to Sect. 2.1, in polynomial time, one can decide if there exists a hitting set of depth one for the *MMHS* problem with rectangles anchored on a horizontal line from one side (*MMHSRAHL*), as follows. Define the weight of a point as the number of rectangles it stabs. Now, apply the algorithm of [1] to compute a maximum weight set of points (no two of them share a rectangle). Then, to see if there is a hitting set of rectangles having depth exactly 1, check if the total weight of the points is equal to the number of rectangles.

3.2   Axis-Parallel Strips

We prove that the *MMHS* problem with axis-parallel strips (*MMHSS*) is NP-hard using a reduction from the *P1in3SAT* problem. We generate an instance $Z(S,P)$ of the *MMHSS* problem from ϕ, an instance of the *P1in3SAT* problem.

The gadget for a variable x_i includes $2m$ horizontal strips $\{1, 2, \ldots, 2m-1\}$ and $2m$ points $\{p_1, p_2, \ldots, p_{2m}\}$. The j-th strip contains the points p_j and p_{j+1}, for $1 \leq j \leq 2m-1$ (see Fig. 6(a)). The points are on a vertical line. However, we move some of the points to the right to some clause gadgets at later stage. It is observed that there are exactly two different sets of points, either all even indexed or all odd indexed, which stab all the strips with depth exactly 1. We stack the variable gadgets vertically from top to bottom.

The gadget for a clause C_ℓ is a vertical strip v^ℓ. The clause gadgets are placed one after another to the right of the points corresponding to the variable gadgets.

For each variable, we order the clauses that contains it. Let C_ℓ be a clause that contains x_i, x_j, x_k, then according to this ordering we say that C_ℓ is a ℓ_1-th, ℓ_2-th, and ℓ_3-th clause for $x_i, x_j,$ and x_k respectively. Now for the clause C_ℓ we move the three points $p_{2\ell_1}, p_{2\ell_2},$ and $p_{2\ell_3}$ in the vertical orientation from $x_i, x_j,$ and x_k respectively to inside v^ℓ.

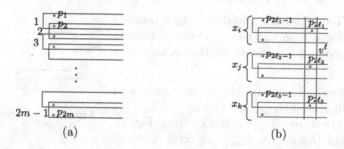

(a)   (b)

Fig. 6. (a) Variable gadget. (b) Clause gadget and its interaction with variable gadgets.

Clearly, the number of strips and points is polynomial with respect to the number of variables and clauses in ϕ. Hence the construction can be done in polynomial time. We now prove the following theorem.

Theorem 4. *The MMHSS problem is NP-hard.*

Proof. We prove that exactly one literal is true in each clause of ϕ if and only if Z has a hitting set with depth exactly 1. For variable x_i, we choose even indexed points if x_i is true, else choose odd indexed points. This clearly stabs all variable strips with depth 1. Since exactly one literal is true in each clause of ϕ, exactly one point will stab a clause strip. On the other hand assume that there is a hitting set of points with depth exactly 1. Now stabbing all the variable strips with depth 1 requires either all even or all odd indexed points. So we set x_i to be true if even indexed points are selected, otherwise, set x_i to be false. Since the depth of the hitting set is 1, exactly one point in a clause strip is selected. $\square$

3.3 Rectangle Intersecting a Horizontal Line

We show that the *MMHS* problem with rectangles intersecting a horizontal line (*MMHSRIHL*) is NP-hard using a reduction from the *PP1in3SAT* problem.

The variable gadget is similar to the variable gadget defined in Sect. 3.2, but now the strips are vertical and they are intersecting a horizontal line. The clause gadget is similar to that in Sect. 2.3, but now, for each clause, the rectangular box of Sect. 2.3 is itself a rectangle. Next, using a process as in Sect. 3.2, we shift (vertically) points from the variable gadgets to these clause rectangles. Hence, as in the proof of Theorem 4, we conclude the following.

Theorem 5. *The MMHSRIHL problem is NP-hard.*

Similar to Theorem 5, we prove that the *MMHSRATHL* problem is NP-hard.

4   Generalized Minimum Membership Hitting Set

NP-Hardness: We prove that the *GMMHS* problem of stabbing horizontal unit segments by vertical unit segments (*GMMHSUSeg*) is NP-hard. The reduction is from the *PP1in3SAT* problem.

Variable Gadget: Each variable gadget consists of a *variable chain* and at most $2m$ *clause chains*, each corresponding to a clause leg that connects to a variable.

Fig. 7. A variable gadget.

Variable Chain: Each variable chain consists of $8m+2$ unit horizontal segments $\{h_1, h_2, \ldots, h_{8m+2}\}$ positioned like a rectangular fashion (see Fig. 7). The segments $\{h_1, h_2, \ldots, h_{4m}\}$ are on a horizontal line and are responsible for connecting the clause chains to the variable chain from above. Similarly, the segments $\{h_{4m+2}, h_{4m+3}, \ldots, h_{8m+1}\}$ are on another horizontal line and are responsible for connecting the clause chains to the variable chain from below.

Clause Chains: Let C_ℓ be a clause in $\mathcal{C}_{above}$ that connects the variables x_i, x_j, and x_k through left, middle, and right legs respectively. Then for a left or middle, or right leg, we construct a *left* or *middle*, or *right* chain respectively. The left and middle chains are depicted in Fig. 8(a) and (b) respectively. The right chain is similar to the left chain but flipped vertically.

Let us consider a clause $C \in \mathcal{C}_{above}$ that is a ℓ-th clause for the variable x_i. In the variable chain of x_i, we shift the $h_{4\ell-2}$-th segment slightly left and the $h_{4\ell-1}$-th segment slightly right (see Fig. 8(c)). Place the chain for C above these two segments such that h' and $h_{4\ell-2}$ are stabbed by a vertical segment and h'' and $h_{4\ell-1}$ are stabbed by another vertical segment. Note that for each variable at most $2m$ chains are connected with its variable chain, at most m from either above or below. The variable chain and at most $2m$ left, middle, or right chains together form a big circular like arrangements of segments, called *big-cycle*. Note that, this big-cycle contains an even number of both horizontal and vertical segments and along the cycle at most 2 consecutive horizontal segments are stabbed by a vertical segment. We now have the following observation.

Observation 1. *For each variable gadget, there are two optimal solutions, either all red or all blue vertical segments each of size half of the total number of vertical segments present in a big-cycle.*

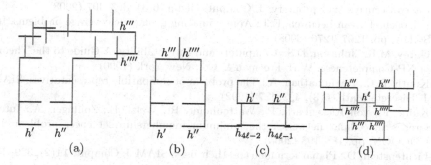

Fig. 8. (a) A left chain. (b) A middle chain. (c) Attaching a clause chain to a variable chain. (d) Clause gadget and connection with the three variable gadgets.

Clause Gadget: Let $C_\ell \in \mathcal{C}_{above}$ be a clause that contains x_i, x_j, and x_k. The gadget for C_ℓ is a single horizontal segment h^ℓ. The position of h^ℓ with respect to the three chains corresponding to x_i, x_j, and x_k is shown in Fig. 8(d).

This completes the construction. Note that this construction can be done in polynomial time with respect to the number of the variables and clauses in ϕ. An argument similar to that in the proof of Theorem 4 leads to the following theorem.

Theorem 6. *The GMMHSUSeg problem is NP-hard.*

Approximation for the GMMHSUSeg Problem: First we convert this problem to the *MMHS* problem with unit squares. Let H and V be given sets of unit horizontal and vertical segments. For each horizontal segment $h \in H$, take a unit square $t_h \in T$ such that the bottom boundary of t_h coincides with h and for each vertical segment $v \in V$, take the top endpoint, $p_v \in P$ of v. Clearly, finding a set $V' \subseteq V$ that stabs all the horizontal segments in H while minimizing the number of times a segment in H is stabbed by segments in V' is equivalent to finding a set of points $P' \subseteq P$ that stabs all the unit squares in T while minimizing the number of points in P' that is contained in a unit square in T.

Because the *GMMHSUSeg* problem is NP-hard, in another way we can say that the *MMHS* problem with unit squares is also NP-hard. Since for unit squares the *MMHS* and *MMSC* problems are dual to each other, the result of [3] ensures the following theorem.

Theorem 1. *There exists a 5-approximation for the GMMHSUSeg problem where the optimal objective value is bounded by a constant.*

References

1. Chan, T.M., Grant, E.: Exact algorithms and APX-hardness results for geometric packing and covering problems. Comput. Geom. **47**(2, Part A), 112–124 (2014)
2. Dom, M., Guo, J., Niedermeier, R., Wernicke, S.: Red-blue covering problems and the consecutive ones property. J. Comput. Geom. **6**(3), 393–407 (2008)
3. Erlebach, T., van Leeuwen, E.J.: Approximating geometric coverage problems. In: SODA, pp. 1267–1276 (2008)
4. Garey, M.R., Johnson, D.S.: Computers and Intractability; A Guide to the Theory of NP-Completeness. W.H. Freeman & Co., New York (1990)
5. Knuth, D.E., Raghunathan, A.: The problem of compatible representatives. SIAM J. Discrete Math. **5**(3), 422–427 (1992)
6. Kuhn, F., von Rickenbach, P., Wattenhofer, R., Welzl, E., Zollinger, A.: Interference in cellular networks: the minimum membership set cover problem. In: COCOON, pp. 188–198 (2005)
7. Lichtenstein, D.: Planar formulae and their uses. SIAM J. Comput. **11**(2), 329–343 (1982)
8. Mulzer, W., Rote, G.: Minimum-weight triangulation is NP-hard. J. ACM **55**(2), 11:1–11:29 (2008)
9. Nandy, S.C., Pandit, S., Roy, S.: Covering points: minimizing the maximum depth. In: CCCG (2017)
10. Narayanaswamy, N.S., Dhannya, S.M., Ramya, C.: Minimum membership hitting sets of axis parallel segments. In: Wang, L., Zhu, D. (eds.) COCOON 2018. LNCS, vol. 10976, pp. 638–649. Springer, Cham (2018). https://doi.org/10.1007/978-3-319-94776-1_53
11. Raz, R., Safra, S.: A sub-constant error-probability low-degree test, and a sub-constant error-probability PCP characterization of NP. In: STOC, pp. 475–484 (1997)
12. Schaefer, T.J.: The complexity of satisfiability problems. In: STOC, pp. 216–226 (1978)

Capacitated Discrete Unit Disk Cover

Pawan K. Mishra, Sangram K. Jena, Gautam K. Das[(⊠)], and S. V. Rao

Indian Institute of Technology, Guwahati, India
{pawan.mishra,sangram,gkd,svrao}@iitg.ac.in

Abstract. Consider a capacitated version of the discrete unit disk cover problem as follows: consider a set $P = \{p_1, p_2, \cdots, p_n\}$ of n customers and a set $Q = \{q_1, q_2, \cdots, q_m\}$ of m service centers. A service center can provide service to at most $\alpha(\in \mathbb{N})$ number of customers. Each $q_i \in Q$ $(i = 1, 2, \cdots, m)$ has a preassigned set of customers to which it can provide service. The objective of the capacitated covering problem is to provide service to each customer in P by at least one service center in Q. In this paper, we consider the geometric version of the capacitated covering problem, where the set of customers and set of service centers are two point sets in the Euclidean plane. A service center can provide service to a customer if their Euclidean distance is less than or equal to 1. We call this problem as (α, P, Q)-covering problem. For the (α, P, Q)-covering problem, we propose an $O(\alpha mn(m+n))$ time algorithm to check feasible solution for a given instance. We also prove that the (α, P, Q)-covering problem is NP-complete for $\alpha \geq 3$ and it admits a PTAS.

Keywords: Geometric covering · NP-complete · PTAS

1 Introduction

The geometric set cover problem is one of the extensively studied optimization problem in computational geometry. In a geometric set cover problem, range space is defined as $S = (X, \mathscr{R})$, where X is a set of points (finite or infinite) in $\mathbb{R}^2$ and $\mathscr{R}$ is a (finite or infinite) family of subsets of X which is called as ranges. These ranges are defined by the intersection of X and geometric objects such as unit disk, unit square, axis parallel rectangles and in general any convex pseudo-disks. The objective of the geometric set cover problem is to find a minimum cardinality $R' \subset \mathscr{R}$ of ranges such that all points in X are covered, i.e., for all $p \in X$, there exist $\mathbf{r} \in R'$, such that, $p \cap \mathbf{r} \neq \phi$. This problem is a special case of the general *set cover* problem. Though the general set cover problem is NP-hard to approximate within a factor of $\Omega(\log n)$ [10], some geometric set cover problems admit a PTAS.

Let $P = \{p_1, p_2, \cdots, p_n\}$ be a set of n red points and $Q = \{q_1, q_2, \cdots, q_m\}$ be a set of m blue points on the Euclidean plane. For a given integer α, a subset $Q' \subseteq Q$ is said to be α-cover of P if the point set P can be partitioned into $P_1, P_2, \cdots, P_\ell$ such that $|P_i| \leq \alpha$ for each $i = 1, 2, \cdots, \ell$ and there exist points

© Springer Nature Switzerland AG 2019
G. K. Das et al. (Eds.): WALCOM 2019, LNCS 11355, pp. 407–418, 2019.
https://doi.org/10.1007/978-3-030-10564-8_32

$q'_1, q'_2, \cdots, q'_\ell \in Q'$ such that each point in P_i is covered by the unit disk centered at q'_i. In this article, for a given α, a red point set P and a blue point set Q, the objective is to find minimum cardinality α-cover of P with respect to the point set Q. We denote it as (α, P, Q)-covering problem.

Our interest in (α, P, Q)-covering problem arose from the coverage problem in wireless software defined networking (SDN). In SDN, we decouple the control plane (controls the traffic routing) from the data plane (packet forwarding). The *switches* are responsible for data plane and *controllers* for control plane. SDN controller gathers information from the switches which falls in the coverage area of the controller. Depending on the price of installation of a controller, there is a limitation on the number of switches a controller can communicate irrespective of the number of switches falls in controller's range. This constraint on the number of switches controlled by a controller inspired us to study at the (α, P, Q)-covering problem.

2   Related Work

The (n, P, Q)-covering problem, known as DUDC problem in the literature, is a well-studied problem as it has wide application in wireless networks and facility location problem [8]. The DUDC problem is NP-complete [8]. Brönnimann and Goodrich [4] proposed the first constant factor algorithm for the DUDC problem. They made an interesting connection between the DUDC and ϵ-net. Exploiting this connection, they proposed an $O(1)$ approximation algorithm in the 2-dimensional Euclidean space. Briefly, given a range space $S = (X, \mathcal{R})$, an ϵ-net is a subset $P \subseteq X$ such that $P \cap R \neq \phi$ for all $R \in \mathcal{R}$ with $|R| \geq \epsilon n$. They used the theorem of Haussler and Welz which states that for a range space with VC dimension d, there exists an ϵ-net of size $O(\frac{d}{\epsilon} log \frac{d}{\epsilon})$[13]. The constant factor approximation algorithm proposed by Brönnimann and Goodrich depends on the constant in the size of ϵ-net.

Subsequently, many results published with different constant factor approximation algorithms using different techniques. In 2004, Călinescu et al. [5] proposed a 108-approximation algorithm which runs in $O(m^2)$. Ambuhl et al. [1] improved the approximation factor to 72 by keeping the running time as $O(m^2)$. Carmi et al. [6] further enhanced the approximation factor to 38 with the running time of $O(m^6)$. Claude et al. [7] improved the result and gave a 22-factor approximation algorithm with the running time of $O(m^2n^4)$. Using the idea of Ambuhl [1] and Claude [7], Das et al. [8], proposed an 18-factor approximation algorithm, which runs in $O(n \log n + m \log m + mn)$ time. A year after, Fraser et al. [11] proposed a 15-factor approximation algorithm using the result of Das et al. [8]. More recently, Manjanna et al. [2] proposed a $(9 + \epsilon)$-factor approximation algorithm, which runs in $O(\max(m^6 n, m^{2(1+6/\epsilon)+1}))$ time. Based on the local search algorithm, Mustafa and Ray [15] proposed a PTAS which runs in $O(mn^{O(\epsilon^{-2})})$ time.

2.1   Our Contribution

In this article, we studied (α, P, Q)-covering problem. We proposed an algorithm to check the feasibility of a given instance in $O(\alpha mn(m + n))$ time where $m = |Q|$, $n = |P|$. Using the feasibility algorithm, the optimal solution for $(1, P, Q)$-covering problem can be obtained. We proved the problem is NP-complete for $\alpha \geq 3$ and proposed a PTAS for the problem.

2.2   Organization

The remainder of the paper is organized as follows. In Sect. 3, we propose an algorithm to check the feasibility of (α, P, Q)-covering problem. In Sect. 4, we prove the decision version of (α, P, Q)-covering problem is NP-complete for $\alpha \geq 3$. In Sect. 5, we propose a PTAS for the problem.

3   Feasibility Test

In this section, we discuss a polynomial time algorithm to check whether there exists an α-cover of the red point set P with respect to blue point set Q.

The feasibility checking algorithm is based on the maximum matching algorithm in a bipartite graph. Here, we construct the bipartite graph from any arbitrary instance of (α, P, Q)-covering problem which leads us to check the feasibility in polynomial time. Given a set $P = \{p_1, p_2, \ldots, p_n\}$ of n red points, a set $Q = \{q_1, q_2, \ldots, q_m\}$ of m blue points and an integer α, we construct a bipartite graph $G = (V_1 \cup V_2, E)$, where $V_1 = \{v_{ij} \mid 1 \leq i \leq m \text{ and } 1 \leq j \leq \alpha\}$ is a set of vertices corresponding to set of points in Q such that for each point $q_i \in Q$ we considered α vertices in V_1, namely, $v_{ij}(j = 1, 2, \cdots, \alpha)$ and V_2 is the set of vertices corresponding to the points in P and $E = \{e = (v_{ij}, v_\ell) \mid v_{ij} \in V_1 \text{ and } v_\ell \in V_2, \text{ and unit disk centered at } q_i \text{ covers } p_\ell\}$ (see Fig. 1). The total number of vertices in the bipartite graph is $|V| = |V_1| + |V_2| = \alpha m + n$ and the maximum possible number of edges $|E| = \alpha mn$. The construction of the bipartite graph takes $O(\alpha mn)$ time. The procedure to check the feasibility is described in Algorithm 1.

Lemma 1. *Algorithm 1 computes correctly in* $O(\alpha mn(m + n))$ *time.*

Proof. Observe that if the cardinality of the maximum matching is n, then there exists an edge from each vertex of V_2 to one of the vertex of V_1. That means each point of P is covered by at least one disk. As α copies of each point of Q taken in V_1, it ensures that each point in P is covered with given disks without violating the capacity constraint.

The time complexity of the algorithm depends upon computing maximum matching in the bipartite graph $G = (V_1 \cup V_2, E)$, which takes $O(\alpha^2 m^2 n + \alpha mn^2)$ time [14]. Therefore, the overall time complexity of algorithm is $O(\alpha mn(m+n))$.

Algorithm 1. Feasibility_Test(α, P, Q)

    Input : A set P of n red points, a set Q of m blue points and an integer α.
    Output: True and a subset $Q' \in Q$ if there exists an α-cover otherwise False.

1: Construct a bipartite graph $G = (V_1 \cup V_2, E)$ as described above.
2: Compute maximum matching in G and store the result in $MaxMatching$ [14].
3: Store all the maximum matching edges in M'.
4: **for** each edge $e = (u, v) \in M'$ **do**
5:     Set $N \leftarrow u \in V_1$
6: **end for**
7: Set $A \leftarrow$ points corresponding to each vertex in N.
8: **if** $MaxMatching == n$ **then**
9:     Report $True$ and A.
10: **else**
11:     Report $False$.
12: **end if**

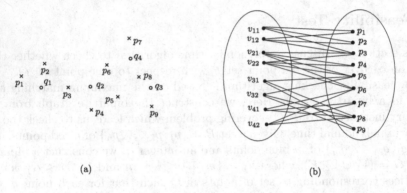

(a)   (b)

Fig. 1. (a) An instance of (α, P, Q)-covering problem, (b) Construction of bipartite graph for $\alpha = 2$, here vertex v_{ij} represent j^{th} copy of disk i.

4 Hardness of the (α, P, Q)-covering Problem

In this section, we show that the $(3, P, Q)$-covering problem is NP-complete. Using the NP-complete proof for $(3, P, Q)$-covering problem, we can conclude (α, P, Q)-covering is NP-complete for $\alpha \geq 4$ also. The $(3, P, Q)$-covering problem is in NP since for a given certificate, we can verify it in polynomial time (see Algorithm 1). To complete the prove, next we prove $(3, P, Q)$-covering problem is NP-hard by showing a special case $(3, P, P)$-covering is NP-hard.

The vertex cover problem on planar graph of degree at most 3 is known to be NP-complete [12]. To prove NP-hardness of (α, P, P)-covering problem, we provide polynomial time reduction from the decision version of the vertex cover (VC) problem on planar graph of degree at most 3 to the decision version of (α, P, P)-covering problem.

Decision version of the VC problem on planar graph (Vc-Pla)
Instance: An undirected planar graph $G = (V, E)$ with maximum degree 3 and a positive integer k.
Question: Does there exist a vertex cover $V'(\subseteq V)$ of G such that $|V'| \leq k$?

Decision version of (α, P, P)**-covering problem**
Instance: A set P of n points, an integer α, and a positive integer k.
Question: Does there exist an α-cover $P'(\subseteq P)$ of P such that $|P'| \leq k$?

Lemma 2 ([17]). *Consider a planar graph $G = (V, E)$ with maximum degree 4. The graph G can be embedded on the plane such that its vertices are at integer coordinates and its edges are line segments of the form $x = i$ or $y = j$, for integers i and j.*

The embedding in Lemma 2 can be done in linear time with at most two bends along each edge [3]. See Fig. 2(b) for an example.

Lemma 3. *Any planar graph $G = (V, E)$ with maximum degree 3 and $|E| \geq 2$ can be embeded on the Euclidean plane with each of its vertices is at $(3i, 3j)$ and edges as a sequence of line segments on the lines $x = 3i$ or $y = 3j$ for integers i and j.*

Proof. Follows from Lemma 2.

Let $G = (V, E)$ be an instance of Vc-Pla with $|E| \geq 2$. An instance of $(3, P, P)$-covering problem can be constructed from G in polynomial-time as follows:

We construct an instance of (α, P, P) in four steps.

Step 1: Embedding
The instance of G is embedded in the plane using the algorithms proposed in [3].

In this embedding, each edge is a sequence of connected line segment(s). The length of line segments used in embedding is of length three units. If ℓ is the number of line segments in the embedding, then 3ℓ is the sum of the length of the line segments. We name *node points* to the points in the embedding correspond to each vertex of G.

Step 2: Adding extra points to the embedding
In the embedding, we add a point at each coordinate $(3i, 3j)$ along every path between two nodes other than the node points. We call these points as bend points. The line segments in the embedding are classified into two catagories, called as *proper* and *improper*. The *proper* line segments are the line segments which none of the end points are node points. All the line segments other than *proper* line segments are named as *improper* line segments.

For each edge (p_i, p_j) of length 3 units (here p_i, p_j are node points) we add two points at distances 0.72 and 1.22 units from p_i and p_j, respectively (thus adding four points in total, see the edge (p_1, p_2) in Fig. 2(c)).

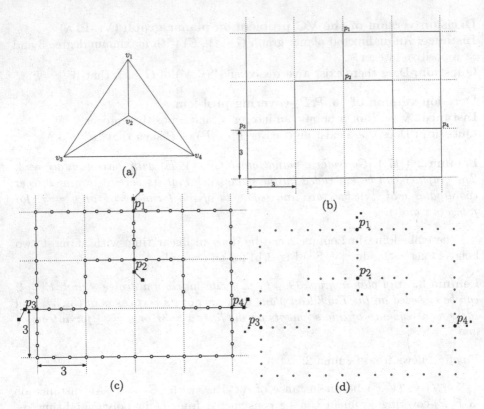

Fig. 2. (a) A planar graph G of maximum degree 3, (b) its embedding G' on a grid of cell size 3×3, (c) adding of extra points to G', (d) obtained instance of the (α, P, P)-covering problem.

For each edge of length greater than 3 units, we also add points as follows: for each improper line segment we add three points at distances 0.75, 1.5, and 2.25 units from the end point corresponding to a node point in G, and for each proper line segment we add two points at distances 1 and 2 units from its end points, i.e., bend points (see the edge (p_2, p_3) in Fig. 2(c)). We name the points added in this step along with the bend points as *joint points*.

Step 3: Adding extra line segments and points
For each node point p_i add a line segment of length 0.70 units (on the lines $x = 3i$ or $y = 3j$ for some integers i or j), without coinciding with the already drawn line segments. Adding these line segments on the $x = 3i$ or $y = 3j$ lines is always possible without losing the planarity of the graph G, as the maximum degree of G is 3. Now, add one point (say x_i) on these line segments at distance 0.70 units from point p_i and add another point (say y_i) at distance 0.32 units from x_i touching the line at distance 0.99 units from p_i and add a line segment from x_i to y_i (see the support point added with respect to (p_1) in Fig. 4(c)). As per the construction of y_i, the distance of y_i from all the points excluding

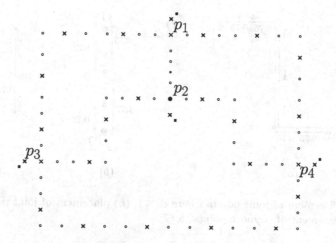

Fig. 3. Set of cross points are in the solution.

x_i is greater than one unit. The points added in this step are named as *support points*.

Let N, J, and S be the set of node points, joint points, and support points respectively. Let $N = \{p_i \mid v_i \in V\}$, $J = \{q_1, q_2, \ldots, q_{3\ell+1}\}$, and $S = \{x_i, y_i \mid v_i \in V\}$.

Step 4: Construction of P

Let $P = N \cup J \cup S$. Observe that, $|N| = |V|(= n)$, $|J| = 3\ell + 1$, where ℓ is the total number of line segments in the embedding, and $|S| = 2|V|(= 2n)$. Hence, $|P| = 3(n + \ell) + 1$. Therefore, P can be constructed in polynomial-time.

Theorem 1. (α, P, P)-*covering problem is NP-complete.*

Proof. For any given set $P' \subseteq P$ and a positive integer k, we can verify whether P' is an α-cover of P such that $|P'| \leq k$ in polynomial-time (see Algorithm 1). We prove the hardness of $(3, P, P)$-covering problem by reducing VC-PLA to it. Let $G = (V, E)$ be an instance of VC-PLA. Construct an instance P of $(3, P, P)$-covering problem as discussed above. We now prove the following claim.

Claim. G has a vertex cover of size at most k if and only if P has an α-cover of size at most $k + \ell + n$.

Necessity: Let $D \subseteq V$ be a vertex cover of G such that $|D| \leq k$. Let $N' = \{p_i \in P \mid v_i \in D\}$, i.e., N' is the set of points in P that correspond to the vertices in D. From each support point associated with a point belongs to N, we select the nearest support point (x_i) in the solution. Let this set is S'. From each set of points corresponding to a line segment in the embedding we choose 1 point as follows: Initially $J' = \emptyset$. As D is a vertex cover, every edge in G has at least one of its end vertices in D. Let (v_i, v_j) be an edge in G and $v_i \in D$ (the tie can be broken arbitrarily if both v_i and v_j are in D). Note that, the edge (v_i, v_j) is

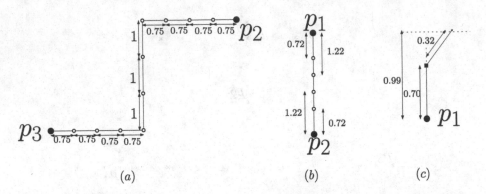

Fig. 4. (a) Placement of joint points where $\ell' > 1$, (b) placement of joint points where $\ell' = 1$, (c) placement of support points to G'.

represented as a sequence of line segments in the embedding. Start traversing the points (of (v_i, v_j)) from p_i, where p_i corresponds to v_i, and add every third point to J' encountered in the traversal without including p_j (see (p_3, p_4) in Fig. 3. The cross points are part of J' while traversing from p_3). Apply the above process to each edge in G. Observe that the cardinality of J' is ℓ as we choose 1 point from each set of points on a segment in the embedding. Let $P' = N' \cup J' \cup S'$. Now, we argue that P' is a 3-cover of P. Each $p_i \in N$ is covered by x_i in S'. If $p_i \in N'$ (i.e., the corresponding point $v_i \in D$ in G), then it covers all its neighbor points in J and all the other points in J is covered by atleast one point $q_j \in J'$. The existence of q_j is guaranteed by the way we constructed J'. If $p_i \notin N'$, then note that p_i is covered by one of the points in S. Therefore, every point in P is covered by at least 1 point in P' and no points cover more than 3 points. Thus, P' is a 3-cover of P and $|P'| = |N'| + |J'| + |S'| \le k + \ell + n$.

Sufficiency: Let $P' \subseteq P$ be a 3-cover of size at most $k + \ell + n$. We prove that G has a vertex cover of size at most k with the aid of the following claims.

Claim(i): At least one of the support points corresponding to each node point belongs to P'.

Proof of Claim (i): The claim follows from the fact that support point y_i corresponding to node p_i is covered only by the support points x_i and/or y_i.

Claim(ii): The points corresponding to each segment in G' in the embedding must contribute at least 1 point to P', i.e., $|J \cap P'| \ge \ell$, where ℓ is the total number of segments in the embedding.

Proof of Claim (ii): If ℓ' is the number of segments between p_i and p_j, then the total number of points between p_i and p_j is $3\ell' + 3$ including p_i and p_j. Now if both p_i and p_j are in P', then p_i and p_j can cover its neighbor points. So, in the worst case $(3\ell' + 3 - 4) = 3\ell' - 1$ number of points has to cover. It needs at least $\left\lceil \frac{3\ell' - 1}{3} \right\rceil = \ell'$ number of points. Thus, the claim follows.

Claim(iii): If p_i and p_j correspond to end vertices of an edge (v_i, v_j) in G, and both p_i, p_j are not in P', then there must be at least $\ell' + 1$ points in P' from the points corresponding to segment(s) representing the edge (v_i, v_j), where ℓ' is the number of segments representing the edge (v_i, v_j) in the embedding.

Proof of Claim (iii): Let (v_i, v_j) be an edge in G such that p_i and p_j are not in P'. By Claim (ii) $|J \cap P'| \geq \ell$. Hence, the points corresponding to each segment between p_i and p_j representing the edge (v_i, v_j) contributes at least ℓ' points in P'. We argue that if both p_i and p_j are not in P', then the number of points in P' from the points corresponding to each segment representing the edge (v_i, v_j) is at least $\ell' + 1$.

As per our construction of points from the graph G', if there exist ℓ' segments between points p_i and p_j then we consider exactly $3\ell' + 1$ points between them. Observe that one point can cover at most 3 points. So, to cover $3\ell' + 1$ points at least $\left\lceil \frac{3\ell'+1}{3} \right\rceil = \ell' + 1$ points are required.

We shall show that, by removing and/or replacing some points in P', a set of k points from N can be chosen such that the corresponding vertices in G is a vertex cover. The vertices in S' account for n points in P' (due to Claim (i)). Let $P' = P' \setminus S'$ and $D = \{v_i \in V \mid p_i \in P' \cap N\}$. If any edge (v_i, v_j) in G has none of its end vertices in D, then we do the following: consider the sequence of points corresponding to segments representing the edge (v_i, v_j) in the embedding. Since, both p_i and p_j are not in P', there must exist a segment having two of its points in P' (due to Claim (iii)). Consider the points corresponding to that segment having two points in P'. Delete any one of the point on the segment and introduce p_i (or p_j). Update D and repeat the process till every edge has at least one of its end vertices in D (due to Claim (ii)). D is a vertex cover in G with $|D| \leq k$. Therefore, (α, P, P)-covering problem is NP-hard. We have already shown that it is in NP. Therefore, (α, P, P)-covering is NP-complete. $\square$

5   A PTAS

We apply the local search algorithm to find an α-cover of the red point set P with respect to the blue point set Q for a given integer α. We prove that the local search algorithm produces a PTAS (see Algorithm 2 for the detailed pseudocode).

Algorithm 2. Local_Search(α, P, Q)

Input: A red point set P of size n, a blue point set Q of size m and an integer α.
Output: An α-cover subset $Q' \subseteq Q$.

1: $Q' \leftarrow Q$. (Assume the given instance has a feasible solution.)
2: **while** there exist $B \subseteq Q'$ of size at most k and $B' \subseteq Q$ of size at most $k - 1$ such that $(Q' \setminus B) \cup B'$ is a feasible solution for (α, P, Q)-cover problem, i.e., call Feasibility_Test (α, P, Q').
3: set $Q' \leftarrow (Q' \setminus B) \cup B'$.
4: **endwhile**
5: Report Q'.

Lemma 4. *The time complexity of the Algorithm 2 is* $O(\alpha m^{2k+1} n(m+n))$.

Proof. The number of local improvement steps is bounded by the number of blue points. Hence, there is a scope of at most m local-improvement steps. In each step, it is required to verify at most $\binom{m}{k}\binom{m}{k-1} \leq m^{2k-1}$ different local improvements. The time to check if certain local improvement is possible takes $O(\alpha mn(m+n))$ time. So, the overall time complexity of the algorithm is $O(\alpha m^{2k+1} n(m+n))$.   □

A subset $Q' \subseteq Q$ is called k-locally optimal if it is not possible to perform local improvement step. Now, we prove that Algorithm 2 produces $(1+\epsilon)$-factor approximation result.

Locality Condition: Let $Q_{opt} \subseteq Q$ be an optimal solution of the (α, P, Q) covering problem and $Q' \subseteq Q$ be an α-covering set returned by the local search algorithm. It is possible to construct a planar bipartite graph $G = (Q' \cup Q_{opt}, E)$ such that for each $p \in P$, there exists two vertices $u \in Q_{opt}$ and $v \in Q'$ sharing an edge $(u, v) \in E$. Note that $u \in Q_{opt}$ and $v \in Q'$ sharing an edge if their Euclidean distance is less than or equal to 1.

The locality condition for the range space consisting of points and disks is established in [16]. The locality condition for (α, P, Q)-covering problem can be established with the aid of same kind of arguments as in [16]. We define neighborhood of the vertices in the graph G, i.e., $N_G(u)$ is the set of neighbors of u in the planar bipartite graph and neighborhood function for the subset Y of the vertices of graph G, $N_G(Y) = \bigcup_{u \in Y} N_G(u)$.

Lemma 5. *Let* $Q_{opt} \subseteq Q$ *be an optimal solution and* $Q' \subseteq Q$ *be returned by Algorithm 2. Assume* $Q_{opt} \cap Q' = \phi$ *and if there exists a planar bipartite graph* $G = (Q' \cup Q_{opt}, E)$, *then for every subset* $Q'' \subseteq Q'$ *of size almost* k, $|N_G(Q'')| \geq |Q''|$.

Proof. Let $G = (Q' \cup Q_{opt}, E)$ be a bipartite graph. Since both Q' and Q_{opt} are α-cover sets for the point set P, then for each point $p \in P$ there exist a point $q' \in Q'$ and $q_{opt} \in Q_{opt}$ such that the Euclidean distance between (i) p and q' and (ii) p and q_{opt} are less than or equal to 1.

Claim: For any $Q'' \subseteq Q'$, $(Q' \setminus Q'') \cup N_G(Q'')$ is a feasible α-cover.

Proof of the claim: If there is a point $p_i \in P$ which is covered by the unit disk centered at a point in Q'', then one of the neighbors in $N_G(Q'')$ also covers the point p_i because of the locality condition. Therefore, $N_G(Q'')$ covers all the points which are covered by Q''. So, $(Q' \setminus Q'') \cup N_G(Q'')$ is a feasible α-cover.

The above claim implies that if $Q'' \subseteq Q'$ is a set of at most k unit disks, then $|N_G(Q'')| \geq |Q''|$, otherwise there is a scope of local improvement step.   □

Without loss of generality, we can always assume that $Q_{opt} \cap Q' = \phi$. If not, let $I = Q_{opt} \cap Q'$, $Q^* = Q \setminus I$, $Q_{opt}^* = Q_{opt} \setminus I$, $Q''' = Q' \setminus I$ and let P' be the set of points which are not covered by the disks centered at I. Q_{opt}^* and Q''' are disjoint. Also Q_{opt}^* is an α-cover of minimum size for the point set P'. Now assume $Q_{opt} = Q_{opt}^*$, $Q' = Q'''$ and $P = P'$.

Theorem 2. *[9] For any planar graph $G = (V, E)$ of n vertices, there is a set $X \subseteq V$ of size at most $\frac{c_1 n}{\sqrt{r}}$, such that $V \setminus X$ can be partitioned into n/r sets $V_1, V_2 \ldots V_{n/r}$ satisfying*

   i. $V_i \leq c_2 r$
   ii. $N(V_i) \cap V_j = \phi$ *for* $i \neq j$, *and*
   iii. $|N(V_i) \cap X| \leq c_3 \sqrt{r}$

where $c_1, c_2, c_3 > 0$ and $N(.)$ defines the neighborhood function.

Lemma 6. $|Q'| \leq (1 + c/\sqrt{k})|Q_{opt}|$ *for some constant c.*

Proof. Lemma 6 follows from Lemma 5 and planar separator theorem of Federickson [9].

If we assume $r = k/c_2$ in Theorem 2, then $|V_i| \leq k$. Let $Q_i' = Q' \cap V_i$ and $Q_{opt_i} = Q_{opt} \cap V_i$.

From Lemma 5, $|Q_i'| \leq |Q_{opt_i}| + |N(V_i) \cap X|$, for all i. Otherwise, $Q' \cap N'(V_i)$ can be replaced by Q_{opt_i}, which contradicts the fact that Q' is a k-locally optimal subset. Now,

$$|Q'| \leq |X| + \sum_i |Q_i'|$$

$$\leq |X| + \sum_i |Q_{opt_i}| + \sum_i |N(V_i) \cap X| \quad \text{(See above discussion)}$$

$$\leq \frac{c_1 n}{\sqrt{r}} + |Q_{opt}| + \frac{n}{r} c_3 \sqrt{r} \quad \text{(See Theorem 2)}$$

$$\leq \frac{c_1 n}{\sqrt{r}} + |Q_{opt}| + \frac{n}{\sqrt{r}} c_3$$

$$\leq |Q_{opt}| + c\frac{n}{\sqrt{r}}$$

$$\leq |Q_{opt}| + c\frac{|Q_{opt}| + |Q'|}{\sqrt{r}} = (1 + c/\sqrt{k})|Q_{opt}|$$

Thus, $|Q'| \leq (1 + c/\sqrt{k})|Q_{opt}|$, where c is a constant.   □

Theorem 3. *Algorithm 2 produces $(1 + \epsilon)$-factor approximation result in $O(\alpha m^{2k+1} n(m + n))$ time.*

Proof. The time complexity result follows from the Lemma 4. And the approximation result of the theorem follows from the Lemma 6 by putting $k = O(\epsilon^{-2})$.   □

References

1. Ambühl, C., Erlebach, T., Mihalák, M., Nunkesser, M.: Constant-factor approximation for minimum-weight (connected) dominating sets in unit disk graphs. In: Díaz, J., Jansen, K., Rolim, J.D.P., Zwick, U. (eds.) APPROX/RANDOM -2006. LNCS, vol. 4110, pp. 3–14. Springer, Heidelberg (2006). https://doi.org/10.1007/11830924_3

2. Basappa, M., Acharyya, R., Das, G.K.: Unit disk cover problem in 2D. J. Discrete Algorithms **33**, 193–201 (2015)

3. Biedl, T., Kant, G.: A better heuristic for orthogonal graph drawings. Comput. Geom. **9**(3), 159–180 (1998)

4. Brönnimann, H., Goodrich, M.T.: Almost optimal set covers in finite VC-dimension. Discrete Comput. Geom. **14**(4), 463–479 (1995)

5. Călinescu, G., Mandoiu, I.I., Wan, P.J., Zelikovsky, A.Z.: Selecting forwarding neighbors in wireless ad hoc networks. Mobile Netw. Appl. **9**(2), 101–111 (2004)

6. Carmi, P., Katz, M.J., Lev-Tov, N.: Covering points by unit disks of fixed location. In: Tokuyama, T. (ed.) ISAAC 2007. LNCS, vol. 4835, pp. 644–655. Springer, Heidelberg (2007). https://doi.org/10.1007/978-3-540-77120-3_56

7. Claude, F., et al.: An improved line-separable algorithm for discrete unit disk cover. Discrete Math. Algorithms Appl. **2**(01), 77–87 (2010)

8. Das, G.K., Fraser, R., López-Ortiz, A., Nickerson, B.G.: On the discrete unit disk cover problem. Int. J. Comput. Geom. Appl. **22**(05), 407–419 (2012)

9. Federickson, G.N.: Fast algorithms for shortest paths in planar graphs, with applications. SIAM J. Comput. **16**(6), 1004–1022 (1987)

10. Feige, U.: A threshold of ln n for approximating set cover. J. ACM (JACM) **45**(4), 634–652 (1998)

11. Fraser, R., López-Ortiz, A.: The within-strip discrete unit disk cover problem. Theor. Comput. Sci. **674**, 99–115 (2017)

12. Garey, M.R., Johnson, D.S.: Computers and Intractability: A Guide to the Theory of NP-Completeness. Freeman, New York (1979)

13. Haussler, D., Welzl, E.: ε-nets and simplex range queries. Discrete & Computational Geometry **2**(2), 127–151 (1987)

14. Kleinberg, J., Tardos, E.: Algorithm Design. Addison-Wesley Longman Publishing Co. Inc, Boston (2005)

15. Mustafa, N.H., Ray, S.: Improved results on geometric hitting set problems. Discrete Comput. Geom. **44**(4), 883–895 (2010)

16. Mustafa, N.H., Ray, S.: PTAS for geometric hitting set problems via local search. In: Proceedings of the Twenty-fifth Annual Symposium on Computational Geometry, pp. 17–22. ACM (2009)

17. Valiant, L.G.: Universality considerations in VLSI circuits. IEEE Trans. Comput. **100**(2), 135–140 (1981)

Author Index

Printed in the United States
By Bookmasters